Lecture Notes in Computer Science 5601

José Mira José Manuel Ferrández
José R. Álvarez Félix de la Paz
F. Javier Toledo (Eds.)

Methods and Models in Artificial and Natural Computation

A Homage to Professor Mira's Scientific Legacy

Third International Work-Conference on the Interplay Between Natural and Artificial Computation, IWINAC 2009
Santiago de Compostela, Spain, June 22-26, 2009
Proceedings, Part I

Springer

Volume Editors

José Mira
José R. Álvarez
Félix de la Paz
Universidad Nacional de Educación a Distancia
E.T.S. de Ingeniería Informática
Departamento de Inteligencia Artificial
Juan del Rosal, 16, 28040 Madrid, Spain
E-mail: info@iwinac.org

José Manuel Ferrández
F. Javier Toledo
Universidad Politécnica de Cartagena
Departamento de Electrónica, Tecnología de Computadoras y Proyectos
Pl. Hospital, 1, 30201 Cartagena, Spain
E-mail: info@iwinac.org

Library of Congress Control Number: 2009928632

CR Subject Classification (1998): F.1, F.2, I.2, G.2, I.4, I.5, J.3, J.4, J.1

LNCS Sublibrary: SL 1 – Theoretical Computer Science and General Issues

ISSN 0302-9743
ISBN-10 3-642-02263-4 Springer Berlin Heidelberg New York
ISBN-13 978-3-642-02263-0 Springer Berlin Heidelberg New York

springer.com

Printed in Germany

Typesetting: Camera-ready by author, data conversion by Scientific Publishing Services, Chennai, India
Printed on acid-free paper SPIN: 12693443 06/3180 5 4 3 2 1 0

Preface

Continuing Professor Mira's Scientific Navigation

Professor José Mira passed away during the preparation of this edition of the International Work-Conference on the Interplay Between Natural and Artificial Computation. As a pioneer in the field of cybernetics, he enthusiastically promoted interdisciplinary research. The term cybernetics stems from the Greek *Κυβερνήτης* (kybernetes), which means steersman, governor, or pilot, the same root as government. Cybernetics is a broad field of study, but the essential goal of cybernetics is to understand and define the functions and processes of systems that have goals, and promote circular, causal chains that move from action to sensing to comparison with a desired goal, and again to action. These definitions can be applied to Prof. Mira. He was a leader, a pilot, with a visionary and extraordinary capacity to guide his students and colleagues to the desired objective. In this way he promoted the study and understanding of biological functions for creating new computational paradigms able to solve known problems in a more efficient way than classical approaches. But he also impressed his magnificent and generous character on all the researchers and friends that worked with him, imprinting in all of us high requirements of excellence not only as scientists, but also as human beings.

We all remember his enthusiastic explanation about the domains and levels in the computational paradigm (CP). In his own words, this paradigm includes not only the physical level, but also the meaning of calculus passing over a symbolic level (SL) and a knowledge level (KL), where percepts, objectives, intentions, plans, and goals reside. In addition, in each level it is necessary to distinguish between the semantics and the causality inherent to that level phenomenology (own domain, OD) and the semantics associated to phenomenologies in the external observers domain (EOD). It is also important to note that own experiences, which emerge from neural computation in a conscious reflexive level, only match partially with what is communicable by natural language. We want to continue Prof. Mira's scientific navigation by attaining a deeper understanding of the relations between the observable, and hence measurable, and the semantics associated to the physical signals world, i.e., between physiology and cognition, between natural language and computer hardware.

This is the theme of the IWINAC meetings the "interplay" movement between the natural and artificial, addressing this problem every two years. We want to know how to model biological processes that are associated with measurable physical magnitudes and, consequently, we also want to design and build robots that imitate the corresponding behaviors based on that knowledge. This synergistic approach will permit us not only to build new computational systems based on the natural measurable phenomena, but also to understand many of the observable behaviors inherent to natural systems.

The difficulty of building bridges over natural and artificial computation was one of the main motivations for the organization of IWINAC 2009. These two books of proceedings contain the works of the invited speakers, Profs. Maravall and Fernández, and the 108 works selected by the Scientific Committee, after a refereeing process. In the first volume, entitled *Methods and Models in Artificial and Natural Computation: A Homage to Professor Mira's Scientific Legacy*, we include some articles by Prof. Mira's former disciples, who relate the relevance of their work with him from a scientific and personal point of view, the most recent collaborations with his colleagues, and the rest of the contributions that are closer to the theoretical, conceptual, and methodological aspects linking AI and knowledge engineering with neurophysiology, clinics, and cognition. The second volume entitled *Bioinspired Applications in Artificial and Natural Computation* contains all the contributions connected with biologically inspired methods and techniques for solving AI and knowledge engineering problems in different application domains.

An event of the nature of IWINAC 2009 cannot be organized without the collaboration of a group of institutions and people, whom we would like to thank, starting with *UNED* and *Universidad Politécnica de Cartagena*. The collaboration of the *Universidade de Santiago de Compostela*, and especially its rector Senen Barro, has been crucial, as has the efficient work of Roberto Iglesias and the rest of the Local Committee. In addition to our universities, we received financial support from the Spanish *Ministerio de Educación y Ciencia*, the *Programa de Tecnologías Futuras y Emergentes (FET) de la Comisión Europea*, the *Xunta de Galicia*, *APLIQUEM s.l.*, *I.B.M.*, *Fundación Pedro Barrié de la Maza* and the *Concello de Santiago de Compostela*. Finally, we would also like to thank the authors for their interest in our call and the effort in preparing the papers, a condition *sine qua non* for these proceedings, and to all the Scientific and Organizing Committees, particularly the members of these committees that have acted as effective and efficient referees and as promoters and managers of pre-organized sessions on autonomous and relevant topics under the IWINAC global scope.

Our deep gratitude goes to Springer and Alfred Hofmann, along with Anna Kramer and Erika Siebert-Cole, for the continuous receptivity and collaboration in all our editorial joint ventures on the interplay between neuroscience and computation.

All the authors of papers in this volume as well as the IWINAC Program and Organizing Committees dedicate this special volume to the memory of Prof. Mira as a person, scientist and friend. We will greatly miss him.

June 2009 Organizing Committee

Organization

General Chairman

José Mira, Spain

Honorary Committee

Roberto Moreno Díaz, Spain
Senén Barro Ameneiro, Spain
Roque Marín Morales, Spain
Ramon Ruiz Merino, Spain
Emilio López Zapata, Spain
Diego Cabello Ferrer, Spain
Francisco J. Ríos Gómez, Spain
José Manuel Ferrández Vicente, Spain

Organizing Committee

José Manuel Ferrández Vicente, Spain
José Ramón Álvarez Sánchez, Spain
Félix de la Paz López, Spain
Fco. Javier Toledo Moreo, Spain

Local Organizing Committee

Senén Barro Ameneiro, Spain
Roberto Iglesias Rodríguez, Spain
Manuel Fernández Delgado, Spain
Eduardo Sánchez Vila, Spain
Paulo Félix Lamas, Spain
María Jesús Taboada Iglesias, Spain
Purificación Cariñena Amigo, Spain
Miguel A. Rodríguez González, Spain
Jesús María Rodríguez Presedo, Spain
Pablo Quintía Vidal, Spain
Cristina Gamallo Solórzano, Spain

Invited Speakers

Dario Maravall, Spain
Javier de Lope Asiaín, Spain
Eduardo Fernandez, Spain
Jose del R. Millan, Switzerland
Tom Heskes, The Netherlands

Field Editors

Dario Maravall, Spain
Rafael Martinez Tomas, Spain
Maria Jesus Taboada Iglesias, Spain
Juan Antonio Botia Blaya, Spain
Javier de Lope Asiaín, Spain
M. Dolores Jimenez Lopez, Spain
Mariano Rincon Zamorano, Spain
Jorge Larrey Ruiz, Spain
Eris Chinellato, Spain
Miguel Angel Patricio, Spain

Scientific Committee (Referees)

Andy Adamatzky, UK
Michael Affenzeller, Austria
Igor Aleksander, UK
Amparo Alonso Betanzos, Spain
Jose Ramon Alvarez-Sanchez, Spain
Shun-ichi Amari, Japan
Razvan Andonie, USA
Davide Anguita, Italy
Margarita Bachiller Mayoral, Spain
Antonio Bahamonde, Spain
Alvaro Barreiro, Spain
Juan Botia, Spain
Giorgio Cannata, Italy
Enrique J. Carmona Suarez, Spain
Joaquin Cerda Boluda, Spain
Enric Cervera Mateu, Spain
Antonio Chella, Italy
Eris Chinellato, Spain
Erzsebet Csuhaj-Varju, Hungary
Jose Manuel Cuadra Troncoso, Spain
Felix de la Paz Lopez, Spain
Javier de Lope, Spain

Gines Domenech, Spain
Jose Dorronsoro, Spain
Richard Duro, Spain
Patrizia Fattori, Italy
Eduardo Fernandez, Spain
Antonio Fernandez-Caballero, Spain
Jose Manuel Ferrandez, Spain
Kunihiko Fukushima, Japan
Jose A. Gamez, Spain
Vicente Garceran-Hernandez, Spain
Jesus Garcia Herrero, Spain
Juan Antonio Garcia Madruga, Spain
Francisco J. Garrigos Guerrero, Spain
Charlotte Gerritsen, The Netherlands
Marian Gheorghe, UK
Pedro Gomez Vilda, Spain
Manuel Graña Romay, Spain
Francisco Guil-Reyes, Spain
Oscar Herreras, Spain
Juan Carlos Herrero, Spain
Cesar Hervas Martinez, Spain
Tom Heskes, The Netherlands
Fernando Jimenez Barrionuevo, Spain
M. Dolores Jimenez-Lopez, Spain
Jose M. Juarez, Spain
Joost N. Kok, The Netherlands
Elka Korutcheva, Spain
Markus Lappe, Germany
Jorge Larrey-Ruiz, Spain
Maria Longobardi, Italy
Maria Teresa Lopez Bonal, Spain
Ramon Lopez de Mantaras, Spain
Vincenzo Manca, Italy
Riccardo Manzotti, Italy
Dario Maravall, Spain
Roque Marin, Spain
Rafael Martinez Tomas, Spain
Jose Javier Martinez-Alvarez, Spain
Jesus Medina Moreno, Spain
Victor Mitrana, Spain
Jose Manuel Molina Lopez, Spain
Juan Morales Sanchez, Spain
Ana Belen Moreno Diaz, Spain
Arminda Moreno Diaz, Spain

Douglas Mota, Brazil
Isabel Navarrete Sanchez, Spain
Nadia Nedjah, Brazil
Taishin Y. Nishida, Japan
Manuel Ojeda-Aciego, Spain
Jose T. Palma Mendez, Spain
Juan Pantrigo, Spain
Gheorghe Paun, Spain
Juan Pazos Sierra, Spain
Jose M. Puerta, Spain
Carlos Puntonet, Spain
Alexis Quesada Arencibia, Spain
Luigi M. Ricciardi, Italy
Mariano Rincon Zamorano, Spain
Victoria Rodellar, Spain
Camino Rodriguez Vela, Spain
Ramon Ruiz Merino, Spain
Angel Sanchez Calle, Spain
Jose Luis Sancho-Gomez, Spain
Jose Santos Reyes, Spain
Andreas Schierwagen, Germany
Jordi Solé i Casals, Spain
Antonio Soriano Paya, Spain
Maria Taboada, Spain
Settimo Termini, Italy
Fco. Javier Toledo Moreo, Spain
Jan Treur, Netherlands
Ramiro Varela Arias, Spain
Marley Vellasco, Brazil

Table of Contents – Part I

Table of Contents – Part II

A Model of Low Level Co-operativity in Cerebral Dynamic*

J. Mira and A.E. Delgado

Dpto. de Inteligencia Artificial, ETS Ing. Informática, UNED, Madrid, Spain

Abstract. We present a conceptual model of cerebral dynamics based on the neuropsychological findings of Lashley, Luria, and J. Gonzalo on the residual function after traumatic and surgical lesions in animals and man. The model proposes a co-operative structure with polifunctional modules distributed on the same anatomical substratum. The level of co-operation depends closely the language used to describe the neural dialogue, analogic, logic or symbolic. At low level we can use a nonlinear convolution-like formulation with adaptive kernels to explain some of the experimental results, although the more relevant properties of cerebral dynamics need more sophisticated formulations.

1 Introduction

In this paper we propose a conceptual model of cerebral dynamics based mainly on the hypothesis and neuropsichological findings of Lashley [1], Luria [12], and J. Gonzalo [9] concerning the high residual function after traumatic and surgical lesions in animals and man. Studying this data we find the followings points:

1. Lashley, in 1929, stated that practically any intact part of a functional area has the capacity to carry out the functions that have been lost by the lesion.
2. Lashley, in 1937, proposed a Field Theory with dynamic and gradual localization of cortical functions.
3. Lashley, in his paper "In search of the engram" (1951) he proposed that all the cells must be constantly active and participating in every activity, so that there are multiple representation of sensorial information and co-operative neural processes.
4. Luria, in 1974, explored the theory of "Functional Systems" where each anatomical area contributes with one or more "factors" in terms of which are synthesized the global functions of cortical tissues. So the lesion eliminates "factors" but the function remains, although depressed.
5. J. Gonzalo, in 1952, introduces the concepts of cerebral dynamics and gradual distribution of sensorial functions on fields described by parameters such as excitability, permeability and differential sensibility.

* This paper was written in 1981, but has not been published until now. It was presented at the "Fifth International Congress of Cybernetics and Systems" which took place in Mexico D.F., and only its summary was published in the proceedings.

J. Mira et al. (Eds.): IWINAC 2009, Part I, LNCS 5601, pp. 1–20, 2009.

6. J. Gonzalo, in 1978, introduced the concepts of sensorial reinforcement and facilitation.

These neurophysiological findings suggest that neural processes (at least the cortical ones) are co-operative in such a way that neurons and neural nets are functionally connected and working together towards a common goal. The behavior of individual neurons is not relevant, otherwise, the effect of lesion would be catastrophic since standard lesion eliminates more than 8×10^4 cortical units.

The suggestions of W.S McCulloch [13], Beurle [2], and Wilson and Cowan [35], to name but few, also point in this direction. All the above researchers have attempted to shed some light on the basis of the high reliability and functional stability of the cortical tissue.

On the other hand, the importance of error in logic has been one of the uppermost topics of discussion since J. von Neumann's paper [34] "on reliable computation from unreliable components" to the present proposals on fault tolerant multiprocessors by Rennels [27,26], Katsuki [11], Hopkins [10] or Siewiorek [31]. Therefore the problem of reliability is of great interest not only in neurocybernetics but in computer sciences. The study of Brain reliability-computer reliability relationships can be of mutual convenience when looking for analogies of distributed computing structures and computing structures with properties similar to the neural tissue.

What we are seeking are abstract structures, algorithms and languages with a built-in capability of circumventing areas damaged by physical lesion which can continue with the overall function, although in general somewhat reduced. The subsequent physical realization is irrelevant at this level to the problem of understanding. If it is man-made we are in the field of fault tolerant computing. If we speak of neural correlates we are making a step forward in the understanding of information processing strategies in the nervous system.

2 A Conceptual Model of Neural Co-operativity

If we were obliged to design a lesion-tolerant reliable structure, the following properties would be included:

- Distributed processing.
- Modularity, with very complex modules.
- Coupling, with a high degree of connectivity and cross-references.

These properties are easily embodied in neural nets [18]. Thus, for example, let us consider an area of the sensorial cortex that receives specific and unspecific inputs and generates outputs to the white matter. We can formulate its global function, including the previously cited properties, introducing the following hypotheses [6,7,8,15,16]:

1. The global anatomical substratum (functional space) operates as a *modular layered computer* over which are the different sensorial modalities, specific functions are allocated to specific areas but not exclusive to that area alone, almost each are having a potential response to any other modality.

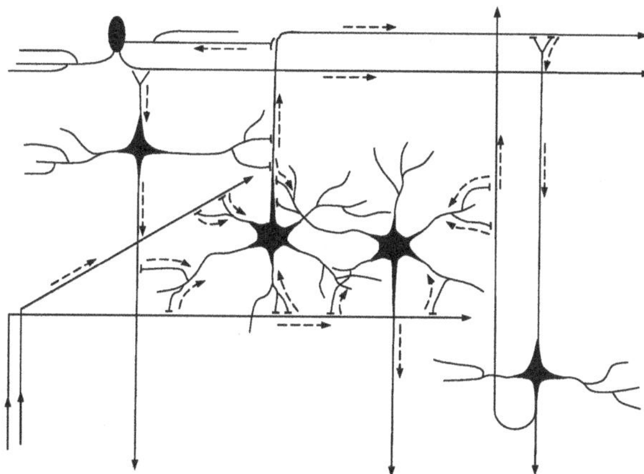

Fig. 1. Cajal arch for the stellate layer

2. The layers can be associated to the horizontal, stellate and pyramidal cells and the modules can be obtained from the anatomical generalization of *Cajal archs*. Figure 1 shows a possible functional module for the stellate layer. The specific neural realizations of the functional modules however, are not relevant at this level of conceptual modeling.
3. The information entering the layered computer is diffused among the functionally connected nets. Each net co-operates in all the sensorial functions in an unspecific way, "conversing" in such a way that we need to invent several languages to cope with the neurophysiological data and the observed behavior. The selection of the known languages in physics and mathematics (analogic, logic, and symbolic), determine the possible outcomes of the model. We will deal with this in more detail later.
4. At this conceptual level, the co-operative processes performed by a layer of nets, $\mathfrak{N}$, can be characterized by the following parameters:

$$\mathfrak{N} \triangleq \{\overrightarrow{x} \in S, \mu_k(\overrightarrow{x}), L(\overrightarrow{x}), R(L), E_k(\overrightarrow{x}), F_k, \mathfrak{L}\}$$

where:

S is the area of calculus (the physical functional space)

$\mu_k(\overrightarrow{x})$ is the density of the kth functional modality distributed on the same area of calculus (S)

$L_k(\overrightarrow{x})$ represent the position and size of a possible damaged area (physical lesion)

$R(L)$ is the measure of the residual function after the lesion $L(\overrightarrow{x})$. It depends on $\mu(\overrightarrow{x})$, $L(\overrightarrow{x})$ and $E(\overrightarrow{x})$

$E_k(\overrightarrow{x})$ is *excitability* and it measures the sensibility and permeability to the spatial summation

F_k represents the global function corresponding to the kth sensorial modality without lesion

$\mathfrak{L}$ is the language used for dialogue between the modules (slow potentials, spikes or neurophysiological symbols)

5. *Anatomical measures.* A lesion, $L(\overrightarrow{x})$, eliminates an area of calculus in the functional space, so that the residual tissue carries out the global function in a depressed way (residual function), but maintaining the organization (functional stability).

To represent the lesion at the anatomical level in our conceptual model we use the following measures:

Rendundancy, (R). Measures the overlap of the generalized dendritic fields (see figure 2).

$$R = \sum_{i,j=1}^{n,m} \mu_{ij} \frac{D_{ij}}{\bar{\mu} \cdot S}$$

where S is the area of calculus, μ_{ij} the density of function, and D is the generalized dendritic field of a neural net.

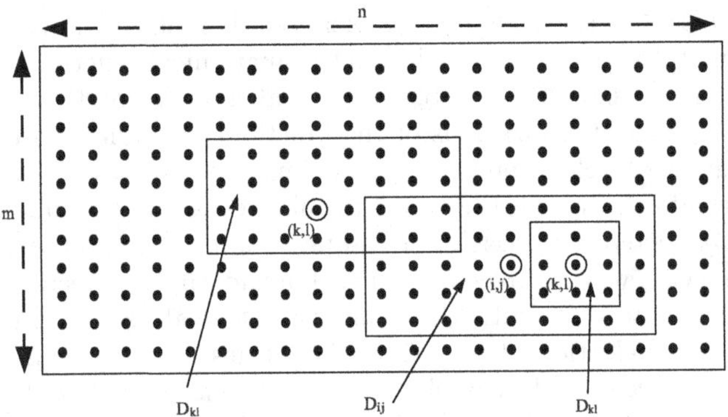

Fig. 2. Area of calculus

Connectivity, (C). Measures how many neurons (or neural nets) receive information from a specific one

$$C = \frac{A + D}{S_0}$$

where A is the axonic field and S_o the elementary area.
Residual Function, (RF). Measures the function that remains after a lesion.

$$(RF)_k = \int_{S-L(\overrightarrow{x})} \mu_k(\overrightarrow{x}) \cdot d\overrightarrow{x}$$

For $\mu_k(\overrightarrow{x}) = \text{cte}$, adding the residual function $R(S-L)$ to the functional deficit (FD), we obtain the redundancy. Thus the greater the redundancy and the connectivity of a layer, the greater its reliability.

In figure 3 we illustrate the two extreme cases of redundancy. For minimum redundancy, $(R = 1)$ there is no overlapping between neighboring dendritic fields and

$$\sum_{i,j}^{n,m} D_{ij} = S$$

For maximum redundancy, $(R = n \cdot m)$ the local areas of calculus are the same as the total ones and

$$D_{ij} = S\,,\; R = n \cdot m$$

Clearly, the redundancy is closely related to the lesion tolerance. A layered computer with maximum redundancy could be practically destroyed and preserve the global function.

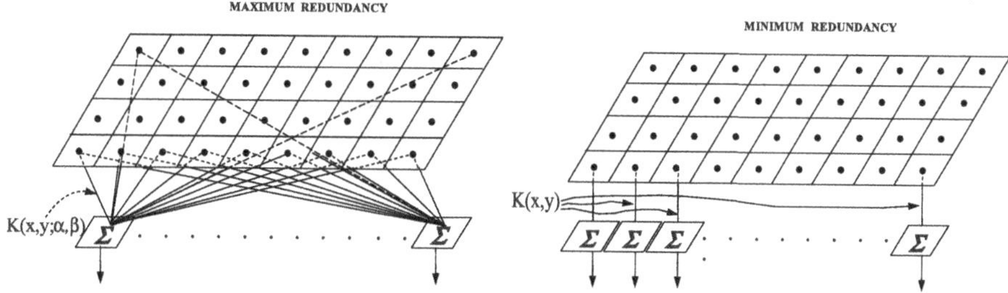

Fig. 3. Redundancy degree

The most we can conclude from the anatomical data at this point is illustrated in the table of figure 4 for the cat's visual cortex. This table has been obtained applying our model to the experimental data of Sholl [30]. We show the neural densities by layer for any kind of neuron as well as for the dominant type (pyramidal, stellate or horizontal). Assuming a lesion of $10^3 D_{ij}$, we obtain the corresponding functional deficit (FD) and residual function (RF). Comparatively, the most reliable layer is the basal part of the pyramidal one.

We can say nothing more about co-operativity at this conceptual level. We need to postulate the possible languages of the brain to go on with the model.

LAYER	NUM. of NEURONS	DOMINANT	S (cm^2)	So (cm^2)	n=m	D_{ij} (μ^2)	A_{ij} (μ^2)	M(D)	R	C(%)	L (μ^2)	FD	RF
II	55	50 P_2	10	14	8370	$D^A=8\cdot10^4$	$3\cdot10^4$	$8\cdot10^{-5}$	5.600	$C_{II-II}=43\cdot10^{-5}$ $C_{II-III}=70\cdot10^{-5}$ $C_{II-IV}=43\cdot10^{-5}$	$8\cdot10^7$	449	$5{,}6\cdot10^3\,(1-9\cdot10^{-2})$
						$D^B=3\cdot10^4$		$3\cdot10^{-5}$	2.100		$3\cdot10^7$	63	$2{,}1\cdot10^3\,(1-3\cdot10^{-5})$
III	62	42 E	10	17	7570	$8\cdot10^4$	$8\cdot10^4$	$8\cdot10^{-5}$	4.720	$C_{III-II}=C_{III-IV}=65\cdot10^{-5}$ $C_{III-III}=94\cdot10^{-5}$	$8\cdot10^7$	337	$4{,}7\cdot10^3\,(1-8\cdot10^{-2})$
IV V VI	207	192 P	10	4	16398	$D^A=8\cdot10^4$	$3\cdot10^4$	$8\cdot10^{-5}$	21.512	$C_{IV,V,VI-II}=150\cdot10^{-5}$ $C_{IV,V,VI-III}=277\cdot10^{-5}$ $C_{IV,V,VI-IV}=150\cdot10^{-5}$	$8\cdot10^7$	1.721	$21{,}5\cdot10^3\,(1-8\cdot10^{-2})$
						$D^B=3\cdot10^4$		$3\cdot10^{-5}$	8.070		$3\cdot10^7$	242	$8{,}1\cdot10^3\,(1-3\cdot10^{-2})$

Fig. 4. Co-operative parameters to the cat's visual cortex

3 Inventing the Languages of the Brain

Let us considered the minimum co-operative structure made up of the two neurons of figure 5, mutually connected and sharing the same afferent information. What kind of messages are they interchanging?. That is the question we need to answer in order to go on with the operational formulation of the co-operative processes in cerebral dynamics.

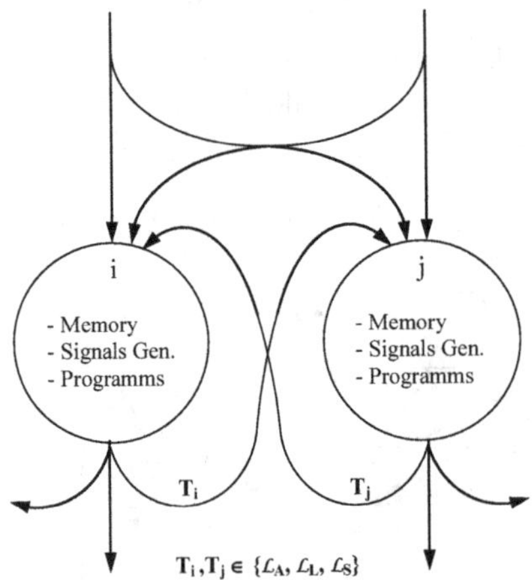

Fig. 5. Two neurons dialogue

According to Pribram [25] from the experimental point of view, the work in neurology is oriented towards the identification of the set of brain codes involved in the different phase of information processing. Thus we can speak of chemical, biophysical (slow potentials), or logical (spikes) languages. However, the more genuine properties of cortical controlled behavior cannot be explained through these languages.

We need to invent more sophisticated languages to describe the machine that has invented so many natural and artificial languages as the brain has done. In fact, it seems a plausible argument to consider that no machine can devise a more complicated language than its own. The neurons of the brain generate the natural languages. So their own language must be as complicated or more complicated than the natural one. Communication between brains takes place by means of signs and symbols, and probably the same is true of the communication between neurons with the interchange of neurophysiological symbols.

Perhaps we haven't dared to propose such complex languages at neuronal level because we have not been able to come to terms with our own inability to recreate a machine of such unimaginable complexity. Thus we forget Kant's principles: "Our representation of things, as these are given to us, does not conform to these things as they are in themselves, but these objects, as appearances, conform to our mode of representation" [23]. Kenneth Craik, in his book "The nature of explanation", also points in the same direction. Even Pitts and McCulloch makes a criticism of this: "That language in which information is communicated to the homunculus... neither needs to be nor is apt to be built on the plan of those languages men use toward one another" [24].

Coming back to he dialogue between the two neurons of figure 5, we can make operational formulations on at least three coexistent languages:

1. *Analogic.* Integro-differential equations linear or nonlinear.
2. *Logic.* Deterministic and/or probabilistic automata and fuzzy logic.
3. *Symbolic.* Something like very complex programming language, with at present unknown mathematical counterparts. Neurons, like human beings in music and painting, probably convey information by "shared agreement".

In modeling, what we are making are formal representations of experimental data and theories in terms of these concrete formal tools. Then the properties and limitations of these mathematical tools are imposed on the models. In fact, to be more precise, the conclusions of a model are always implicit in the nature of the mathematics used in its formulation.

Let us illustrate this argument in the case of our models of co-operativity in neural nets.

4 Operational Formulation in the Analogic Language

In the field of sensorial functions we can use the conceptual model of paragraph two, substituting the modules for effective procedures of calculus. In the analogic case each layer acts as a nonlinear-adaptive-spatio-temporal filter followed by a threshold function and specified by the quintuple:

$$F_i = \{*,\, K_i\,(r,\, S - L,\, t)\,;\, S,\, L,\, \theta\,(L)\}$$

where $*$ is the convolution operator, K_i the local kernel, S and L the functional space and the area of lesion respectively, and $\theta\,(L)$ is the adaptive threshold depending on the size of the lesion.

In figure 6 we show the more frequent connections between layers with the multiple feedback loops included from the Cajal arches.

The function of each layer can now be specified in terms of the corresponding kernel, K_i, in a formally equivalent way to the classical models proposed for the information processing in the retina, as for example suggested by Moreno-Díaz et al (1979, 1980) [21,22]. The kernel characterizes completely to the layer. Thus a periodic kernel detects periodicities, and a short one (stellate layer) detects local properties.

We have some clues that can help us to postulate the kernels corresponding to the different layer:

- *Mean value.* The volume under the kernel must be finite and of value related to the threshold.

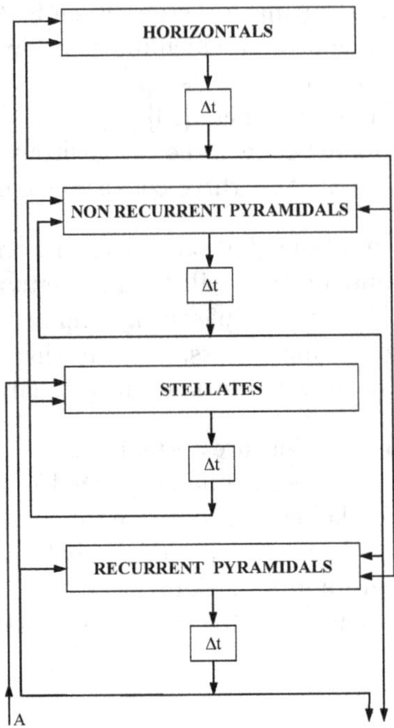

Fig. 6. Cortical layered "computer"

- *Economy* (from Barlow). Data strings without changes are not biologically significants. So the volume under the kernel for a certain interval must be null. This suggests modulated kernels.
- *Inhibition.* In many neurons, the synapses of the soma are inhibitory. So in general it is convenient to assume an inhibitory area around the center of the kernel.
- *Stability.* In the layers where the dominant neuron possesses an axon with collateral branches into the basal dendritic field. There always exists feedback. If this feedback is positive the system would oscillate. So in general, we postulate negative feedback.

From these clues and the anatomical evidence we propose a kernel of the form

$$K_H(x,y) = C_1 e^{-\pi\left[(x-\alpha_1)^2+(y-\beta_1)^2\right]} + C_2 e^{-\pi\left[(x-\alpha_2)^2+(y-\beta_2)^2\right]}$$

The kernel is wide and centered far away from the area where the processed information is discharged. This layer receives information from the stellate layer (III), the C fibres and the recurrent pyramidal layers (IV and VI). Its axon discharge in the pyramidals of all the layers. Figure 7 shows the neural net corresponding to the functional module whose repetition makes up the spatial filter corresponding to the horizontal layer.

To illustrate the dynamic significance of the form of this kernel we have carried out a digital simulation of the processes of spatial convolution threshold, delay

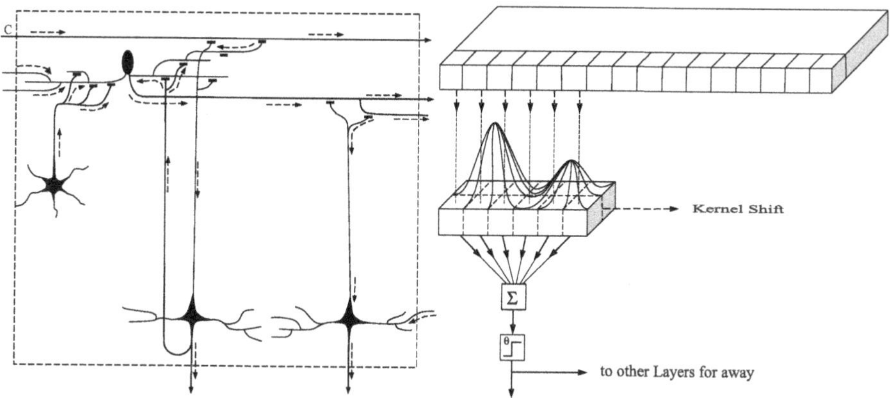

Fig. 7. Horizontal layer arch and filter

and feedback. We stimulate it with a spatial impulse and look for the temporal evolution of the output under constant input, with and without lesion. The corresponding results are showed in figures 8, 9, and 10.

Figure 8 shows the input with a 7×7 lesion. Figure 9 shows the transient and stable outputs for a threshold of $U = 250$, and in figure 10 the same for $U = 50$. We can observe that, although the lesions remain, the elimination of an inhibitory area generates a "false" stimulation in such a way that the threshold is overcome and there is a new area of firing.

In the digital simulation we have approximated the direct and feedback kernels by the following matrices:

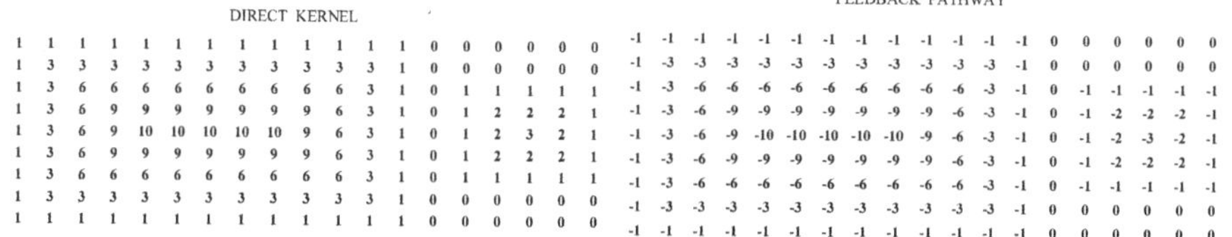

DIRECT KERNEL																		
1	1	1	1	1	1	1	1	1	1	1	1	1	0	0	0	0	0	0
1	3	3	3	3	3	3	3	3	3	3	3	1	0	0	0	0	0	0
1	3	6	6	6	6	6	6	6	6	6	3	1	0	1	1	1	1	1
1	3	6	9	9	9	9	9	9	9	6	3	1	0	1	2	2	2	1
1	3	6	9	10	10	10	10	10	9	6	3	1	0	1	2	3	2	1
1	3	6	9	9	9	9	9	9	9	6	3	1	0	1	2	2	2	1
1	3	6	6	6	6	6	6	6	6	6	3	1	0	1	1	1	1	1
1	3	3	3	3	3	3	3	3	3	3	3	1	0	0	0	0	0	0
1	1	1	1	1	1	1	1	1	1	1	1	1	0	0	0	0	0	0

FEEDBACK PATHWAY																		
-1	-1	-1	-1	-1	-1	-1	-1	-1	-1	-1	-1	-1	0	0	0	0	0	0
-1	-3	-3	-3	-3	-3	-3	-3	-3	-3	-3	-3	-1	0	0	0	0	0	0
-1	-3	-6	-6	-6	-6	-6	-6	-6	-6	-6	-3	-1	0	-1	-1	-1	-1	-1
-1	-3	-6	-9	-9	-9	-9	-9	-9	-9	-6	-3	-1	0	-1	-2	-2	-2	-1
-1	-3	-6	-9	-10	-10	-10	-10	-10	-9	-6	-3	-1	0	-1	-2	-3	-2	-1
-1	-3	-6	-9	-9	-9	-9	-9	-9	-9	-6	-3	-1	0	-1	-2	-2	-2	-1
-1	-3	-6	-6	-6	-6	-6	-6	-6	-6	-6	-3	-1	0	-1	-1	-1	-1	-1
-1	-3	-3	-3	-3	-3	-3	-3	-3	-3	-3	-3	-1	0	0	0	0	0	0
-1	-1	-1	-1	-1	-1	-1	-1	-1	-1	-1	-1	-1	0	0	0	0	0	0

In a similar way we have simulated the pyramidal and stellate layer. Thus, in figure 11 we show the kernel used for the stellate layer according to the Cajal arch shown in figure 1 to illustrate the conceptual model. The analytic expression corresponds to the rotation of the curve of the dendritic tree density.

In the non recurrent pyramidal layer there are two dendritic fields, the apical and the basal one. The apical part picks up information from the horizontal layer and the C fibres. The corresponding kernel could be represented by a double periodic function modulated by the dendritic tree density, such as is shown in figure 12. In the same figure is shown a possible basal kernel (OFF-ON type), centered in the soma and picking up information from B fibres, and collaterals of the axon.

After the simulation of single layers, we connect the whole as in figure 6, as a layered computer composed by four nonlinear spatial filters in cascade.

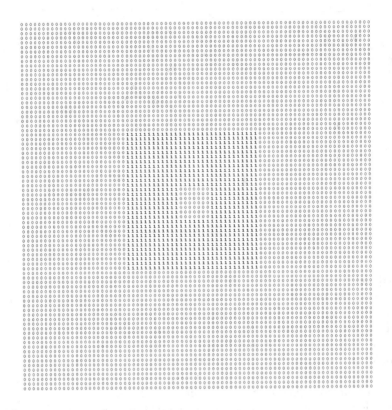

Fig. 8. Stationary input and 7 × 7 lesion to the horizontal layer

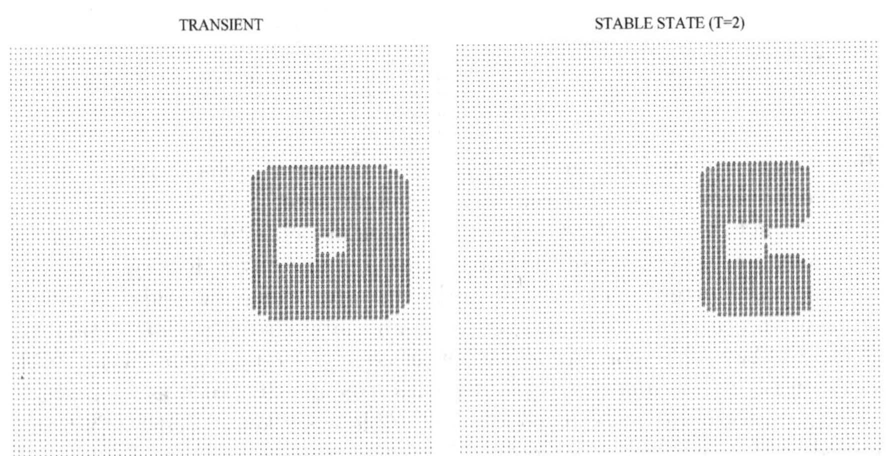

Fig. 9. Output of the horizontal layer (lesion 7 × 7 and $U = 250$)

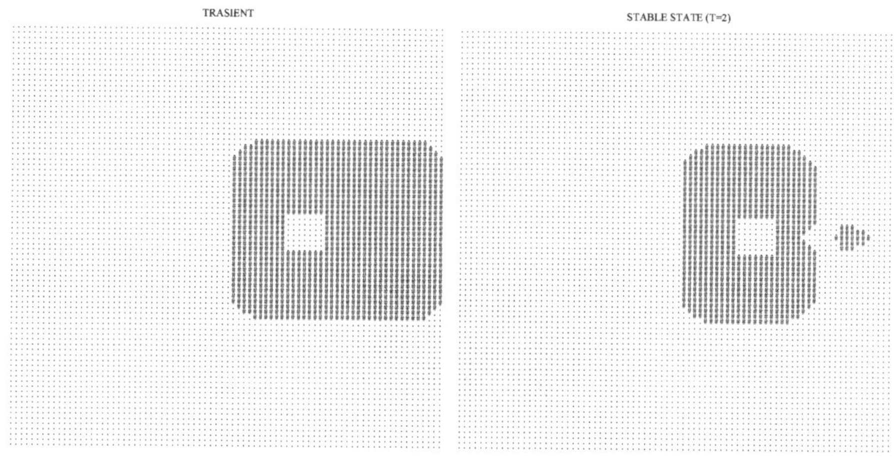

Fig. 10. Output of the horizontal layer (lesion 7 × 7 and $U = 50$)

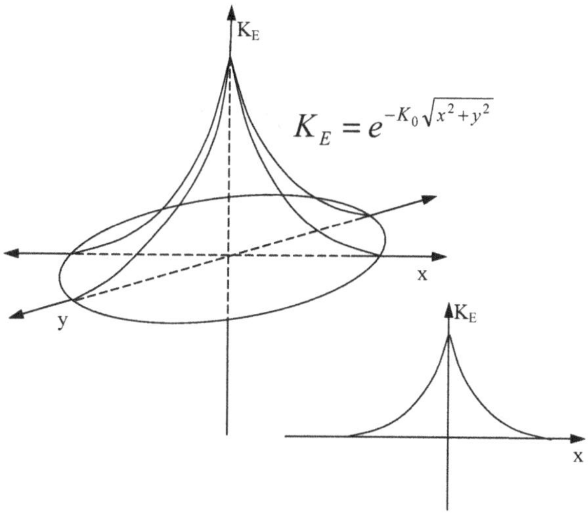

Fig. 11. Kernel for the stellate layer

The feed-forward and feedback connections in the layer are such that each layer can receive information from all the others with axonic fields in the area and send its own information to the same and other layers. Thus, for example, from the stellates goes to pyramidals and then to horizontal coming back to stellates after four synaptic delays. The information is then picked up by the apical dendritic field and processed with the new afferences. Thus, after five synaptic delays the transitory behavior ends and there is stable interaction among all the layers.

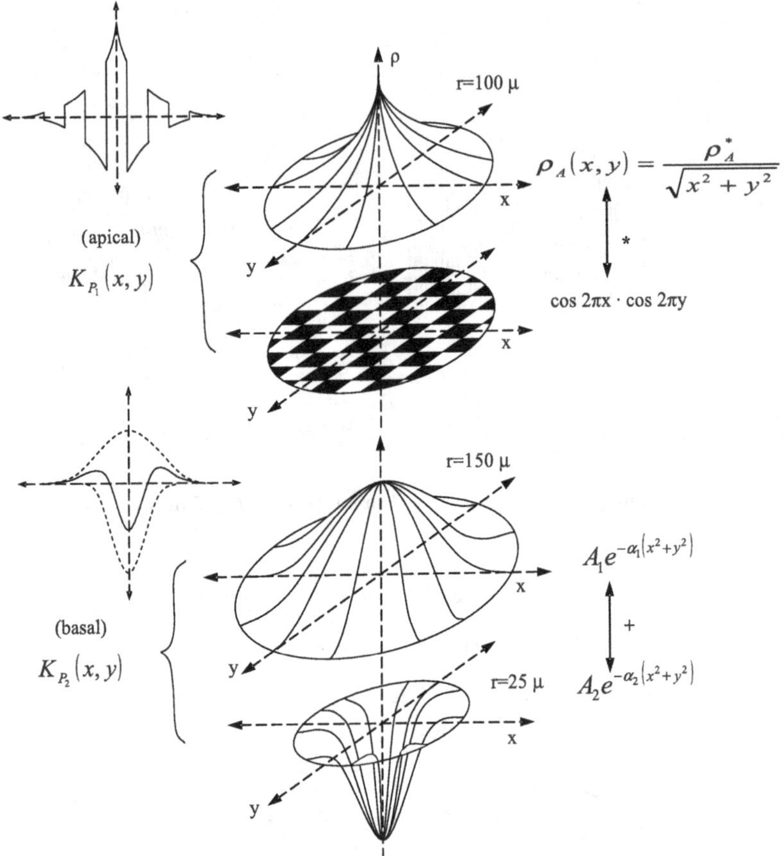

Fig. 12. Kernel for the pyramidal layer

For the sake of simplicity let us suppose that the general input to the layered computer is through the stellate layer (A fibres) and the output is through the recurrent and nonrecurrent pyramidal layers. As in the digital simulation of a single layer, we only consider the spatio-temporal evolution of the outputs under constant input (spatial impulse).

To analyze the lesion tolerance on this model we introduce a 7×7 damaged area in a specific layer and observe its evolution through the others. Thus a lesion 7×7 in the stellate layer is absorbed by the other layers, although it depends on the threshold value. A first defense of a neural net against lesion is the adaptability of its threshold.

Finally, let us illustrate the possible complexity of the model, lowering the thresholds and observing the diversity of firing patterns after stationary spatial impulse input. Figure 13 shows some of the outputs of the stellate layer. The excitatory field is extended oscillating between excitation and inhibition in a non predictable way.

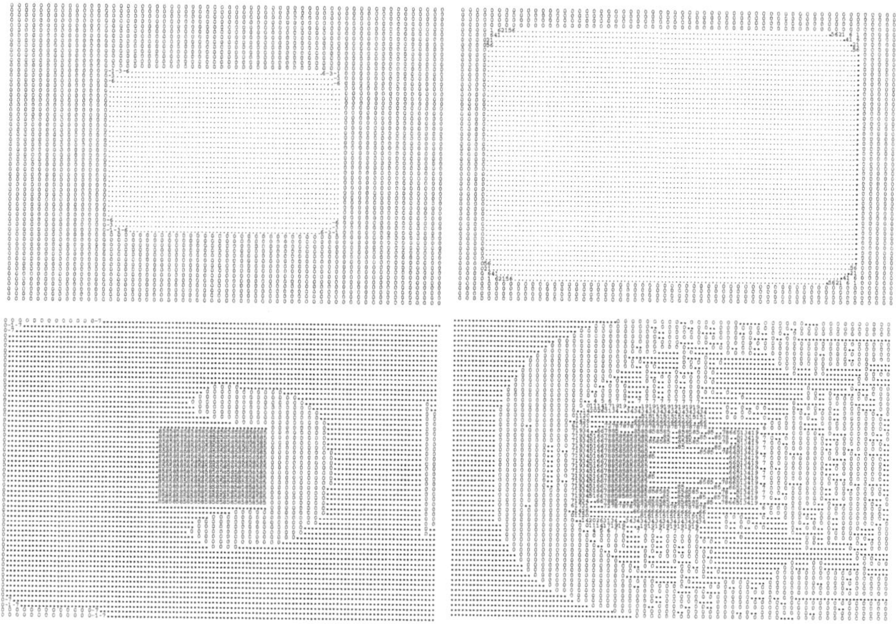

Fig. 13. Temporal evolution of firing patterns

5 The Limits of the Analogic Level

The limits of the analogic language are inherent to mathematical tools used. So we can only speak of the spatio-temporal evolution of excitation, transient behaviors, stability and so on.

To extend the possibilities of this level of description we can make the following improvements:

1. Assume some processing previous to the input to the cortex.
2. Assume plasticity and adaptability of the threshold and kernels.
3. Generate a new theory on nonlinear cortical processing.
4. Elevate to higher languages (logic and symbolic).

The first two points are developed here in an abridged form, the general nonlinear statement is under research in collaboration with professor Moreno-Díaz and for the jump to higher linguistic levels we offer some suggestions in the next paragraphs.

With regard to the preprocessing problem, let us consider the meaning of the lesion on the visual pathway where, from retina to visual cortex, it is plausible to suppose a preprocessing. For example, from the neurophysiological findings of Campbell [3,4,28] concerning the harmonic analysis in spatial frequencies the ganglion cells with small receptive fields will project probably a long way from

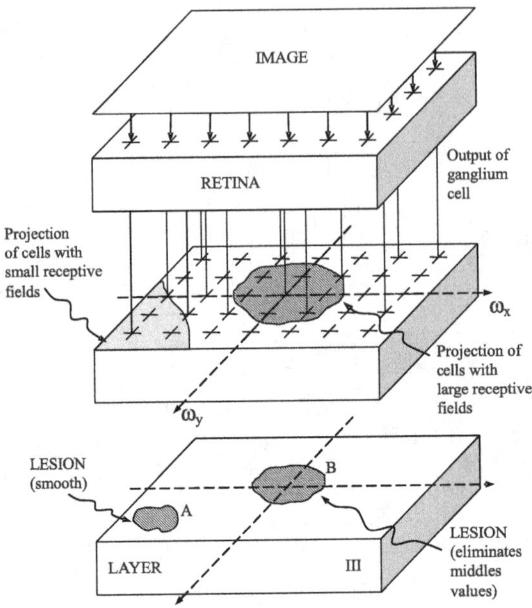

Fig. 14. Lesion on a space of properties

the centre (low selectivity). On the other hand, ganglion cells with large receptive fields will project on the centre (high selectivity).

Thus, we can consider the cortex as a space of properties in such a way that after a lesion such as A, (figure 14), high frequency information is lost (low pass filter) but the global function remains although without the finest details. After a lesion such as B, middle values (high pass filter) are eliminated and in any case the functional deficit is proportional to the eliminated mass tissue (Lashley).

The inclusion in this low-level cooperativity model of the concepts and neurophysiological findings of Luria (1974) and J. Gonzalo (1952) needs a fuzzy formulation of the filter. However, the proposals of J.Gonzalo (1978) on the sensorial reinforcement and facilitation to the spatial summation as a defense of the neural tissue against lesion can be formulated by rendering our model more adaptable.

All the damaged layers reform themselves equivalent to the remainder which have not suffered lesion, obtaining a new kernel by means of any one of the elastic transformations shown in figure 15, where:

T_{C1} = *Excitability* (degree of flatting)
T_{C2}= *Permeability* (expansion of the kernel)
$T_{C1}{\cdot}T_{C2}$= *Compensation* of the excitability decrease by permeability increase.

The coefficients, $C_0\left(L\right), C_1\left(L\right), C_2\left(L\right)$ correspond to the "scale factors" in the residual function. That is to say, any area of the sensorial cortex is characterized by a kernel and a set of scale factors, corresponding to the physiological

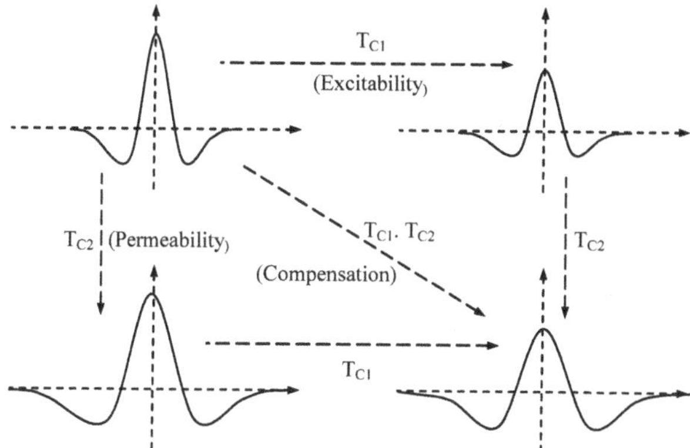

Fig. 15. Elastic transformations group

parameters, excitability, differential sensibility and permeability to the spatio-temporal summation. After lesion the neural tissue holds the global structure and function but works on a different scale:

$$\theta\,(S-L) = C_0\,(L)\cdot\theta\,(S)$$

$$K\,(S-L) = T_{C_1}\cdot K\,(S)$$

That is, the threshold increases and the kernel is flattened. Before this decrease in excitability, the tissue reacts "expanding" the kernel, T_{C_2}, facilitating the spatio-temporal summation,

$$K^*\,(S-L) = T_{C_2}\cdot T_{C_1}\cdot K\,(S)$$

Finally, the differential sensibility is also included in this model. Mach kernel detects the properties inherent in its shape. So is it expands so does its capacity to detect, according to the experimental results of J. Gonzalo [9].

6 Low Level Co-operativity with Logic Language

The logic level in the description of cerebral dynamics started with McCulloch and Pitts classical paper on "A logical calculus of the ideas immanent in nervous activity" [13]. Since then, the initial physiological and psychological motivations have been progressively lost [17] to a great extent, while the theory of these new mathematical beings, the automata, was being built.

At this logical level of description the more elaborate formulation belongs to Moreno-Díaz [20,19,29] and collaborators, who consider the neuron to be a probabilistic automaton result of a measure on the state variables of a deterministic one.

Co-operativity in formal neural nets develops because all the neurons participate in all the microstates of the net contributing a state-variable. The outputs of the net define its observable macroscopical states as measures on the microstates in such a way that co-operativity can be formulated in terms of microscopic and macroscopic descriptions in the physical sense. After lesion there is a deep alteration in the microstates but the global function is associated to the macroscopical one and so, although depressed, it survives. We have dealt with this problem elsewhere [7,5].

7 Towards Symbolic Languages

If we use the theory of automata to represent neural behavior it is clear again that the only questions we can ask are of the following king [17]. When defining any logical function is it always possible to specify an automaton (neural net) which realizes it? When specifying an automaton (neural net) is it always possible to determine the function which defines it? When using unreliable modules is it possible to build a reliable modular automaton? Is it possible to predict the cycles of oscillation as well as the stable states?. In fact these were the questions asked and solved by McCulloch-Pitts, Kleane, von Neumann, Arbib, da Fonseca, Moreno-Díaz, Winograd and Cowan, Minsky and Paper, and many others.

The genuine properties of cerebral dynamics however stand cut to these questions and answers and there is a shared feeling among some researchers who are looking for more sophisticated formal tools. It is our personal belief that we need to invent a new symbolic language to attain a more thorough understanding of the brain.

McCulloch's work on triadic relations [14], the introduction of a Calculus of Signification and Intention by J, Simoes da Fonseca and Mira [33], and the proposal of Santesmases et al [29], to name a few, points in this direction.

Let us consider a possible symbolic language. The elementary symbols would designate things, relations between things and basic concepts relevant to survival strategies. The accumulation of processes resulting from cultural evolution also add to these primary necessities. At a primitive level, the neurophysiological symbols could be related to

1. Alert signals.
2. Upkeep (maintenance) of basic variables and control of the internal milieu (homeostasis).
3. Emotive evaluation of the sensorial world.
4. Programs of automatic behavior (such as conditioned reflex).
5. Any other strategies concerning the stability of the species (sexual behavior and so on).

These symbols must have as referents very specific Physic-Neurophysiologic Patterns, as there is clearly an unequivocal neural coding of physiological variables such as temperature, arterial pressure, respiratory rhythm, and levels of glucose and urea in blood.

From these symbols, the neural nets would generate messages to control elementary behaviors of

1. Eating and drinking.

2. Sleeping.
3. Inhibition-alert.
4. Search (a state of alert with manipulation of the environment).
5. Escape aggression.

Formally, according to the J. da Fonseca interpretation of C. Morris's work [32], we need to specify:

1. When and why a specific symbol is emitted?
2. How it is identified in reception.
3. What kind of behavior it induces?
4. How they are manipulated (syntax and rules of transformation).

Thus, for example, in a function of co-operative decision, the neurons would interchange structural messages in a certain number of fields

$$[Basic\ Information] * [Individual\ Decision] * [Power] * [Identification]$$

Each neuron or neural net receives a lot of these messages and generates a new one as a synthesis or a local decision, which is transmitted to many other units. Let us invent some of the possible elementary symbols from which messages are generated. Each one has the corresponding contrary symbol. Thus, survival importance (SI), has the corresponding (NSI), whose operative meaning in reception, is the inhibition of the meaning corresponding to SI.

Symbols	Meaning in Emission	Meaning in Reception
SI (NSI)	Input or input operated with memory is survival import	1. $EI(\uparrow)$ N = Alert 2. $EI(\uparrow$or $\downarrow)$ AN=Alert T AD
P (M)	Positive emotive valoration of stimulus	P (search) M (alert, escape, aggression)
N (AN)	Input normal or unexpected	N (nothing) AN (alert and memory search)
IN	Internal necessities satisfied (S) or not (N)	INS (nothing) INN (search)
TH TL TM	Auto-evaluation tone of the relative authority TH(high), TL (low), TM (medium)	Compare and select maximum

Thus for example, possible messages are:

$$*SI,\ INN,\ AN,\ M*$$

$$*NSI,\ INS,\ AN,\ P*$$

In the first case, an important to survival situation is described. In the second case a more relaxed situation is present.

From this basic information individual decision are generated ordering the alternatives:

$$*D6,\ D5, D1*$$

where $\{D_i\}$ is the set of possible decisions:

$D0$ (stand in repose)
$D1$ (alert)
$D2$ (search)
$D3$ (satisfaction of internal needs)
$D4$ (approximation)
$D5$ (aggression)
$D6$ (escape)

Possible complete message could be:

$$**SI,\ INN,\ AN,\ M*D6,\ D5,\ D1*TH*NET1**$$

$$**NSI,\ INS,\ AN,\ P*D1,\ D2,\ D0*TL*NET2**$$

where each net includes its power and identification.

From this kind of information representation the neural information processing starts. Giving rise to new formally equivalent messages.

The problem of inventing symbolic languages whose neurophysiological counterparts are the neural codes is really formidable and comes back to the original motivations that were behind the birth of Cybernetics. So let us consider the later proposal in the McCulloch sense: "When I point don't look at my finger!. Look where I point!".

References

1. Beach, F.A., et al. (eds.): The Neuropsychology of Lashley. McGraw Hill, New York (1960)
2. Beurle, R.I.: Properties of mass of cells capable of generating pulse. Phil. Trans. 240, 55–94 (1956)
3. Campbell, F.W., Fulikowsky, J.J.: Orientational selectivity of the human visual system. J. Physiol. 187 (1966)
4. Campbell, F.W., Green, G.G.: Optical and retinal factors affecting visual resolution. J. Physiol. 181, 576–593 (1965)
5. Delgado, A.E.: Modelos neurocibernéticos de dinámica cerebral. PhD thesis, ETSI Telecomunicación. Univ. Politécnica Madrid (1978)
6. Delgado, A.E., Mira, J.: Bases para un modelo de dinámica cerebral. Inf. y Automática (37), 5–16 (1978)
7. Delgado, A.E., Mira, J.: Residual function after the lesion in formal neural nets. In: Lainiotis, D.G., Tzannes, N.S. (eds.) Applications of Information and Control System, pp. 209–216. Reidel (1980)
8. Delgado, A.E., Mira, J., Moreno-Díaz, R.: Un modelo conceptual de dinámica cerebral. Inf. y Automática (39), 5–15 (1979)

9. Gonzalo, J.: Las funciones cerebrales humanas según nuevos datos y bases fisiológicas. Trabajos del Instituto Cajal de Inv. Bio. (XLIV) (1952)
10. Hopkins, A.L., et al.: FTMP-A highly reliable fault-tolerant multiprocessor for aircraft. Proceedings of IEEE 66 (October 1978)
11. Katsuky, D., et al.: Pluribus-an operational fault-tolerant multiprocessor. Proc. of the IEEE 66, 1146–1159 (1978)
12. Luria, A.R.: El cerebro en Acción. Fontanella, Barcelona (1974)
13. McCulloch, W.S.: Embodiments of Mind. MIT Press, Cambridge (1965)
14. McCulloch, W.S., Moreno-Díaz, R.: On a calculus for triadas. In: Caianielo, E. (ed.) Neural Networks, pp. 78–86. Springer, Berlin (1968)
15. Mira, J., Delgado, A.E., Moreno-Díaz, R.: Co-operative processes in cerebral dynamic. In: Lainiotis, D.G., Tzannes, N.S. (eds.) Applications of Information and Control System, pp. 273–280. Reidel (1980)
16. Mira, J., et al.: On the lesion tolerance problem for co-operative processes. In: Richter, L., et al. (eds.) Implementing Functions: Microprocessors and Firmware, pp. 71–80. North-Holland Pub. Comp., Amsterdam (1981)
17. Mira, J., Simoes da Fonseca, J.: Neural nets from the view point of signification and intention. In: Simoes da Fonseca, J. (ed.) Signification and Intention, pp. 13–30. Gulbenkian Found. Pub., Lisboa (1970)
18. Moreno-Díaz, R.: Personal communication. We are indepted to Professor Moreno-díaz by his feelings on the biological simulation of man made machines (1961)
19. Moreno-Díaz, R.: Deterministic and probabilistic neural nets with loops. Mathematical Biosciences 11, 129–236 (1971)
20. Moreno-Díaz, R., McCulloch, W.S.: Circularities in nets and the concept of functional matrices. In: Proctor, L. (ed.) Biocib. of the CNS, pp. 145–150. Littel and Brown, Boston (1969)
21. Moreno-Díaz, R., Rubio, E.: A theoretical model for layered visual processing. Int. J. Bio-medical Computing 10, 231–243 (1979)
22. Moreno-Díaz, R., Rubio, E., Núñez, A.: A layered model for visual processing in avian retina. Biol. Cybernetics 38, 85–89 (1980)
23. Paper, S.: Introduction to the McCulloch book. In: Embodiment of mind. MIT Press, Cambridge (1965)
24. Pitts, W., McCulloch, W.S.: How we know universals, the perception of auditory and visual forms. In: Embodiment of mind, pp. 46–66. MIT Press, Cambridge (1965)
25. Pribram, K.H.: Languages of the Brain. Prentice-Hall, Inc., Englewood Cliffs (1971)
26. Rennels, D.A.: Reconfigurable modular computer networks for spacecraft on-board processing. Computer, 49–59 (July 1976)
27. Rennels, D.A.: Distributed fault-tolerant computer systems. Computer, 55–56 (March 1980)
28. Santesmases, J.G., et al.: Harmonic analysis in the visual pathway of the higher mammals. In: Rose, J. (ed.) Progress of Cybernetics, vol. II, pp. 403–418. Gordon Breach, London (1970)
29. Santesmases, J.G., et al.: Formal neural nets. In: Lainiotis, D.G., Tzannes, N.S. (eds.) Proc. of the First Int. Conf. on Inf. Sc. And Syst., vol. 2. Hemisphere Publishing Corporation, London (1977)
30. Sholl, D.A.: Organización de la corteza cerebral. EUDESA, Buenos Aires (1962)
31. Siewiorek, D.P., et al.: A case study of C.mmp, Cm and C.vmp: part. i and ii. Proceedings of IEEE 66, 1176–1220 (1978)

32. Simoes da Fonseca, J.: Bases neuronais da vida psiquica. PhD thesis, Lisboa (1971)
33. Simoes da Fonseca, J., Mira, J.: A calculus of signification and intention. In: Simoes da Fonseca, J. (ed.) Signification and Intention, pp. 5–11. Gulbenkian Found. Pub., Lisboa (1970)
34. von Neumann, J.: The general logical theory of automata. In: Jefferson, R.A. (ed.) Cerebral Mechanisms in Behavior, Hixon Symposium, New York (1951)
35. Wilson, R., Cowan, J.D.: Excitatory and inhibitory interactions in localized populations of model neurons. Biophysical Journal 12, 1–24 (1972)

On Bridging the Gap between Human Knowledge and Computers: A Personal Tribute in Memory of Prof. J. Mira*

R. Marin

Dept. de Ingeniería de la Información y Comunicaciones, Facultad de Informática, Universidad de Murcia, Spain
roquemm@um.es

Abstract. This paper is an attempt at analyzing the scientific trajectory of Prof. Mira. It is also an attempt at understanding how Prof. Mira's scientific contributions arose from the different pathways of study along which he advanced during his professional lifetime.

1 Introduction

One of the favorites phrases of Prof. José Mira during his last years of life was: "When I grow up I want to be my own fellowship student". He said this to me many times, in fact every time we met in a conference or in a PhD commission. Indeed, this sentence was not a joke, but a piece of advice.

During recent decades, research has become over-bureaucratized and subject to innumerable constraints, especially in Spain. The old cliché "Publish or perish" has now become "Publish, write thousands of project reports, follow the trend, don't leave the beaten track, don't think too much, develop applications and transfer it to business, or perish".

When Prof. Mira said that he wanted to become his own fellowship student, what he meant was that he would prefer to allocate his valuable time to freely thinking about fundamental problems in Artificial Intelligence (AI), as well as to producing contributions that had the potential to push forward the state of the art in this field. However, to achieve this end, he would first have to free himself from the burdens and hindrances of the predominant research system.

This was one of many good pieces of advice that I received from him. But, while I have still not managed to put it into practice, Prof. Mira could and did during his professional life. And I am convinced that it was during his last decade of professional activity that he developed the scientific contributions of greatest value and achieved the most novel and ambitious milestones of his wide scientific career.

He devoted his whole life to understanding the essential differences between human knowledge and machine knowledge, as well as to building bridges to help

* This work was supported by the Spanish Ministerio de Ciencia e Innovación and the European Regional Development Fund of the European Commission (FEDER)under grants TIN2006-15460-C04-01, PET2006-0406 and PET2007-0033.

J. Mira et al. (Eds.): IWINAC 2009, Part I, LNCS 5601, pp. 21–30, 2009.

fill, step by step, the gap between both sides of this divide. In accordance with his free nature, he did not rule out any way of constructing this bridge and advanced from both sides of the gap. He worked passionately on the connectionist perspective of AI from the beginning of his scientific activity. But he also fought on the symbolic perspective front, helping to put into their true context the advances made in Knowledge-based Systems (KBS) and himself proposing significant methodological contributions in the Knowledge Engineering (KE) field.

The IWINAC conference, now in his 3rd edition, was a successful creation of Prof. Mira and is the greatest evidence of his constant concern on devising bridges between natural and artificial intelligence. Allow me to remind you what the IWINAC acronym means: International Work-conference on the *Interplay* between Natural and Artificial Computation.

This paper is an attempt, necessarily brief and incomplete, at analyzing the scientific trajectory of Prof. Mira. It is also an attempt at understanding how Prof. Mira's scientific contributions arose from the different pathways of study along which he advanced during his professional lifetime: pathways that he frequently covered in parallel and simultaneous ways. And, most important for me, this is an attempt to understand the scientific and personal affinities that arose between us during the many years that I was lucky enough to spend working and learning as a researcher alongside to him.

This article is just a partial, personal and inevitably one-sided view of his work. My perspective is constrained by the small fraction of his wide activity in which I took part. Moreover, in my opinion, the main reference for understanding his scientific heritage is the opening talk that the President of UNED (the Spanish Open University) asked him to prepare for the start of the academic year in 2003. To my knowledge, this talk entitled "From human knowledge to machine knowledge" is only published in Spanish [1]. I hope it will soon be translated into English and included in a compilation of his best papers. However, I will borrow some of the most enriching sentences from this talk and use them here for the benefit of the non-spanish reader.

2 Contributions and Trajectory

If I had to choice the most relevant scientific contributions of Prof. Mira, especially from among those belonging to his last stage, I would emphasize three of them.

First, I would stress his work on bio-inspired neural networks. This was his great passion, and one that lasted throughout his scientific life. But this was only his second passion. As we all know, his greatest and most ever-lasting passion was his wife, Prof. Ana Delgado. She played a privileged role in this research line. This was a successful combination if ever there was one. An intense mutual admiration, which everybody could see when they were together, was combined with the outstanding intellectual power of two scientists hand in hand for many years. I am convinced that if fate had not brought together that successful combination, the world would have lost much of what they, together, produced.

Second, I would emphasize his Theory of Two Domains and Three levels. Allow me to name it TD^2L^3, although Prof. Mira preferred to define it as a "methodological building where the different components of knowledge lie, including computable and non-computable parts". Years ago he talked to me about his ideas on this new view of KE.Uninitiated, I did not fully understood the power of this apparently vague theory. But, I could later see its value and I moved from my initial skepticism to admitting that he was right, as usual. Indeed, my intense critical awareness was inculcated in me by Prof. Mira during my research training stage.

In fact, in this third IWINAC conference we can read a posthumous paper of Prof. Mira and collaborators, in which he clearly shows the practical application of this theory in the domain of Intelligent Visual Surveillance. I fervently recommend that you read this paper, in which, we can all admire the lucidity of his proposal. That is indeed a work that contributes to clearing up the mists that have hung over KE for decades.

And third, I would like to mention his interest in modularizing and standardizing design work in KE. He was looking for a method inspired in electronic design. We can implement complex electronic functions by combining elementary building blocks encapsulated in integrated circuits, and these elementary functions are perfectly characterized, predictable and repeatable. Some years ago he wanted to engage me in this daring task, but I am afraid I was not much help. Now I regret not having allocated more time to collaborating in this idea. As I said before, I have not yet been able to follow his advice on becoming my own fellowship student, and to allocating more time to the truly important things that interest all of us.

This last idea on modularizing the KE design task was perhaps his unfinished legacy. If fate had given him more time, this research line and his theory of TD^2L^3 would finally have converged. Of that I am convinced.

To fully understand how he achieved all these astonishing results, we need to reveal the intellectual sources in which he drank. We also need to emphasize his cleverness to integrate these knowledge sources all over his scientific trajectory. An in deep analysis exceed the limits of this paper. And anyway I do not feel qualified to carry out that impossible task. But a few brushstroke should suffice to devise a general overview.

The primary sources for Prof. Mira were undoubtedly his readings in Cybernetics and especially Neurocybernetics. His research training began with two projects headed by Prof. Santesmases and Prof. Moreno-Díaz, the latter of whom would some years later change his status from mentor to collaborator and friend. In these initial projects the topics included neural networks, analysis and control of animal behavior, as well as electronic systems related to neural networks.

Through Prof. Moreno-Diaz, he was influenced by the work of Warren McCulloch. Together with Prof. Da Fonseca, they sought an operational definition of basic concepts of meaning and intention. They studied their relationships with learning, language abilities and other cognitive processes, as well as the relationship with pathological dysfunctions in these processes.

In their contributions to those early projects, Prof. Mira and his colleagues began to sketch out an idea of what later was known as computable knowledge. All this happened around 1968, more than 40 years ago, and it is fascinating to see how the intuition of those pioneers led them to the fundamental problems of AI, which even today are the subject of study and debate.

Later, at the University of Granada, Prof. Mira began his upward progress and became leader of his own research projects. In his previous research training, the line in Neurocybernetics was dominant. Indeed, he maintained a commitment to this line throughout his career. But the electronic implementation of neural networks gradually led him to the development of electronic instrumentation for Biology and Medicine.

In fact, his first funded projects were entitled: "Electronic instrumentation for the rehabilitation of the spasmofemics" (1976) and "Electronic analysis of dysarthria: application to diagnosis and rehabilitation of certain types of mentally disabled" (1979).

Interestingly, in these initial projects, the development of electronic systems was not an end in itself. There was already an interplay between Electronics and Neurocybernetics. In some way, this is similar to what happened later, when he was not only interested in Artificial Intelligence and Natural Intelligence, but primarily in the mutual interactions between them. This conference bears witness to this persistent pursuit.

For example, in his first project he proposed a neurocybernetics model of voluntary motor activity governed by feedback loops. Then, he analyzed the neurophysiological bases of speech and apraxic disorders. And finally this led him to develop electronic prototypes that, according to the previously proposed model, would contribute to the rehabilitation of speech disorders.

But, in his early projects Electronics was also a means to model, simulate and analyze social and cognitive processes. In this way, his interest in distributed architectures for fault-tolerant computing was born, an idea based on biological tolerance to brain damage.

A second example in this line was his interest in studying the temporal evolution of social behavior in animals. In this case, the main assumption was that, in those aspects of animal behavior that are of a social nature, animals are indistinguishable from each other, and react to stimuli as a whole. I think these ideas have their roots in the deep influence of his training as a physicist. In essence, statistical mechanics is a tool to obtain macroscopic descriptions of clusters of indistinguishable particles, and Prof. Mira transferred these ideas to the study of social behavior.

Later, at the University of Santiago, he began a new project entitled "Electronic instrumentation for detection, analysis and rehabilitation of hypoacusic patients" (1982). Part of the project was aimed at the detection and monitoring of deafness from its genesis, i.e. during pregnancy. Prof. Mira directed my doctoral thesis work on this line. During this work, he clearly saw the potential of the then emerging knowledge-based systems (Expert Systems, as we called them then) and their applicability to this specific problem.

I feel that I was a privileged witness of the beginning of what was his third line of research, Knowledge Engineering. But I will speak about this in more detail in the next section, because of the deep implications it had for my own personal research training.

As a summary of this brief description of his early scientific career, we can say that there were three basic axes to the research of Prof. Mira: Neurocybernetics, Electronic Instrumentation in Medicine and Biology and Knowledge Engineering. However, these research lines were never isolated in his mind, and he could see their interactions and their mutually enriching potential.

3 My Own Research Training

I met Prof. Mira in October 1979, when I was a 4th year student of Physical Sciences at the University of Granada. I was beginning my specialization in Electronics. His lessons about General Electronics fascinated me from the beginning. He had a special ability to catch our attention and make clear the most complex concepts.

From the onset, he always encouraged us to think critically, both then as students and after as trainee researchers. He provided solid fundamentals and then we were encouraged to build on them ourselves. Prof. Mira was a very keen agriculturalist, but he not only grew trees. He also grew students and researchers. I like to remember him today as a grower of minds.

Somehow, he opened our minds, implanted original and innovative ideas, and encouraged us to grow. He nurtured us. He watched our growth and channelled any progress we made, in the right direction. In this task, he showed tireless dedication and a profound involvement. But the nutrients that made us grow were simply those tremendous doses of enthusiasm with which we were imbued and the affection he lavished on us. Unavoidably, we quickly became lasting friends.

My degree dissertation, directed by him, was about the design and implementation of a biofeedback system for electromyographic signals. Its aim was to provide the subject with a path that would enable an artificial proprioceptive loop for the voluntary control of muscle contraction and relaxation.More specifically, it was applied to controlling the tone of the facial muscles involved in speech, contributing to the rehabilitation of the spasmophemic subject.

By mid-1981, I was already an assiduous guest at his research laboratory in the University of Granada, but then Prof. Mira became a Professor at Santiago de Compostela University, so we did not have the opportunity to develop closer links at that stage. However, Prof. Mira left a well trained group in Granada, which provided me with valuable help during the final stage of my master thesis work.

Some time later, Prof. Mira told me about a new research project that he was launching in Santiago University and he invited me to join their team as a fellowship student. In September 1982 I arrived in Santiago and started working on my PhD. The goal, focused on the medical domain of Obstetrics, was to develop tools for monitoring and diagnosing the maternal-fetal state in high-risk pregnant patients. There was special emphasis on the antepartum detection of congenital deafness, since the work fell within the framework of the research project entitled "Electronic instrumentation for detection, analysis and rehabilitation of hypoacusic patients" (1982), to which I referred earlier .

One of the tools we developed was a system for the implementation and interpretation of Fetal Acoustic Stress Test, a test whose clinical value was well established, but for which there were no devices permitting the systematic, precise and reliable interpretation of results. Basically, the device provided acoustic stimuli applied to the fetus through the abdominal wall. A fetus in good neurophysiological state usually responds to stimuli with an acceleration of the fetal heart rate and with movements. These signals could be acquired through standard cardiotocographic monitors. Our system acquired the signals through an A/D converter. Then, an old Apple-IIe microcomputer implemented signal processing tasks that detected significant events and interpreted the results [2].

In this work, I worked closely with Dr. Francisco Ríos, who was also doing his PhD work in this area, focusing primarily on electronics and associated signal processing techniques. In these years I also shared offices and weekends with Prof. Emilio Zapata, Prof. Ramón Ruiz, Prof. Diego Cabello, and many others, including Prof. Senén Barro and Dr. María Taboada, all of whom were gradually absorbed into the group. All of us, encouraged by Prof. Mira, quickly became friends.

My contribution in this stage of the project focused on techniques for fetal heart rate signal processing based on Finite State Automata, and later on fuzzy classification. But the core of my thesis followed a very different path.

One day, Prof. Mira came into the lab and gave me a book entitled "Computer-Based Medical Consultations: MYCIN". We had often discussed the need to take in account clinical data in order to correctly interpret signal analysis results. We were aware of the need to use this contextual information to assess the maternal-fetal state. With his usual lucidity, Prof. Mira thought that the emerging Expert Systems could be the key to solving the problem. He looked for the book that could best help me in this new task. The book in question, written by Prof. Edward Shortliffe, described in great detail the design of one of the first successful expert systems and I learnt a lot from it.

Prof. Mira encouraged me to get down to the task immediately, and in a couple of years we had designed and implemented an expert system for diagnosing the antepartum maternal-fetal status. The system managed both data from biomedical signals and contextual data from the clinical history. I named this system CAEMF. He was not very happy with this acronym. Indeed, he was much more skilled in seeking acronyms but he accepted my proposal.

He always maintained that CAEMF was the first expert system implemented in Spain. It is probably true, but it is also true that there were other groups in the country pursuing similar goals and they were starting to publish their results simultaneously. Anyway, a new line of research in the group was launched, focusing first on Knowledge Based Systems and later on Knowledge Engineering. Over the years, this line grew slowly under the tutelage of Prof. Mira, who gradually lent more weight to work in this field. With KBS, Prof. Mira was beginning to build bridges from the other side of the gap, the symbolic perspective. For him this was nothing new. The novelty was the methodology and the kind of representation and inference that KBS provided, clearing the way to practical application in increasingly complex tasks.

Other projects followed, all led by Prof. Mira, and the group gradually shifted its focus from Electronics to KBS. The most immediate consequence was the formal incorporation of three members of the group in the newly born Spanish area of Computer Science and Artificial Intelligence: Prof. Mira himself, his wife Prof. Delgado and myself. As you may well expect, this decision had a profound impact on my future career.

These later research projects were: "Fault-tolerant multiprocessor system for monitoring in Coronary Care Units and Artificial Intelligence programs for emergency pre-diagnosis" (1984), "TAO: A therapy adviser for oncology" (1986) and "Therapy advising in Oncology " (1988). I was not involved in the first one because I was finishing my PhD work. However, I was deeply involved in the other two, especially in the TAO project, funded by the European Commission. This was one of the first ESPRIT projects obtained by Spain after joining the European Union. In fact, this call, ESPRIT-II, was the first to which our country had access. Prof. Mira gave the group a major boost by internationalizing our research.

As a passing anecdote, during one of the demonstrations of the prototypes in Brussels, we had as neighbors the team that worked on the KADS project. KADS later became an important referent in KE. Years later, my future research group at the University of Murcia established a collaboration with one of the youngest members of the team that developed KADS and CommonKADS, Prof. Richard Benjamins. Through him, I met Prof. Ameen Abu-Hanna, from the University of Amsterdam, with whom we have kept up a very fruitful collaboration in Temporal Data Mining.

4 Scientific and Personal Heritage

In this section, I will try to describe how, through my training as a researcher, Prof. Mira had a profound impact on my future career, and most importantly, on my own personal development.

While working on the expert system CAEMF it soon became apparent that the representation language and inference engine that was used in early expert systems, such as MYCIN, were not sufficient to solve the problem. While medical KBS were geared to a single consultation, in our domain we had to follow-up the maternal-fetal state throughout the pregnancy. Data collected in multiple consecutive consultations had to be managed and compared throughout time. It was clear that our system needed some temporal reasoning capability.

At this time, Allen was about to publish a work that introduced the first model for temporal reasoning based on constraints propagation. In contrast, temporal logics lacked the necessary expressiveness in the domain, or were openly untractable. For several weeks, coinciding with a summer holiday, I was in a quagmire. Upon my return, Prof. Mira encouraged me to explore one of the possible ways. The idea was to broaden the knowledge representation language with temporal expressions and extend the inference engine with simple primitive functions to handle temporal data through a kind of date arithmetic. This solution required an implementation from scratch, but after a couple of years, we had a expert system fully implemented in CommonLISP, providing simple but effective temporal reasoning capabilities.

Some time later, I had access to the work of Allen and further temporal models. I saw the possibility of integrating formal models of temporal reasoning based on Constraint Satisfaction Problems with KBS. At this stage, Prof. Mira was about to move to the UNED in Madrid and distance made our collaboration slow and difficult. So, I opened my own line of research in constraint-based temporal reasoning, while Prof. Mira worked in Madrid on the field of temporal logics. For some years, every time we had that opportunity to meet we would exchange information and discuss scientific matters .

Then I moved to the University of Murcia, in 1993, where I set up my own research group and went deeper into this line of temporal reasoning, while Prof. Mira returned to his roots and focused on the field of bio-inspired neural networks. The newly established group of Murcia worked on medical applications of temporal reasoners and, later, on Temporal Data Mining and Temporal Case Based Reasoning. Undoubtedly, all this work stems from the interest in the subject that Prof. Mira instilled in me. Some papers in this field are [3], [4] and [5].

But the influence of Prof. Mira went much further. For example, when working in the fetal acoustic stimulation system, we found a case of congenital deafness. The mother was deaf and the fetus did not respond to acoustic stimuli, even after the completion of several tests to rule out a false positive. That early diagnosis of congenital hearing loss would allow us to begin rehabilitation and education tasks as soon as possible after the birth. As you will all know, the key to success in the rehabilitation of deaf children is an early start.

That case made me feel that the endless hours I had spent working on my thesis had a sense of purpose beyond the purely professional. I realized that well-directed research (and any research directed by Prof. Mira was) could produce tangible benefits, even though it was might be a "one-off". In scientific terms, the result was not very valuable as it was based on a single case, and only led to a rather low quality paper. But the personal impact was deep.

Any subsequent research lines on my own group have always been aimed at applications that could have a potential social benefit. We are now working on AI applications in Medicine and, to a lesser extent, we have worked the fields of Agriculture and Ecology. This is what allows us to feel we are working on something that could be useful from a social point of view and has the potential to improve, even slightly, the world in which we live. This lesson, one of the most important of my life, I learnt from Prof. Mira.

I would like to conclude by quoting some sentences extracted from Prof. Mira's writings [1]. In my opinion they provide a good summary of his fundamental ideas, and were often repeated by word of mouth, or in his papers. I fully support them and they may partially explain the scientific and personal affinities that existed between us.

In accordance with his critical spirit, Prof. Mira was convinced that the pompous terminology used time after time in AI almost since its inception has hindered rather than facilitated an in-depth understanding of the gap between AI's potential and what we are really able to model. Once he wrote [1]:

> "An important part of the language of AI has been imported from cognitive science and the language has created a sense of reality from the repetition of terms, without taking in consideration their change in meaning".

As politicians know well, a lie repeated a thousand times becomes the truth. Chomsky himself once wrote on this basic principle of political propaganda, an idea that has a sad Goebbelian origin. In AI, this unreal terminology is not needed. Prof. Mira knew it, and did not was fooled by this:

> "The world of computing in general, and of AI in particular, has the technological usefulness, complexity and beauty enough for it not to need any added features that it does not possess. The task of interdisciplinary understanding, modeling, programming and formalizing the most genuine human knowledge is fascinating."

As a student of physics, I had learned to appreciate the value of mathematics in physics. Being a student of Prof. Mira I discovered that we both shared the same passion. He taught me to apply mathematics and formal modeling correctly, first in the field of Electronics and later in AI. During my training as a researcher at his side, he continually insisted, not only on the value of formal modeling, but also on its limits. I can think of no better quote than this one that he liked to use:

> "I owe to F. Bombal my knowledge of the work of the Nobel Prize in Physics E. P. Wigner, entitled "The unreasonable effectiveness of mathematics in the natural sciences", where he textually say: "The miracle of the appropriateness of the language of mathematics for formulating physical laws is a wonderful gift that we neither understand nor deserve. We should be grateful for that, and hope it will continue to be valid in the future, and [...] that it may spread to other branches of knowledge".

But Prof. Mira was perfectly aware of the long road ahead, and the uncertainties that the field of AI still hides. He often wrote about the need of new formal tools to understand and describe the cognitive processes. In fact, he was continuing a long tradition of science. Almost 400 years ago, in 1623 Galileo published his book *Il saggiatore* (The Assayer), describing what we would today call the scientific method. It was in this book that we find his famous quotation: "But this grand book [... the Universe ...] cannot be understood unless one first learns to comprehend the language and read the characters in which it is written. It is written in the language of mathematics, and its characters are triangles, circles, and other geometric figures without which it is humanly impossible to understand a single word of it; without these one is wandering in a dark labyrinth". Exactly twenty years after this quotation, Sir Isaac Newton was born and the work of formalizing Physics began.

This thought was applied by Prof. Mira to this new frontier of the knowledge that is the research in AI. Therefore, I would like to end this paper with a last quote from his writings:

> "We are still pursuing the new mathematics that will allows us to model cognitive processes with the same clarity and precision with which differential-integral calculus has allowed us to formulate physical theories [...] This is one of the basic problems of AI, which is obliged to inherit from mathematical logics and mathematical physics the only formal tools available to describe mental processes [...] We do not believe, however, that a logical-formal language suffices to describe the most genuine aspects of human knowledge. Nor are we sure that perception, reasoning and decision processes are essentially formal [...] A computer is based on formal models using logic and mathematics and there is no evidence that these mathematics are powerful enough (or adequate enough) to formalize cognitive processes".

I understand this sentence as a message of hope that addresses the pathway to follow in the future. Perhaps we will see the development of this new language during the next decades. And then, we will remember to Prof. Mira as a pioneer that clearly pointed up toward the future steps in AI. Whatever happens in the future, the memory of an outstanding scientist and an exceptional human being will remain forever in our minds.

References

1. Mira, J.: Del conocer humano al conocer de las máquinas. UNED (2003)
2. Marín, R.: Automatic Acoustic Stimulation System for Nonstress Foetal State Evaluation. Medical and Biological Engineering and Computing 25, 147–154 (1987)
3. Marín, R., Cárdenas, M.A., Balsa, M., Sánchez, J.L.: Obtaining Solutions in Fuzzy Constraint Networks. International Journal of Approximated Reasoning 16(3-4), 261–288 (1997)
4. Navarrete, I., Sattar, A., Wetprasit, R., Marín, R.: On Point-Duration Networks for Temporal Reasoning. Artificial Intelligence 140, 39–70 (2002)
5. Juarez, J.M., Campos, M., Palma, J., Marin, R.: Computing Context-Dependent Temporal Diagnosis in Complex Domains. Expert Systems with Applications 35(3), 991–1010 (2008)

Personal Notes about the Figure and Legate of Professor Mira

Ramón Ruiz-Merino

Dpto. Electrónica, Tecnología de Computadoras y Proyectos,
Universidad Politécnica de Cartagena
ramon.ruiz@upct.es

Abstract. This contribution pretends to be a retrospective look, inevitably partial and subjective, to the scientific legate that Professor José Mira has left us. Pepe Mira has been not only the alma mater of this Conference but also a maestro for many of us. This is my personal tribute and my most felt remembrance for him.

1 Introduction

Those who had the fortune of share a part of our life with Pepe Mira had the opportunity of perceiving his dazzling personality, his humanity and his enormous professional and personal integrity. He was able to face both the profession and the life with a healthy optimism and some ethical principles in which he firmly believed. I have not forgotten some of his more valuable pieces of advice, and I have tryed to put them in practice along my professional life. For decades I had the hope to resemble Professor Mira and to be recognized as a disciple of him, which has been a mark of pride for me. Pepe used to say that profession and life is a race through separated lanes, where each one is free to help the one who is running beside him; however we should never cross the lines to obstruct the way of anybody. This principle defined his personal and ethic way of life.

After our stage at the University of Santiago de Compostela, when the group of young researchers who had been working with Professor Mira for more than a decade began to disseminate among different universities, each one of us discovered that universities and departments were varied, each one having its own positive values, but also its internal problems and miseries, sometimes very evident. We realized that the atmosphere of cooperation, scientific productivity and friendship we enjoyed during this stage was not very usual but rather a privilege. And Professor Mira constituted the true reference for the group and the main responsible of this privileged environment.

I must dedicate a special mention to Professor Ana Delgado, Pepe Mira's wife, his inseparable mate and his most valuable collaborator. I can not remember Pepe without the presence and the support of Ana, without her quiet and self-sacrificing work, her balance and generosity. Ana remains as the principal trustee of Professor Mira's legate, and she is now the only depositary of our estimation and admiration.

J. Mira et al. (Eds.): IWINAC 2009, Part I, LNCS 5601, pp. 31–37, 2009.

In the next, I will briefly review the two periods I had the privilege of having Professor Mira as advisor, the first one at the University of Granada and the second one at the University of Santiago de Compostela. I will try to clarify how this years contributed to my personal and professional growth. I am convinced that in no other group I would have had the opportunities I enjoyed during these years, and surely my professional career would have been less productive and rewarding.

2 1978-1981: Instrumentation for Rehabilitation of the Spasmofemics

There has been a long time from the last years of the seventies, when I met the research team in charge of José Mira at the University of Granada. Then I was a student of Physics, and the research of the group was centered around models of language production from a cybernetic perspective, as well as the development of instrumentation to diagnose and treat language disorders. The interest and deep knowledge of José Mira in relation to physical models describing the central nervous system behaviour was the reason that led the group to this field, and consequently my first research period was strongly determined by this choice. There were two funded projects granted by the Spanish Comisión Asesora de Investigación Científica y Técnica which brought the necessary funds to develop this work.

The start point was the hypothesis that the language, like any other voluntary motor activity, is the result of a multiplicity of feedback loops which are simultaneously active, according to Luria's and Muñoz-Sotés' concepts [1]. The goal was twofold. First, this research pretended to advance in the development of conceptual schemes and language production models. Second, we were involved in designing and implementing electronic instruments whose aim was both to analyse some speech motor dysfunctions through the measure of a set of speech fluency features such as phonemic speed, intensity and rhythmicity, and to allow the rehabilitation of apraxic disorders of spasmofemic nature. The role of this instrumentation was to interfere the kinestesic-auditive feedback loop, by means of the introduction of masking acoustic signals within this loop (noise, delayed speech and low frequency tones), as well as electrical and visual stimuli depending on the patient's response. This constitutes a classical scheme of reinforcement learning, which forced to patients to execute in a voluntary way the motor patterns that were defectively recorded during the language acquisition stage or evoked in conflictive situations.

I remember the fascination that this research field produced on myself. I was a student who was familiar only with the terminology and conceptual schemes of Physics, and in a lesser scale of Electronics, since I was just beginning my speciality cycle in Electromagnetism and Electronics. And suddenly I had to face neurocybernetic models, cerebral function schemes and a wide variety of psychological and physiological concepts and terminology. Even more, in this period I hardly knew a few words of English because my training in foreign language had been done in French. However, I felt I was safe at the extent that José Mira was my advisor, and this feeling convinced myself to enthusiastically initiate my career in this field.

My personal contribution to this project was focused from the beginning in the development of microprocessor-based systems. These devices were components still not well studied in this period, at least in Spanish academic institutions, and of course they were absent at all of the university courses. Documentation about microprocessor was an authentic nightmare, and the existing documents were intricate datasheets and reference manuals. However, Pepe Mira knew that integrated processors had to play a fundamental role in the future, and he employed a substantial part of his time and effort in providing us with a privileged material. Our group at the University of Granada enjoyed one of the scarce microprocessor development systems sold in Spain, and in these privileged conditions I started my knowledge of microprocessor systems together with a group of young researchers who constituted an unrepeatable team gathered by José Mira among his best students. This singular group would remain joined for more than a decade.

Development of microprocessor-based systems posed the need of knowing these systems from different sides, including the support electronics and peripheral circuits, operation cycles, programming in assembler language or the development systems and the corresponding tools to allow high-level programming and debugging. This comprehensive knowledge on microprocessors allowed me to face the design and debugging tasks involved in the implementation of a special purpose microprocessor-based system intended to analyse the speech rhythmicity in terms of three different fluency patterns, a system including a hand-made data acquisition system. The skills I acquired during this first stage were decisive for my future work, and this training would suppose a very valuable starting point in order to face the research topic that constituted my PhD Thesis.

This initial research activity, closely monitored by Pepe Mira, gave place to my MsC dissertation which allowed me to obtain a degree in Physics [2]. This kind of work was not very usual in the context of a Faculty of Physics, more involved in theoretical research and less interested in technological and applied issues. This work, as well as other works aimed to the development of instrumentation which were undertaken in this period, had a common and simple requirement that Pepe Mira imposed: they had to work, to operate correctly before they could be presented. Once this requirement was satisfied, the formal description of the work was a responsibility of Pepe. He dictated us on the fly paragraphs we took to write our reports, and his words were an astonishing portent of structure and clarity, more eloquent and descriptive than many of the texts we used. Also, I still remember with horror the condition that the José Mira's students had to satisfy before a dissertation could be approved: a practical operation test showing the correctness of the instrument in front of the examination board. Nobody else was subjected to such a test; this was the Mira's imprint and an additional proof of his demanding honesty. Fortunately my work satisfied this test and I finally reached my degree in Physics.

My stay at the University of Granada paradoxically ended when I obtained a research fellowship to initiate my doctoral Thesis in this University, being José Mira my doctoral advisor. This grant coincided with his appointment as Professor at the University of Santiago de Compostela and I did not have to think much about the dilemma that came in relation to my future: I decided to follow Professor Mira to his new destination, leaving behind the city of Granada and its University.

3 1981-1989: Electronic Systems for Monitoring Patients in Coronary Care Units

This new stage at the University of Santiago de Compostela led to the opening of new research lines, focusing our interest in applications where the instrumentation gained a new dimension. This new generation of instrumentation was equipped with more sophisticated processing techniques and signal abstraction mechanisms which in some cases were clearly within the new domain of the artificial intelligence.

The aim of my doctoral thesis was to build a monitoring system for patients recovering from acute myocardial infarction within Coronary Care Units [3]. This work was posed according to a two-sided scope. On the one hand, we pretend to construct an information processing system having enough computing power to address this task, which forced us to consider as computing platform a multiprocessor architecture, given the existing integrated processors were unable to individually cope with this computational load. This architecture would need also to implement robust fault-tolerance mechanisms in order to guarantee its correct operation even in the event of failures in any of its modules [4]. On the other hand, we had to provide the system with programs to process the signals coming from the patients, paying special attention to electrocardiogram, so that alarms and warnings could be activated in the event that certain conditions would occur during this monitoring [5].

The previous work of the group in the field of development of microprocessor systems, and specially a multiprocessor architecture designed by Emilio López Zapata as part of his doctoral Thesis, proved to be a very valuable basis to undertake the work related to the processing platform. This multiprocessor architecture was endowed with an operating system designed ad-hoc for this system which implemented a rather complex structure of fault-tolerance strategies. In my case, the training I acquired in the field of microprocessors during my previous stage at Granada allowed me to interpret the key aspects of this multiprocessor architecture, and to undertake its redesign to adapt it to this new application.

The role of Professor Mira in this period was progressively experiencing a change as his young collaborators were growing up. He was no more the ubiquitous tutor whose continued attendance we needed to go on, as it was at Granada.

More and more he let us the autonomy which was essential for our professional development. His interventions acquired a strategic profile, so that he intended most of the time only to guide our work towards where he considered we would have greater possibilities of having success. His proven ability to look at the research lines from a global and interdisciplinary perspective, his criticism and his proverbial eloquence were sufficient to convince us of the convenience or the transcendence of certain approaches. For example, the idea of providing the multiprocessor system with fault-tolerant mechanisms was his, considering different degradation levels for the constituent modules and implementing test and functional replacement strategies in order to permit the survival of some critical functions. Again we may recognize in this idea the interest of Professor Mira in applying neurocybernetic models and mechanisms of functional redundancy, which are present in the nervous system, to artificial multiprocessor systems, bridging the gap between natural and artificial computing structures, and offering genuine and effective solutions in the artificial world. Even the name given to this multiprocessor architecture had evident neurological reminiscences (Lesion-Tolerant Multiprocessor - LTMuP).

This period Professor Mira was not only an inexhaustible source of ideas. He assumed also the role of research manager, given his experience and academic status. And he did take over this unpleasant task absolutely convinced this was a part of his duties within the group. A significant part of his time was destined to obtain the funds which made possible the continuity of our research lines, using his knowledge and wisdom in preparing research proposals which were regularly funded. Those were hours stolen to sleep that Professor Mira devoted to the bureaucracy, of which we were only marginal participants. Anyone of those who shared these years with him can remember the weeks when Pepe Mira had to prepare the documentation for a new proposal, and those endless days in which we saw Pepe and Ana filling forms and writing reports, feeding a bureaucracy to which we have later succumbed inexorably.

This years meant for me the beginning and my growth as university teacher, and I feel compelled to refer to the academic stature of Professor Mira despite the main purpose of this session is to highlight his scientific trajectory. Those who have been fortunate to have him as a teacher when we were students, and later to have his tutelage when we faced our first years as lecturers, have felt amazed and jealous of his extensive knowledge, his endless ability to communicate and his dominance of the language. But above all Professor Mira had a special ability to motivate, to find the seductive aspects of the subjects he taught. These educational values have meant for me a whole reference model throughout my career, a stimulus in the pursuit of the academic excellence that Professor Mira had got.

In this stage at the University of Santiago, the groundwork of our scientific trajectories was established. Professor Mira was able to diversify and open new paths, while maintaining a cohesion and shared goals that stimulated us to know the other research lines and to cooperate with the other group members. Seen in this way, it is not surprising that twenty years after the departure of most of us to other destinations, after having got our professional emancipation, we still continue to seek common ground to cooperate and participate in coordinated projects from the healthy diversity that the time has led us.

4 The Maturity: Inheritance from Professor Mira

It can not be said that those who collaborated with Professor Mira those years at Granada and Santiago de Compostela have followed similar trajectories. The geographical dispersion led to an adjustment of our academic and research interest in line with both the new destinations and the personal experience acquired during the previous years. In fact, the members of the "group of Santiago" we are located in different areas of knowledge, ascribed to a wide variety of faculties and university degrees and our research lines have evolved following very different paths.

May be illustrative to review the research topics which have matured since then within the academic groups leaded by Professor Mira's collaborators, throughout the PhD Thesis which have been presented within these groups during the last fifteen years. Without pretending to cover exhaustively all lines of work, and only as a sample, we can enumerate the following fields: systems for intelligent monitoring in medical applications, hardware architectures and microelectronic systems for speech signal processing with clinical purposes, electronic architectures intended for speech encryption, time-frequency techniques for biosignal variability analysis, thermographic signal analysis to detect landmines, algorithms for temporal data mining, algorithms and hardware architectures for efficient computation in fuzzy systems, neural computation architectures based on adaptive resonance theory, solid modelling of biological structures from computer tomography images, application of ANN to process and interpret medical images, mobile robotics, microelectronic design of systems devoted to deformable active contour processing, efficient hardware architectures for computing of discrete transforms, techniques and tools for the automation of the high-level synthesis of mixed-signal microelectronic systems, electronic solutions in telemedicine, design and modelling of high performance mixed-signal microelectronic systems, etc. And this enumeration represents just a sample of the enormous diversity of topics addressed by researcher who belonged to the group of Santiago, and a significant part of Professor Mira's heritage.

But surprisingly this diversity has not prevented we have looked for opportunities for cooperating in common areas, working together in coordinated research projects. And more important, we have looked for occasions to meet each other, enjoying each of the opportunities we have been able to meet. And always with the figures of Pepe and Ana as reference and unifying link.

Since the years at the University of Santiago, fellowship and friendship ties have survived time, thanks at a great extent to the atmosphere of connivance in an environment of demanding work that Professor Mira established in our Department. These ties have persisted over time, despite the physical distance that separates us at the moment. We all remember with particular affection the celebration of 25th anniversary of the creation of the specialty of Electronics at the Faculty of Physics in the University of Santiago de Compostela, whose responsible was Professor Mira. It was the last time we had the opportunity to meet those who were members of that Department, to pay tribute to the "boss" and to spend some unforgettable days together with those who shared with us a really exciting time. We were all older, but essentially the same friends that years ago shared a common academic project.

Finally, some of us have placed their work in research fields closer to that Professor Mira has devoted his attention the last years. Other of us have addressed research fields conceptually more distant, but certainly consistent with the path determined by our previous work within the group which leaded Professor Mira. Anyway, I am convinced that we are what we are thanks to those years when we grew in this profession, and thanks in a large part to the influence of Professor Mira. He fought for each and every one of us, and provided us with the best background any one can wish for: the training and the professional independence which allowed us to face our future with ambition, confidence and self-assurance in ourselves.

Thank you, Pepe.

Acknowledgements

This work was supported by the Spanish Ministry of Ciencia e Innovación (MICINN) and the European Regional Development Fund of the European Commission (FEDER) under grant TIN2006-15460-C04-04.

References

1. Mira, J., Delgado, A.E., Zapata, E.L., Ríos, F., Ruiz, R.: Electronics Aids in the Analysis of Speech Fluency Disorders. In: 5th International Congress on Cybernetics and Systems, México, Julio (1981)
2. Ruiz, R.: Un Sistema Microprocesador para el Tratamiento Digital de la Voz, Tesina de Licenciatura. Universidad de Granada (1981) (in Spanish)
3. Ruiz, R.: Sistema Multiprocesador para Monitorización Interactiva en UCC, Tesis Doctoral. Universidad de Santiago de Compostela (1986) (in Spanish)
4. Ruiz, R., Zapata, E.L., Mira, J.: Reconfiguration in LTMP (Lesion-tolerant Multimicroprocessor System). In: Albertos, P., de la Puente, J.A. (eds.) IFAC LCA 1986: Components, Instruments and Techniques for Low Cost Automation and Applications, Valencia, November 1986, pp. 563–567 (1986)
5. Ruiz, R., Hernández, C., Mira, J.: Method for Mapping Cardiac Arrhythmias in Real Time using Microprocessor Based Systems. Medical and Biological Engineering and Computing 22, 160–167 (1984)

Intelligent Patient Monitoring: From Hardware to *Learnware*

Senén Barro

Dept. Electrónica e Computación, Universidade de Santiago de Compostela, 15782 Santiago de Compostela, Spain
senen.barro@usc.es

1 Introduction

Some time ago, in response to Proust's musings on the role of the writer in language, the French philosopher Gilles Deleuze described literary creation as the result of a series of events that take place on the edge of language [11]. This figure is easily recognisable by anyone who, at some point, has decided to move to that territory in permanent tension which is the edge of knowledge; that fine line which many reach seeing naught but the path by which they have arrived, and which very few are capable of taking further away. Prof. José Mira knew that place well. And those fortunate enough to accompany him had the opportunity to learn through him part of what makes up the essential baggage for undertaking that journey: simple things, such as honesty, creativity and enthusiasm.

Throughout his scientific career he conducted research which, taking cybernetics as a starting point [15], would engage in the study of those physical or formal systems capable of performing some form of computation and which, inspired by human beings, would be incorporated into the, then, still incipient science which we call Artificial Intelligence.

One of the constant themes running throughout his academic life was the need to conduct research that would repay the society that sponsors this research with a benefit to improve it, and which he wished to direct towards the domain of medicine, which, conversely, would also enable him to set the human being as the object of study, and thus aspire to an *"understanding of the living being and its dialectic interaction with the environment of the both the individual and of the species"*, which would ultimately need to be conducted *"with the same precision and formal beauty as physics and mathematics"*. It is this compulsion which underlies the research projects: "ELECTRONIC INSTRUMENTATION FOR THE REHABILITATION OF SPASMOPHEMIA", "ELECTRONIC ANALYSIS OF DYSARTHRIA: APPLICATION TO THE DIAGNOSIS AND REHABILITATION OF CERTAIN TYPES OF THE MENTALLY DISABLED " and "ELECTRONIC INSTRUMENTATION FOR THE DETECTION, ANALYSIS AND REHABILITATION OF HYPACOUSIA". Projects which share two fundamental goals: the understanding of the phenomena underlying certain pathologies specific to neurology, and the desire to obtain results to aid in the rehabilitation of specific groups of patients.

This study soon evidenced the need to monitor a set of biopotentials - basically, the Electrocardiogram (ECG), the Electromyogram (EMG) and the

J. Mira et al. (Eds.): IWINAC 2009, Part I, LNCS 5601, pp. 38–45, 2009.

Electroencephalogram (EEG)- and to develop to that end specific technology and methods and techniques for the digital processing of bio-signals. In keeping with the aforementioned line of work, Prof. Mira undertook the research Project entitled "FAULT-TOLERANT MULTIPROCESSOR SYSTEM FOR THE MONITORING OF A CORONARY UNIT AND ARTIFICIAL INTELLIGENCE PROGRAMS FOR URGENT PRE-DIAGNOSES", thus providing a specific nature to a line of research that would occupy the majority of my career as a scientist, and which provides the title for this conference: intelligent patient monitoring.

2 First Stage: Monitoring Hardware

The problem posed was relevant: the monitoring and detection of cardiac arrhythmias in patients suffering from some type of ischemic cardiopathy and admitted to an Intensive Care Unit (ICU). Confirmation of the importance of this type of monitoring is to be found in the fact that in these types of patients any form of rapid ventricular arrhythmia can easily degenerate into episodes of ventricular fibrillation, resulting in cardiac arrest in a matter of seconds.

The solution proposed consisted in analysing those intervals associated with ventricular electrical activity, over a single-channel ECG signal. After characterising the ventricular activity associated to each beat by means of a set of descriptors, a classification was made. With this in mind, a specially designed multiprocessor system was developed to ensure the processing of the ECG signal in real time, and to demonstrate a resistance to faults which had to endow the system with the margin of reliability essential for the critical nature of the monitoring tackled. This solution, developed on the basis of the architecture proposed by Prof. Emilio Zapata, was the subject matter of the PhD thesis of Prof. Ramón Ruiz [17], who, along with Prof. Mira, would be the supervisor for my own PhD thesis.

In 1985, after obtaining a degree in Physics, I joined the research group led by Prof. Mira in the Department of Particle Physics and Electronics in the University of Santiago de Compostela. My contribution to the research conducted into patient monitoring would lead to the creation of the first version of the SUTIL system, an intelligent monitoring system for intensive and exhaustive follow up of patients in ICU, which was the aim of my PhD thesis [1].

SUTIL was conceived as an evolution of the previous system. With regard to its physical structure, SUTIL's hardware was selected on the basis of criteria of flexibility in the definition and expansion of the system. This condition determined the selection of a multiprocessor architecture for the acquisition and process of multiple low- and medium-level signals, which would allow the incorporation of other signals of importance in the ICU, starting with the pressure in various cavities. This architecture comprised modules compatible with the widely used VME standard. The mass storage of information, its high-level treatment and user-system interaction were resolved using a compatible PC connected via a serial communication channel. This initial version of SUTIL was installed in the University Hospital of Elche where, under the supervision of Dr Francisco

Palacios, it proved to be a useful tool from the perspectives of both healthcare and research.

The rapid evolution of electronics would result in the swift appearance of standard monitoring equipment, from a number of commercial companies, with a drastic reduction in implementation costs, and which currently forms part of the basic instrumentation of any ICU. Since the first version of SUTIL, we have witnessed a great deal of technological progress, above all in the field of telecommunications, which has rendered both monitoring itself and access to the information it supplies ubiquitous. The telemonitoring of patients is now commonplace [8], and the availability of ever-smaller devices, along with the development of new sensors, means that we will soon have sophisticated monitoring devices incorporated into our clothing and, without noticing, into the setting where we conduct our daily activity.

3 Second Stage: Knowledge-Based Monitoring

Here it should be noted that the technological development referred to in the previous section has had little effect on the functions supplied by the commercial monitoring systems over the last thirty years. Nevertheless, the objectives proposed in patient monitoring by Prof. Mira were already highly ambitious from the outset: monitoring could not simply be limited to rendering a set of patients' physiological variables visible, in a more or less elaborated manner. In order for SUTIL to be of clear clinical value, it would need to endow information with greater semantic content, taking it closer to the language that the physician uses in decision-making.

The solution was found in Artificial Intelligence; in his own words: *"the human capacity for abstraction, symbolic processing, the handling of uncertainty and the intensive use of the different types of knowledge characteristic of the medical domain mean that the interpretation [...] of biomedical signals will always more reliable than that obtained by conventional digital processing techniques"*. This premise, which guided the work conducted in the aforementioned project, has always been the same initial hypothesis from which all the projects that I have directed throughout my scientific career have stemmed.

In that early project, the structuring based on the levels of processing which was performed on the acquired signals added a final level, where a processing procedure was implemented using an acute myocardial infarct knowledge base in a distinguished manner [16].

This initial approach to the problem of intelligent patient monitoring enabled us to grasp the extent of the problem, and to characterise the drawbacks that we would encounter, which years later Prof. Roque Marín and I would summarise in [9]: 1) the complexity of the human body and of the physio-pathological processes that take place in it; 2) the enormous quantity of knowledge available on the human being and, which is worse, the still greater lack of knowledge; 3) the nature of a large extent of this knowledge, which is characteristic of what is usually referred to as "common sense" knowledge, and it is expressed by means of natural language; 4) the great degree of variability that is shown by different patients, even with the same diagnoses and similar therapeutic actions, and even within the same patient over time; 5) the vast amount of data which it is necessary to handle, leading to an overload of data and information for medical staff, which, on occasion, may hinder more than help in the decision-making process. This characterisation has enabled us to determine the nature of the formal and methodological tools with which to tackle the problem of monitoring.

One of the most salient features of medical knowledge is its imprecision and uncertainty, a trait directly related with the fact that the habitual medium for transmitting this knowledge is natural language. The need to render its representation and subsequent reasoning computable has accounted for a substantial part of the efforts of the research group which I have been leading since 1990. An early work in this setting dealt with the classification of heartbeats described linguistically by a medical specialist [2,4,5,6,10]. The formal framework chosen was Fuzzy Set Theory, within the bounds of which we have conducted research, both theoretical and applied, which has given rise to a set of models, techniques and tools that have proved to be effective in intelligent patient monitoring. Here, specific mention should be made of the problem of representing and reasoning over temporal information [7,14], as these aspects are absolutely essential in the monitoring domain, where everything of relevance for the patient occurs over time, and it is its specific evolution and temporal layout, with regard to the ordered set of events that define its context, which requires an interpretation or diagnosis, and a response or therapy. Subsequent studies dealt with problems such as the identification of ischemic episodes, hypovolemic shock patterns and obstructive apnea, among others, with the aim of reducing the workload of the clinical staff by means of the generation of a set of alarms with a great semantic content through temporal abstraction techniques [12,18].

Another of the problems which proved to be highly important in the patient-monitoring field was that of computationally handling knowledge from the domain, which Prof. Mira identified at an early stage with expert knowledge [16]. He soon came to understand that the handling of knowledge required a systematic approach, which would give rise to what would later be referred to as Knowledge Engineering, and to which he made numerous contributions. In this matter, patient monitoring posed certain specific challenges: the tasks to be performed would need to be organised, on one hand, according to the level of abstraction that they occupied in terms of the processing of the knowledge available, and on the other, the real-time requirements posed by the problem. Thus arose the need to experiment with different architectures, and to research the modelling of knowledge in the monitoring domain [19].

4 Third Stage: *Learnware*

One of the problems posed by intelligent patient monitoring is related with some of the abovementioned drawbacks: the still greater lack of medical knowledge, and the great degree of variability that is shown by different patients, even with the same diagnoses and similar therapeutic actions, and even within the same patient over time. The result of this is that an approach based solely on the use of the knowledge available on a domain must necessarily be incomplete; coupled with this is the difficulty inherent in the elicitation of certain forms of knowledge which are not always easily recognisable or expressible, if not through the use made of them.

The solution would have to be found in *"a change in the computational model, where external programming is substituted by learning via training, using supervised and unsupervised algorithms"*. The inspiration would come from nature: *"biological computation is an unlimited source of inspiration, and nervous systems [...] are the most advanced alternatives to von Neumann machines and to the syntactic-semantic rigidity of conventional programming"*. Artificial Neural Networks constituted a research setting, within the framework of Artificial Intelligence, where Prof. Mira made contributions at the very onset of their development in Spain. Nevertheless, he was always wary of the euphoria and opportunism that surrounded the publication of the results he was obtaining, which he always wished to evaluate on merit: *"neural computation is currently understood as all computation that is modular, distributed, with small-grained, autonomous processes and/or processors generally organised into a multi-layer architecture, and in which part of the programming is substituted by learning, understood as the adjustment of parameters by correlational or supervised procedures."*

The application of neural computation to patient monitoring seemed appropriate, particularly to the morphological characterisation of the heart beat, understood as the assignation of each beat to the different morphological classes. Morphological characterisation is usually a step prior to determining the origin of cardiac activation and the detection of arrhythmias. This problem responds to the drawbacks referred to above: normal heart beat morphologies vary greatly between different patients, and even within the same patient the morphological classes may by very different, depending on the derivation of EGC being considered; new morphological classes appear during monitoring, and these may evolve over time, depending on a number of factors, such as nervous tension, mood state and digestive activity, among others. Moreover, signals may be contaminated during monitoring with noise from a number of origins (muscular, electrical, respiratory, etc.); this increases the morphological variability of the classes present, and can even introduce artefacts that may appear to be new classes. This requires continuous learning throughout processing and a certain capacity to adapt to the changes in morphologies by means of updating of the classes learned; this must also selectively evaluate the different signal channels on the basis of criteria of flexibility, quality or interest applied to each of them. Our proposal consisted of the design of a new multi-channel neural computation architecture based on ART network architecture, thus satisfying the requirements stipulated above [13].

Another of the drawbacks highlighted above has now, at the same time, been converted into an opportunity: the vast amount of data which it is necessary to handle, leading to an overload of data and information for medical staff. The development of new, ever more sophisticated sensors has increased the availability of data which show different perspectives from the physio-pathological process that we aim to analyse, providing medical research with hitherto unexploited resources. Moreover, the development of telecommunications technology has enabled us to extend the setting of medical research, which was limited to hospital units but now incorporates the very homes of patients, enabling us to ascertain what occurs in the initial phases of the development of an illness, or of a subsequent deterioration. Here Artificial Intelligence's vocation is to play a key role in developing data analysis tools, to find unsuspected correlations and describe data arising from new types of monitoring, supplying useful, comprehensible information to improve our knowledge of the illness. This resulted in the inception of a new discipline, Data Mining, in which we are working actively at the present. Even though Data Mining has an extensive recognised set of tools for tackling data analysis, patient monitoring poses a number of problems which require new approaches. More specifically, the temporal nature of the data under analysis, wherein resides extremely valuable information for better understanding the processes that underlie the illness. Thus was also born the research field which has come to be known as Temporal Data Mining, and which is still in its infancy.

5 Epilogue

Among the characteristics that best defined Prof. Mira were his outstanding scientific intuition, his capacity to conceive ideas from where others find themselves out of their depth, while many look on from the shore, and his talent in transforming them into reality, to make them possible in front of our very eyes.

One example of Prof. Mira's exceptional scientific vision was witnessed by those of us lucky enough to attend the Summer Course organised in the Spanish Open University (UNED) in 1990, a short time after his transfer from the University of Santiago de Compostela. The title of the course was *Artificial Intelligence in Medicine*, and the majority of those who had written theses under his supervision were invited as lecturers. It was a wonderful opportunity to witness the scope of the research that Prof. Mira had embarked upon, since each lecture corresponded broadly to what are now the most important lines of investigation into Artificial Intelligence followed in Spain and, in particular, their application in the field of medicine.

Allow me to provide a further example of Prof. Mira's clear-sightedness. In the Introduction, I mentioned the project which provided the impetus for my research into intelligent patient monitoring: " FAULT-TOLERANT MULTIPROCESSOR SYSTEM FOR THE MONITORING OF A CORONARY UNIT AND ARTIFICIAL INTELLIGENCE PROGRAMS FOR URGENT PRE-DIAGNOSES". A successful application for funding for this project was submitted in 1984; 25 years down the line,

I would not hesitate to apply for funding for a research project of the same name. That project proved to be a true scientific programme, evidencing the human referent that has always inspired the research of Prof. Mira: that fault-tolerant system which is still a challenge today.

These brief and hastily gathered notes pay tribute to the man who was my mentor. His account of the research that we have shared will no doubt find future scientists in more than one of their readers. The writer, Javier Marias, once said that certain writers imbue us with enthusiasm, the desire and even the need to write; others are so intimidating that you never want to pick up even a pencil ever again. As a scientist, Prof. Mira clearly belonged to the former group. His passion for science rubbed off on you, as was evident in each and every one of the lectures he gave, especially at that fleeting, enlightening moment that occurs when the lecture comes to an end and the spirit of the audience is revealed. Today, those of us privileged to have been touched by his presence and his teaching are deeply saddened by the knowledge that this silence will never again be filled by his resounding voice.

References

1. Barro, S.: SUTIL: Sistema cuasi-integral de monitorización inteligente en UCC. PhD thesis, Universidad de Santiago de Compostela (1988)
2. Barro, S., Ruiz, R., Mira, J.: Fuzzy beats labelling for intelligent arrhythmia monitoring. Computers and Biomedical Research 23, 240–258 (1990)
3. Barro, S., Ruiz, R., Mira, J.: A multi-microprocessor for on-line monitoring in a CCU. Medical & Biological Engineering & Computing 28, 339–349 (1990)
4. Barro, S., Ruiz, R., Presedo, J., Cabello, D., Mira, J.: Arrhythmia monitoring using fuzzy reasoning. In: 12th Annual International Conference of the IEEE Engineering in Medicine and Biology Society (1990)
5. Barro, S., Ruiz, R., Presedo, J., Mira, J.: Grammatic representation of beat sequences for fuzzy arrhythmia diagnosis. International Journal of Biomedical Computing, 245–259 (1991)
6. Barro, S., Ruiz, R., Marín, R., Presedo, J., Bugarín, A., Mira, J.: Knowledge-based fuzzy classification of signal events. Lecture Notes in Medical Informatics 45, 428–433 (1991)
7. Barro, S., Marín, R., Mira, J., Patón, A.R.: A model and a language for the fuzzy representation and handling of time. Fuzzy Sets and Systems 61, 153–175 (1994)
8. Barro, S., Castro, D., Fernández-Delgado, M., Fraga, S., Lama, M., Presedo, J., Vila, J.A.: Intelligent Telemonitoring of Critical-Care Patients. IEEE Engineering in Medicine and Biology Magazine 18, 80–88 (1999)
9. Barro, S., Marín, R.: A call for a stronger role for fuzzy logic in medicine. In: Barro, S., Marín, R. (eds.) Fuzzy Logic in Medicine. Studies in Fuzziness and Soft Computing, pp. 1–17. Springer, Heidelberg (2002)
10. Cabello, D., Salcedo, J.M., Barro, S., Ruiz, R., Mira, J.: Fuzzy k-nearest neighbor classifier for ventricular arrythmia detection. International Journal of Biomedical Computing, 77–93 (1991)
11. Deleuze, G.: Critique and Clinics, Paris (1993)
12. Félix, P., Barro, S., Marín, R.: Fuzzy constraint networks for signal pattern recognition. Artificial Intelligence 148, 103–140 (2003)

13. Fernández-Delgado, M., Barro, S.: MART: A multichannel ART-based neural network. IEEE Transactions in Neural Networks 9(1), 139–150 (1998)
14. Marín, R., Barro, S., Bosch, A., Mira, J.: Modelling the representation of time from a fuzzy perspective. Cybernetics and Systems 25, 217–231 (1994)
15. Mira, J.: Modelos cibernéticos del aprendizaje. PhD thesis, Universidad de Madrid (1971)
16. Mira, J., Otero, R.P., Barro, S., Barreiro, A.: On knowledge based system in CCUs: monitoring in patients follow-up. Lecture Notes in Medical Informatics 45, 174–192 (1991)
17. Ruiz, R.: Sistema multiprocesador para monitorización interactiva en UCC. PhD thesis, Universidad de Santiago de Compostela (1986)
18. Presedo, J., Vila, J., Barro, S., Ruiz, R., Palacios, F., Taddei, A., Emdin, M.: Fuzzy modelling of the expert's knowledge in ECG-based ischaemia detection. Fuzzy Sets and Systems 77, 63–75 (1996)
19. Taboada, M.J., Lama, M., Barro, S., Marín, R., Mira, J., Palacios, F.: A problem-solving method for 'unprotocolised' therapy administration task in medicine. Artificial Intelligence in Medicine 17, 157–180 (1999)

A Look toward the Past of My Work with the Professor José Mira

Francisco Javier Ríos Gómez

Departamento de Electrónica, ETS de Ingenieros Industriales
Universidad de Málaga, Spain
fjrios@uma.es

Abstract. The José Mira's projection has been very important in my personal and professional life. I was their pupil in the faculty and in the development of my doctoral thesis, co-worker in their first investigation projects and, however, his friend. I have a great regret due to their sudden disappearance, and I want that these notes are of acknowledgment like homage toward their work and dedication in the scientific aspects and in their human quality, their generosity and capacity to integrate to people and ideas in a friendly and family environment. I present a summary of the topics in those that I worked with him, the investigation lines that opened up and that they affected me directly among the years 1981 to 1989 in their stages of Granada and Santiago de Compostela and what their work projected on mine and in my professional later career. I include some anecdotes that I have remembered while was writing.

1 Introduction

I met the Professor José Mira in the year 1978 being student of the specialty of Electronic of the Sciences Faculty (Physics section) of the University of Granada. I was fascinated by their peculiar way of imparting the Subject of Electronic as a closed and self-sufficient science in their bases (he said). I discovered my ability for the electronics design systems and he also discovered it, because it proposed me to make the work of Degree related with the analysis of the biological signals. He also intended me to accompany him in the adventure of going to Santiago of Compostela with the purpose of creating the specialty of Electronic in the grade of Physical Sciences. I was nine years in Santiago being formed as educational and investigator and it directed me the Doctoral Thesis that I defended in the 1987. Being already professor of university, he understood fulfilled their formative mission with me and, like he had made with other collaborators, it allowed flying my interests so that I traced my future. He left in the 1989 to UNED in Madrid. And I left Santiago two years later toward the University of Málaga where I reside currently. From the 1989 up to 1995 I maintained a relationship with him in the distance, sharing impressions and making him participant as consultant in some of my initiatives. In 1995 he Presided the Tribunal of the first Thesis that I directed and, although my main line of work began to move away from his, I completed some of their proposals in connection with the development of

J. Mira et al. (Eds.): IWINAC 2009, Part I, LNCS 5601, pp. 46–52, 2009.

Computer Integral Systems of help to the diagnosis and of acquisition of clinical parameters, the analysis of the speech signals and the physical implementation of some algorithms in Applications Specific Integrated Circuits (ASICs), some of them suggested by himself in the Granada stage. I saw him in the 2007, in the celebration of the 25 anniversary of the electronic specialty in the grade of physical sciences of Santiago's University. He gave me a great happiness, seeing that he hadn't changed their aspect, nor their appeased character, kind and cheerful.

2 Granada Stage. Instrumentation for the Analysis of Biological Signals

In Granada city, the Professor Mira devoted in three investigation projects routed to the development of electronic instrumentation and applications for the diagnosis of stuttered, disartrics and sensorineural hearing loss [1][2][3], physiologic alterations that had a direct relationship with the nervous system. He had conscience that the signals that we analyzed: the speech production, the electromyogram or the electroencephalogram, were the macroscopic representation of thousands of electric global process associated to the brain activity, acting on itself or in excitable muscles. We observed to a full soccer stadium with spectators and we sought to know what one said. But, the demonstrable fact that from time to time all agreed to say something in common, it justified the interest of the analysis.

The state of the Technology at the end of the 70 was faulty. The electronics of integrated functions was incipient and they began to appear some commercial microprocessors but of scarce benefits. Of this way, we defended the idea of combining the analogical electronics with the digital one to get the wanted objective [4]. For example, if it was necessary to calculate a logarithm, we used as previous stage a logarithmic amplifier and later, we sampled the result to treat it on a microprocessor. I participated with him in the following lines:

- The development of a spectrum analyzer for biopotentials and audio signals (my Grade work): An instrument that carried out the spectral analysis of signals of very low frequency based on an analog recycling voltage controlled filter with a constant and programmable quality factor [5]. With this analyzer we could study all those signals spectrally with a very little computational cost. I still conserve that analyzer as a museum piece that reminds me that time. And it works!
- Development of amplifiers for EMG, EEG and ECG: We design differential amplifiers of low noise and high rejection factor [6]. We learned on electronic and on the behavior of the biological signals. We discover that it was necessary to use a common earth for the instrumentation amplifier and the analysis (uP) system, and we check it in the head of D. Alejandro Vega, collaborator that nearly is electrocute in one of the experiments.
- Development of portable prototypes for the analysis of the rhythm of production syllabic: It interested us that the patient carried the instrumentation

with the purpose of to extract and to analyze the behavior of the parameters outside of the clinical environment [7]. We design a syllabic analyzer based on the calculation of the half value of the speech signal. An additional logic based in the determination of the temporary distance among syllabic packets was adjusted to a normal or anomalous production rhythm. Our portable prototype was built totally with CMOS logic devices.
– Development of acoustic, luminous and electric actuators: an alternative path of feedback with the patient analysis scope. We close the feedback loop using actuators in the patient. We tried to modify their behavior, by means of advertisements with remuneration or punishment signals (biofeedback). The biofeedback and conductive methods as solutions to certain types of disartrics were not of our pleasure, but I must say that in some patients, the technique worked enough well, especially with the penalty signals. In the experiments we requested volunteers by means of the local radio in Granada. We required people with problems in the production rhythm (stammerers). Not we know the reason for which the request of patients passed to the news EFE service of national scope. The Professor Mira and his co-workers were two or three weeks attending telephone calls of all Spain. The notice was altered by the media and they believed that we had solved the stammerers problem. Inside the development plan of devices of electric stimulation to highlight the devices of stimulation of facial muscles and neck. They were electric stimuli with wave in spike form (square signals filtered in different orders), low frequency (between 80 and 300Hz), and variable intensity. These signals forced in the patients a pleasant sensation of relaxation. The prototypes were applied in patients of the section of diseased neurological of the "San Juan de Dios" Hospital in Granada.

3 Santiago Stage. End of Instrumentation Projects and Tendency toward the Computational Techniques of the Artificial Intelligence

With their wife, Ana (always next to him), the Professor José Mira moves to Santiago in 1982 being accompanied by a small group of collaborators: Emilio Zapata, Diego Cabello, Roque Marín, Ramón Ruiz and Me. The work of the professor goes much more there of its activity like investigator. He builds the Electronic Specialty of Physics Faculty, and in little time (up to 1987), he gives labour stability to their entire group. The group had been enlarged with own people formed there under their guides. The students of the Electronic Specialty were very numerous and they overcame in a factor 3 or 4 to those of other specialties. The Mira Professor had capacity to integrate and to build so much in the investigate facet as in the educational one. It also integrated in the affective plane: It liked to organize collective periodic meetings in which we enjoyed a good food. I remember those that we did in the Market of Cattle of Santiago or the excursions to the Noia's bay as long as it allowed the time it.

I was collaborating with him in educational and research tasks in the Projects of Disartrics and Sensorineural Hearing Loss that we maintained from Granada. It swept all the phases of the human's growth for the study of these dysfunctions. A relationship arose this way with the Department of Gynecology of the 'General Hospital of Galicia' with the purpose of studying the fetal behavior during the pregnancy. We develop methods to analyze the state of the foetus auditory system and the group of physiologic variables that were susceptible of being obtained by means of non invasive methods. My doctoral Thesis it tried on the analysis of the ventricular fetal function starting from three signals: The Fetal Abdominal Electrocardiogram, Fetal Phonocardiogram and Fetal Doppler Cardiogram [8]. These signals are obtained from the maternal abdomen and are contaminated by multiple noise sources. Our work was the treatment of these signals to investigate on the foetus heart function and its alterations. The contribution that we made was the development of techniques for the elimination of the maternal electrocardiogram, and techniques of coherent averaging to obtain fetal averaged QRS complexes (10-20 beats), as well as the measurement of the different temporary heart intervals of opening and closing cardiac valves. We continued making use of analogical preprocessing techniques combined with the digital treatment of the signals [9]. To understand the difficulties of digital processing in that time, let us say that we use a computer AppplelIe with 8 bits Rockwell processor: 6502 and a memory of 64Kbytes.

The electronics devices appears present in the initial Mira's Projects, but as the technology allows increasing the computational capacity, D. José spreads to take advantage of that circumstance to approach projects but related with the software development. This way, the idea of the use of computers to help to the medical diagnosis and to develop models that allowed to explain the operations of the nervous system as a system of decision and calculation (I believe that the obsession of all their life), makes him to go aboard their following projects, guiding them toward that work line [10][11][12] ... In little time it demonstrates that their main investigation was in the named 'Artificial intelligence'. In the year 1989, he changes the area of knowledge of 'Electronics' to the knowledge area 'Computer Science and Artificial Intelligence' and he leaves to UNED where their third stage begins.

Following their line and already as principal investigator, I get two Projects related with Integral Computerization of Clinical Histories, including the analytic information through the bus MIB (Medical Information Bus) [13][14]. The technology of 90's allowed the possibility to integrate the medical diagnosis with the analytic results of the patient's clinic in a Database that allowed the agility of the consultation and the ruled classification of all the clinical variables (analytic chemistry, images, parameters,...)[15][16][17].The collaboration with medical specialized teams was essential to approach certain problems related with the prediagnosis, diagnosis and treatment throughout the disease. I found difficulties to join efforts in the integral computerization of Hospitals. By one hand, the health national system was interested in transferring competencies to the autonomous communities. By other hand, the physicians that I knew were also averse to capture their decisions and diagnoses in computer systems that appeared as competitors and correctors of their own decisions.

4 Influences in My Later Work

In 1992 I marched to the University of Malaga and found new surroundings of fellow workers and a Microelectronics Lab in progress. I created my own investigation group and I remained in the electronics area separating to me of the main line of Jose Mira. However, their influence made that it approached some of the Mira's positions and ideas applying them to microelectronics devices. In special to the design of ASICs for the real-time execute of certain algorithms. I returned again to work with the speech signals (from my Degree work), enlarging my knowledge on her considering it as biologic signal and information coded source. I deepened in issues that we had not played as the analysis of the wave glottal result of the exclusive movement of the vocal chords without the action of the buccal cavity. We measure the glottal wave by the variations of impedance that takes place when an electric constant current is injected to the neck during the phonation process. We build ASICs for the analysis of Pitch, the glottal wave and the syllabic production. Circuits that gave place to my first Doctoral Theses directed [18][19][20]. José Mira participated as member of the Tribunal of my first Thesis directed.

These works related with the speech analysis interested to a foniatrics and otorhinolaryngologist group in 'Carlos Haya' Hospital of Malaga with those we established relationship and carried out a work inside environment of a clinical foniatrics. We study the possibility to diagnose some larynx pathologies by analysis of speech signals taken place by the patients and after classifying to patient normal and with some concrete pathology. We analyze some such parameters as Jitter or variability Pitch, Shimmer or variability of the emitted energy and the Glottal Noise or level of bottom noise during the phonation. These parameters were shown good candidates to detect the adolescent maturity, and the existence of some types of nodules, hiatuses and incipient polyps. The resulting effort was enough wide as to defend another Doctoral Thesis that I directed in a shared way [21].

It interested me the speech code for their efficient transmission in reduced band channels and guided my work toward the field of the communications. Together with my collaborators, we developed efficient speech codecs with transmission rates smaller than 300Bps [22]. On the other hand, keeping in mind that the hearing is the organ of perception of the speech, we was interested in its study to apply the behaviour from this to the codecs based on the auditory perception. By means of the cochlear dynamic, we discovered that the auditory persistence, and the lateral inhibition of the nervous cells of the cochlea, both phenomena similar to the visual one, you could build a very efficient codec modifying the intervals of analysis of the perceived energy (masking principle), and reduced tonal excitation (inhibition principle) to a very small number of tones gave the development of a codec of an extremely low rate transmission (less than

100Bps), perfectly intelligible for the hearing [23]. Natural consequence of the process of cochlear excitation was the development of a microelectronic model of cochlear prosthesis based on these principles [24].

All these works went them supplementing with the general line of my group guided toward the design of circuits and systems in the field of the communications. Today, my line of main work is centered in the optical communications so that I have left to the margin my interest for the biological signals. Nevertheless, I must recognize that my work during the decade of the 90 came marked in great measure by the influence of Mira's Professor and of its instrumentation stage.

5 Gratefulness

I have presented some small brushstrokes of my work shared with the Mira's Professor during the stages of Granada and Santiago in those that had the privilege and the pleasure of working with him, to learn of him and to enjoy with him moments that I will never forget and that they are part of me.

I must also express my gratitude and recognition to their wife Ana E. Delgado. She has always been beside José Mira's Professor in a quiet and patient way and she has been their affective and adviser support in all the stages of their life.

References

1. Mira, J.: Instrumentación Electrónica para la Rehabilitación de Espasmofemias. Ref.: 2656/76. Proyecto de Investigación Básica de Presidencia del Gobierno. Universidad de Granada (1977-1979)
2. Mira, J.: Análisis Electrónico de Disartrias: Aplicación al Diagnóstico y Rehabilitación de Algunos Tipos de Disminuidos Psíquicos. Ref.: 3803/79. CAICYT. Universidad de Granada (1980-1982)
3. Mira, J.: Instrumentación Electrónica para la Detección, Análisis y Rehabilitación de Hipoacúsicos. Ref.: 995/82. CAICYT. Universidad de Granada (1983-1986)
4. Mira, J., Ríos, F., Zapata, E.L., Delgado, A.E., Cabello, D.: On the Use of Hybrid Architecture for Biosignal Processing. In: World Congress on Medical Physics and Biomedical Engineering. Proceedings, 8.39. Hamburg (1982)
5. Ríos, F.: Un Analizador Armónico para Biopotenciales y señales de Audio. Memoria de Licenciatura. Universidad de Granada (1981)
6. Mira, J., Marín, R.L., Ríos, F., Túnez, S.: Amplificadores de Instrumentación para Biopotenciales. In: 5° Congreso de Informática y Automática, Madrid, pp. 203–207 (1982)
7. Mira, J., Ríos, F., y Delgado, A.E.: Circuitos Electrónicos para el Control del Habla Rítmica. Revista de Informática y Automática 48, 14–19 (1981)
8. Ríos, F.: Evaluación Automática de la Función Ventricular Fetal. Tesis Doctoral. Universidad de Santiago de Compostela (1987) ISBN: 84-398-9625-5
9. Ríos, F., Mira, J., Delgado, A.E., Varela, R., Tojo, R.: Analog Processor Implementation of a Real Time Harmonic Profile Estimator for use in Clinical Paediatrics. In: Mediterranean Electrotechnical Conference, Madrid (1985)
10. Mira, J.: Sistema Multiprocesador Tolerante a Fallos para la Monitorización de una Unidad Coronaria y Programas de IA para Prediagnóstico de Urgencia. Ref.: 0615/84. CAICYT. Universidad de Santiago de Compostela (1984-1986)

11. Mira, J.: A Therapy Advisor for Oncology. Ref.: 1592/86. ESPRIT (CE), Universidad de Santiago de Compostela, Universidad de Leeds, Royal Freed Hospital London and Hospital le Peroní (Montpellier) (1986-1991)
12. Mira, J.: Consejeros de Terapia en Oncología. Ref.: 88/0315 CICYT. Universidad de Santiago y UNED (1989-1991)
13. Ríos, F.: Desarrollo de una Base de Datos Relacional Integral para un Servicio de Tocoginecología. Ref.: 6090225025/89. Universidad de Santiago de Compostela (1989-1990)
14. Ríos, F.: Un Sistema Informático Integral para un Servicio de Toco-Ginecología. Desarrollo de Interfaces con el BUS M.I.B. para Instrumentación Clínica. Xunta de Galicia Ref.: XUGA20607B90 (1990-1992)
15. Ríos, F., Dono, L., Montesinos, I., González, O.: A Entity-Relation Model for a Tcho_Gynecology Service. Rev. Medical Informatics 17, 269–278 (1992)
16. Dono, L., Ríos, F., Beiro, J.M., Montesinos, I., González, O.: Towards the Concept of Automatic Intelligent Hospitals: An Interconexion Model for a Distributed Data Base in a Tocho-Gynecology Service. In: First International Workshop on Mechatronics in Medicine and Surgery, Torremolinos (1992)
17. Dono, L., Ríos, F., Beiro, J.M., Montesinos, I., González, O.: Data Model Management in a Clinical Service: Application in an Echography Unit for Tocho-Gynaecology. Rev. Medical Informatics 19, 331–346 (1994)
18. Escaño, R.: SINAP: Un sistema integral de análisis del Pitch. Universidad de Málaga (1995) ISBN: 84-7496-476-8
19. Rodríguez, J.: Un sistema de análisis de voz basado en la onda glotal. Universidad de Málaga (1998) ISBN: 84-7496-548-9
20. Martín, J.F.: AS: Un ASIC para la valoración de secuencias silábicas en el Habla continua. Universidad de Málaga (1998) ISBN: 84-7496-547-0
21. Diez-de-Baldeón, F.: Valoración Diagnóstica de Patologías de Laringe Mediante Técnicas no Invasivas de Procesado de señales de voz. Universidad de Málaga (1999) ISBN: 84-7496-816-X
22. Ríos, F., y Romero J.: Circuito Integrado Monolítico Codec-Encriptador de Baja Tasa para Señales de Voz. Patente. Solicitud No.: 9800213. Publicación No.: 2143396 (1998)
23. Vargas, S.: Codec de Voz de Baja Tasa Basado en Modelo de Cóclea Implementado en DSP Mediante Técnicas de Diseño Independientes de la Tecnología. Proyecto Fin de Carrera. ETS de Ingenieros Industriales. Universidad de Málaga (2001)
24. Ríos, F., Fernandez, R., Romero, J., Martín, J.F.: Corti's Organ Physiology-Based Cochlear Model: A Microelectronic Prothestic Implant. Rev. Bioengineered and Bioinspired Systems 5119, 46–53 (2003)

Remembering José Mira

Isabel Gonzalo-Fonrodona

Departamento de Óptica. Facultad de Ciencias Físicas,
Universidad Complutense de Madrid, Ciudad Universitaria s/n, 28040-Madrid, Spain
igonzalo@fis.ucm.es

Abstract. Here I emphasize the importance of the help, advice, suggestions and backing that I received from José Mira in the task of dissemination and formalization of the research of Justo Gonzalo (1910-1986). This survey intents to be an expression of my profound gratitude to José Mira.

My relationship with José Mira was based on the common interest on the research of Justo Gonzalo (1910-1986).

I had the opportunity to get in closer contact with José Mira when he and Ana Delgado contacted Justo Gonzalo some years before 1978. Both Mira and Delgado were working in neurocibernetic models of cerebral dynamics (this is the title of the Ph. D. thesis of Delgado [1] directed by Mira) and therefore they were very interested in cerebral processes, brain models and cerebral dynamics. Gonzalo had developed a functional model of cerebral dynamics [3,4] based on the study of patients from the Spanish Civil War (1936-39) [5], put it in relation to many other cases in the literature and also to civil cases at hand. Such a model reinterpreted fundamental concepts introduced by Lashley [6], as the mass action, and introduced the definition of functions in gradation through the cerebral cortex, which account for the multisensoriality of the anatomic areas as well as for their mutual influence. Some aspects of this model together with data and interpretations from Lashley and Luria [7] were deeply analyzed and discussed in the Ph. D. thesis of Delgado [1] where Lasheley, Luria and Gonzalo were considered the three basic authors concerning the organization of the nervous tissue in relation with behavior. In this thesis, operational formulations (analogical, logical and symbolic) of basic concepts were proposed in order to understand some aspects of the laws of the cerebral dynamics. An extension of the concepts developed in the thesis was published later in [2], where reference to previous works is made.

Some years after the death of Gonzalo, I translated into English a part of his work (his published work was in Spanish) in order to disseminate it more widely. I then came into knowledge of the "Inernational Conference in Honor of W.S. McCulloch 25 Years after His Death", which was going to take place in Canary Islands (Spain) in 1995. When I contacted Mira for this event, he encouraged me to present a summary of the work of Gonzalo. The help, advice and support that I received from Mira were decisive. Since then, I kept in close contact with him as well as with his wife and coworker Delgado. Thanks to him I could present in that conference a summary that, with his help, could be placed in the appropriate

J. Mira et al. (Eds.): IWINAC 2009, Part I, LNCS 5601, pp. 53–56, 2009.

context. This was the first time that the work of Gonzalo was presented actively in an international event (only a summarized translation of the reference [4] was disseminated in the "International Conference on Alpha Processes in the Brain" in Lübeck (Germany) in 1994). The contribution to the Conference on McCulloch was focused on general aspects of Gonzalo's research [8]. One of them was the dynamic effects that appear under a type of cerebral lesion and that are due to changes in the central nervous excitation, these effects being dependent on the position of the lesion and on the active neural mass (then on the neural mass lost). These effects give place to striking phenomena in perception (e.g., inverted or tilted perception), observed in several patients under careful measures, since the phenomena can be easily overlooked. Similar phenomena were described later, even recently, in the bibliography (reviewed in [9]). Another remarked aspect was the functional model where functions in gradation were defined through the cortex, highlighting the heterogeneity and unity of the cortex and accounting for multisensory integration in the cerebral process.

For further works, Mira proposed me a more active approach: It was desirable to present original contributions formalizing and opening new perspectives of the previous findings. This idea made me to change the way I was approaching the work and has had very fruitful consequences until nowadays.

For the contribution to the "International Work-Conference on Artificial and Natural Neural Networks" in 1997 (IWANN'97), the help of Mira was crucial once more. That contribution was the first work where the theory of similitude and allometry developed by Gonzalo in relation to the cerebral system was exposed [10], apart from unpublished notes of doctoral courses [11]. The allometry arises here from the similitude between two cerebral systems related between them by a scale change. In the cerebral system with a lesion, the same organization as in a normal man is kept but the sensory functions become altered differently (allometricaly) depending on their excitability demands, thus revealing in part the functional behavior of sensorial structures. The perception splits or decomposes into stages of incomplete perception depending on the stimulus intensity. More complex or higher functions, which demand more cerebral excitability, become reduced in a greater degree than simpler functions with a lower excitability demand. An essential continuity, based on the quantity of cerebral excitation, was then established between higher and lower sensory functions. The analogy established by Gonzalo between the biological growth and the sensory growth led him to establish power functional relations between different sensory functions, verified by his observations. My small contribution in this work was to find the particular power functions from the fitting of available data of visual acuity and direction versus visual field amplitude.

For the contribution to the IWANN'99, the support from Mira was decisive once more. The contribution was devoted to one of the striking phenomena found by Gonzalo in the decomposition of perception: the inverted or tilted perception under low stimulus intensity in patients with lesions in the so called central zone. Visual, tactile and auditive inversion were reported. Another remarkable phenomenon described was the improvement of perception by multisensory facilitation, for example, by muscular stress. The experimental data were fitted with functions according to Fechner law in order to present the results in this Work-Conference [12].

In the communication presented in the IWANN 2001, and following the suggestions of Mira, an effort was made to interpret and expose in a simple mathematical form some of the results obtained by Gonzalo. A time-dispersive linear model was used to describe basic aspects of the temporal dynamics of cerebral excitation, characterized by a permeability to the excitation and a reaction velocity [13]. This work was already much closer to the type of work suggested by Mira. This model was extended to other aspects in the contribution to the IWANN 2003: The remarkable phenomena of reinversion of the visual perception by increasing the stimulus or by facilitation by other different stimulus were reinterpreted under the previous time-dispersion formulation of the cerebral excitation, suggesting that the multisensory facilitation would be essentially a nonlinear effect [14].

The work presented in the "Second international Work-Conference on the Interplay Between Natural and Artificial Computation", IWINAC 2007, arose also from the continuous backing and enthusiastic interest of Mira. The work focused on the physiological Stevens power law of perception (more accepted than Fechner law) and on scaling power laws, in order to reinterpret data obtained by Gonzalo in cases with lesions leading to the syndrome he characterized. Very good agreement with scaling power laws was found for the improvement of visual perception (including the reinversion of a perceived arrow) with facilitating stimulus, with the remarkable result that the dominant exponents are 1/4, as in the case of many biological observables with respect to the mass. We suggested that Stevens' law in the cases analyzed is a manifestation of universal allometric scaling power laws of the biological neural networks [15]. From these findings, we planned together with Mira and Delgado to develop these ideas in the near future. Mira then suggested me to made a review, to be published in Neurocomputing, on the results, model and formalization of Gonzalo's research up to that time. The review was made [16] connecting it to the recent and increasing number of works on multisensory and cross-modal effects, where transitional multisensory zones, even multisensory interactions in the primary sensory cortices are proposed, as was already proposed by Gonzalo by means of the functional cerebral gradients through the cortex.

Finally, the work for the present IWINAC 2009 is an example of the successful and fruitful way of working that Mira suggested. The work is an extension of that presented in the IWINNAC 2007. It is found that for high enough stimulus, and for different sensory systems, the perception scales as a quarter power law, as in the case of facilitating stimulus. The analysis supports assertions made by Gonzalo based on his observations, that the integrative cerebral process initiated in the projection path reaches far regions of less specificity.

I must also stress that thanks to the generous and altruistic support from Mira, I have got in contact with numerous scientists interested in these problems, this opening to me new and promising perspectives.

We expect to continue working along these lines, remembering the suggestions and advices of Mira to whom I express my acknowledgement and gratitude.

References

1. Delgado, A.E.: Modelos Neurocibernéticos de Dinámica Cerebral. Ph.D.Thesis. E.T.S. de Ingenieros de Telecomunicación, Univ. Politécnica, Madrid (1978)
2. Mira, J., Delgado, A.E., Moreno-Díaz, R.: The fuzzy paradigm for knowledge representation in cerebral dynamics. Fuzzy Sets and Systems 23, 315–350 (1987)
3. Gonzalo, J.: La cerebración sensorial y el desarrollo en espiral. Cruzamientos, magnificación, morfogénesis. Trab. Inst. Cajal Invest. Biol. 43, 209–260 (1951)
4. Gonzalo, J.: Las funciones cerebrales humanas según nuevos datos y bases fisiológicas: Una introducción a los estudios de Dinámica Cerebral. Trab. Inst. Cajal Invest. Biol. 44, 95–157 (1952)
5. Gonzalo, J.: Investigaciones sobre la nueva dinámica cerebral. La actividad cerebral en función de las condiciones dinámicas de la excitabilidad nerviosa. Publicaciones del Consejo Superior de Investigaciones Científicas, Inst. S. Ramón y Cajal, Madrid, vol. I (1945), II (1950) (available in: Instituto Cajal, CSIC, Madrid) (1945)
6. Lashley, K.S.: Brain mechsnisms and intelligence. Univ. of Chicago Press, Chicago (1929)
7. Luria, A.R.: Restoration of Function after Brain Injury. Pergamon Press, Oxford (1963)
8. Gonzalo, I., Gonzalo, A.: Functional gradients in cerebral dynamics: The J. Gonzalo theories of the sensorial cortex. In: Moreno-Díaz, R., Mira, J. (eds.) Brain Processes, Theories and Models. An Int. Conf. in honor of W.S. McCulloch 25 years after his death, pp. 78–87. MIT Press, Massachusetts (1996)
9. Gonzalo-Fonrodona, I.: Inverted or tilted perception disorder. Revista de Neurología 44, 157–165 (2007)
10. Gonzalo, I.: Allometry in the J. Gonzalo's model of the sensorial cortex. In: Mira, J., Moreno-Díaz, R., Cabestany, J. (eds.) IWANN 1997. LNCS, vol. 1240, pp. 169–177. Springer, Heidelberg (1997)
11. Gonzalo, J.: Unpublished notes of doctoral courses on Fisiopatología Cerebral, Madrid (1956-1966)
12. Gonzalo, I.: Spatial Inversion and Facilitation in the J. Gonzalo's Research of the Sensorial Cortex. Integrative Aspects. In: Mira, J., Sánchez-Andrés, J.V. (eds.) IWANN 1999. LNCS, vol. 1606, pp. 94–103. Springer, Heidelberg (1999)
13. Gonzalo, I., Porras, M.A.: Time-dispersive effects in the J. Gonzalo's research on cerebral dynamics. In: Mira, J., Prieto, A. (eds.) IWANN 2001. LNCS, vol. 2084, pp. 150–157. Springer, Heidelberg (2001)
14. Gonzalo, I., Porras, M.A.: Intersensorial summation as a nonlinear contribution to cerebral excitation. In: Mira, J., Álvarez, J.R. (eds.) IWANN 2003. LNCS, vol. 2686, pp. 94–101. Springer, Heidelberg (2003)
15. Gonzalo-Fonrodona, I., Porras, M.A.: Physiological Laws of Sensory Visual System in Relation to Scaling Power Laws in Biological Neural Networks. In: Mira, J., Álvarez, J.R. (eds.) IWINAC 2007. LNCS, vol. 4527, pp. 96–102. Springer, Heidelberg (2007)
16. Gonzalo-Fonrodona, I.: Functional grsdients through the cortex, multisensory integration and scaling laws in brain dynamics. In: Mira, J., Ferrández, J.M. (Guest eds.) Neurocomputing, vol. 72, pp. 831–838 (2009)

Revisiting Algorithmic Lateral Inhibition and Accumulative Computation*

Antonio Fernández-Caballero[1,3], María T. López[2,3], Miguel A. Fernández[2,3], and José M. López-Valles[1]

[1] Universidad de Castilla-La Mancha
Escuela de Ingenieros Industriales de Albacete, 02071 - Albacete, Spain
caballer@dsi.uclm.es

[2] Universidad de Castilla-La Mancha
Escuela Superior de Ingeniería Informática, 02071 - Albacete, Spain

[3] Instituto de Investigación en Informática de Albacete, 02071 - Albacete, Spain

Abstract. Certainly, one of the prominent ideas of Professor Mira was that it is absolutely mandatory to specify the mechanisms and/or processes underlying each task and inference mentioned in an architecture in order to make operational that architecture. The conjecture of the last fifteen years of joint research of Professor Mira and our team at University of Castilla-La Mancha has been that any bottom-up organization may be made operational using two biologically inspired methods called "algorithmic lateral inhibition", a generalization of lateral inhibition anatomical circuits, and "accumulative computation", a working memory related to the temporal evolution of the membrane potential. This paper is dedicated to the computational formulations of both methods, which have led to quite efficient solutions of problems related to motion-based computer vision.

1 Introduction

One of the most important problems in Artificial Intelligence (AI) and in Computational Neuroscience (CN) is to find effective calculation procedures that enable connecting the analytic models of the behavior of individual neurons, typical of the Neurodynamics [1], [2], with the formulations in natural language of the concepts and inferences associated to the high-level cognitive processes [23], [27], [28], [30]. Three years ago,the research community has celebrated the fiftieth anniversary of AI and it is evident that there is still no general and satisfactory solution to this problem, as clearly stated by Professor José Mira in one of his last papers [29]. The reasons for this lack of links between the models belonging to Physics and AI reside in the intrinsic complexity of the cognitive processes and in the lack of adequate methodological approaches.

The search for architectures of the cognition (the"logic of the mind") is a long-haul task with very limited results that has roots in the ancient Greece.

* This article is dedicated to the memory of our close master and friend, Professor José Mira.

J. Mira et al. (Eds.): IWINAC 2009, Part I, LNCS 5601, pp. 57–66, 2009.

Since then the task has essentially been of interest to philosophers, neurophysiologists, psychologists, mathematicians and, more recently, professionals of the field of computation in general and of Artificial Intelligence (AI) in particular. We are in front of a partial, fragmented and non-structured knowledge and look for an abstract structure that allows us to order adequately these pieces of knowledge. The architecture needs organization and structure, easy indexation and search and, finally, efficient use of this knowledge in inference and reasoning.

In the robotics field the name architecture is kept explicitly and such terms as "reactive", "situated" or "representational" are used to specify the organization of the software and hardware of an autonomous robot. In AI these organizational approaches are usually called paradigms and it is again distinguished between connectionist, situated, symbolic or representational and hybrid. In this work our architecture of hybrid character, which combines a bottom-up part of connectionist type with a top-down one of symbolic type is presented. To fix ideas we focus on the visual path but we also give out the conjecture of the potential validity in other sensory modalities and in the understanding and synthesis of other tasks in which reactive components are combined with others of intentional nature.

The ideas underlying the work are: (1) In order to theorize and to solve problems that go beyond the retinotopic projection step it is necessary to propose neurophysiologically plausible synthetic architectures (knowledge models), but without the precision by which the more peripheral structures are known. (2) It is not sufficient to model and interpret the neuronal function at physical (physiological) level of registrations of slow potentials or spikes trains. It is necessary to re-formulate the neuronal mechanisms at symbolic level in terms of inferential rules and frames, closer to the natural language used by an external observer when describing the visual processes, or when performing psychophysical experiments [13], [14].

2 Description of the Two Neural Processes

In order to make operational any architecture it is necessary to specify the mechanisms and/or processes underlying each subtask and inference mentioned in that architecture. The conjecture of our team is that any bottom-up organization may be made operational using two biologically inspired methods called (1) "algorithmic lateral inhibition" (ALI), a generalization of lateral inhibition (LI) anatomical circuits [3], and, (2) "accumulative computation" (AC). Firstly, ALI is considered in the following section and AC in the next one.

2.1 Algorithmic Lateral Inhibition

Lateral Inhibition at the Physical Level. There are some neuronal structures that are repeated at all the integration levels in the nervous system and that emerge again in the global behavior of human beings. This leads us to

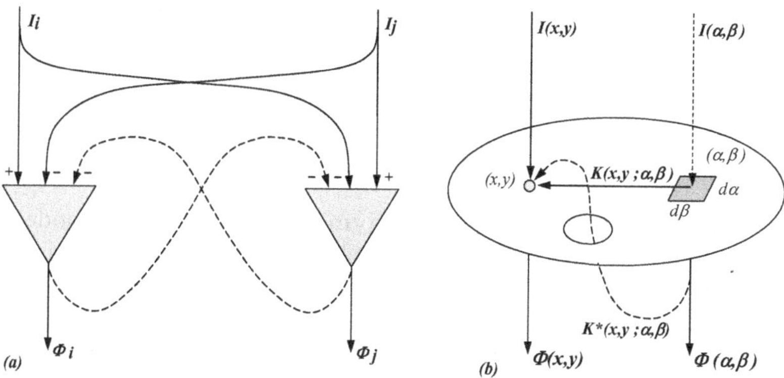

Fig. 1. Non-recursive (continuous line) and recursive (discontinuous line) LI connectivity schemes at the physical level. (a) Discrete case. (b) Continuous case.

think that they have endured the evolution process because they were adaptively useful for interacting with the environment. If we look at these structures from an electronic and computational perspective we could say that they are the "basic functional modules" in terms of which evolution has designed the best architecture for the nervous system (NS) to process information. This is the case of the lateral inhibition circuits, due to the wide range of functions which they may synthesize, to their capacity of explaining in neurophysiology, and their facility of abstraction when accepting to maintain invariant their structure in front of semantic level changes.

LI schemes may be found in levels such as neurogenesis, dendro-dendritic contacts, neuronal circuits in the retina, lateral geniculate body and in cerebral cortex, in the interaction between groups of neurons (ocular and orientation dominance columns) [24]. At the physical level, in terms of interconnected circuits that are described using a language of signals, there are two basic connectivity schemes: (1) non-recursive LI, and (2) recursive LI, with feedback (as outlined in Fig. 1). In non-recursive LI, the modification of a unit response depends on the inputs to the neighboring units, and in recurrent LI it depends on the outputs of the neighboring units.

If extending the formulation to the continuous case, (Fig. 1), the terms of interaction become nuclei of a convolution integral, so that the output, $\Phi(x,y)$, is the result of accumulating the direct excitation of each unit, $I(x,y)$, with the inhibition from modulating the excitation received by the connected neighboring units, $I(\alpha,\beta)$,via the weight factors $K(x,y;\alpha,\beta)$. The difference nucleus, $K(x,y;\alpha,\beta)$, is now responsible for the specific form of the calculation which, in all cases, acts as a detector of contrasts. In the recursive LI , it is the direct output, $\Phi(x,y)$, which accumulates the inhibition from the responses of the neighboring units, $\Phi(\alpha,\beta)$, weighted by an interaction coefficient $K^*(x,y;\alpha,\beta)$, basically different from that of the direct path (K).

Here again, in recursive LI, the shape and size of the receptive field (the interaction nucleus K^*) specifies the network connectivity (cooperation-competition area) and the calculation details (syntony, orientations, shapes, speeds, and so on). The most usual shape in K and K^* is obtained from subtracting two Gaussians. Thus the calculation structures inherent in the entire LI network are obtained: (1) a central area (ON or OFF), (2) a peripheral area (OFF or ON), and, (3) an excluded region (outside the receptive field).

Algorithmic Lateral Inhibition at Symbol Level. The LI model at the physical level is limited to a language of physical signals as functions of time. The first possible abstraction, which passes from a circuit to an algorithm, is obtained by rewriting the accumulation and inhibition processes in terms of rules (conditionals "if-then"), as a generalization of the weighted sum and the threshold. The condition field of a rule generalizes the weighted sum and the conditional generalizes the non-linearity of the threshold. Furthermore, the structure of the LI model is maintained. In other words, now also receptive fields (data fields in a FIFO memory) with an excitatory center and an inhibitory periphery, both for the input space (non-recursive LI) and for the output space (recursive IL) are defined. However, the nature of these input and output spaces is changed drastically, to be spaces of representation; that is, they are multidimensional spaces, where each "axis" represents possible values of an independent variable which measure an independent possible property of inputs and outputs. For the input space, the properties are those pertinent to the description of the external and internal "world", as seen by the previous neural nets. For the output space, properties are "decisions" of the net under consideration in the proper time axis. In the simplest case, the axons of each neuron of the net are proposing a sequence of "decisions" that fill in the output FIFO memory. Additionally, the decision rules in ALI need not be limited to analytic but can include also logical-relational operators as components of the calculus carried out by its condition fields [3], [24].

With this interpretation of ALI, each element of calculus samples its data in the center and periphery of the volume specified by its receptive field in the input space, and also samples in the center and periphery of the volume which specifies its receptive field in the output space. By specifying the nature of the decision rules the different types of calculus attainable by a network of algorithmic lateral inhibition (ALI) are obtained.

Algorithmic Lateral Inhibition as a Method at Knowledge Level. To complete the possibilities of the LI network, we do a new abstraction process and we pass from the symbol level to the knowledge level by generalizing the rules in terms of inferences [4], [24]. Through this abstraction process it is possible to consider ALI circuits as the anatomical support of an inferential scheme (Fig. 2).

2.2 Accumulative Computation

Usually the time evolution of the neuron membrane potential is modeled by a first order differential equation known as the "leaky integrator model". If we

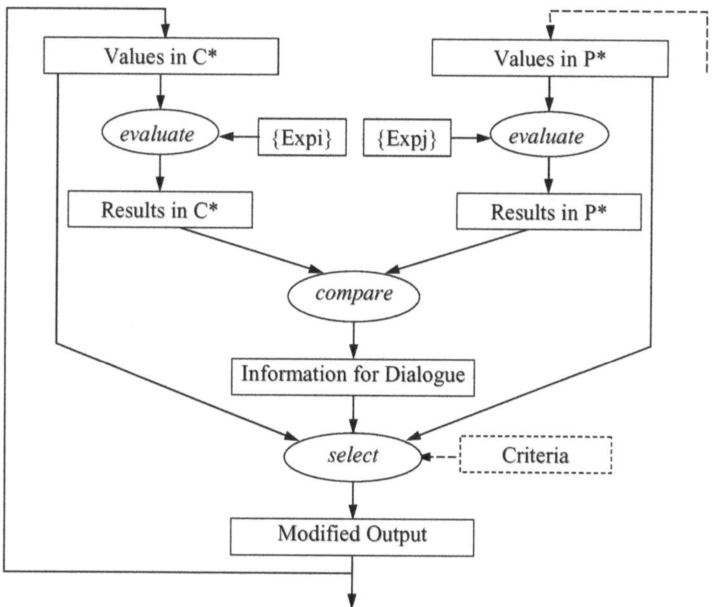

Fig. 2. Recurrent ALI inferential scheme. The results of the evaluation in the central (C*) and peripheral (P*) parts of the feedback receptive field are first compared and the result of this comparison is used to select the updated output.

move from differential equations to equations in finite differences then we get the temporal ALI model previously described, in which each element of calculus samples its input and feedback data from two FIFO memories. A different way of modeling time evolution of membrane potential is consider the membrane as a local working memory in which neither the triggering conditions nor the way in which the potential tries to return to its input-free equilibrium value, needs to be restricted to thresholds and exponential increases and decays. This type of working memory was named "accumulative computation" (AC), [5], [6], [25] and is characterized by the possibility of controlling its charge and discharge dynamics in terms of:

1. The presence of specific spatio-temporal features with values over a certain threshold.
2. The persistency in the presence of these features.
3. The increment or decrement values ($\pm\delta Q$) in the accumulated state of activity of each feature and the corresponding current value, $Q(t)$.
4. The control and learning mechanisms.

The upper part of Fig. 3 shows the accumulative computation model's block diagram. The model works in two time scales, a macroscopic, t, associated to the external data sequence to be processed by the net and a microscopic one, τ,

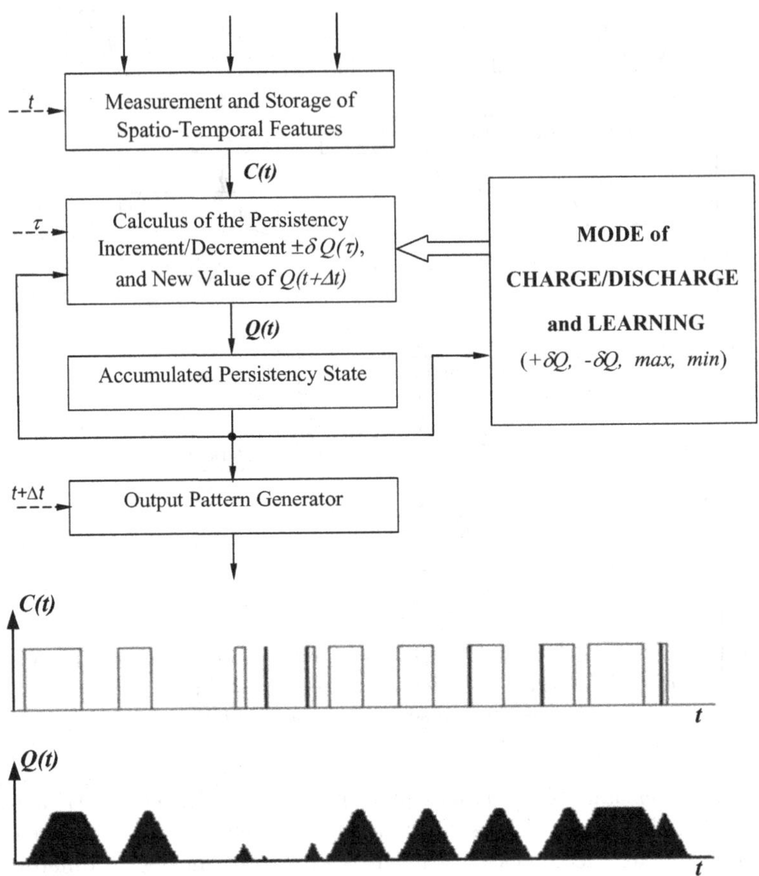

Fig. 3. The AC working memory model (upper part) and an example of the temporal evolution of the accumulated persistency state, Q(t), in response to an specific sequence of values of a detected feature, C(t), (lower part)

associated to the set of internal processes that take place while the external data (an image, for instance) remain constant. The lower part of Fig. 3 illustrates the temporal evolution of the state of the charge in an AC working memory in front of a particular one-dimensional stimuli sequence.

The control rules used to calculate the persistency of motion through time are:

$$Q[x,y;t+\Delta t] = \begin{cases} \max(Q[x,y;t] - \delta Q, min), \text{ if } C[x,y;t] = 1 \\ \min(Q[x,y;t] + \delta Q, max), \text{ otherwise} \end{cases} \quad (1)$$

where $C[x,y;t] = 1$ means that motion has been detected on pixel $[x,y]$ at t.

3 Applications in Computer Vision

In this paper we have re-explored the fact that with only two synthetic mechanisms, ALI and AC, it is sufficient to do computational an important part of visual processes. Some of the works of our group support the validity of this methodological approximation, as described in this section.

Firstly, AC was applied to the problem of the classification of moving objects in long image sequences [5], showing its capacity to be implemented in real-time [7]. Later on the combination of AC and ALI was used in the resolution of the problem of segmenting moving silhouettes in video sequences [8], [9]. Its neuronal nature was described in detail [10], as well as the model for motion detection [11] and the influence of each parameter of the combination between AC and ALI [12]. In all these papers the novel model based in neural networks combining AC and ALI was denominated "lateral interaction in accumulative computation" (LIAC) [11].

From the good results obtained by means of these methods in computer-vision-based motion analysis, the following step was the challenge of facing selective visual attention (dynamic) by means of a research line [15], [16] where the importance of the incorporation of new parameters appeared. This research aims in describing a method for visual surveillance based on biologically motivated dynamic visual attention in video image sequences. The system is based on the extraction and integration of local (pixels and spots) as well as global (objects) features. Our approach defines a method for the generation of an active attention focus on a dynamic scene for surveillance purposes. The system segments in accordance with a set of predefined features, including gray level, motion and shape features, giving raise to two classes of objects: vehicle and pedestrian. The solution proposed to the selective visual attention problem consists of decomposing the input images of an indefinite sequence of images into its moving objects, defining which of these elements are of the user's interest at a given moment, and keeping attention on those elements through time. Features extraction and integration are solved precisely by incorporating the mechanisms described, AC and ALI.

Among others, we also introduced velocity to improve the capture of the attention on the objects in movement [17]. In this case, we highlight the importance of the motion features present in our algorithms in the task of refining and/or enhancing scene segmentation in the methods proposed. The estimation of these motion parameters is performed at each pixel of the input image by means of the AC method. "Motion presence", "module of the velocity" and "angle of the velocity", all obtained from AC computation method, are shown to be very significant to adjust different scene segmentation outputs in our dynamic visual attention method [18], [19].

In parallel with the previous works, again mainly AC was used, and ALI in a minor degree, to improve the segmentation of moving objects, by introducing stereovision, with the purpose of adding a parameter that can contribute enormously in any tracking task in the three-dimensional world; this is parameter depth [20]. In this work a method that turns around the existing symbiosis between stereovision and motion is used; motion minimizes correspondence ambiguities, and stereovision enhances motion information. The central idea behind our approach is to transpose the spatially-defined problem of disparity estimation into the spatial-temporal domain. Motion is analyzed in the original sequences by means of AC and the disparities are calculated from the resulting two-dimensional motion charge maps [22].

To date the first results applied to the potential usefulness in mobile robotics have also been introduced [21]. Finally, the LI model at the knowledge level, the so-called ALI [26], seems to be a promising method to solve problems of cooperation and dialogue in the theory of multi-agent systems.

4 Conclusions

Neurally-inspired computer vision involves at least memory processes, alert, surveillance, selective attention, divided attention, reinforcement learning, intentionality and conscience. The complexity of these mechanisms is the reason for the difficulty found when trying to synthesize them computationally. A satisfactory solution, close to biology, is still a long term objective.

In this work we have revisited the works of the joint research of Professors José Mira and Ana E. Delgado with our team in computer vision by using a description framework based in the use of two basic neuronal mechanisms, the lateral inhibition and the working memory related to the temporal evolution of the membrane potential, which was denominated accumulative computation. In order to extract the maximum performance the two mechanisms at the three levels used in Artificial Intelligence to describe a calculation (the physical level, the level of the symbols and the knowledge level) have been formulated. We have proposed a method to interpret the function performed by a neural circuit, not only at the level of neurophysiologic signals but also in terms of symbolic rules and inferential schemes.

In the previous section, we have also commented some of the results obtained by the research team from using the conceptual scheme described in the previous sections. These results although reasonable in AI, are far away from approaching biological solutions for at least two reasons. The first one is the enormous constituent difference between biological systems and synthesis elements used in computer vision. The second reason is the ignorance on the architectures of cognition, starting from the doubt on the suitability of the term architecture to describe perception in humans.

Nevertheless, when seeking to construct solutions able to be computed with programs, cameras and robots, there is no other way than trying to equip the knowledge that we have on the biological solution to the problem of attention with structure and organization. Anything that we can not describe in a structured, declarative, clear, precise and formal way is not computable. That is why we think that our approach of firstly looking into neuroscience as an inspiration source, then proposing real computational models, and at the same time evaluating its efficiency in problem solving, independently of its biological origin, is useful.

Acknowledgements

This work was partially supported by the Spanish Ministerio de Ciencia e Innovación under project TIN2007-67586-C02-02, and by the Spanish Junta de Comunidades de Castilla-La Mancha under projects PII2I09-0069-0994, PII2I09-0071-3947 and PEII09-0054-9581.

References

1. Deco, G., Rolls, E.T.: A neurodynamical cortical model of visual attention and invariant object recognition. Vision Research 44(6), 621–642 (2004)
2. Deco, G., Rolls, E.T.: Short-term memory, and action selection: A unifying theory. Progress in Neurobiology 76(4), 236–256 (2005)
3. Delgado, A.E., Mira, J., Moreno-Díaz, R.: A neurocybernetic model of modal cooperative decision in the Kilmer-McCulloch space. Kybernetes 18(3), 48–57 (1989)
4. Delgado, A.E., Mira, J.: Algorithmic lateral inhibition as a generic method for visual information processing with potential applications in robotics. Computational Intelligent Systems for Applied Research, 477–484 (2002)
5. Fernández, M.A.: Una arquitectura modular de inspiración biológica con capacidad de aprendizaje para el análisis de movimiento en secuencias de imagen en tiempo real, PhD Thesis, UNED, Madrid, Spain (1995)
6. Fernández, M.A., Mira, J., López, M.T., Alvarez, J.R., Manjarrés, A., Barro, S.: Local accumulation of persistent activity at synaptic level: Application to motion analysis. In: Sandoval, F., Mira, J. (eds.) IWANN 1995. LNCS, vol. 930, pp. 137–143. Springer, Heidelberg (1995)
7. Fernández, M.A., Fernández-Caballero, A., López, M.T., Mira, J.: Length-Speed Ratio (LSR) as a characteristic for moving elements real-time classification. Real-Time Imaging 9(1), 49–59 (2003)
8. Fernández-Caballero, A.: Modelos de interacción lateral en computación acumulativa para la obtención de siluetas, PhD Thesis, UNED, Madrid, Spain (2001)
9. Fernández-Caballero, A., Mira, J., Fernández, M.A., López, M.T.: Segmentation from motion of non-rigid objects by neuronal lateral interaction. Pattern Recognition Letters 22(14), 1517–1524 (2001)
10. Fernández-Caballero, A., Mira, J., Fernández, M.A., Delgado, A.E.: On motion detection through a multi-layer neural network architecture. Neural Networks 16(2), 205–222 (2003)
11. Fernández-Caballero, A., Mira, J., Delgado, A.E., Fernández, M.A.: Lateral interaction in accumulative computation: A model for motion detection. Neurocomputing 50, 341–364 (2003)
12. Fernández-Caballero, A., Fernández, M.A., Mira, J., Delgado, A.E.: Spatio-temporal shape building from image sequences using lateral interaction in accumulative computation. Pattern Recognition 36(5), 1131–1142 (2003)
13. Heinke, D., Humphreys, G.W.: Computational models of visual selective attention: A review. Connectionist Models in Cognitive Psychology 1(4), 273–312 (2005)
14. Heinke, D., Humphreys, G.W., Tweed, C.L.: Top-down guidance of visual search: A computational account. Visual Cognition 14(4-8), 985–1005 (2006)
15. López, M.T.: Modelado computacional de los mecanismos de atención selectiva mediante redes de interacción lateral, PhD Thesis, UNED, Madrid, Spain (2004)

16. López, M.T., Fernández-Caballero, A., Mira, J., Delgado, A.E., Fernández, M.A.: Algorithmic lateral inhibition method in dynamic and selective visual attention task: Application to moving objects detection and labeling. Expert Systems with Applications 31(3), 570–594 (2006)
17. López, M.T., Fernández-Caballero, A., Fernández, M.A., Mira, J., Delgado, A.E.: Motion features to enhance scene segmentation in active visual attention. Pattern Recognition Letters 27(5), 469–478 (2006)
18. López, M.T., Fernández-Caballero, A., Fernández, M.A., Mira, J., Delgado, A.E.: Visual surveillance by dynamic visual attention method. Pattern Recognition 39(11), 2194–2211 (2006)
19. López, M.T., Fernández-Caballero, A., Fernández, M.A., Mira, J., Delgado, A.E.: Dynamic visual attention model in image sequences. Image and Vision Computing 25(5), 597–613 (2007)
20. López-Valles, J.M.: Estereopsis y movimiento. Modelo de disparidad de carga: Un enfoque con inspiración biológica, PhD Thesis, Universidad de Castilla-La Mancha, Albacete, Spain (2004)
21. López-Valles, J.M., Fernández, M.A., Fernández-Caballero, A., López, M.T., Mira, J., Delgado, A.E.: Motion-based stereovision model with potential utility in robot navigation. In: Ali, M., Esposito, F. (eds.) IEA/AIE 2005. LNCS, vol. 3533, pp. 16–25. Springer, Heidelberg (2005)
22. López-Valles, J.M., Fernández, M.A., Fernández-Caballero, A.: Stereovision depth analysis by two-dimensional motion charge memories. Pattern Recognition Letters 28(1), 20–30 (2007)
23. Marr, D.: Vision. A Computational Investigation into the Human Representation and Processing of Visual Information. W.H. Freeman and Company, New York (1982)
24. Mira, J., Delgado, A.E.: What can we compute with lateral inhibition circuits? In: Mira, J., Prieto, A.G. (eds.) IWANN 2001. LNCS, vol. 2084, pp. 38–46. Springer, Heidelberg (2001)
25. Mira, J., Fernández, M.A., López, M.T., Delgado, A.E., Fernández-Caballero, A.: A model of neural inspiration for local accumulative computation. In: Moreno-Díaz Jr., R., Pichler, F. (eds.) EUROCAST 2003. LNCS, vol. 2809, pp. 427–435. Springer, Heidelberg (2003)
26. Mira, J., Delgado, A.E., Fernández-Caballero, A., Fernández, M.A.: Knowledge modelling for the motion detection task: The algorithmic lateral inhibition method. Expert Systems with Applications 27(2), 169–185 (2004)
27. Mira, J., Delgado, A.E.: On how the computational paradigm can help us to model and interpret the neural function. Natural Computing 6(3), 207–209 (2006)
28. Mira, J.: The Semantic Gap, Preface of Bio-inspired Modeling of Cognitive Tasks. In: IWINAC 2007. LNCS, vol. 4528. Springer, Heidelberg (2007)
29. Mira, J.: Symbols versus connections: 50 years of artificial intelligence. Neurocomputing 71(4-6), 671–680 (2008)
30. Newell, A.: The Knowledge Level. AI Magazine, 1–20 (Summer 1981)

Detection of Speech Dynamics by Neuromorphic Units

Pedro Gómez-Vilda[1], José Manuel Ferrández-Vicente[2], Victoria Rodellar-Biarge[1], Agustín Álvarez-Marquina[1], Luis Miguel Mazaira-Fernández[1], Rafael Martínez-Olalla[1], and Cristina Muñoz-Mulas[1]

[1] Grupo de Informática Aplicada al Tratamiento de Señal e Imagen, Facultad de Informática, Universidad Politécnica de Madrid, Campus de Montegancedo, s/n, 28660 Madrid

[2] Dpto. Electrónica, Tecnología de Computadoras, Univ. Politécnica de Cartagena, 30202, Cartagena

pedro@pino.datsi.fi.upm.es

Abstract. Speech and voice technologies are experiencing a profound review as new paradigms are sought to overcome some specific problems which can not be completely solved by classical approaches. Neuromorphic Speech Processing is an emerging area in which research is turning the face to understand the natural neural processing of speech by the Human Auditory System in order to capture the basic mechanisms solving difficult tasks in an efficient way. In the present paper a further step ahead is presented in the approach to mimic basic neural speech processing by simple neuromorphic units standing on previous work to show how formant dynamics -and henceforth consonantal features-, can be detected by using a general neuromorphic unit which can mimic the functionality of certain neurons found in the Upper Auditory Pathways. Using these simple building blocks a General Speech Processing Architecture can be synthesized as a layered structure. Results from different simulation stages are provided as well as a discussion on implementation details. Conclusions and future work are oriented to describe the functionality to be covered in the next research steps.

1 Introduction

Neuromorphic Speech Processing is an emerging field attracting the attention of many researchers looking for new paradigms helping in better understanding the underlying brain processes helping in speech perception, comprehension and production [8] [14]. This can also be extended to audio processing in general when aspects as emotion or speaker recognition are concerned or in scene analysis [13] [17]. The present paper is aimed to extend previous work on Neuromorphic Speech Processing using a layered architecture of artificial Neuron-like Neurons derived from the functionality of the main types of neurons found in the Auditory Pathways from the Coclhea to the Primary and Secondary Auditory Cortex [7]. In early work the typology a General Simple Unit was defined

J. Mira et al. (Eds.): IWINAC 2009, Part I, LNCS 5601, pp. 67–78, 2009.

under the paradigm of mask Image Processing. It was shown how one of these Mask Units can be adapted to model different neuronal processes as Lateral Inhibition to enhance Formant Detection. Besides it was shown how using different process masks the general simple structure could be configured to detect formant dynamics, as the ascending or descending patterns found in consonants or approximants. On these premises the present work is intended to show how based on this General Simple Unit a general layered architecture can be defined for the detection of phonemes from formant positions and dynamics, advancing one step in the definition of a fully Bio-inspired Speech Processing Architecture. The paper is organized as follows: A brief description of formants and formant dynamics is given in section 2. In section 3 the different neurons in the Auditory Pathway are defined accordingly to their functionality, and the structure of a General Simple Unit is presented to mimic the main functionalities found with use in Speech Processing, and the Neuromorphic Speech Processing Architecture based on these simple units is presented. In section 4 some results are given from simulations, accompanied by a brief discussion. Conclusions and future work lines are presented in section 5.

2 Perceiving the Dynamic Nature of Speech

Speech can be defined as the result of a complex interaction of the sound produced by different organs as the vocal folds or the constrictions of the vocal tract (lips, tongue, teeth, palate, velum, pharinx, nose) either in pseudo-periodic vibration (voiced speech) or in turbulent noise-like behaviour (unvoiced speech). The Vibration and Articulation Organs reduce or enhance the frequency contents of the resulting sound, which is perceived by the human Auditory System as a flowing stream of stimuli distributed accordingly with the dominant frequencies present in it, when properly separated by the Cochlea. An injection of complex spike-like neural stimuli is released to the Auditory Fibres in the Auditory Nerve which are then processed at the level of the Brain Stem and distributed to the Auditory Primary and Secondary Areas over the Cortex. Two important functions may be highlighted from this rather brief description: That speech sounds are dominated by certain enhanced bands of frequencies called formants in a broad sense, and that the assignment of meaning is derived both from dominant frequency combinations as well as from the dynamic changes observed in these combinations in time. Therefore speech perception can be seen as a complex parsing problem of time-frequency features. The most meaningful formants in message coding are the first two ones, designated classically as F_1 and F_2 in order of increasing frequency, F_1 being the lowest, in the range of 250-700 Hz. F_2 sweeps a wider range, from 700 to 2400 Hz. As the present study is focussed in the dynamic features of speech, a rather illustrative example has been presented in Fig. 1 where the spectrogram of a sentence with rapid formant changes has been reproduced. Formants are characterized in this power envelope spectrogram by brighter levels or envelope peaks, whereas the darker ones indicate envelope valleys. Envelope peaks are especially interesting because the

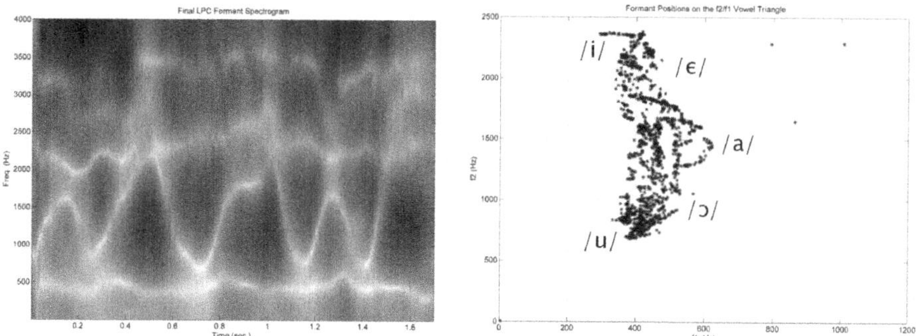

Fig. 1. Left: Adaptive Lineal Prediction (ALP) Spectrogram corresponding to the syllables utterance of -Where were you while you were away- phonetically $[\mathrm{h}^{\omega}\epsilon^{I}\omega\epsilon^{I}\mathrm{ju}^{h}\omega\mathrm{ailju}^{\omega}\epsilon^{I}\mathrm{a}\omega\epsilon\mathrm{j}]$ uttered by a male speaker. The IPA has been used for annotation 2. Right: Vowel triangle showing the five reference vowels in English framing the formant trajectories of the utterance.

perceptual processing of the auditory pathways work with dominant frequencies related to formants. What can be observed in the figure is that the first formant is oscillating between 350 and 650 Hz, whereas the second formant experiences abrupt fluctuations between 700 and 2200 Hz. High positions of the second formant point to front vowel-like sound, as [ϵ, i, j], whereas low ones correspond to back vowel-like sounds as [u, ω]. The positions of [ϵ, i, a, u] correspond to the zones where the formant positions are stable or slightly changing, as around the peaks of F_2 (0.15-0.17, 0.45-0.50, 0.7-0.75, 0.85-0.95, 1.15-1.17, 1.25-1.30, 1.38-1.42) whereas the positions of [j, ω] correspond to the complementary intervals where strong dynamic changes of formant positions can be observed.

When plotting F_2 vs F_1 formant trajectories appear as clouds of dots showing the vowel-like structure of the message. The vertices mark the positions of the extreme front [i], back [u] and middle [a] vowels. Stable positions produce places where formant plots are denser, whereas dynamic or changing positions produce trajectories, appreciated in the figure as bead-like lines. Formant transitions from stable Characteristic Frequencies (CF) to new CF positions (or virtual loci) are known as FM (frequency modulation) components.

3 Perceiving the Dynamic Nature of Speech

The Auditory System described in Fig. 2 as a chain of different sub-systems integrated by the Peripheral Auditory System (Outer, Middle and Inner Ear) and the Higher Auditory Centers is the structure responsible for Speech Perception. The most important organ of the Peripheral Auditory System is the Cochlea (Inner Ear), which carries out the separation in frequency and time of the different components of sound and their transduction from mechanical to neural activity, based on the hair cells [1]. Electrical impulses propagate to higher neural centers through auditory nerve fibers of different characteristic frequencies

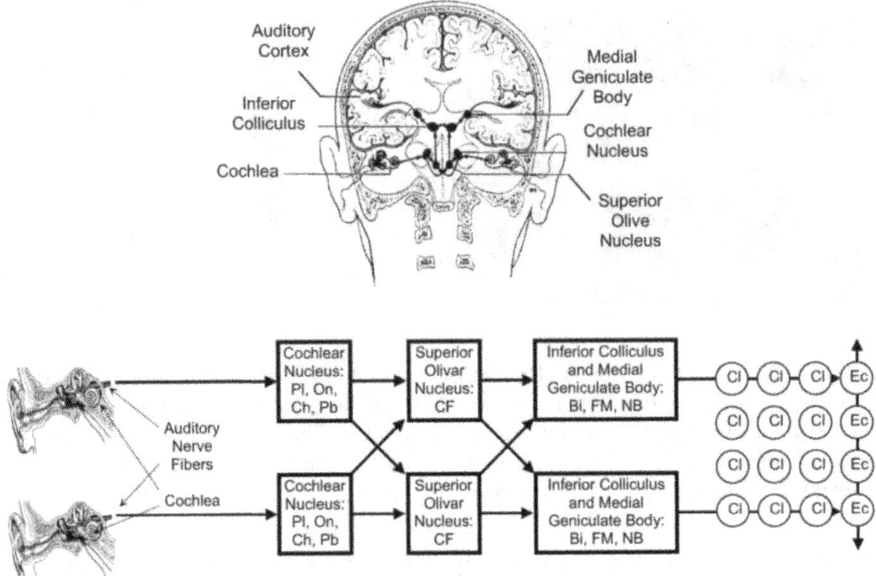

Fig. 2. Speech Perception Model. Top: Main neural pathways in the Peripheral and Central Auditory Centers (taken from [5]). Bottom: Simplified main structures and specialized neurons found in them.

(CF) responding to the spectral components (or harmonics f_0, f_1, f_2...) of speech. Within the cochlear nucleus (CN) different types of neurons are specialized specific processing as described below. The Cochlear Nucleus feeds information to the Olivar Complex, where sound localization is derived from inter-aural differences, and to the Inferior Colliculus (IC) organized in spherical layers with orthogonal isofrequency bands. Delay lines are found in this structure to detect temporal features in acoustic signals. The thalamus (Medial Geniculate Body) acts as a last relay station, and as a tonotopic mapper of information arriving to the Primary Auditory cortex as ordered feature maps.

The functionality of the different types of neurons is the following:

- Pl: Primary-like. Reproduce the firing stream found at its input as relay stages.
- On: Onset. Detect the leading edge of a new firing stream or package, discriminating the background firing rate from a new stimulus rate.
- Ch: Chopper. Specialized in dividing a continuous stimulus into slices.
- Pb: Pauser. Act as delay lines, firing sometime after the stimulus onset.
- CF: Characteristic Frequency. Tonotopically organized, responding to a narrow band of frequencies centred in a specific one.
- FM: Frequency Modulation Sensitive. Specialized in detecting changes in frequency their role is crucial in detecting dynamic speech features.

- NB: Noise Burst Sensitive. React to broadband stimuli, as those specific of turbulent noise found in unvoiced consonants.
- Bi: Binaural Processors. Specific of binaural hearing by contrasting phase-shifted stimuli. They are found mainly in the Inferior Colliculus.
- Cl: Columnar. Organized linearly in narrow columns through the layers of the Auditory Cortex. Their function may be related with short-time memory [12], although they are still under controversy.
- Ec: Extensive Connectors. The outer layers of the Auditory Cortex seem dominated by extensive connections among distant columns.

To simulate frequency separability in time some choices are available, as filter-banks, gammatones, FFT or LPC, among others. The present work is based in formant-like pattern detection on LPC spectrograms [4] as the one in Fig. 1, seen as a $1 \leq m \leq M$, frequency channels giving the envelope of the power spectral density of speech as:

$$X(m,n) = 20 \cdot log_{10} \left| \sum_{p=1}^{P} a_{k,n} e^{-jmp\Omega\tau} \right|^{-1} \tag{1}$$

where $a_{p,n}$ is the coefficient set of a P-order predictor obtained at the time instant n from a speech signal *x(n)*, and and τ and Ω are the resolutions in time and frequency. One of the m frequency-separated channels may be seen as the cumulative activity of an auditory fibre associated to a characteristic frequency f_m. The matrix *X(m,n)* can be seen as a two-dimensional Auditory Image, describing the activity in time of a linear layer of CF units in frequency. Many tools devised for image processing can be used for the detection of time-frequency features, as CF or FM patterns [9], using for instance simple masks or submatrices *w(i,j)* as:

$$\tilde{X}(m,n) = \sum_{i=-I}^{I} \sum_{j=0}^{J} w_{i,j} X(m-i, n-j) \tag{2}$$

where w_{ij} is a *(2I+1)x(J+1)* mask with a specific set of weights defined to mimic a specific neuron. The activity of this matrix may be represented by a generalized Unit structure as the one shown in Fig. 3, where the output activity $\tilde{X}(m,n)$ is the result of applying a weight mask *w(i,j)* to the selected inputs from *X(m,n)*, adding or subtracting the incoming stimuli (depending on their excitatory or inhibitory nature coded in the specific weight) and applying a threshold nonlinear function. These outputs will constitute a new layer of channels coded as, where m is now a positional channel index (Unit number) and n is the time index. The lateral-inhibition filtering active in certain neuron associations in the Inferior Colliculus is a special case of (2) where the weights of columns *j=1,2* are *zeros* and weights $w_{-1,0}=w_{1,0}$=*-1/2* and $w_{0,0}$=*1*. Different neurons as the one defined in (2) organized in consecutive layers will mimic some of the speech processes of interest in the study.

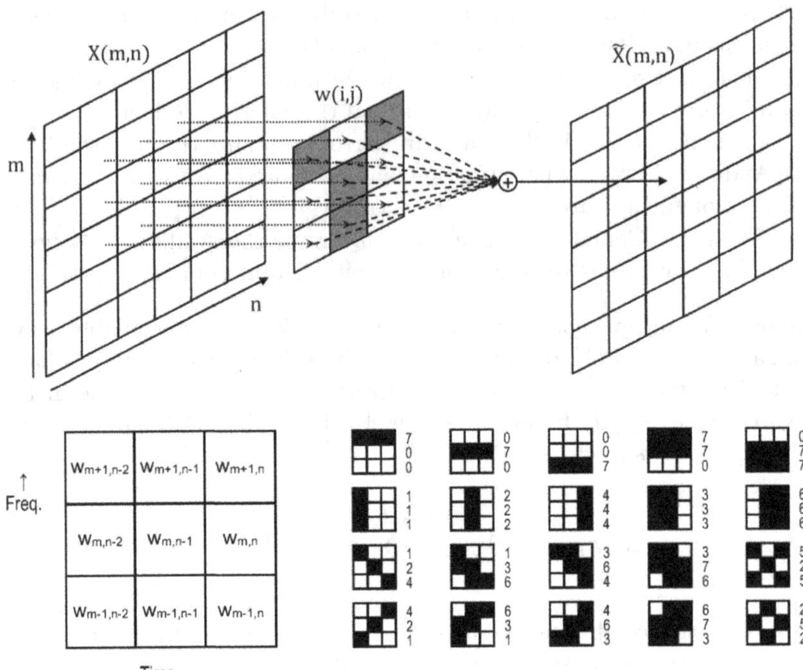

Fig. 3. Mask-based Neuromorphic Units. Top: Structure of a General Unit. Bottom Left: 3x3 weight mask. Bottom Right: 3x3 Masks for feature detection on the formant spectrogram. Each mask is labelled with the corresponding octal code (most significant bits: bottom-right).

In the bottom part of the figure some examples of weights for 3x3 masks are shown. To produce unbiassed results, the weight associated to each black square is fixed to $+1/s_b$ and the weight associated to white squares is fixed to $-1/s_w$, s_b and s_w being the number of squares in black or white found in a 3x3 mask, respectively. The Neuromorphic Speech Processing Architecture proposed for dynamic formant tracking may be then described by the structure presented in Fig. 4 using these Units.

This architecture is composed by different layers of specific neuromorphic units mimicking the physiological units found in the Auditory Pathways accordingly with the description given above as follows:

- LIFP: Lateral Inhibition Formant Profilers, as already described.
- $+f_{M1-K}$, $-f_{M1-K}$: Positive and Negative Slope Formant Trackers (K bands) detecting ascending or descending formant activity using masks 124-376 and 421-673.
- $f1_{1-K}$, $f2_{1-K}$: First and Second Energy Peak Tracker, intended for formant detection mimicking CF neurons, using masks 700-077.

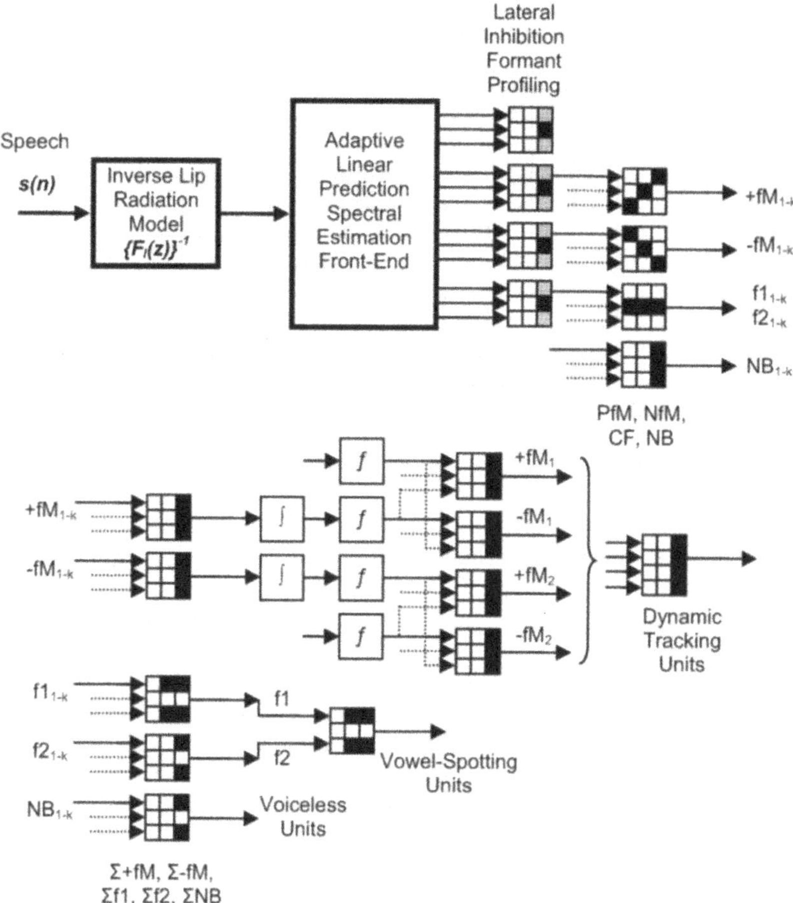

Fig. 4. Neuromorphic Speech Processing Architecture for a mono-aural channel

- S+fM, S-fM: These are integrators or accumulators working on the inputs of previous Formant Tracker Integration Units on certain specific bands (350-650 Hz for the first formant, or 700-2300 Hz for the second formant).
- $+fM_1$, $-fM_1$, $+fM_2$, $-fM_2$: First and Second Formant Integration Units (positive and negative slopes), estimating the features FM1 and FM2 in Table 1.
- SNB: Noise Burst Integration Units (111-666) for wide frequency activity.
- VSU: Voiceless Spotting Units. These integrate the outputs of different SNB's acting in separate bands to pattern the activity of fricative consonants.
- WSU: Vowel Spotting Units. These integrate the activity of Sf1 and Sf2 units to detect the presence of vowels and their nature.
- DTU: Dynamic Tracking Units. These integrate the activity of different dynamic trackers on the first two formants to detect consonant dynamic features.

An example of a consonant feature map showing the nuclear set of consonant phonemes of Spanish, is given in the following Table.

IPA Char	p	t	c	k	b	d	ɟ	g	f	θ	ʃ	x	β	ð	ʒ	γ
Alpha	p	t	c	k	b	d	J	g	f	T	S	x	B	D	Z	G
O/N	o	o	o	o	o	o	o	o	o	o	o	o	o	o	o	o
V/U	u	u	u	u	v	v	v	v	u	u	u	u	v	v	v	v
DC	s	s	s	s	s	s	s	s	f	f	f	f	f	f	f	f
AP	bl	a	p	v	bl	a	p	v	ld	d	p	v	bl	da	pa	v
O/R	-	-	-	-	-	-	-	-	-	-	-	-	-	-	-	-
FM1	a	a	a	n	a	a	a	n	a	a	a	n	a	a	a	n
FM2	a	n	d	n	a	n	d	n	a	n	d	n	a	n	d	n
CF	-	-	-	-	-	-	-	-	-	-	-	-	-	-	-	-
NB	a	mh	h	ml	-	-	-	-	a	mh	h	ml	-	-	-	-

Fig. 5. Table. Nuclear set of consonant phonemes shared by the Spanish Dialects.

The first row of the table shows the IPA code of the corresponding phoneme, whereas the second row gives the corresponding ASCII-IPA equivalent, also known as *Kirshenbaum code* [3]. The third row gives the Oral/Nasal feature, the fourth the Voiced/Voiceless quality, the fifth describes the degree of closure (stop/fricative), the sixth expresses the Articulation Place (bl: bilabial, a: apical, p: palatal, v: velar, ld: labiodental, d: dental, da: dento-alveolar, pa: palato-alveolar). The seventh feature is the quality of Oval/Round and is applied only to vowels. The eighth is the dynamic movement of F_1, (a: ascending, d: descending). The ninth is the respective feature to F_2. The tenth is the stable CF condition (for vowels). The eleventh codes the presence of broadbands (a: all-band, l: low-band, m: middle-band, h: high-band).

4 Results and Discussion

In what follows some of the capabilities of the proposed structure will be shown with emphasis in the detection of the most meaningful dynamic consonantal features. For such a typical example as is the sentence -Where were you while you were away- will be modelled using mainly FM+ and FM- units. The details of the architecture are the following: M=1024 units are used as characteristic frequency outputs from LPC, defining a resolution in frequency of barely 8 Hz for a sampling frequency of 8000Hz. These are sampled each 5 msec. to define a stream of approximately 200 pulses per second. The dimensions of the +FM and -FM units are 7x7, which means that the connectivity in frequency extends from +3 to -3 neighbour neurons, whilst the delay lines in the pauser units go from 0 to 30 msec. As the frequency distribution is linear, the number of channels integrated for the first formant (350-650 Hz) is around 38, whereas for the second formant (700-2300) is around 204.

The templates in Fig. 6 reproduce the behaviour of a Second Energy Peak Tracking Unit (Positive Slope) on the output of LIFP neurons, integrating the

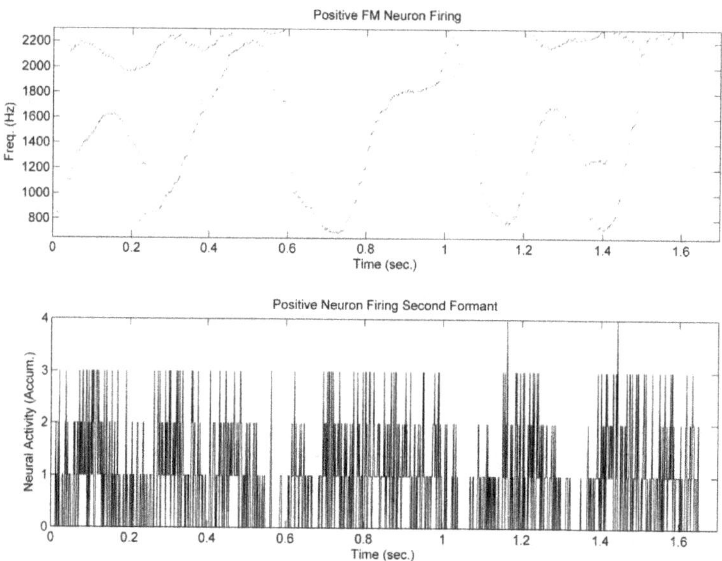

Fig. 6. FM Unit firing. Top: Activity of $+f_{M700-2300}$ Units coding the band of the second formant F_2. Bottom: Activity of Second Formant Integration Units (positive slope $+f_{M2}$).

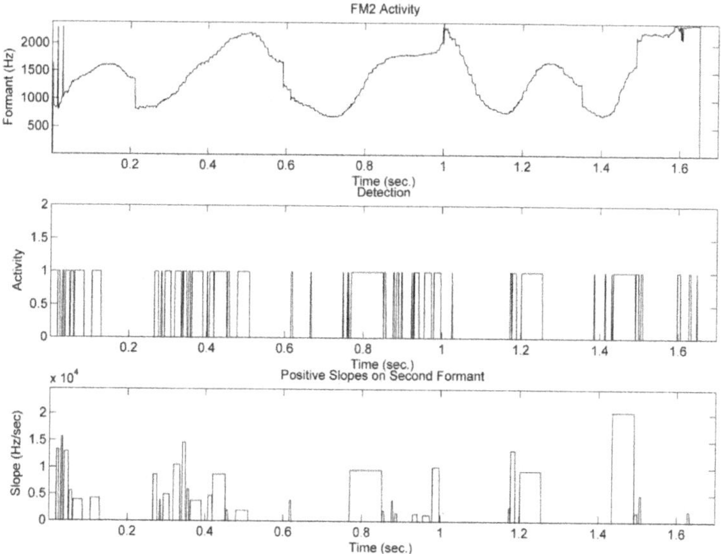

Fig. 7. Top: Output of Activity at the output of LIFP Units in the band of the second formant. Middle: Output of the Second Formant Integration Unit $+f_{M2}$ reproducing the positive slope intervals in the second formant. Bottom: Value of the slope detected on the given interval.

activity of up 204 units, most of which are not firing simultaneously. A correspondence among the firing histograms and the positive slopes of the second formant is clearly appreciated. This activity is integrated ($\int$) and thresholded by a saturation function (f) to produce the plots in Fig. 7 (middle). Finally in the bottom template of the figure the values of the different intervals detected with positive slope and the value of the average slope are given. These values are calculated integrating the firing histograms on the interval detected by the specific Integration Unit ($+f_{M2}$ in this case). The utility of these results is to be found in automatic phonetic labeling the speech trace, following maps as the one in Fig. 5, as well as in the detection of the speaker's identity [15].

5 Conclusions

Through the present it has been shown that formant-based speech processing may be carried well-known bio-inspired mask processors (CF and FM units) The structures studied correspond roughly to the processing centres in the Olivar Nucleus and the Inferior Colliculus. The study of short-time memory-like structures found in the upper levels of the brain, and especially the columnar structures of the Auditory Cortex [12] using low order regressors, fundamental for phonemic parsing deserve an extensive further attention. The lower and mid auditory pathways have been intensively and extensively researched, and a good deal of helpful and useful knowledge of use in Neuromorphic Speech Processing has been produced [14]. Nevertheless the cortical structure functions in speech processing lacks a similar description, their functionality being a great challenge even nowadays. Through the works of Mountcastle [12] some kind of functionality could be inferred related mainly with the short memory capability of bidirectional linear structures reproducing cortical columns, essential for the parsing of phonetic sounds leading to the emergence of the word as a semantic unit. This idea was discussed during the celebration of IWINAC07 at La Manga after the presentation of our work [6] and during a subsequent colloquium directly with Prof. Mira and some other colleagues. In that occasion Prof. Mira insisted in that the leading work to be revisited on the topic of short memory phonetic parsing should be that one of Rafael Lorente de No [11]. This researcher was one of the most outstanding disciples of Santiago Ramón y Cajal, and lived and worked in the USA for some five decades till his death in Tucson, AZ in 1990. It seems that it was Lorente de No's work what eventually leaded to Mouncastle's. During two other occasions in the early spring of 2008 Prof. Mira insisted in that Lorente de No's should be brought forth again to the interest of modern Neuromorphic Computing for the proposal of new mechanisms in Speech Processing and Understanding. Shortly after he passed away leaving the challenge open for the new generations. In memoriam: Cajal, Lorente de No, Mira.

Acknowledgements

This work is being funded by grants TIC2003-08756, TEC2006-12887-C02-00 from the Ministry of Education and Science, CCG06-UPM/TIC-0028 from

CAM/ UPM, and by project HESPERIA (http.//www.proyecto-hesperia.org) from the Programme CENIT, CDTI, Ministry of Industry, Spain.

References

1. Delattre, P., Liberman, A., Cooper, F.: Acoustic loci and transitional cues for consonants. J. Acoust. Soc. Am. 27, 769–773 (1955)
2. Deller, J.R., Proakis, J.G., Hansen, J.H.: Discrete-Time Processing of Speech Signals. Macmillan, New York (1993)
3. Gómez, P., Godino, J.I., Alvarez, A., Martínez, R., Nieto, V., Rodellar, V.: Evidence of Glottal Source Spectral Features found in Vocal Fold Dynamics. In: Proc. of the ICASSP 2005, pp. 441–444 (2005)
4. Hermansky, H.: Should Recognizers Have Ears? In: ESCA-NATO Tutorial and Research Workshop on Robust Speech Recognition for Unknown Communication Channels, Pont-à-Mousson, France, April 17-18, 1997, pp. 1–10 (1997)
5. Ferrández, J.M.: Study and Realization of a Bio-inspired Hierarchical Architecture for Speech Recognition, Ph.D. Thesis, Universidad Politécnica de Madrid (1998) (in Spanish)
6. Gómez, P., Martínez, R., Rodellar, V., Ferrández, J.M.: Bio-inspired Systems in Speech Perception: An overview and a study case. In: IEEE/NML Life Sciences Systems and Applications Workshop (by invitation), National Institute of Health, Bethesda, Maryland, July 13-14 (2006)
7. Haykin, S.: Neural Networks - A comprehensive Foundation. Prentice-Hall, Upper Saddle River (1999)
8. Irino, T., Patterson, R.D.: A time-domain, level-dependent auditory filter: the gammachirp. J. Acoust. Soc. Am. 101(1), 412–419 (1997)
9. Jahne, B.: Digital Image Processing. Springer, Berlin (2005)
10. Mendelson, J.R., Cynader, M.S.: Sensitivity of Cat Primary Auditory Cortex (AI) Neurons to the Direction and Rate of Frequency Modulation. Brain Research 327, 331–335 (1985)
11. Mountcastle, V.B.: The columnar organization of the neocortex. Brain 120, 701–722 (1997)
12. Ojemann, G.A.: Organization of language cortex derived from investigation during neurosurgery. Sem. Neuros. 2, 297–305 (1990)
13. O'Shaughnessy, D.: Speech Communication. IEEE Press, Park Avenue (2000)
14. Rauschecker, J.P., Tian, B., Hauser, M.: Processing of Complex Sounds in the Macaque Nonprimary Auditory Cortex. Science 268, 111–114 (1995)
15. Sams, M., Salmening, R.: Evidence of sharp frequency tuning in human auditory cortex. Hearing Research 75, 67–74 (1994)
16. Schreiner, C.E.: Time Domain Analysis of Auditory-Nerve Fibers Firing Rates. Curr. Op. Neurobiol. 5, 489–496 (1995)
17. Secker, H., Searle, C.: Study and Realization of a Bio-inspired Hierarchical Architecture for Speech Recognition. J. Acoust. Soc. Am. 88(3), 1427–1436 (1990)
18. Sejnowski, T.J., Rosenberg, C.R.: Parallel networks that learn to pronounce English text. Complex Systems 1, 145–168 (1987)
19. Suga, N.: Cortical Computational Maps for Auditory Imaging. Neural Networks 3, 3–21 (1990)

20. Suga, N.: Basic Acoustic Patterns and Neural Mechanism Shared By Humans and Animals for Auditory Perception: A Neuroethologists view. In: Proceedings of Workshop on the Auditory bases of Speech Perception, ESCA, July 1996, pp. 31–38 (1996)
21. Waibel, A.: Neural Network Approaches for Speech Recognition. In: Furui, S., Sondhi, M.M. (eds.) Advances in Speech Signal Processing, pp. 555–597. Dekker, New York (1992)

Spatio-temporal Computation with Neural Sensorial Maps

J.M. Ferrández-Vicente[1], A. Delgado[2], and J. Mira[2]

[1] Dpto. Electrónica, Tecnología de Computadoras, Univ. Politécnica de Cartagena,
[2] Departamento de Inteligencia Artificial, UNED
jm.ferrandez@upct.es

Abstract. In this work, the conexionist computational paradigm used in Artificial Intelligence (AI) and Autonomous Robotics is revised. A Neural Computational paradigm based on perceptual association maps is also analysed regarding his close relation with the conexionist paradigm. We explore the Singer hypothesis about the evidence of the same neural mechanism previous to retinotopic projections. Finally, the implications of this computational paradigm with maps in the context of the definition, design, building and evaluation of bioinspired systems is discussed.

1 Introduction

When we speak about Bioinspired Computing, we refer to computational models based on biological phenomenology in general, and particularly to neural computation thorough the Computational Paradigm (CP), showed in Figure 1. This paradigm includes not only the physical level, but also the meaning of the calculus passing over a symbolic level (SL) and a knowledge level (KL), where the percepts, objectives, intentions, plans, and goals reside. In addition, in each level it is necessary to distinguish between the semantics and the causality inherent to that level phenomenology (Own Domain, OD) and the semantics associated to that phenomenologies by the External Observers Domain (EOD). It is also important to note that the own experiences, that emerge from the neural computation in a conscious reflexive level, only matches partially with the communicable by natural language.

Assuming that a Neural Network is the anatomical and physiological structure of a calculus is evident if we observe the external medium (initial phases of sensor modules and latest phases of motor modules). Then, we are speaking about electrical measurable signals (currents, voltages) associated to a physical semantic (amplitude, frequency, waveform, derivatives, integralsetc).

The problem of the computational interpretation of neural functions begins while ascending in the abstraction level, from receptors to cortex, with the identification of neural and anatomical structures with a semantic level more complex than sensorial and motor neural networks signals. At that level we are speaking about spatio-temporal patterns of specific signals, that present unitary and autonomous sensorial patterns with an unitary and autonomous meaning associated to the medium characteristics (objects, events and activities) of relevant

J. Mira et al. (Eds.): IWINAC 2009, Part I, LNCS 5601, pp. 79–86, 2009.

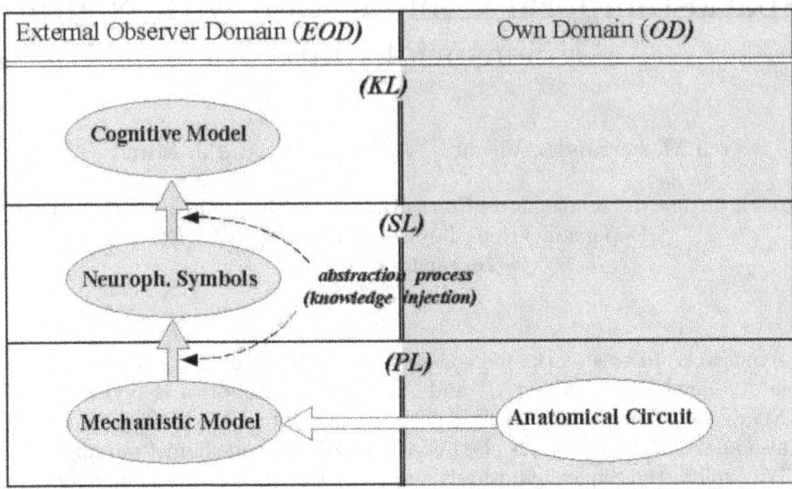

Fig. 1. The figure of the external observer enables us to establish a clear distinction between causal (OD) and descriptive (EOD) entities. This wider view of the CP, with three levels (PL, SL, KL) and two domains for each level (OD, EOD), enhances the explanatory facilities of the CP.

interest for the survival of an animal structurally couple to that environment. Different types of retinal ganglion cells in rana pipiens and the concept of affordances in a environment (the echological aproximation to perception proposed by [1]) are clear paradigmatic examples of the neurophysiologic symbolic proposal and the semantic sep concept between complex patterns and the sensorial signals that constute them. The extension to semantics is at ht e complex pattern level, and the relational structure to the low semantic level signals Singer conjecture [2] extends retinotopy to to central neural networks, and it serves as basis to this paper proposal, that pretends to define the neural computation as constructions, and associations between two map hierarchies (sensory and motor). These two hierarchies are organized according the ascendent pathway of the computational paradigm (signal, symbols and percepts, concepts and intentions). This proposal requires identifying the anatomical support of these entities, or locating the neurons or neural ensembles tuned to that spatio-temporal signals pattern, identified by the animal with an unitary and autonomous meaning, independent and overlapped to individual signals meaning that constitute the ensemble. These symbolic neurons (objects, images) with invariant perception will have a maximal response when the external code reference is presented or remembered. In the same way the electrical stimulation of this neuronal ensemble (for instance during neurosurgery) will evoke the subject to perceive the corresponding meaning. Since human beings birth, they begin to build their own neural representations of the environment, and from our own experience with this environment we predict, plan, decide and act as proposed by K. Craik [3].

A similar framework appears when the knowledge level is reached, where the concepts, percepts, intentions, objectives and plans reside. The objects at the conscious levels are directly related with the specific activity patterns from sensorial, motor or psycho-emotive limbic areas (prefrontal, areas 10-11 Broadmann) At these levels , physical, symbolic and intentional, the resulting computation consists in te spatio-temporal activation of the axonal structures, while the semantics of this computation is intrinsic to the network architecture (with local connectivity, lateral inhibition, positive and negative reinforcement, and convergent and divergent schemes). The paper is structured by revising the classical conexionist paradigm beginning from its origins (Neurocybernetics y Bionics) and linking with its reborn in 1986. Finally we revise a computational paradigm based on association maps, the auditory neural hierarchy is described for proposing spatio-temporal neural maps as new paradigm for computation in predominant real time sensory systems. We conclude by discussing the connection of this neural computation with the new bioinspired systems design.

2 The Conexionist Paradigm

In the conexionist paradigm, the so-called Artificial Neural Networks (ANN), the problem of knowledge representation is dealt by using numerical labels as inputs/outputs to the network. The inference problem is then solved with a numerical classifier with numerical parameters adjusted by means of a learning algorithm, supervised or not.

The architecture of a conexionist agent is in form of modules, organize in layers, with multiple units, elemental processors or neurons, strongly coupled, that compute a single local function, in general a weighted addition followed by a sigmoid or other non linear function, that limits the response dynamical range of each unit.

The relevant characteristics of this knowledge modeling and programming are the following:

1. The problems solved with ANN are characterized by an adaptive numerical classifier, that associates a set of acquired values, represented by numerical label lines, with the values of other limited set, represented by the last layer of neurons output, that are also numerical labeled lines.
2. An important part of the available knowledge corresponds to the data analysis phase. The user must decide what are the input/output variables, the most proper local function, the number of units per layer, and the amount of layers, the weights initialization...etc.
3. It is also important to know the balance between the data and the available knowledge, and also the kind of data, if they are labeled or not. Labeled data, with a priori known network response are used for supervised learning, and in later stages for validating and evaluating the network behavior. Non-labeled data are used for preprocessing and for autoorganizative learning, where the network automatically organizes all the knowledge inherent to data statistics.

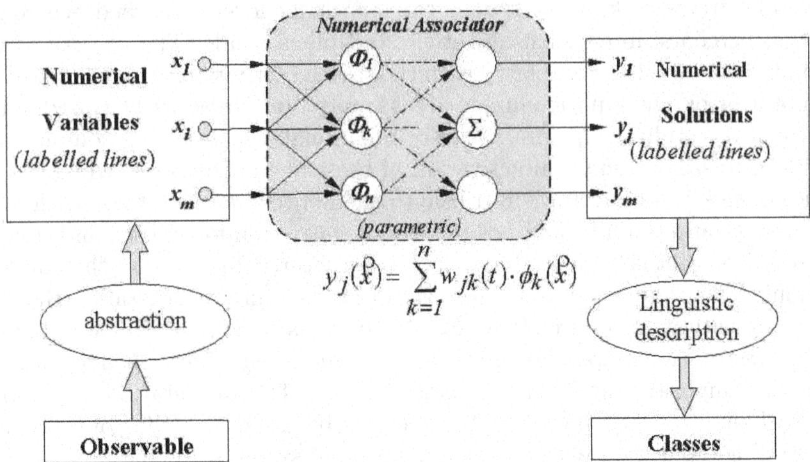

Fig. 2. ANNs general architecture like adjustable parametric associators beween two numerical representations

Figure 2 shows the basic architecture of a conexionist model, that corresponds to a parallel and oriented graph where the nodes are the neurons, and the arcs are the adjustable weights.

Finally the numerical solutions of the ANNs, the outputs of the end layer units are interpreted in terms of the labels associated to that classes.

It is important to remark the strict numerical profile of the conexionist paradigm. The network only associates numbers in term of the functions used in the first and second layers. These functions can be lineal (Fourier, eigenvalues pricipal component analysis) or non-linear (radial-base functions, polynomials, Wiener-Volterra series).

Conexionist has two main orientations. The first one and more usual is Knowledge Engineering described previously. The network has a fixed architecture and it behaves as a general purpose classifier. There exist other approximation more oriented to biology, because it incorporates the knowledge about the neural anatomy and physiology. It is called bioinspired conexionist. Some example are the lateral inhibition networks, the reflex arcs, habituation and sensatization circuits and the response pattern generators, where the specific connectivity, the feedback loops and the excitatory and inhibitory synaptic contacts are determinant of the function computed by the network. This ascendent approximation to intelligence is more modest than the symbolic approach and it uses adaptation mechanisms of an agent to its environment. This is equivalent to assume that intelligence may be considered as a superior form of adaptation, built over other more elemental forms of adaptation and shared with the rest of the phylogenetic scale.

In the context of computational maps, classic conexionist can suggest: (1) to look for the distiction between observable and cluster maps, and to take into account the recursive behaviour of the relation during all the process of afferent categorization. (2) To consider the idea of reference base transformation ("hidden layer"). There exist maps that behave as functional maps that code in a more efficient and economic way the external world. Posterior Neural Networks modules only "see" at the environment through these internal coded representations.

While supervised learning does not have a clear biological similarity, the unsupervised algorithms (Hebbian) have and also the reinforcement learning.

3 Real Neural Maps

It is known that sensory surface exists in the cortex and other parts of the brain, and that inputs are regularly ordered building maps with the provided sensory representations. The evidence for these maps has been proved using electrical stimulation by Penfield [4], and by neurological impairments due to stroke or another cortical damage. There have been described somatosensory maps [Nelson.80], olfactory maps [5], visual maps [6], auditory maps [7]. The meaning of the somatotopic, retinotopic, or tonotopic ordered organization is not completely understood. Why similar stimuli are represented in close locations, while very different inputs tend to occupy distant spaces. It is reasonable that neurons with most common connections group in a closely neighborhood in order to avoid metabolic costly distant connections. However, this organization of simple stimuli is less evident in higher centers of the neural hierarchy. This is possible because these higher centers are specialized in detecting spatio-temporal combinations of these basic inputs, making possible in this way to detect complex patterns.

Neural sensory maps can also be defined as systematic spatio-temporal distributions of sensory spaces in a bidimensional surface, that is bidimensional representations of peripheral receptors. The main function is in this way to cluster the n-dimensional spatio-temporal receptor space in an ordered bidimensional topographical organization. These spatio-temporal ordered neural maps offer many computational advantages over random distribution representations. The efficiency comes from minimizing connectivity, reducing redundancy and enhancing computational cost.

3.1 Auditory Maps

Auditory system has been better described than visual system because of the unidimensional characteristic of sounds. We perceive combinations of frequencies varying in the temporal domain, while images are bidimensional representations moving over time.

In cats, the auditory connections that arrive to auditory cortex are ordered by frequencies. This tonotopy is preserved at primary auditory cortex. There exist neurons en auditory area AI that respond only specific frequency and are organized in an ordered and binaural way by increasing frequencies and intensities axes. On the other hand, secondary auditive cortex has not a tonotopic organization, finding neural cells that responds to different frequencies or combinations of them. This organization is nearly the same in primates, around sylvian fissure. In human beings, cerebral cartography like PET, or fMRI has shown evidence of a ordered tonotopy in primary auditive area.

In cortical auditive area, like in visual areas, or somatosensory areas, functional columns have been located. All the neurons in a column, the cortex has six layers, have similar responses to stimulus. These responses differ a lot even in adjacent columns. That means that each column response to a specific frequency. These columns have also binaural selectivity, been excited by both cochlear inputs (EE bands) or only by contralateral inputs and inhibited by ipsilateral sounds (EI bands). As conclusion, like visual system that is organized in dominant bands, auditory systems is also organized in frequency characteristic bands in an ordered maps in primary areas.

In auditory cortex there have been detected neurons that respond only to a specific frequency, as mentioned, but also to combinations of frequencies and delayed frequencies, that vary from 20 to 50 ms, in Posterior areas, while in anterior auditive cortex the delays were detected were until 20 ms. The functional organizing structure of this area is the most relevant aspect for proposing a computational paradigm. As mentioned in the previous paradigm, the animal organizes its inputs depending on their significance for its survival or behaviour. In humans, the auditory cortex has evolved for providing oral communication between human beings, so it must be organized according the speech components.

The basic speech components are four: Characteristic Frecuency (CF) units, that are composed by stationary frequencies, Frequency Modulated (FM) elements, that consists in transitions between frequencies, Noise Burst (NB) components, bands of noise around certain frequency and Voice Onset Time (VOT) units, that code certain sound delays. The CF elements are detected by the neurons that response to specific frequencies, found and many animals, and in human beings. There has been also detected FM neurons that fire in response to certain FM rate, that is detect specific slopes in the spectrogram. Some of them detect the slope orientation while others response both to the same positive and negative slopes. The frequency transitions cover the range from 0.05 to 0.8 kHz/ms in some units, with positive or negative selectivity. The NB elements have been detected in macaques by Rauschecker . There exist units that fire when certain noisy stimuli is provided, with specific central frequency and bandwidth. In certain areas the noisy stimuli evoke neural responses 90 per cent bigger that pure tones. The neural selectivity detected also the characteristic bandwidths for each neuron, and this bandwidths were also ordered in a map within a specific arrangement, perpendicular to the frequencies axe. Finally the VOT elements are detected by neurons specialized in detecting specific delays, varying from 0.4 to 18 ms. In an ordered way. Summarizing, there exists neurons that respond to the specific speech components and they organized in ordered maps.

3.2 Maps Isomorphism

Each map has its properties and it is necessary to define precisely the pairs Property $i->$ Neuron j, and the values associated to each neuron. One of the main arguments for the functional importance of neural spatio-temporal maps is the close relation between map activation and perception. Some studies show the relationship between map neural discharges and the sensation, and the separate stimulation in the map evokes separate perceptions, but it is important to note that the representation in the spatio-temporal neural map is not ordered the same way the receptors structure. For instance, there no exist a retinotopic map in visual primary cortex with the same arrangement of retinal ganglion cells organization. The may account for inputs invariant in any parameter of the stimuli dimension.

Neural spatiotemporal maps may also provide a substrate for a local reorganization in response of experience, training or damage in any sensory input or region in the map. This map plasticity provides flexibility to the sensory representation during the whole life of the structure.

4 Discussion

In this article we have two main conclusions. The first one is more oriented to Artificial Intelligence and Robotics, while the second one focuses Neurophysiology, medicine, and neuroprosthetics.

The conexionist computational paradigm with biologic inspiration in AI and robotics have been revised, looking in all cases their relation with computational maps. Classic connexionist methods provide the concept of adaptive classifier that associates different classes of observable classes to other specific classes (concepts). In this way if we want to find the corresponding relation with a maps computation, we need to follow the ascendant perception organization, identifying the anatomical structures that support its properties, and relations. We also need to distinguish between unimodal maps, crossmodal maps, multimodal maps.

The real neural maps are organized in hierarchical functional maps, no matter where the inputs come. The regions that handle auditory inputs are similar to the regions that handle touch, and they look like the areas that process visual scenes. The brain works with uniform neural maps in their cortical surface. And these maps are ordered in specific ways, giving much area to relevant stimuli. So probably they are performing the same parallel computational spatio-temporal operation, and this simple structure may be used as a new computational paradigm in Artificial Intelligence and Neuroprosthetics.

During the last year, I have been discussing with Professor Mira, the work of Libet [8], Luria [9], Singer [2], Marr [10], Braitenberg [11], Gonzalo, trying to find the answer to the semantic gap. We had many projects in common, the sensorial summation, the non-invasive deep brain stimulation, the National Network on Natural and Artificial Computation. I will try to continue, the path you began, and I will try to do it my best, I owe you, Pepe.

Acknowledgements

Two the authors (Mira, J and Delgado, A.E.) are grateful for the support from projects TIN2004-07661-C02-01 and TIN2007-07586-C02-01, in whose context this work has been developed. J.M. Ferrandez is supported by TIN2008-06893-C03, and Fundación Séneca 08788/PI/08.

References

1. Gibson, J.J.: The ecological approach to visual perception. Houghton-Mifflin, Boston (1979)
2. Singer, W.: The Observer in the Brain. In: Understanding Representation in the Cognitive Sciences, pp. 253–256. Kluwer Academic/Plenum Publishers, New York (1999)
3. Craik, K.: The Nature of Explanation. Cambridge University Press, Cambridge (1943)
4. Penfield, W., Rasmussen, T.: The cerebral cortex of man. Macmillan, New York (1954)
5. Uchida, N., et al.: Odor maps in the mammalian olfactory bulb: domain organization and odorant structural features. Nature Neuroscience 3(10), 1035–1042 (2000)
6. Merzenich, M.M., Kaas, J.H.: Principles of organization of sensory-perceptual systems in mammals. Progress in psychobiology and physiological psychology (1980)
7. Suga, N.: Cortical Computational maps for auditory imaging. Neural Networks 3, 3–21 (1990)
8. Libet, B.: Unconscious cerebral initiative and the role of conscious will in voluntary action. Behavioral and Brain Sciences 8(4), 529–566 (1985)
9. Luria, A.: The working brain. Basic Books, New York (1973)
10. Marr, D.: Vision. Freeman, San Francisco (1982)
11. Braitenberg, V.: Vehicles, experiments in synthetic psychology. In: Sober, E. (ed.) (1984)
12. Gonzalo, J.: Las funciones cerebrales humanas segun nuevos datos y bases fisiologicas, Instituto Cajal. CSIC (1954)

Knowledge-Based Systems: A Tool for Distance Education

Pedro Salcedo L.[1], M. Angélica Pinninghoff J.[2], and Ricardo Contreras A.[2]

[1] Faculty of Education
[2] Faculty of Engineering
University of Concepción, Chile
psalcedo@udec.cl

Abstract. This work describes how, starting from the PhD thesis *A Knowledge-Based System for Distance Education*, under the supervision of Professor José Mira Mira, different ideas have evolved opening new research directions. This thesis has emphasized the usefulness of artificial intelligence (AI) techniques in developing systems to support distance education. In particular, the work concerning the modelling of tasks and methods at knowledge level, and in the use of hybrid procedures (symbolic and connectionist) to solve those tasks that recurrently appear when designing systems to support teaching-learning processes.

Keywords: Prediction, Academic Performance, Neural Networks.

1 Introduction

This paper is a summarized text that presents the doctoral thesis *A Knowledge-Based System for Distance Education*, developed under the supervision of professor José Mira Mira, that shows how different artificial intelligence techniques can be used in developing support systems for distance education. Key aspects are the modelling of tasks and methods at knowledge level, and the use of hybrid procedures, symbolic and connectionist, for solving those tasks that recurrently appear when designing systems to support the teaching-learning process. In particular, in the developed platform we have used Bayesian networks (BN), artificial neural networks (ANN) and rule systems (RS). The decision in choosing one or another technique for implementing a subtask or inference step takes into account the balance between the data and the available knowledge. We use symbolic methods (rules, tables, frames, graphs) in all those tasks in which clear, complete, precise and unequivocal descriptions are available. On the other hand, we use BN when tasks are not precise or when tasks present some uncertainty. Finally, ANN is used in subtasks in which supervised learning is required (adapting contents to students' profile).

The validation and evaluation of the proposal is achieved through the development of a specific platform; MISTRAL [5,6], that incorporates these artificial intelligence techniques including adaptation and evaluation mechanisms. This platform has been used to develop a set of courses with different contents and

J. Mira et al. (Eds.): IWINAC 2009, Part I, LNCS 5601, pp. 87–96, 2009.

taking into account a variety of students profiles, in an attempt to highlight the strengths and weaknesses that it presents.

Additionally, we present different works that have arisen from the research topic that the doctoral thesis proposes. It includes a platform for distance training, using PLCs remote control and considers AI techniques for adapting the most appropiate contents to different students. In recent years some efforts have been devoted to predict academic performance in students belonging to university level as well as to high school level. The effort can also be illustrated by considering methods at knowledge level in the design and adaptation of a platform for teaching architecture students.

AI Techniques in Education

The main goal of the doctoral thesis was to study systems that can help in distance teaching from the point of view of Knowledge Engineering, knowledge-based system in particular. In doing so, we followed the usual stages in current methodologies: knowledge level modelling, the use of methods from a reusable components library, inference processes by using symbolic and connectionist techniques depending on available data and knowledge, operator implementation, and finally, system validation and evaluation.

Table 1. AI techniques classification

AI Technique	Example	Description
Basic Rules	CAI of Uhr	Uses Rule-based systems
Knowledge-based system	SCHOLAR	Uses solving strategies from AI
Natural Language and rules	CALCHEM	Uses natural language (interface)
Production system and natural language	SOPHIE	Interacts through natural language
T-rules and production system	GUIDON	Uses backward chaining
Backward and forward chaining	PTA	Uses an expert module
Models declarative knowledge	LISP, GEOMETRY, GEOMETRY ALGEBRA	Uses tutoring strategy
Internet rules	TANGOW, APHID, AHA	Adapts teaching strategies
Case based reasoning	ARTHUR	Adapts strategy based on cases
Bayesian networks	I-PETER	Uses learning diagnosis in an English course
Neural networks	NBSU	Adapts the learning environment

Table 2. AI techniques to be used when designing support systems in the teaching-learning process

AI Technique	Task in which it is used
Semantic networks	For modelling and structuring conceptual knowledge
Frames	For modelling knowledge related to different entities in the knowledge layer for each task
CommonKADS CML	General methodology for modelling knowledge of tasks, methods, inferences, roles and concepts
Rules	For implementing inferences when available descriptions are clear and complete
Neural networks	Tasks that require supervised learning
Bayesian networks	Tasks that present uncertainty

The initial step was to consider different techniques that have been used in different modules belonging to Intelligent Tutoring Systems (ITS), Hypermedia Adaptive Systems (HAS) and systems for distance education. In this way various attempts were made in order to create computational systems that could supply complementary solutions or alternatives considering education in general, and later evolving towards Internet and onto distance education. A classification of techniques is shown in Table 1.

Studying these techniques and the way in which they have been used in different systems, was the starting point for proposing that the modelling of tasks and methods could be combined with hybrid procedures, symbolic rules and connectionist networks in order to solve tasks that recurrently appear when designing systems for supporting teaching-learning processes. In particular, it is necessary to use AI techniques and knowledge engineering techniques. These techniques are shown in Table 2.

2 Available Knowledge for the Conceptual Model

Taking into account the state-of-the-art in distance teaching, and considering the tasks that appear in a recurrent way as necessary for its synthesis as well as the methods used for its synthesis, it is clear that tasks that appear frequently are those summarized in Figure 1, and Table 3.

All the tasks mentioned in Table 3 were considered when designing and implementing MISTRAL, the platform used in order to validate the proposal. It is important to notice that there exists a different complexity degree of the different tasks, and that there is a strong need to use AI methods and techniques in evaluating, in selecting personalized teaching-learning strategies for different students' profiles (it considers modelling) and in didactic material preparation. As it can be noticed, to completely automate these tasks is far from a definite solution.

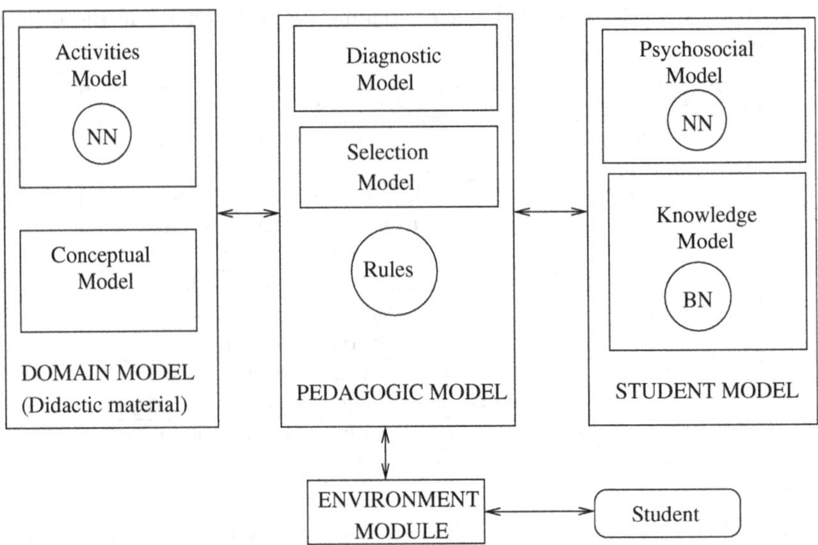

Fig. 1. Interaction among knowledge models

Table 3. Global task and subtasks for a distance education system

Global task	Distance teaching and learning
Subtasks	
Material preparation	Previous knowledge tutorial Theoretical content Activities Evaluation Complementary activities
Teaching/Learning Strategies	Adaptation of user model Strategy configuration Sequential contents presentation Feedback
Dialogues	On line (public, private) Off line (public, private)
Evaluation	Contents (text, images, formulas) Portfolio of activities (Teacher evaluation, automatic evaluation)
Management	Class list, attendance, dialogues

With the previous idea in mind, it was established that the usual approach in knowledge engineering could help, by proposing solutions to the problem of uncertainty and to incomplete data. In this work it was summarized the available knowledge for constructing a conceptual model for all the three tasks: domain model, pedagogic model and student model.

3 The Model of the Course Model

Modelling the knowledge for a course is one of the first tasks to be solved in order to put into the system what contents and activities are more suitable for a specific student's profile. In structuring the didactic material, the human expert must establish, as a first step, the set of concepts to be used in constructing the structure to support that knowledge. These concepts and their static relationships are called the knowledge layer of the domain.

These relationships are considered by the human expert when structuring the material. A general goal is defined through a set of specific objectives, which in turn are defined as a set of activities that are to be evaluated in different ways.

Semantic networks are an adequate tool in the task of modelling the static knowledge; because they allow to determine how the specific contents in a particular course are related. Additionally, semantic netwoks allow to make explicit the previous knowledge required for a specific content, that can be reflected as state transition graph. This allows us to model the knowledge so as to establish the way in which the system may be automated. These graphs can help to go from the static (semantic) network to a set of possible paths. This is analogous to the way in which automata theory is used to describe the dynamic behavior of a system. To reach a state of knowledge (a node in the graph) it is necessary to accomplish all the previous activities and evaluations required to access that state.

4 The Diagnostic Model

Adapting a teaching strategy is associated to the available tools that allow to measure learning. The knowledge required to successfully achieve a goal is closely related to the previous task of generating the didactic material. The knowledge of the diagnostic task domain is modelled, same as with the former, in a graph where each node is related to one or more nodes belonging to the conceptual model (didactic material). The evaluation graph is called the dual graph of the graph containing the teaching material.

The diagnosis provides the necessary information for the student's model and for the teaching strategy, which is explained as follows: It can be separated into two stages; the first one for determining the psychosocial characteristics of a student, and the second one that is iteratively in charge of the interaction process between the student and the platform. In order to obtain psychosocial data, a set of test can be used; these data can be used to diagnose the student's success through neural networks. In general, this technique allows the generation of a personalized teaching strategy to foresee the future performance of a student [5].

On the other hand, it is possible to periodically diagnose the student's performance through a variety of techniques like portfolios or adaptive tests. This last technique based on Bayesian networks allows the possibility to chose the best activity or question for the knowledge the student needs [6].

5 The Student Model and the Teaching Strategy

The modelling of the student can be as having two parts, (1) the psychosocial model, that considers the psychological and social characteristics to be taken into account during the teaching process, and (2) the knowledge model, that allows us to establish what the conceptual nodes in a course are, that the students learn, and what the activities involved in that process are. The learning experiences should be consistent with the student's learning style. Learning styles include cognitive, affective and physiologic traits, that help to understand how individuals acquire, interact and respond to their learning environments.

The learning style is the way in which people learn. It should be noticed that different people have different learning styles. Different authors characterize the learning styles in different ways, but they often agree in choosing a point in a continuous line. There is not a *better* style, and there is not a relationship between learning style and intelligence. Although it is usual to recognize a dominant learning style in a person, non dominant components always appear; i.e., nobody presents an exclusive learning style.

Different researchers have presented models and theories to deal with and measure learning styles. That is the case of [2,3], who has identified four different styles using the terms Diverging, Assimilating, Converging and Accommodating, and has developed a procedure for measuring them. This procedure has been used to test the methodology we implemented.

From Kolb's point of view, when learning an individual has to be involved in new experiences, and has to be able to observe and think about those experiences, integrating observation and thinking to concepts, models and logic theories. In other words, he/she should be able to take decisions, solve problems and challenge these theories when new conditions arise.

It is very important the way in which the student acquires and relates to nodes of knowledge. The method that a student follows can be represented as an analog procedure to graph searching as we can usually find it in AI techniques. But, the representation of knowledge as a space of states and the representation of learning as a searching procedure through the space of states, is still an open issue.

6 Model for the Teaching Strategy

A teaching strategy is a set of steps aimed at changing the student behavior, in a structured way that uses didactic materials. In particular, when considering a distance teaching/learning process the strategy will help to structure didactic material in various ways, in an attempt to search for the best strategy for a particular student, depending on his/her specific features.

At this point, a semantic network appears as a natural tool for structuring the didactic material, and as a helpful representation for generating a strategy which is adequate to the student's knowledge.

Students presenting different levels of knowledge need different sequences of contents that consider different profiles. In this manner, the teaching strategy for a specific learning style (and for a particular student), consists of a sequence of activities that are to be specified in the didactic material model. Rules derived from pedagogy or psychopedagogy will act as a guide to generate these sequences.

Activities that determine a strategy as adequate belong to a closed and predefined set, which in turn implies that the main effort for a strategy is a selection process. The above mentioned MISTRAL platform, additionally allows the possibility to manage different and not previously defined activities.

7 System Design

The design stage is associated to techniques and other issues coming from AI that allowed the resolution of the tasks identified through the analysis of the target system. In this stage a set of techniques belonging to knowledge engineering (KE) were used for linking the computerized system to the model generated in the previous stage.

KE is a set of techniques (and knowledge) that allow the application of the scientific knowledge to the use of the knowledge [1]. By using techniques and tools that belong to KE it is possible to design, develop, produce and manage those learning environments currently requested by enterprises and educational institutions for complementing the classic presential education.

One important methodology is CommonKADS, used as the European standard for developing knowledge-based systems [7]. CommonKADS supports the development cycle for software through the use of a set of related models that capture the fundamental features of a system and his environment.

For modelling knowledge (diagnostic, student and strategy) in the design stage, we used CommonKADS and frames. A frame is a collection of attributes, usually called slots, that contain values describing some entity in the world. CommonKADS and frames are necessary tools to accomplish the following stage: the system implementation.

8 System Implementation

Before implementing the system it was necessary to choose the most adequate tools and languages to solve each one of the different tasks specified in the previous stages.

We have been working from the point of view of an external observer and at knowledge level in such a way that all the considered entities maintain their semantic, using a non restrictive notation to denote them. The idea is to keep a set of words close to the description an expert could make concerning distance teaching/learning. This description should consider distance teaching, distance learning, selection and organization of didactic material and, finally, how the evaluation proceeds.

In this stage we carried out the step that goes from the models corresponding from the observer domain to the proper domain (PD), both at the knowledge level. In doing so, it was necessary to formally describe all the inferences derived from the schemes that reflect the different subtasks. We used a general operator library, and a general criteria based on the balance between data and knowledge. It was noticed that in some cases operators are analytic in nature, while in other cases are algorithmic. It was observed that it is possible to get a particular set of inferences that, based on the knowledge they use, cannot be directly applied. That is the case in which during the inference process we deal with uncertainty. To manage this particular issue we used macro operators (Bayesian networks and neural networks).

In this stage, a platform, MISTRAL, was developed, modelling the knowledge and the tasks just described, specifying a number of primitive inferences that are necessary to implement by using tools and techniques that belong to AI. For example, neural networks (macro operator) through NeuroSolutions, Bayesian networks (macro operator) through SMILE.

9 Architecture and Core of MISTRAL

The architecture system is based on the standard Web model, in which the server receives requests from students, professors or an administrator through browsers. These requests are transmitted to a process manager program. The program after delivering the received data, remains waiting for HTML pages generated by MISTRAL, and that are to be shown to users. If the information received by the process administrator program corresponds to a user request not connected to the system, the process administrator starts a new process. In case the user is connected, data is directly set to that user.

10 Derived Directions for Research

An e-learning platform, ATENEA, was implemented to manage continuous education programs. It can adapt the teaching strategy to students' psychosocial profiles. Additionally, this platform controls PLCs (Programmable Logic Controllers) remotely. Currently, ATENEA is being used for continuous education in Chilean forestry enterprises.

It has been established a research direction in students' performance prediction, at university level as well as at high school level, using a neural network that considered a training set containing more than 200 thousand real cases collected around the world, through the PISA project [4].

Another interesting topic is the consideration of the lexical availability for predicting success/failure in schooling performance and the adaptability of remedial activities.

The design of e-learning courses for architecture students in the University of Bio Bio (Chile) is another interesting initiative, it used the models proposed in this research.

11 Conclusions

The most important conclusions are related to the difficulty to automate the tasks identified during the analysis of the system, and the usefulness of using artificial intelligence and knowledge engineering in distance education, from two different key aspects. First, in showing, explicitly through knowledge level modelling, the difficulty to reduce to a computation the basic processes for giving a structure to the didactic material, for adapting this material to a student profile and automatically evaluate the learning for a particular student and for a variety of topics. Secondly, in integrating distance learning as a specific application field of AI. In this way, allowing the reuse of the components for tasks, methods, inferences, roles and domain knowledge modelling.

More precisely, the following results were obtained:

A wide comparative study has been varried out on different reasoning techniques used in intelligent tutoring systems, adaptive hypermedia systems, and distance education platforms. The tasks involved in the teaching/learning process have been modelled by using CommonKADS and UML. The inference process has been implemented in MISTRAL to illustrate the capabilities an weaknesses of the considered models. It has been developed a procedure for using the platform that allows: (i) to create, modify and manage courses in a friendly way, for non specialist users; (ii) to reduce time and effort in structuring course contents, because of the way in which modules and activities can be included; (iii) to adapt the teaching strategy to the student's profile that is represented in MISTRAL as acquired knowledge and learning style, and (iv) to include usual facilities related to synchronous and asynchronous dialogues, attendance control, level of participation and general management of the system. It has been shown the usefulness of applying artificial neural networks and Bayesian networks for adapting the teaching process to the student's profile and to specific topics to be covered. Finally, it has been possible to get promising results in evaluation, adapting test activities to the automata states that represent the level of knowledge of a student.

None of these results could have been possible without the guidance, support and enthusiasm of professor José Mira Mira.

References

1. Alonso, A., Guijarro, B.: Ingeniería del Conocimiento. Aspectos Metodológicos. Pearson Prentice Hall, London (2004)
2. Kolb, D.: The Learning Style Inventory. Technical Manual. McBer, Boston (1976)
3. Kolb, D.: Learning Style and Disciplinary Differences. In: Chickering, A.W. (ed.) The Modern American College. Jossey-Bass, San Francisco (1981)
4. Pinninghoff, M.A., Salcedo, P., Contreras, R.: Neural Networks to Predict Schooling Failure/Success. LNCS. Springer, Heidelberg (2007)
5. Salcedo, P., Pinninghoff, M.A., Contreras, R.: MISTRAL: A Knowledge-Based System for Distance Education that Incorporates Neural Networks for Teaching Decisions. In: Mira, J., Álvarez, J.R. (eds.) IWANN 2003. LNCS, vol. 2687, pp. 726–733. Springer, Heidelberg (2003)

6. Salcedo, P., Pinninghoff, M.A., Contreras, R.: Computerized Adaptive Tests and Item Response Theory on a Distance Education Platform. In: Mira, J., Álvarez, J.R. (eds.) IWINAC 2005. LNCS, vol. 3562, pp. 613–621. Springer, Heidelberg (2005)
7. Schreiber, G., Akkermans, H., Anjewierden, A., de Hoog, R., Shadbolt, N., Van de Velde, W., Wielinga, B.: Knowledge Engineering and Management: The CommonKADS Methodology. MIT Press, Cambridge (1999)

ANLAGIS: Adaptive Neuron-Like Network Based on Learning Automata Theory and Granular Inference Systems with Applications to Pattern Recognition and Machine Learning

Darío Maravall and Javier de Lope

Perception for Computers and Robots
Universidad Politécnica de Madrid
dmaravall@fi.upm.es,javier.delope@upm.es

Abstract. In this paper the fusion of artificial neural networks, granular computing and learning automata theory is proposed and we present as a final result ANLAGIS, an adaptive neuron-like network based on learning automata and granular inference systems. ANLAGIS can be applied to both pattern recognition and learning control problems. Another interesting contribution of this paper is the distinction between pre-synaptic and post-synaptic learning in artificial neural networks. To illustrate the capabilities of ANLAGIS some experiments on knowledge discovery in data mining and machine learning are presented. Previous work of Jang *et al.* [1] on adaptive network-based fuzzy inference systems, or simply ANFIS, can be considered a precursor of ANLAGIS. The main, novel contribution of ANLAGIS is the incorporation of Learning Automata Theory within its structure.

Keywords: Artificial Neural Networks, Granular Computing, Learning Automata Theory, Pattern Recognition, Machine Learning.

1 Knowledge-Based Production Rules with Self-learning and Adaptive Capabilities

Knowledge-based production rules of the type *If* {*condition*} *Then* {*Action*} are very powerful and widely used computational instruments to encode the knowledge needed for the implementation of intelligent systems. Not surprisingly they are the basic building blocks of expert systems, probably the most significant and successful contribution of the early "symbolic Artificial Intelligence" [2]. However, production rules-based expert systems have been criticized of being brittle [3] in the sense that they are unable to generalize from narrow, specialized domains, where they usually perform quite satisfactorily. Therefore, big efforts have been devoted to endow production rules with self-learning and adaptive capabilities [4,5] to cope with this generalization/induction problem.

In this paper we propose an artificial Neuro Granular Network with self-learning and adaptive stochastic connections which is based on the fusion of

J. Mira et al. (Eds.): IWINAC 2009, Part I, LNCS 5601, pp. 97–106, 2009.

Learning Automata Theory with knowledge-based production rules, implemented by a Neuro Granular Network, with stochastic connections that is able to learn and adapt to novel domains and environments. These learning and adaptation capabilities are based on Learning Automata Theory [6].

2 Adaptive Neural Network Based on Learning Automata and Granular Inference Systems: ANLAGIS

Summarizing, our proposed Adaptive Neuron-like network of Granular and Learning Automata System (ANLAGIS) is composed of two principal parts or subnets: (1) the granulation part or classification subnet which plays the role of partitioning the state space into indistinguishable atomic information granules or states or, as we propose to call them, *ballungen*[1]; (2) the second part of ANLAGIS is the action-selection subnet which is in charge of choosing for each particular information granule or state or ballungen, the appropriate available action. This part of ANLAGIS is based on a learning automaton [9] that chooses each possible action according to a probability distribution that varies as a function of the responses (either punishment or reward) of the environment. This learning automaton governs the self-learning and self-adaptive stochastic connections of ANLAGIS. In Fig. 1 it is shown the block diagram of ANLAGIS.

As commented a few lines above and in the caption of Fig.1, ANLAGIS consists of two main parts: the first one is in charge of partitioning the state space in the appropriate information granules or basic states or clusters or *ballungen* and the second part of ANLAGIS (the right side of Fig. 1) is devoted to choose the most appropriate action for each state or atomic granule: this choice is based on a learning automaton that updates its probability distribution as a function of the responses given by the environment (either a reward or a punishment). Thanks to the existence of a learning automaton for each state or information granule, ANLAGIS is able to learn in real-time with unsupervised guidance the proper actions. Once again, by "action" we mean either a physical action in the case of a controller or a symbol in the case of a classifier.

[1] In tribute to the Austrian philosopher and outstanding member of the Vienna Circle Otto Neurath who in his master piece philosophical paper *Protokol Sätze* (Protocol Sentences) [7] introduced the term *ballungen* (balloons, clots, conglomerates) to designate *verbal clusters*, a concept precursor to the idea of Wittgenstein's private language and also close to Michael Polanyi's tacit knowledge. Our concept of granulation process in ANLAGIS also attempts to formalize internal numerical sensory information clusters and we think that Neurath's *ballungen* is one of the first attempts in this direction. The final step of anchoring or grounding these internal clusters into symbols in the case of a pattern recognition problem or into actions in the case of a learning control problem we propose to be implemented by means of a probability distribution giving an additional degree of freedom to the self learning capability of ANLAGIS. Another more recent and computationally-oriented similar/related lines of research in this direction are the so-called Computing with Words and the granular computing concept [10,11].

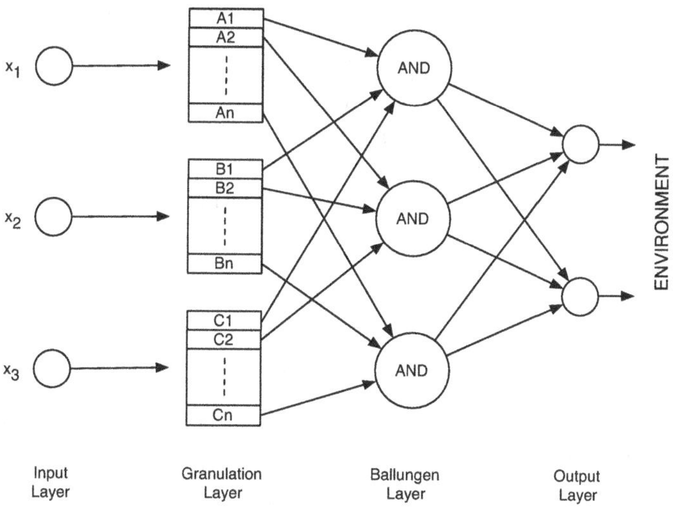

Fig. 1. Block diagram of ANLAGIS. The left-side of the whole network consists of the granulation process of the state variables x_i, so that after the previous interval granulation of each state variable there appears the corresponding AND-neuron that computes the logical conjunction of several interval granulations. Each AND-neuron has $1, 2, \ldots, r$ possible outputs (which are chosen according to a probability distribution that changes in function of the environment responses: see the text for details).

3 The Granulation Process of ANLAGIS

We briefly describe the above mentioned granulation process as formed by the following steps.

Step 1. Selection of the most representative or principal state variables of the system $x_1, x_2, \ldots, x_n$, which are the n principal state variables that optimize the representation of all the indistinguishable states or *ballungen* or clusters.

Step 2. Granulation or partition of all the state variables. In this paper we propose to apply a conventional Boxes-like [8] based granulation of the state variables. Roughly speaking, the interval granulation is a hard open problem in every particular application, which can be solved either by hand by the designers themselves if they have the necessary information or by means of an automated off-line optimization process based on e.g. evolutionary techniques. In this paper we have applied a mixture of both approaches.

4 Self-learning and Adaptive Behavior of ANLAGIS: Reinforcement Learning Based on Learning Automata Theory

Once the principal state variables have been selected and partitioned in the corresponding intervals then the atomic information granules or states or clusters or *ballungen* emerge as a consequence of the interaction with the environment, so that the next step is to map the *ballungen* space $\mathcal{X} = X_1, X_2, \ldots$ to the action space $\mathcal{A} = a_1, a_2, \ldots, a_r$. This mapping process is known in the reinforcement learning parlance as the policy mapping or planning problem. In control theory parlance is known as the controller design problem. Finally in Pattern Recognition problems this mapping corresponds to the classifier design itself.

One of our main contributions in this paper is to propose a probabilistic policy mapping based on learning automata theory so that for each individual state the available actions are selected by means of a probabilistic distribution which is updated depending on the environment responses (reward or punishment). In other words, in this paper we propose to apply a learning automaton for each individual *ballungen*. Taking into account that in ANLAGIS the *ballungen* are implemented by means of an artificial neural network (i.e. the left side of Fig. 1 or the granulation process described above) and more specifically as the *ballungen* are detected by the AND-neurons in Fig. 1 which, at the same time, are connected to every existing output neurons through probabilistic connections (links) then we propose a novel approach to artificial neurons as explained in the sequel.

5 A Novel Approach to Artificial Neural Networks: Pre-synaptic Learning *vs.* Post-synaptic Learning

The conventional "input-oriented" artificial neuron model with pre-synaptic learning is depicted in Fig. 2 below. The activation of neuron's output y_i can be expressed as:

$$y_i = f(\tau, w_i, x_i) \tag{1}$$

where τ is the threshold of neuron i and w_i are its pre-synaptic weights, and x_i are the neuron's inputs; similarly f is the activation function usually implemented as the sigmoid function. As it is well known, in this conventional neuron model, learning is based on the variation of the pre-synaptic weights as a function of the training error.

Now we propose to consider another perspective and to focus instead on the neurons output that in principle can be connected to several target neurons as depicted in Fig. 3.

In the "output-oriented" artificial neuron model the central idea is based on the connections of a generic neuron through its axonal output to what can be

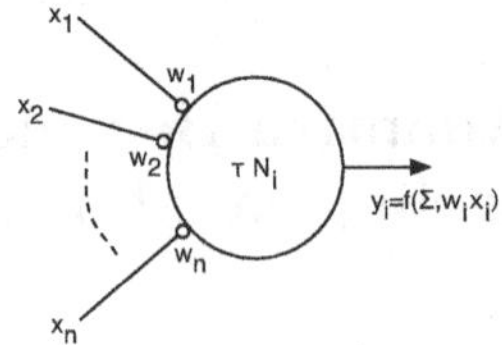

Fig. 2. Input-oriented artificial neuron model

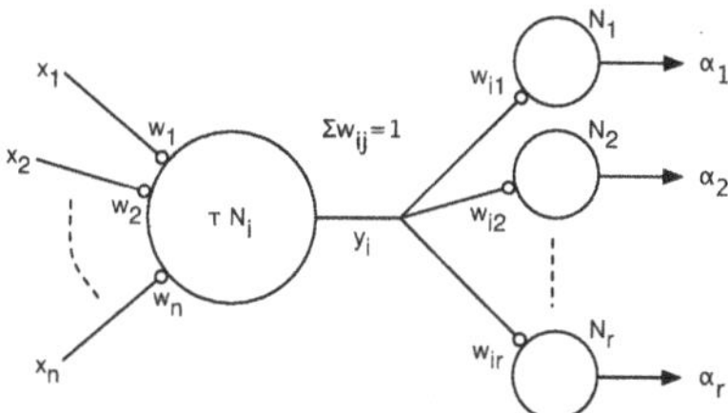

Fig. 3. The "output-oriented" artificial neural network in which a generic neuron n_i is connected to r different target neurons: $N_{i_1}, N_{i_2}, \ldots, N_{i_r}$. In this case learning takes also place in the probabilities of connecting the neuron N_i to the different target neurons (see text for details).

called the target neurons. In this paper we propose to define a probability p_i for each possible connection so that we can use learning automata theory to introduce a learning capability for the output links of a generic neuron. In this novel proposal each neuron has an associated learning automaton in which the action probability vector $P = \{p_i\}$ is defined over the output connections of the corresponding neuron. For this reason we propose the term post-synaptic learning in which, as depicted in Fig. 3, the pre-synaptic weight of target output neuron, w_{i_1}, is also the post-synaptic probability of neuron N_i as regards to its probability connection to output target neuron. In other words, this pre- and post-synaptic weights $w_{ij}(j = 1, \ldots, r)$ are the probabilities of the associated learning automaton so that they hold:

$$0 \leq w_{ij} \leq 1 \quad ; \quad j = 1, \ldots, r \quad ; \quad \sum_{j=1}^{r} w_{ij} = 1 \tag{2}$$

Obviously the post-synaptic learning is based on the reinforcement received from the environment as it happens in learning automata theory. Summarizing, in the global neural network there appear two learning processes: (1) the usual pre-synaptic learning and (2) the above mentioned, a novel concept of post-synaptic probabilistic learning.

6 Pattern Recognition and Machine Learning Using ANLAGIS

Pattern Recognition's objective is to assign a predefined class S_i to an object represented by a features' vector X_j.

Rule-based classifiers that belong to the Machine Learning field are among the most recently and widely used methods in Pattern Recognition and ANLAGIS can be straightforwardly interpreted as a class of rule-based classifiers.

In effect, rule-based classifiers employ a set of classification rules which have the following generic structure:

if {some_specific_conditions_of_the_discriminant_variables/features_hold}
then {class_C_s, with a certain degree of confidence}

More specifically, the antecedent conditions of these rules are usually expressed in Conjunctive Normal Form (CNF):

$$\textit{if} \quad (x_1 \text{ is } A_i \wedge x_2 \text{ is } B_j \wedge x_3 \text{ is } C_k \wedge \ \ldots \ \wedge x_n \text{ is } F_m) \quad \textit{then} \quad \text{Class } C_s$$

where x_1, x_2, x_n are the discriminant variables or features and $A_i(i = 1..n_1)$, $B_j(j = 1..n_2)$, $C_k(k = 1..n_3)$ and $F_j(m = 1..n_m)$ are either fuzzy or Boolean subsets in which the corresponding discriminant variable has been partitioned.

ANLAGIS can be straightforwardly interpreted as a rule based classifier, as the antecedent conditions of the classification rules of ANLAGIS are based on a box-like granulation as previously described in section 3 and also expounded in next section dealing with its application to pattern recognition problems.

To check the capabilities of ANLAGIS as a pattern recognition device we have tested it with two well-known pattern recognition problems: (1) the Iris dataset and (2) the Wisconsin Breast Cancer Diagnostic (WBCD) dataset.

6.1 Iris Dataset

In this well-known and classical multiclass Pattern Recognition (PR) problem, three classes of Iris flowers (*setosa*, *versicolor* and *virginica*) have to be classified according to four continuous discriminant variables measured in centimeters: x_1, sepal length (SL), x_2, sepal width (SW), x_3, petal length (PL), and x_4 petal width (PW).

The Iris dataset [12] was first used and even created by Fisher [13] in his pioneering research work on linear discriminant analysis, and today it is still an up-to-date, standard PR problem for testing discriminant techniques and algorithms.

Granulation of Discriminant Variables of the Iris Dataset for ANLAGIS Classification. As commented previously in several occasions, we have chosen a Box-like granulation of the discriminant variables in order to obtain what we have called *ballungen* in ANLAGIS. Thus, we propose to divide each discriminant variable x_i of the PR problem at hand into three granules or intervals: low, middle, high, as displayed in Fig. 4. where $\bar{x}_i$ is the mean value of the discriminant variable x_i, σ_i is its standard deviation, and k_i is a real parameter to be optimized. For the the Iris PR problem we have obtained an optimum $k_i = 0.7$ by means of a evolutionary algorithm.

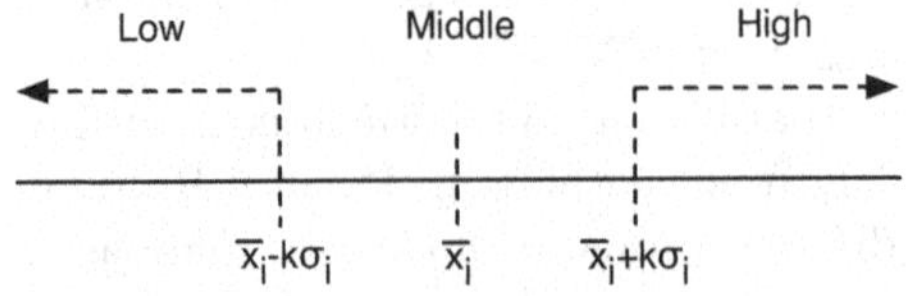

Fig. 4. Granules or intervals in which the discriminant variables have been divided

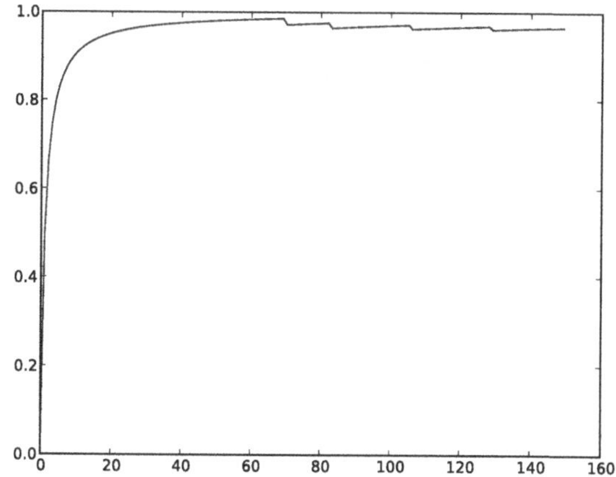

Fig. 5. Normalized true performance ratio obtained by ANLAGIS for the Iris dataset

Table 1. Iris dataset confusion matrix

	Iris Setosa	Iris Versicolor	Iris Virginica
Iris Setosa	50	0	0
Iris Versicolor	0	48	2
Iris Virginica	0	2	48

Experimental Results. As it is a multiclass PR problem we have used confusion matrices for the evaluation of the performance of ANLAGIS as shown in Table 6.1 and we have also obtained the estimation of the true performance ratio (which is the complementary of the true classification error) normalized to unity of ANLAGIS by means of a leave-one-out cross validation technique as displayed in Fig 5.

6.2 WBCD Dataset

The Wisconsin Breast Cancer Diagnostic dataset has been created by medical experts at the University of Wisconsin Hospital for diagnosing breast masses based on an FNA (Fine Needle Analysis) test [12,14]. Five visually assessed discriminant characteristics of an FNA sample are assigned by human specialists, an integer value between 1 and 10 (numbers closer to 1 indicate more possibility of malignant breast mass and, inversely, the closer to 10 the less malignant).

The WBCD dataset consists of 683 cases, corresponding to only two possible classes: benign and malignant diagnostic respectively. As documented in the WBCD dataset itself, there are several published studies about this dataset, thus, for such reason and also for its intrinsic interest the WBCD is an excellent benchmark for any novel PR method or algorithm as ANLAGIS itself.

Table 2. WBCD dataset confusion matrix

	Benign	Malignant
Benign	442	4
Malignant	5	232

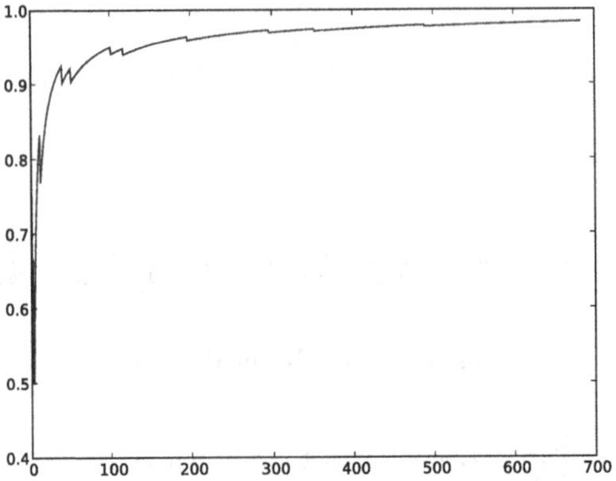

Fig. 6. Normalized performance ratio obtained by ANLAGIS for the WBCD dataset

Granulation of Discriminant Variables of the WBCD Dataset for ANLAGIS Classification. We have also used the same idea of interval granulation than in the Iris dataset (i.e. low, middle and high); in this case the optimum k happens to be 0.6 that has been also obtained by means of an evolutionary algorithm.

Experimental Results. In order to evaluate the performance of ANLAGIS in the WBCD dataset we have used a special case of cross validation technique known as leave-one-out, which is an excellent and widely used technique for computing the almost unbiased estimator of the true error rates in classifiers [15]. However, we have also included the confusion matrix obtained with ANLAGIS (Table 2).

In the leave-one-out cross validation technique, for each particular classifier and using N labelled instances, a classifier is obtained using $N - 1$ cases and tested on the single remaining case. The performance ratio is the number of successes produced on the N test cases divided by N. The normalized performance ratio obtained by ANLAGIS is shown in Fig. 6.

7 Conclusions and Further Work

A novel approach for solving complex pattern recognition and learning control problems based on the fusion of artificial neural networks, granular computing and learning automata theory has been proposed. Also we have also introduced the distinction between pre-synaptic and post-synaptic learning in artificial neural networks. This distinction adds probabilistic terms to the neuron output connections which allow a certain degree of non deterministic results to the output. This fact favors the general system performance.

We have also presented the application of ANLAGIS to Pattern Recognition and Machine Learning problems. More specifically we have tested ANLAGIS on two well-know PR and ML problems: (1) the Iris dataset and (2) Wisconsin Breast Cancer Diagnostic dataset. In both cases, ANLAGIS has proved to obtain excellent classification performance. As it is a rule-base classifier, besides its good recognition performance, an additional advantage of ANLAGIS, is its explanatory capability (i.e. the classification rules used by ANLAGIS can be easily understood by a human interpreter). A future research line is to work in the process of rules extraction of ANLAGIS. Another future work is related to the fundamental granulation process of ANLAGIS, in which we believe that the evolutionary computing may be a kind of interest.

An open and very ambitious future research line is the application of ANLAGIS as a function approximation in adaptive signal filtering and conventional control problems, i.e. in the general field of systems identification.

Concerning the application of ANLAGIS to learning control in robotic systems we refer to our previous work [16].

Acknowledgments

This work has been partially funded by the Spanish Ministry of Science and Technology, project: DPI2006-15346-C03-02.

References

1. Jang, J.S.R.: ANFIS Adaptive network-based fuzzy inference systems. IEEE Trans. on Systems, Man and Cybernetics 23(3) (1992)
2. Kasabov, N.K.: Foundations of Neural Networks, Fuzzy Systems, and Knowledge Engineering, ch. 2. MIT Press, Cambridge (1996)
3. Holland, J.H.: Escaping brittleness: The possibilities of general purpose learning algorithms applied to parallel rule-based systems. In: Michaiski, R.S., Carbonell, J.G., Mitchell, T.M. (eds.) Machine Learning II, pp. 593–623. Morgan Kaufmann, San Francisco (1996)
4. Butz, M.V.: Learning Classifier Systems. In: Proc. GECCO 2008, pp. 2367–2388 (2008)
5. Lanzi, P.L., Stolzmann, W., Wilson, S.W. (eds.): IWLCS 2002. LNCS, vol. 2661. Springer, Heidelberg (2003)
6. Thathachar, M., Sastry, P.: Varieties of Learning automata: An Overview. IEEE Trans. on Systems, Man and Cybernetics, Part B 32(6), 711–722 (2002)
7. Neurath, O.: Protocol sentences, Logical Positivism (The Library of Philosophical Movements), pp. 199–208. A.J. Ayer Free Press (1959)
8. Michie, D., Chambers, R.A.: Boxes: An experiment in adaptive control. In: Dale, E., Michie, D. (eds.) Machine Intelligence 2, pp. 137–152. Oliver & Boyd (1968)

9. Narendra, K., Thathachar, M.: Learning Automata - A Survey. IEEE Trans. on Systems, Man, and Cybernetics 4(4), 323–334 (1974)
10. Zadeh, L.: From Computing with Numbers to Computing with Words — From manipulation of measurements to manipulation of perceptions. IEEE Trans. on Circuits and Systems–II Fundamental Theory and Applications 4, 105–119 (1999)
11. Bargiela, A., Pedrycz, W.: Toward a theory of granular computing for human-Centered information processing. IEEE Trans. on Fuzzy Systems 16(2), 320–330 (2008)
12. Asuncion, A., Newman, D.J.: UCI Machine Learning Repository. University of California, School of Information and Computer Science, Irvine, CA (2007), http://www.ics.uci.edu/~mlearn/MLRepository.html
13. Fisher, R.A.: The use of multiple measurements in taxonomic problems. Annual Eugenics 7, Part II, 179–188 (1936); also in Contributions to Mathematical Statistics (John Wiley, NY 1950)
14. Mangasarian, O.L., Wolberg, W.H.: Cancer diagnosis via linear programming. SIAM News 23(5), 1, 18 (1990)
15. Weiss, S.M., Kapoulas, I.: An empirical comparison of Pattern Recognition, Neural Nets, and Machine Learning Classification Methods. In: Proc. Eleventh International Joint Conference on Artificial Intelligence, pp. 781–787
16. Maravall, D., De Lope, J.: Neuro granular networks with self-learning stochastic connections: Fusion of neuro granular networks and learning automata theory. In: Kasabov, N.K., et al. (eds.) Proc. ICONIP 2008 (2008) (in press)

Neural Prosthetic Interfaces with the Central Nervous System: Current Status and Future Prospects

E. Fernández

Bioingineering Institute, Universidad Miguel Hernández, Alicante
CIBER-BBN (Bioengineering, Biomaterials and Nanomedicine)
Spain
e.fernandez@umh.es

Abstract. Rehabilitation of sensory and/or motor functions in patients with neurological diseases is more and more dealing with artificial electrical stimulation and recording from populations of neurons using biocompatible chronic implants. For example deep brain stimulators have been implanted successfully in patients for pain management and for control of motor disorders such as Parkinson's disease. Moreover advances in artificial limbs and brain-machine interfaces are now providing hope of increased mobility and independence for amputees and paralysed patients. As more and more patients have benefited from these approaches, the interest in neural interfaces has grown significantly. However many problems have to be solved before a neuroprosthesis can be considered a viable clinical therapy or option. We discuss some of the exciting opportunities and challenges that lie in this intersection of neuroscience research, bioengineering and information and communication technologies.

1 Introduction

The development of neuroprosthetic devices can have high potential impact on brain research and brain-based industry and is one of the central problems to be addressed in the next decades. By the year 2040 neurodegenerative diseases, such as Parkinson's disease, Alzheimer's disease, and similar disorders, will overtake cancer as the second most common cause of death in Europe and around the world. This will impose enormous costs upon social systems. As the incidence of these functional deficits increases with aging population, we will be facing an epidemic of neurological problems in the near future. The good news is that we have 20 or 30 years of warning and that at this unique moment in the history of technical achievement, restoration of lost neurological function becomes possible. It is here that implantable neural interfaces march in as a treatment option that could be useful to enhance the quality of life of many patients. Microdevices for cell-electrode interfacing have been available for in vitro and in vivo applications for many years. For example deep brain stimulators have been implanted successfully in patients for pain management and for control of motor disorders such as

J. Mira et al. (Eds.): IWINAC 2009, Part I, LNCS 5601, pp. 107–113, 2009.

Parkinson's disease; cochlear implants are being used for restoring auditory function, micro-array type devices have been implanted in rudimentary artificial vision systems and a wide variety of devices have been developed to control respiration, activate paralysed muscles or stimulate bladder evacuation [1] [2] [3]. Moreover there is preliminary data showing that by using electrophysiological methods it is possible to extract information about intentional brain processes and then translate these signals into models that are able to control external devices [4]. As more and more patients have benefited from this approach, the interest in neural interfaces has grown significantly and individuals interested in pursuing careers in this field continue to increase. However many problems have to be solved before a neuroprosthesis can be considered a viable clinical therapy or option.

2 Neural Prosthetic Interfaces with the Nervous System

Neural prosthetics is an emerging technology that use electrical stimulation of the neural tissue to restore function in individuals with disabilities due to damage of the nervous system [5]. Neuroprosthetic devices commonly record and process inputs from outside the body and transmit this information to the nervous system. A neuroprosthesis, therefore, is a device or a system that communicates with specific groups of neurons to restore as much functionality as possible. The most fundamental requirements of any neural prosthesis are well understood. In order for a device to effectively emulate a neurological system, it has to perform three tasks:

1. First, it must collect the same kind of information that the neuronal system normally collects. Consequently, sensory prostheses always include a processing module that mimics the transfer function of biological sensory receptors so that they can be emulated by their artificial counterparts. For example, an auditory prosthesis should be able to capture acoustic signals and a visual prosthesis has to capture the attributes of the visual scene.
2. Next, it has to process that information.
3. Third, it must communicate the processed information, in an appropriate way, with other parts of the nervous system. Therefore, all neural prostheses need a stable electronic/neural interface that enables selective activation of specific groups of neurons and allow for chronic injection of electrical charge without deterioration of the electrodes or surrounding neural tissue.

As the neural activity evoked by the processing modules is not exactly the same as expected, sensory prostheses must rely on the assumption that the nervous system is able, to a certain degree, adapt to stimuli that differ from what was expected [6].

3 Engineering a Neural Interface

All neural prostheses need a stable electronic/neural interface that enables selective activation of specific groups of neurons and allow for chronic injection of electrical charge without deterioration of the electrodes or surrounding neural tissue. These issues still pose big challenges. Most of the current knowledge about neurons and their functional properties is based on single sequential recordings of their responses using microelectrode techniques. Depending upon the size of the electrode employed, the activity of several neurons was sometimes recorded simultaneously and the resultant record separated into single-unit activity by hardware devices or computer programs. Although these tools have been very useful for understanding many cellular and molecular mechanisms of neurons, it is clear that sensory and motor information is processed in a parallel fashion by populations of neurons working in concert, often embedded in complex multi-layer feedback loops [7] [8]. Thus to consider seriously the use of neuroprosthetic devices to restore motor or sensory pathways, one needs to understand and be able to emulate the dynamic and distributed nature of neural population coding [9]. Furthermore, to optimally record from or excite neurons, it is important to position the microelectrodes very close to the neurons we are trying to interact with (within a few microns). This means that the electrodes must be located deep within the nervous system and have dimensions that are similar to the size of the neurons. Other important attributes are:

- Special design for insertion in any part of cerebral cortex, including sulci of highly folded cortices such as those of humans.
- Reaching neurons at the desired three-dimensional location.
- Bi-directional communication with neurons and ensembles of neurons.
- Providing appropriate and stable electrical interfacing.
- Minimizing tissue damage and scarring.
- Rendering the devices as inert as possible from the biocompatibility, biostability and biofouling standpoints.

Following these basic requirements great efforts have been made to develop penetrating multi-electrodes, with dimensions of the same order of magnitude as the cortical cells that can be used to excite neurons more selectively and with electrical currents much smaller than those used by surface electrodes [10]. In this context the two main dominated approaches are multiple insulated metal microwires [11] [12] [13] and thin-film penetrating microelectrode arrays (Figure 1)

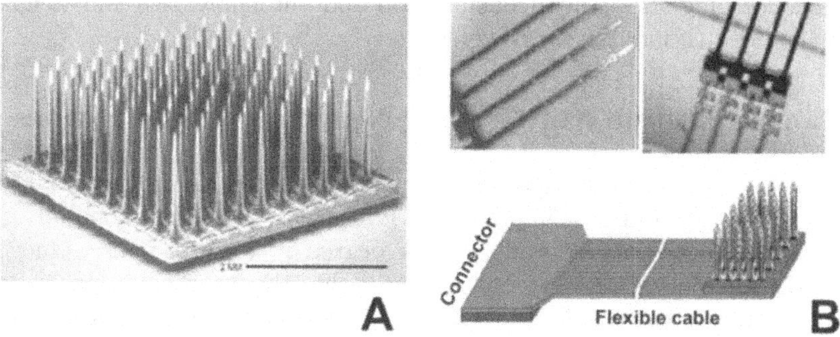

Fig. 1. Examples of 3D multielectrode arrays. A, Utah Electrode Array. B, Neuroprobes Array.

with various substrate materials, insulating dielectrics, and substrate shaping technologies such as the Utah's array [14], the Michigan array [15] and the new NeuroProbes arrays that are being developed in the context of an Integrated European Project (http://neuroprobes.org). Nowadays, only a few of these devices are commercially available.

4 Trends in Instrumentation for Neural Recording and Stimulation

In the field of electronic instrumentation for acquisition and processing of neurophysiological data, a spectacular progress has taken place throughout the last decade thanks to advances in microelectronics and digital processing techniques. Thus it is now possible to find in the market a variety of special computerized multichannel systems that provide amplification and filtering, spike detection and online spike sorting, for up to 128 channels simultaneously. Intermediate solutions are also available, combining parallel amplification, analog-to-digital (A/D) conversion, digital recording, and off-line spike sorting and analysis with powerful software tools running on standard low-cost computers (PC). A number of companies around the world (as *Multichannel Systems, Thomas Recording, Axona, Tucker-Davis Technologies, Plexon, Cambridge Electronic Design, Biomedical Technologies, Cyberkinetics,*) offer a number of hardware and software solutions that are very useful for research laboratories. While these equipments are suitable for either in-vitro or in-vivo acute experiments with the available electrode arrays, they are far from being useful for a chronic use in the neural interfaces of future neuro-prosthetic systems. For neuro-stimulation purposes, the present state is considerably worse since the available neuro-stimulators are able to drive only a reduced number of individually programmable channels simultaneously, and they usually provide only elementary and/or stereotyped stimulation patterns. In order to stimulate sensory areas of the brain, compact neuro-stimulators with considerably higher channel count will be required, and they should provide an effective and flexible functional stimulation of the neural tissue. This represents an important challenge for commercially available electronics in those technical aspects concerning size, power consumption and autonomy, size and weight of battery and coil (if inductively powered), etc.

5 Biocompatibility of Central Nervous System Probes

An important problem reported with all available microelectrodes to date, is long-term viability and biocompatibility. Even those materials generally considered to be biocompatible produce some degree of tissue response. Once implanted the microelectrodes have to remain within the brain environment for a long time. Therefore these devices must not be susceptible to attack of biological fluids, proteases, macrophages or any substances of the metabolism. However although it has been reported that silicon-based shafts, siliconoxide based surfaces and other glass based products are highly biocompatible, there are acute

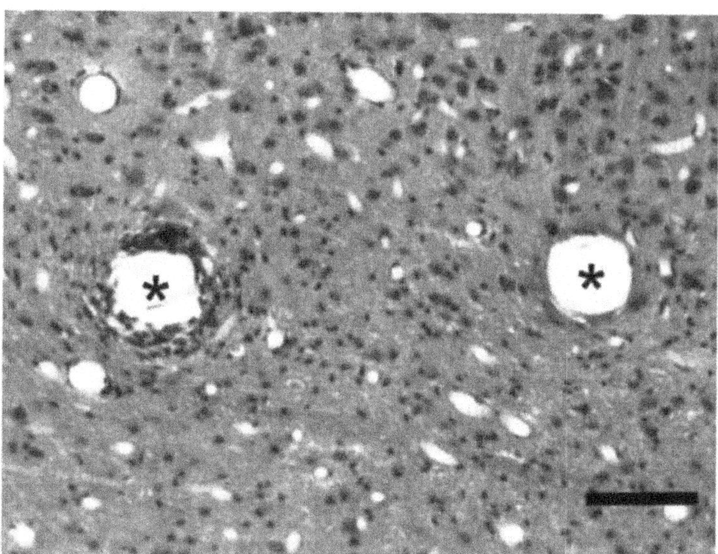

Fig. 2. An example of in vivo variability commonly encountered in implant literature. Asterisks show adjacent tracks from two identical electrodes. The track on the left shows clearly a chronic inflammatory reaction while that on the right seems to have no tissue reaction. Calibration bar = 100 mm.

and chronic inflammatory reactions which affect both the neural tissue and the surface of the microelectrodes [16] [17]. These reactions often result in damage of neurons and microelectrodes and lead to the proliferation of a glial scar around the implanted probes which prevents neurons to be recorded or stimulated (Figure 2). To those who were developing and using the first generation of implantable devices (1940-1980) it was becoming increasingly obvious that the best performance was achieved with materials that were the least reactive chemically [18]. However, while the surface composition of the implant is an important parameter, in some cases physical properties (size, shape, stiffness) are the major determinants of biocompatibility [17]. Furthermore although it is widely accepted to define biocompatibility as "the ability of a material to perform with an appropriate hostresponse in a specific application", nowadays any neural probe is comprised of more than one material, therefore we have to move from a material base to an specific application base definition [19]. Consequently a neural implant can be considered to be biocompatible if,

- It performs its desired function without eliciting any undesirable local or systemic effect in the recipient of the implant or in the own implant materials.
- It remains for a long term within the organism, entirely functional and with the desired degree of incorporation in the host.

This concept is not limited to minimize local lesions, but also encloses the whole behaviour of the implant in its biological environment. Therefore three areas have to be considered, the *"biosafety"*, the *"biofunctionality"* and the *"biostability"*. Biosafety means that the implant does not harm its host in any way, biofunctionality is related with the ability of the device to perform its intended function, and biostability means that the implant must not be susceptible to attack of biological fluids, proteases, macrophages or any substances of the metabolism [17]. In addition it should also be taken into account the *"biotolerability"* or the ability of the implant to reside in the body for long periods of time with only low degrees of inflammatory reaction. All these considerations imply extreme demands on stability and function of the implant and place unique constraints on the architecture, materials, and surgical techniques used in the implementation of central nervous system probes.

6 Challenges and Future Perspectives

The development and realization of neuroprosthetic devices requires extraordinary diverse, lengthy and intimate collaborations among basic scientists, engineers and clinicians. If we can understand more about the fundamental mechanism of neuronal coding, and how to safely stimulate the nervous system, there will be real potential in applying this knowledge clinically. However, we should be aware that there are still a number of important unanswered questions. In the longer term, several neurological pathologies will be alleviated by local microelectrodes, but keeping these devices biologically and electrically viable for many years still remains a difficult problem. Furthermore, a variety of questions regarding safety and non-technological issues need to be addressed. To be able to successfully afford these challenges it is required to consider not only the scientific, technical, engineering and surgical issues that remain to be solved but also take into account ethical and legal issues and develop new educational paradigms. However the majority of science technology education continues to follow along traditional disciplinary boundaries and lacks any effective means to demonstrate the interactions that are needed. Thus one of the main challenges in this field is to get a new generation of scientists, with a truly multidisciplinary formation, and able to work in close collaboration with neuroscientists, engineers, physical scientists, and social and behavioral scientists. It implies a better understanding of the central and peripheral nervous systems and the development of coordinating efforts at the frontier of scientific discovery (i.e. to integrate and converge engineering tools and methods in the areas of neuroscience research, electronics and information and communication technologies) in order to enable people with neural pathways that have been damaged by amputation, trauma, or disease to improve their quality of life.

Acknowledgements

This work was supported by the Spanish Government through grants SAF2008-03694, Cátedra Bidons Egara,and by the European Comission through the project "'NEUROPROBES"' IST-027017.

References

1. Stieglitz, T., Schuettler, M., Koch, K.P.: Neural prostheses in clinical applications–trends from precision mechanics towards biomedical microsystems in neurological rehabilitation. Biomed. Tech. (Berl) 49(4), 72–77 (2004)
2. Fernandez, E., et al.: Development of a cortical visual neuroprosthesis for the blind: the relevance of neuroplasticity. J. Neural Eng. 2(4), R1–R12 (2005)
3. Wolpaw, J.R., et al.: Brain-computer interfaces for communication and control. Clin. Neurophysiol. 113(6), 767–791 (2002)
4. Hochberg, L.R., et al.: Neuronal ensemble control of prosthetic devices by a human with tetraplegia. Nature 442(7099), 164–171 (2006)
5. Loeb, G.E.: Neural prosthetic interfaces with the nervous system. Trends Neurosci. 12(5), 195–201 (1989)
6. Sanguineti, V., et al.: Neuro-Engineering: from neural interfaces to biological computers. In: Riva, R., Davide, F. (eds.) Communications Through Virtual Technology: Identity Community and Technology in the Internet Age, pp. 233–246. IOS Press, Amsterdam (2001)
7. Fernandez, E., et al.: Population coding in spike trains of simultaneously recorded retinal ganglion cells. Brain Res. 887(1), 222–229 (2000)
8. Nicolelis, M.A., Ribeiro, S.: Multielectrode recordings: the next steps. Curr. Opin. Neurobiol. 12(5), 602–606 (2002)
9. Nicolelis, M.A.: Brain-machine interfaces to restore motor function and probe neural circuits. Nat. Rev. Neurosci. 4(5), 417–422 (2003)
10. Normann, R.A., et al.: A neural interface for a cortical vision prosthesis. Vision Res. 39(15), 2577–2587 (1999)
11. Carmena, J.M., et al.: Stable ensemble performance with single-neuron variability during reaching movements in primates. J. Neurosci. 25(46), 10712–10716 (2005)
12. Nicolelis, M.A.: Computing with thalamocortical ensembles during different behavioural states. J. Physiol. 566(Pt 1), 37–47 (2005)
13. Musallam, S., et al.: A floating metal microelectrode array for chronic implantation. J. Neurosci. Methods 160(1), 122–127 (2007)
14. Normann, R.A.: Technology insight: future neuroprosthetic therapies for disorders of the nervous system. Nat. Clin. Pract. Neurol. 3(8), 444–452 (2007)
15. Seymour, J.P., Kipke, D.R.: Neural probe design for reduced tissue encapsulation in CNS. Biomaterials 28(25), 3594–3607 (2007)
16. McCreery, D., Agnew, W.F., Bullara, L.: The effects of prolonged intracortical microstimulation on the excitability of pyramidal tract neurons in the cat. Ann. Biomed. Eng. 30, 107–119 (2002)
17. Heiduschka, P., Thanos, S.: Implantable bioelectronic interfaces for lost nerve functions. Prog. Neurobiol. 55, 433–461 (1998)
18. Williams, D.F.: On the mechanisms of biocompatibility. Biomaterials 29(20), 2941–2953 (2008)
19. Williams, D.: Revisiting the definition of biocompatibility. Med. Device Technol. 14(8), 10–13 (2003)

Analytical Models for Transient Responses in the Retina

Gabriel de Blasio[1], Arminda Moreno-Díaz[2], and Roberto Moreno-Díaz[1]

[1] Instituto Universitario de Ciencias y Tecnologías Cibernéticas
Universidad de Las Palmas de Gran Canaria
gdeblasio@dis.ulpgc.es
rmoreno@ciber.ulpgc.es
[2] School of Computer Science, Madrid Technical University
amoreno@fi.upm.es

Abstract. We propose analytical models for slow potentials transients as they are recorded at different layers of the retina, but mostly as they are integrated to produce ganglion cells outputs. First, two possible pathways from photoreceptors to bipolars and to ganglia are formulated, their formal equivalence being shown for quasi-linear behaviour. Secondly, a linear oscillator analytical model is introduced which is shown also to be equivalent to the first under certain circumstances. Finally, local instantaneous nonlinearities are introduced in the models. Tunning their parameters, the models are very versatile in describing the different neurophysiological situations and responses.

1 Basic Neurophysiological Arguments

Transient local responses to steps and to local point light stimuli have been registered at different retinal layers as the so called slow potentials. Transients also correspond to the generation of the action potentials at the ganglion cells hillock and axon, which are apparently a pulse frequency modulation coding of the slow signals impinging retinal ganglion cells.

In general, for the two cases, the presence of two opposite mechanisms is assumed to explain the process, one provoking departure from and a second provoking return to the resting situation. [1], [2], [3], [4], [5].

Thus, the slow potentials at levels prior to ganglion cells axons, can be explained assuming that transient responses are the result of the action of depolarizing and/or hyperpolarizing effects of two processes, fast and a slow processes, which combinations give rise to all variety of ON and OFF responses. In addition, the spatial origin of the signals corresponding to the processes can be located in the same area of the photoreceptors or on spatially closed areas.

Anatomically, the different retards of local stimuli are produced because signals follow different pathways before interaction. For a two way interaction of fast and retarded processes possibilities are: 1) local stimuli follow different retarded pathways from the beginning (as illustrated in figure 1(a)) for bipolars and ganglia, or 2) one of them undergoes an additional retard provoked by amacrines,

J. Mira et al. (Eds.): IWINAC 2009, Part I, LNCS 5601, pp. 114–120, 2009.

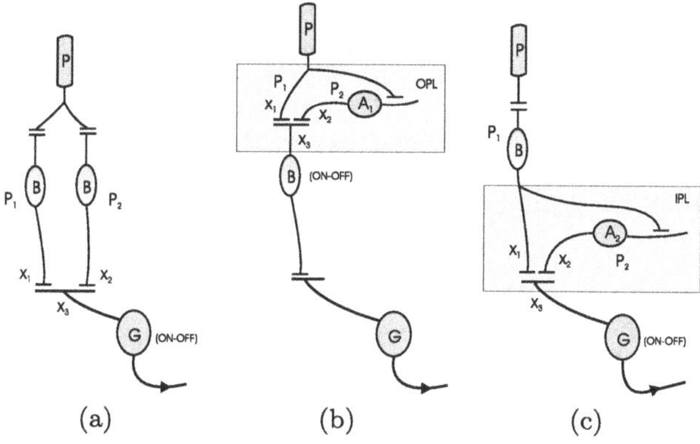

Fig. 1. Two possibilities for the interaction of the fast and retarded signals. (a) The two signals follow different paths from the origin of the signals through two bipolars, B, to interact in a ganglion cell, G. (b), (c) The retarded signals follow different pathways through amacrines A_1 and A_2, to interact in the OPL or IPL with fast signals coming from a photoreceptor, P, or a bipolar, B, respectively.

illustrated in figure 1(b) for the amacrines of the Outer Plexiform Layer (OPL) and in figure 1(c) for the amacrines of the Inner Plexiform Layer (IPL).

In each case, significant non linearities are probably situated at the interaction site, probably presynaptic to ganglion cells.

2 Equivalence between Linear Representations

We assume quasi-linear behaviors from photoreceptors outputs to ganglion synapses in cases of figures 1(a) and 1(c) and up to bipolar synapses in case of figure 1(b).

Introducing a "delay block", D, the two situations can be represented as indicated in figure 2(a) and 2(b) where u is a local stimulus from the photoreceptor and k_1, k_2 and k_3 are synaptic weights. If delays are represented by transport delay modules, the equivalence follows because k_2 and D'_2 commute, being $D_1 = D'_1 + D'_2$; $D_2 = D'_1$.

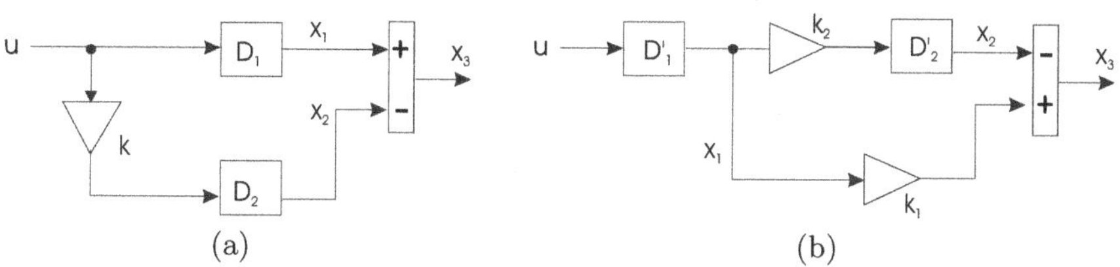

Fig. 2. To represent the delay of the signals of the two situations of figure 1(a) and figures 1(b) and 1(c), a delay block, D, is introduced as shown in figures 2(a) and 2(b)

In the more realistic case of linear retards, as inertial retards for D_1, D_2, D'_1 and D'_2, the state equations corresponding to the diagram of figure 2(a) are:

$$\begin{cases} \dot{x}_1 = a_1(-x_1 + u) \\ \dot{x}_2 = a_2(-x_2 + ku) \end{cases} \tag{1}$$

for presynaptic signals and $a_1 > a_2$. Also:

$$x_3 = x_1 - x_2 \tag{2}$$

for the postsynaptic signal.

In this second order system, we select x_1 and x_3 as state variables obtaining the following state equations:

$$\begin{cases} \dot{x}_1 = -a_1 x_1 + a_1 u \\ \dot{x}_3 = -(a_1 - a_2)x_1 - a_2 x_3 + (a_1 - a_2 k)u \end{cases} \tag{3}$$

In matrix notation, $\dot{x} = Ax + \overrightarrow{r}$, we have:

$$A = \begin{bmatrix} -a_1 & 0 \\ -(a_1 - a_2) & -a_2 \end{bmatrix} \quad \overrightarrow{r} = \begin{bmatrix} a_1 u \\ (a_1 - a_2 k)u \end{bmatrix} \tag{4}$$

Obviously, eigenvalues are $-a_1$ and $-a_2$, the "speeds" of the retarding processes.

For the case of figure 2(b), state equations are:

$$\begin{cases} \dot{x}_1 = -b_1 x_1 + b_1 u \\ \dot{x}_2 = -b_2 x_2 + k_2 b_2 x_1 \end{cases} \tag{5}$$

$$x_3 = k_1 x_1 - x_2 \Rightarrow x_2 = k_1 x_1 - x_3 \tag{6}$$

where $b_1 > b_2$.

Selecting as state variables x_1 and x_3, matrices A' and $\overrightarrow{r}'$ are:

$$A' = \begin{bmatrix} -b_1 & 0 \\ -[k_1(b_1 - b_2) + k_2 b_2] & -b_2 \end{bmatrix} \quad \overrightarrow{r}' = \begin{bmatrix} b a_1 u \\ k_1 b_1 u \end{bmatrix} \tag{7}$$

The equivalence of the two representations require $A = A'$, $\overrightarrow{r} = \overrightarrow{r}'$, that is:

$$a_1 = b_1; \;\; a_2 = b_2; \;\; -[k_1(b_1 - b_2) + k_2 b_2] = -(a_1 - a_2); \;\; k_1 b_1 = (a_1 - a_2 k) \tag{8}$$

which allows us to find the parameters of one representation in terms of the parameters of the other. To obtain representation 2(b) from 2(a) it is additionally required that:

$$k_2 = \left(\frac{a_1}{a_2} - 1\right)(1 - k_1) \tag{9}$$

Figure 3 shows the results of these representations for a sustained-ON response. Parameter values are $a_1 = 5$, $a_2 = 1$, $k = 0.8$ (first representation), which provide values $b_1 = 5$, $b_2 = 1$, $k_1 = 4.2/5$, $k_2 = 3.2/5$ for the second representation.

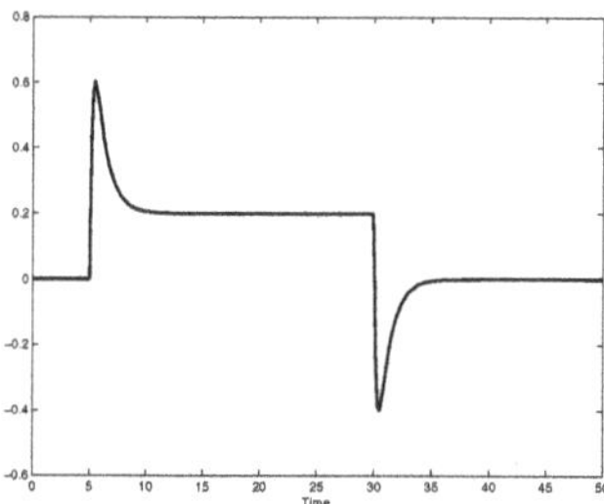

Fig. 3. Identical responses to a step of stimulus by the two representations of section 2 and the LO model of section 3 under the parameters values indicated in the text

3 Linear Oscillator Models

Models for the action potential, which generate the spikes that propagate, for example, in motoneurons or in axon of ganglion cells, follow a general scheme based on the Hodgkin & Huxley equations', which are sometimes generalized to include Voltage Controlled Oscillators Models (VCO) [1], [2].

For slow potentials there is not an equivalent formulation at the level of membrane properties. For this case, it is possible to introduce a model similar to the VCO models for the action potential, after linearization.

The VCO action potential model is given by [2]:

$$\tau \ddot{x} + F(\dot{x}) + A\, sen\, x = g \tag{10}$$

where $\dot{x}$ is the potential in the region that generates the spike, F is a nonlinear function of this potential. Normalizing ($\tau = 1$) and linearizing for slow potentials, equation (10) is reduced to:

$$\ddot{x} = -ax - b\dot{x} + g \tag{11}$$

where g is the stimulus function depending of both the external input u and its derivative $\dot{u}$. Thus, the reduced linear oscillator model is obtained:

$$\ddot{x} = -ax - b\dot{x} + K_1\dot{u} + K_2 u \tag{12}$$

where the resting potential is assumed to be null.

Notice that $a = a_1 \cdot a_2$ and $b = a_1 + a_2$, where a_1 and a_2 are the eigenvalues of equation (12). They correspond to the speeds of the two processes (fast and slow) of the section 2, a_1 associated to a process which follows relatively rapidly the stimulus (probably mediated by Na channels), while a_2 corresponds to a restoring process, probably mediated by K channels.

Notice that equation (12) corresponds to a damped oscillator, where the recovery coefficient, $a = a_1 \cdot a_2$ is something like the probability of interference of the two channels, while the damping is provoked by a joint action of both channels, $a = a_1 + a_2$.

We can compare the Linear Oscillator Model (LO) with the model of figure 2(a) of section 2, by reducing them to a similar state equations form. It is not difficult to show that both representations are equivalent if:

$$\begin{aligned} K_1 &= a_1 - ka_2 \\ K_2 &= a_1 a_2 (1 - k) \end{aligned} \quad (13)$$

However, equivalence between these two models, the LO model and the model with two retards, is not always possible because not only equation (13) is needed but also a_1 and a_2 have to be real numbers. In LO models, since a and b are specified, a_1 and a_2 could be complex and an equivalent model with two retards is not possible. In these cases the LO model provides for overshooting or ringing, as illustrated in figure 4 for $K_1 = 4.2$, $K_2 = 1$, $a_1 \cdot a_2 = 5$ but $a_1 + a_2 = 2$ instead of 6 of figure 3.

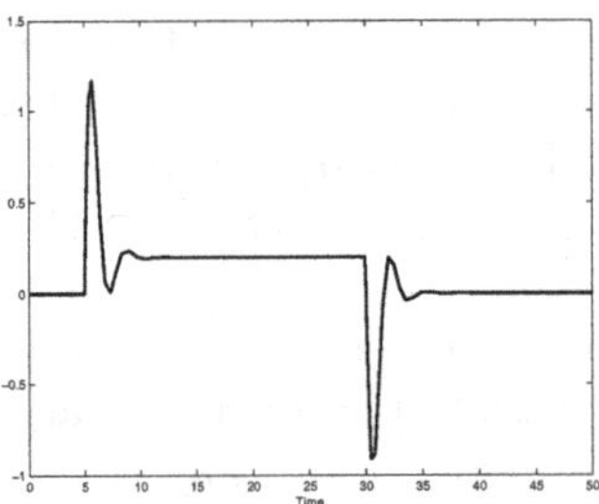

Fig. 4. Overshooting and ringing in the LO model

Slow potentials showing overshooting and ringing have not been found in practice. This makes it plausible for the basic membrane phenomenon to be described linearly by a LO model having real eigenvalues and its model with two retards counterpart.

4 Introducing Non-linearities in the Models

As pointed out before, significant non linearities, apart from the photoreceptors coding of light, occur at the site of the presynaptic synapses and in the process of spike generation coding. In these two cases, only net hyperpolarizing (excitatory) signals remain. Thus, the nonlinear effects can be represented by some type of half wave rectification of the overall input signals, X, to provide some type of resulting activity:

$$A_c = F[X] \quad (14)$$

F being a rectifying local (in space and time) function, as $exp[X]-1$ or $Pos[X]$ (positive part).

We shall illustrate the action of the non linearities by using the model of section 2, given by figure 2(a). The other representations have similar treatment.

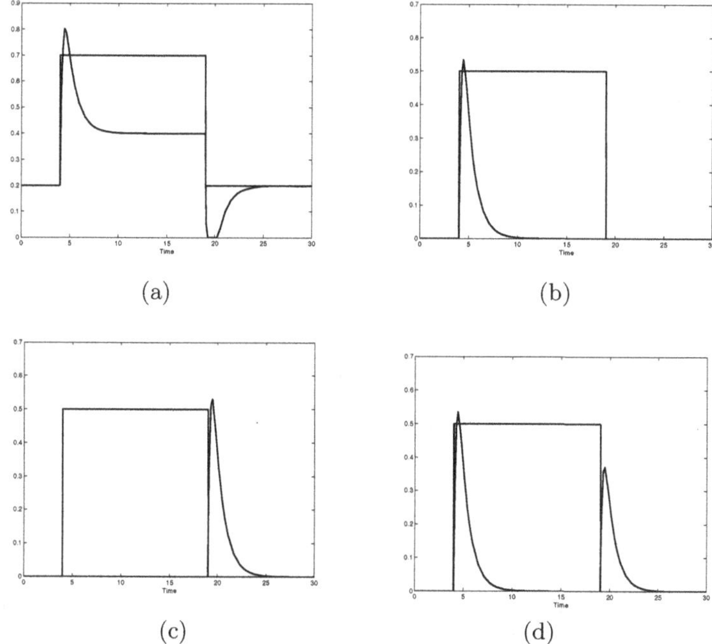

Fig. 5. Several responses for different values of the parameters: (a) Sustained ON, (b) Non sustained ON, (c) Pure OFF and (d) Non-linear ON-OFF

The general postsynaptic interaction of processes given by x_1 and x_2 of section 2 will be a separate rectification of their double linear interaction, that is:

$$A_c = F[x_1 - x_2 + A_1] + F[x_2 - x_1 + B_1] + C \tag{15}$$

A_c will be the total postsynaptic local activity contributing to the output of the retinal cell. A_1, B_1 and C are positive terms which may account for spontaneous responses or background illumination effects.

For the simplest case of an ideal rectification, equation (15) becomes:

$$A_c = A \cdot Pos[x_1 - x_2 + A_1] + B \cdot Pos[x_2 - x_1 + B_1] + C \tag{16}$$

where A and B are positive factors which determine the nature of the process: for $A \neq 0$ and $B = 0$, we have an ON process. For $A = 0$, $B \neq 0$, an OFF process. For $A \neq 0$ and $B \neq 0$, an ON-OFF non-linear process is obtained, with relative weights A and B respectively.

Figure 5 shows the responses to a step of local light stimulus (long enough to allow for stationary regime). Figure 5(a) is the sustained-ON response, for $A = 1$, $B = 0$, $k = 0.8$ and parameters $C = 0$, $A_1 = 0.2$, $B_1 = 0$. Figure 5(b) is the non sustained-ON response, for $k = 1$, $A = 1$, $B = 0$, and the rest of parameters set to zero. Figure 5(c) is the pure OFF response for $k = 1$, $A = 0$, $B = 1$ and the rest of parameters set to zero. Finally, the non-linear ON-OFF response is illustrated in figure 5(d), for $k = A = 1,\ B = 0.7$ and the rest set to zero.

Acknowledgments

This work is supported in part by Spanish Ministry of Science and Innovation (MICINN) under grant TIN2008-06796-C04-02.

References

1. Hodgkin, A., Huxley, A.: A Quantitative Description of Membrane Current and its Application to Conduction and Excitation in Nerve. Journal of Physiology 117, 500–544 (1952)
2. Hoppensteadt, F.C.: An Introduction to the Mathematics of Neurons. Cambridge University Press, Cambridge (1997)
3. Moreno-Díaz, R., de Blasio, G., Moreno-Díaz, A.: A Framework for Modelling Competitive and Cooperative Computation. In: Ricciardi, L.M., Buonocuore, A., Pirozzi, E. (eds.) Retinal Processing, In Collective Dynamics: Topics on Competition and Cooperation in the Biosciences. American Institute of Physics, pp. 88–97 (2008)
4. Moreno-Díaz, R., Moreno-Díaz, A., de Blasio, G.: Local Space-time Systems Simulation of Linear and Non-linear Retinal Processes. In: Quesada, A., Rodríguez-Rodríguez, J.C., Moreno-Díaz Jr., R., Moreno-Díaz, R. (eds.) 12th International Workshop on Computer Aided Systems Theory, EUROCAST 2009, pp. 9–12 (2009)
5. Guttman, R., Feldman, L., Jacobsson, E.: Frequency Entrainment of Squid Axon Membrane. J. Membrane Biol. 56, 9–18 (1980)

Analysis of Retinal Ganglion Cells Population Responses Using Information Theory and Artificial Neural Networks: Towards Functional Cell Identification

M.P. Bonomini[1,3], J.M. Ferrández[1,2], J. Rueda[4], and E. Fernández[1,3]

[1] Instituto de Bioingeniería, Universidad Miguel Hernández, Alicante
[2] Dpto. Electrónica, Tecnología de Computadoras, Univ. Politécnica de Cartagena
[3] CIBER-BBN
[4] Dpto. de Histología y Anatomía, Universidad Miguel Hernández, Alicante
jm.ferrandez@upct.es

Abstract. In this paper, we analyse the retinal population data looking at behaviour. The method is based on creating population subsets using the autocorrelograms of the cells and grouping them according to a minimal Euclidian distance. These subpopulations share functional properties and may be used for data reduction, extracting the relevant information from the code. Information theory (IT) and artificial neural networks (ANNs) have been used to quantify the coding goodness of every subpopulation, showing a strong correlation between both methods. All cells that belonged to a certain subpopulation showed very small variances in the information they conveyed while these values were significantly different across subpopulations, suggesting that the functional separation worked around the capacity of each cell to code different stimuli.

1 Introduction

Our perception of the world, our sensations about light, colour, music, speech, taste, smell is coded in raw data (as a binary data) by the peripheral sensory systems, and sent by the corresponding nerves to the brain where this code is interpreted and coloured with emotions. The raw or binary sensory data consists in sequences of identical voltage peaks, called action potentials or spikes. Seeing consists in decoding the patterns of these spike trains which are sent to the brain, via the optic nerve, by the visual transduction element: the retina. The external world object features, such as size, colour, intensity, are transformed by the retina in a myriad of parallel spikes sequences, which must describe with precision and robustness all the characteristics perceived. Getting insight into this population code is, nowadays, a basic question for visual science.

A considerable number of coding studies have focused on single ganglion cell responses [1] [2]. Traditionally, the spiking rate of aisle cells has been used as an information carrier due to the close correlation with the stimulus intensity in all sensory systems. There are, however some drawbacks when analysing single cell

J. Mira et al. (Eds.): IWINAC 2009, Part I, LNCS 5601, pp. 121–131, 2009.

firings. Firstly, the response of a single cell cannot unequivocally describe the stimulus since the response from a single cell to the same stimulus has a considerable variability for different presentations. Moreover, the timing sequence differs not only in the time events but also in the spike rates, producing uncertainty in the decoding process. Secondly, the same sequence of neuronal events in an aisle cell may be obtained by providing different stimuli, introducing ambiguity in the neuronal response.

New recording techniques arisen from emerging technologies, allow simultaneous recordings from large populations of retinal ganglion cells. At this time, recordings in the order of a hundred simultaneous spike trains may be obtained. New tools for analysing this huge volume of data must be used and turn out to be critical for proper conclusions. FitzHugh [3] used a statistical analyser which, applied to neural data was able to estimate stimulus features. Different approaches have been proposed on the construction of such a functional population-oriented analizer, including information theory [4] [5], linear filters [6], discriminant analysis [7] and neural networks [8].

Analyzing the neural code, in the context of getting useful information for the clustering algorithm, needs to quantify the amount of information each cell conveys. The goal of this study was to quantify their tendency to group themselves in sets of relatives according to their coding performance, using functional clustering of the autocorrelograms and Information Theory and Neural Networks as tools for providing an empirical value for the goodness of a coding capability. Therefore, a functional separation, or classification based on behaviour, was accomplished and the coding abilities of the subsets cells and the whole cluster determined. Finally the strong relationship between stimulus reconstruction using artificial neural networks and mean cell information provided by Information Theory was proved.

2 Methods

2.1 Experimental Procedures

Extracellular recordings were obtained from ganglion cell populations in isolated superfused albino rabbit (Oryctolagus cuniculus) retina using a rectangular array of 100, 1.5 mm long electrodes, as reported previously [7] [9] [10]. Briefly, after enucleation of the eye, the eyeball was hemisected with a razor blade, and the cornea and lens were separated from the posterior half. The retinas were then carefully removed from the remaining eyecup with the pigment epithelium, mounted on a glass slide ganglion cell side up and covered with a Milipore filter. This preparation was then mounted on a recording chamber and superfused with bicarbonate-buffered Ames medium at 35C. For visual stimulation we used a 17" NEC high-resolution RGB monitor. Pictures were focused with the help of lens onto the photoreceptor layer. The retinas were flashed periodically with full field white light whereas the electrode array was lowered into the retina until a significant number of electrodes detected light evoked single- and multi-unit responses. This allowed us to record with 60-70 electrodes on average during each experiment. The electrode array was connected to a 100 channel amplifier (low and high corner frequencies of 250 and 7500 Hz) and a digital signal processor based data acquisition system. Neural spike events were detected by comparing the instantaneous electrode signal to level thresholds set for each data channel using standard procedures described elsewhere [7] [9] [11]. When a supra-threshold event occurs, the signal window surrounding the event is time-stamped and stored for later, offline analysis. All the selected channels of data as well as the state of the visual stimulus were digitized with a commercial multiplexed A/D board data acquisition system (Bionic Technologies, Inc) and stored digitally. Figure 1 shows the experimetal setup. For spike sorting we used a free program, NEV2kit, which has been recently developed by our group [12] and runs under Windows, MacOSX and Linux (source code and documentation is freely available at: http://nev2lkit.sourceforge.net/). NEV2kit loads multielectrode data files in various formats (ASCII based formats, LabView formats, Neural Event Files (.NEV), etc) and is able to sort extracted spikes from large sets of data. The sorting is done using principal component analysis (PCA) and can be performed simultaneously on many records from the same experiment. Different experiments were carried out with albino rabbits.

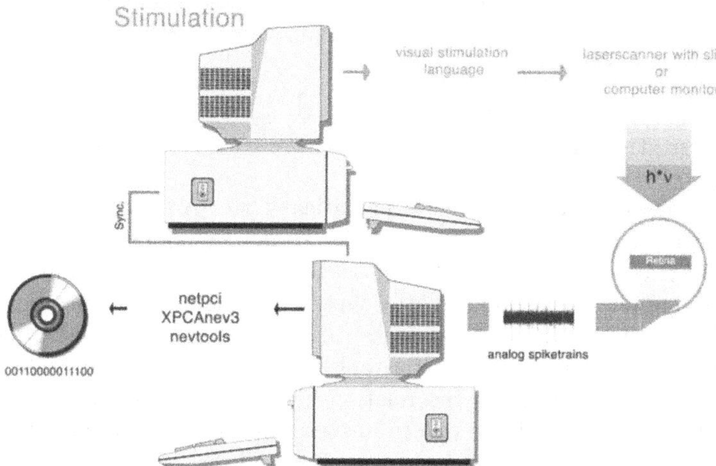

Fig. 1. Retinal ganglion cells stimulation and recording setup

The retinas were stimulated with full field flashes at 16 different light intensities within the gray scale. In order to ensure both the number of trials for each intensity was constant and the probabilities of appearance of each intensity was equal, the following procedure was carried out. Firstly, a lookup table with 16 light intensities equally distributed ranging from black (RGB values: 0, 0, 0) to white (RGB values: 255, 255, 255) was constructed. Afterwards, the elements of a list containing 20 repetitions for each of the intensities from the lookup table were relocated by changing their indexes according to a random entry chosen from an uniform distribution. The list was then loaded by a Python script embedded in VisionEgg for presentation of the flashes. Flashes were 300 ms long so that each trial lasted 96 seconds.

2.2 Separation into Subpopulations

The separation algorithm consists in calculating autocorrelograms [13] [14] on each of the cells in the dataset, a bin size of 10 ms was used and different time shifts such that the complete flash transitions were included in the analysis. The autocorrelograms then fed a non supervised, partitional clustering method for the creation of a varying number of autocorrelograms groups. We will refer to the number of groups with italic k. The same analysis were carried out for an increasing k at every entire population. We will use the terms class, cluster, subset or group interchangeably. The nearest-neighbour or k-means approach was chosen for the clustering method. This approach decomposes the dataset into a set of disjoint clusters and then minimizes the average squared distance from a cluster centroid among the elements within a cluster, while maximizes this distance when regarding the centroids of the different clusters. This defines a set of implicit decision boundaries that separate the clusters or classes of units according to their periodicity. In this way, we end up with groups of relatives that are a subset of the entire array.

2.3 Information Theory

In 1929, Shannon published "The Mathematical Theory of Communication" [15] where thermodynamic entropy was used for computing different aspects about information transmission, it was known as Information Theory. This computation was also applied for computing the capacity of channels for encoding, transmitting and decoding different messages, regardless of the associated meaning. Information theory may be used as a tool for quantifying the reliability of the neural code just by analysing the relationship between stimuli and responses [17] [18]. This approach allows one to answer questions about the relevant parameters that transmit information as well as addressing related issues such as the redundancy, the minimum number of neurons needed for coding certain group of stimuli, the efficiency of the code, the maximum information that a given code is able to transmit, and the redundancy degree that exists in the population firing pattern [19] [20].

In the present work, the population responses of the retina under several repetitions of flashes were discretized into bins where the firing rates from the cells of the population implement a vector n of spikes counts, with an observed probability P(n). The probability of the occurrence of different stimuli has a known probability P(s). Finally the joint probability distribution is the probability of a global response n and a stimulus s, P(s,n).

The information provided by the population of neurons about the stimulus is given by:

$$I(t) = \sum_{s \in S} \sum_{n} P(s,n) log_2 \frac{P(s,n)}{P(s)P(n)} \quad (1)$$

This information is a function of the length of the bin, t, used for digitising the neuronal ensembles and the number of the stimuli in the dataset.

With the purpose of assessing the quality of the subpopulations obtained, the following procedure was carried out. On every subpopulation generated, the information that single cells conveyed about the stimulus as well as the progression of the mutual information values when increasing the number of cells for each subpopulation was calculated. From these, two informational indicators were constructed: the mean cell information, calculated as the sum of the mutual information of each aisle cell divided by the total number of cells in a particular subpopulation, and the subpopulation information, consisting of the overall mutual information for a subpopulation in which all their cells were taken into account.

2.4 Artificial Neural Network

To validate the theory information values as viable descriptors about the capacity of subpopulations at coding different stimuli, A feed-forward, three-layer back-propagation neural network was used [16] to assess the stimulus reconstruction ability of each of the subpopulations and hence, get insight into their coding ability. In order to prevent the number of cells from affecting the subpopulation performances, a fixed number of cells was chosen such that the smallest subpopulation had one surplus cell for permutations. The ANN received the firing rate parameters from either single cells or populations of cells. The network architecture consisted of 20 nodes in the hidden layer, and the same number of neurons as stimuli categories to be recognized in the output layer. The activation function used for all neurons, including the output layer was the hyperbolic tangent sigmoid transfer function given by:

$$f(x) = \frac{2}{1+e^{-2x}} - 1 \quad (2)$$

using as initial momentum and adaptive learning rate the values established by default by Matlab Neural Network Toolbox, the initial weights randomly initialized and the network trained to minimize a sum squared error goal of 0.04, to provide more generality to the recognition stage. Once the network has been trained, the recognition with extended data was executed, and the performance at stimulus reconstruction computed.

3 Results

3.1 Entire Population and Subpopulations Obtained

The generation of subpopulations is illustrated in Figure 2 with an example where the number of clusters was fixed to three. Top panel displays the raster plot of an entire population of ganglion cells while bottom panels (s1, s2 and s3) show the raster plots of the subpopulations obtained by applying the separation method formerly explained using a bin size of 10 ms and a maximum shift of 900 lags. In order to avoid repetitions, the following shortened forms will be defined: the first subpopulation will be referred to as s1, while the second and third one as s2 and s3 respectively. In addition, subpopulations were reordered so that s1 will always account for the subpopulation with fewer cells and subsequent subpopulations will contain an increasing number of cells. Clear differences among the different subpopulations can be perceived. These differences

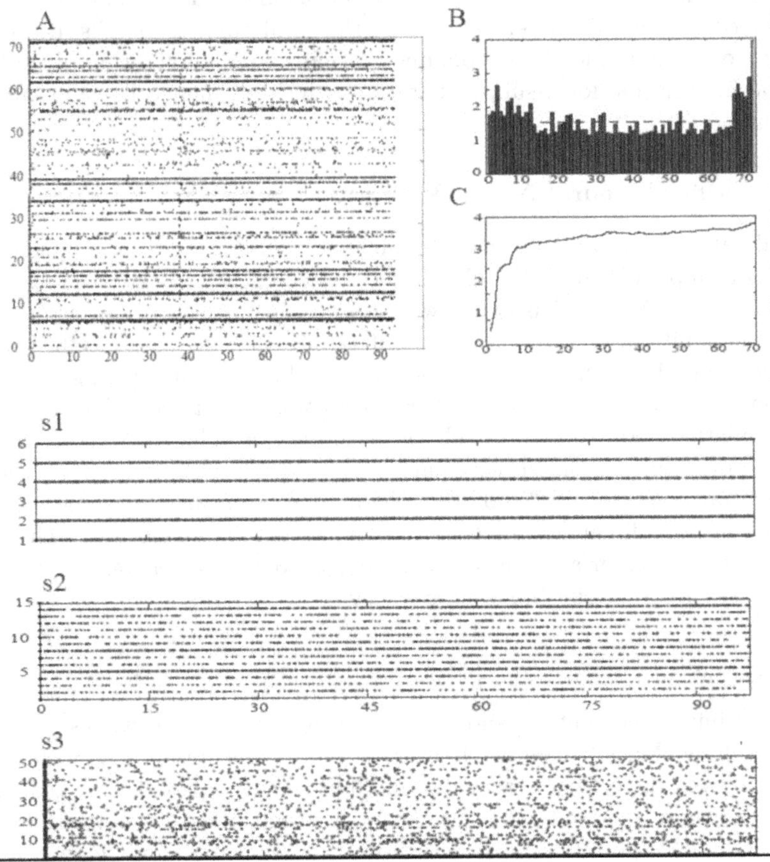

Fig. 2. Example of simultaneously recorded extracellular responses from a population of rabbit ganglion cells to a trial of random full field flashes with 16 different intensities (see methods). (A) Original population raster plot. Each dot represents a single spike. (B) Mutual information values for each cell in the recording (bars) and for the whole population (last gray bar). The overall mean cell information is showed as a dashed line. (C) Accumulative mutual information for an increased number of cells. In this example the number of cluster was fixed to three and the lower panel shows the raster plots for each subpopulation (named s1, s2 and s3).

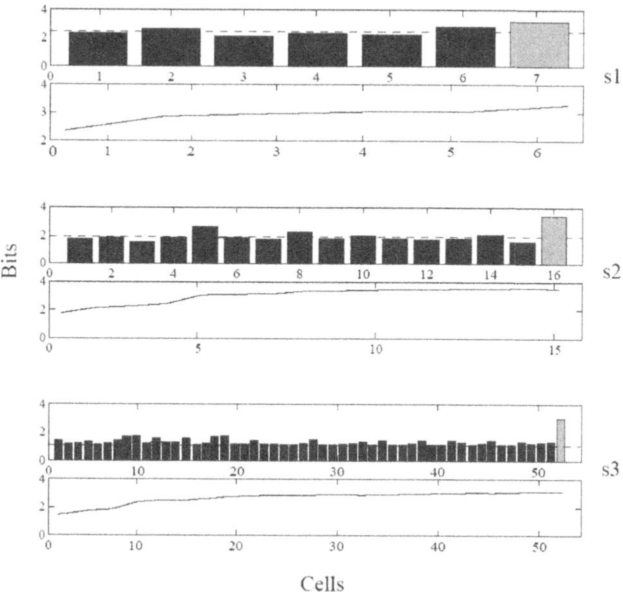

Fig. 3. Information about the stimulus for the subpopulations showed in Fig 2. Upper panels show the mutual information values for each cell in the recording (bars) and for the whole population (last gray bar). The overall mean cell information in each case is showed as a dashed line. Lower panel present the accumulative mutual information for an increased number of cells. Note the relationship between number of cells and mean cell information.

were related mainly with the firing time patterns and the number of cells in each subpopulation. For instance s1, contained very few cells that fired almost constantly during the presentation of the stimuli, s2 contained a considerable number of cells with apparent temporal patterns and s3 was integrated by a higher number of cells which showed a more randomised activity. This behaviour was present for different number of clusters.

3.2 Quality of the Subpopulations: Information Theory Approach

The overall information that each subpopulation accounted for, this is, the subpopulation information, kept similar across classes (one-way ANOVA; $p=0.82$) while the informative value of the individual cells, summarised by the mean cell information significantly varied (one-way ANOVA; $p\leq 0.0005$). Figure 3 shows the information that each cell conveyed about the stimulus. Notice the difference in the MCI (dashed line) and SI (last bar) values.

Notice the difference in the mean cell information value (dashed line) from the subpopulation information (last bar), which is equalized across subpopulations around a similar value to that of the original population. Surprisingly, the subpopulation formed by the very few continuously firing cells gave the higher

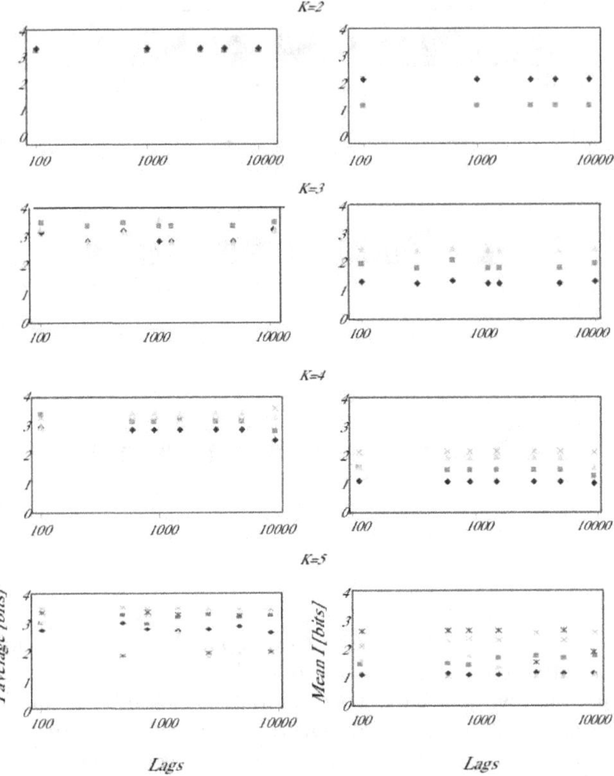

Fig. 4. Information clustering for increasing clusters, k, and all shift lags. Left column: Subpopulation information for k=2 (top panel) and k=5 (bottom panel). Right column: Mean cell information for k=2 (top panel) and k=5 (bottom panel). Abscissa axis: shift lags at which the autocorrelograms for the classes separation was computed. Note the well delimited mean cell information levels for each subpopulation, while the subpopulation information of every subpopulation tends to gather around the mutual information of the original population.

mean cell information (2.420.10 bits; MSE), above the overall value from the whole population (1.480.01 bits; MSE) (Figure 2, right top panel, dashed line). Moreover, s2, kept the moderately informative cells (1.880.07 bits; M SE) and s3 grouped the many worst cells on a mean information basis (1.330.02 bits; MSE). Interestingly, the fewer cells a subpopulation contained, the higher the mean cell information resulted at a certain subpopulation, while the number of cells did not affect the subpopulation information. Figure 4 expands the trends in the generation of subpopulations shown in Fig. 3 for an increasing number of subpopulations at different shift lags. A coherent behaviour is present for all shift lags. Left column figures show subpopulation information whilst mean cell information is represented on the right ones. Once again, it is evident the differences in the mean cell information through out the different subpopulations arisen for

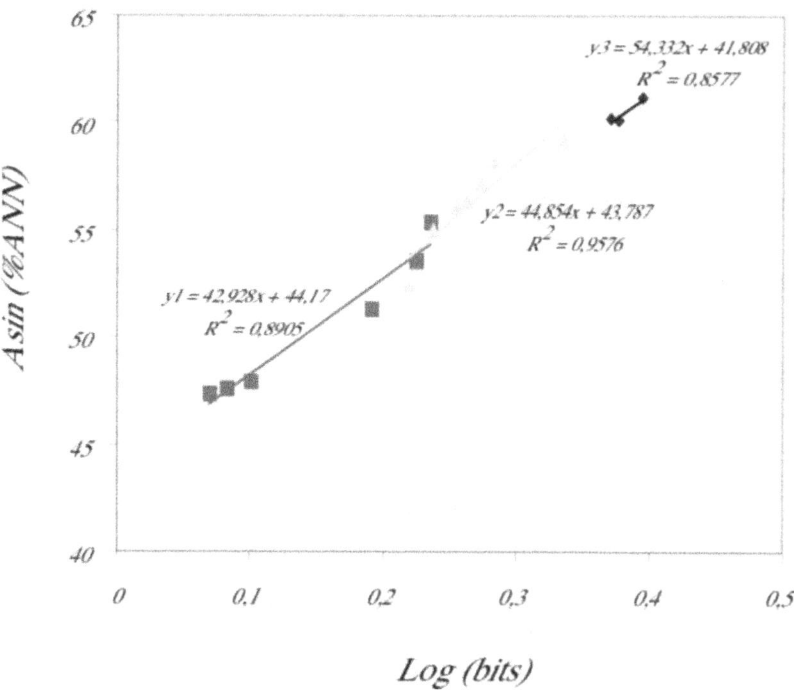

Fig. 5. ANN performance against mean cell information. Line fitting for s1 (squares), s2 (triangles) and s3 (rhombus). Notice that the best correlation coefficient belongs to the subpopulations with the best defined time patterns.

any number of classes, k, while the subpopulation information turned up nearly constant for all the cases. For the case in which four subpopulations are originated, the latter trend starts to suffer whilst for five classes such behaviour is completely vanished. Thus, there is a maximum number of subpopulations which might optimise a clustering strategy that is able to split the entire population on a smart informational basis.

3.3 Artificial Neural Networks and Information Theory

In order to address quality differences across subpopulations, we used a feed-forward back-propagation neural network to study the contribution of the mean cell information indicator to group cells that are better encoders than others with respect to the stimulus applied. Interestingly, the neural network performance obtained with the different sets of cells within each subpopulation was found to be related to the mean cell information. With the aim of quantifying this relationship, ANN performance percentages were transformed to arcsin values and the mean cell information values underwent a logarithmic transformation in order to study the linearity of its relationship. Afterwards, correlation coefficient and regression analysis was applied to the data. Fixing n=5 (see methods), a highly significant positive correlation was found for s1 (r=0.9436, df=4), s2 (r=0,9785, df=4) and s3 (r=0,9261, df=4). Figure 5 shows the lines fitted by means of regression analysis on each of the subpopulations generated from the population shown on Figure 3. Dots represent the samples from which regression analysis was calculated. Equations for the respective line fittings and square r are included in the figures. Here, it is clearly shown the strong relationship between stimulus reconstruction and mean cell information.

4 Discussion

A new method for analysing subsets in a population of neuronal responses has been defined using artificial neural networks and information theory. It permits pruning and classifying the relevant elements of the visual system, getting insight the neural code more accurately. It has been shown that the generated subsets share their own coding behaviour, quantified by information theory and artificial neural nerworks, identifying different subset encoders with different temporal responses. This has been observed for different number of clusters and different time lags. Also, the close relation between the mean information of its cells and the stimulus reconstruction rate has been observed using artificial neural networks. A clear trend in the clustering strategy was present in all the subpopulations generated for certain number of classes, less than five. This is due to the fact that there exists a critic value for which subpopulations with similar sizes but very different behaviours start to originate. This might be explained from a functional point of view. We speculate with a natural number of classes, where every class contributes effectively to coding different elements of the visual scenario such as intensity, colour, texture, orientation [22] or shape. From that number on, the coding process could lose effectiveness, starting to turn up redundant classes or subpopulations. In other words, the classes would represent different kind of cells, like the intensity coders, the colour coders and so forth. Taking into account that the stimulus applied was intensity variation of full field flashes, the best coder subsets in this analysis would effectively code intensity. In fact, they get the best coding capabilities, while the other classes would code other features. This should be addressed in future works by repeating the visual stimuli with other varying parameters, for instance, changing colour or orientation lines, in order to confirm such a behavioral separation and functional cell identification.

Acknowledgements

This work was supported by the Spanish Government through grants TIN2008-06893-C03, TEC2006-14186-C02-02 and SAF2008-03694, Cátedra Bidons Egara, Fundación Séneca 08788/PI/08 and by the European Comission through the project "'NEUROPROBES"' IST-027017.

References

1. Ammermuller, J., Kolb, H.: The organization of the turtle inner retina. I. ON- and OFF-center pathways. J. Comp. Neurol. 358(1), 1–34 (1995)
2. Ammermuller, J., Weiler, R., Perlman, I.: Short-term effects of dopamine on photoreceptors, luminosity- and chromaticity-horizontal cells in the turtle retina. Vis. Neurosci. 12(3), 403–412 (1995)
3. Fitzhugh, R.: A Statistical Analyzer for Optic Nerve Messages. J. Gen. Phyosiol. 41, 675–692 (1958)
4. Rieke, F., et al.: Spikes: Exploring the Neural Code. M. Press, Cambridge (1997)
5. JGolledge, H.D., et al.: Correlations, feature-binding and population coding in primary visual cortex. Neuroreport 14(7), 1045–1050 (2003)
6. Warland, D., Reinagel, P., Meister, M.: Decoding Visual Information from a Population of Retinal Ganglion Cells. J. Neurophysiol. 78, 2336–2350 (1997)
7. Fernández, E., et al.: Population Coding in spike trains of sinultaneosly recorded retinal ganglion cells Information. Brain Res. 887, 222–229 (2000)
8. Ferrández, J., et al.: A Neural Network Approach for the Analysis of Multineural Recordings in Retinal Ganglion Cells: Towards Population Encoding. In: Mira, J., et al. (eds.) IWANN 1999. LNCS, vol. 1607, pp. 289–298. Springer, Heidelberg (1999)
9. Normann, R., et al.: High-resolution spatio-temporal mapping of visual pathways using multi-electrode arrays. Vision Res. 41, 1261–1275 (2001)
10. Ortega, G., et al.: Conditioned spikes: a simple and fast method to represent rates and temporal patterns in multielectrode recordings. J. Neurosci. Meth. 133, 135–141 (2004)
11. Shoham, S., Fellows, M., Normann, R.: Robust, automatic spike sorting using mixtures of multivariate t-distributions. J. Neurosci. Meth. 127, 111–122 (2003)
12. Bongard, M., Micol, D., Fernández, E.: Nev2lkit: a tool for handling neuronal event files, `http://nev2lkit.sourceforge.net/`
13. Bonomini, M.P., Ferrández, J.M., Bolea, J.A., Fernández, E.: RDATA-MEANS: An open source tool for the classification and management of neural ensemble recordings. J. Neurosci. Meth. 148, 137–146 (2005)
14. Bonomini, M.P., Ferrández, J.M., Fernández, E.: Searching for semantics in the retinal code. Neurocomputing 72, 806–813 (2009)
15. Shannon, C.: A Mathematical Theory of Communication. Bell sys. Tech. 27, 379–423 (1948)
16. McClelland, J., Rumelhart, D.: Explorations in Parallel Distributed Processing. M Press, Cambridge (1986)
17. Borst, A., Theunissen, F.: Information Theory and Neural Coding. Nature Neurosci. 2(11), 947–957 (1999)
18. Amigo, J.M., et al.: On the number of states of the neuronal sources. Biosystems 68(1), 57–66 (2003)
19. Panzeri, S., Pola, G., Petersen, R.S.: Coding of sensory signals by neuronal populations: the role of correlated activity. Neuroscientist 9(3), 175–180 (2003)
20. Pola, G., et al.: An exact method to quantify the information transmitted by different mechanisms of correlational coding. Network 14(1), 35–60 (2003)
21. McClelland, J., Rumelhart, D.: Explorations in Parallel Distributed Processing. M. Press, Cambridge (1986)
22. Kang, K., Shapley, R.M., Sompolinsky, H.: Information tuning of populations of neurons in primary visual cortex. J. Neurosci. 24(15), 3726–3735 (2004)

On Cumulative Entropies and Lifetime Estimations*

Antonio Di Crescenzo[1] and Maria Longobardi[2]

[1] Dipartimento di Matematica e Informatica, Università di Salerno
Via Ponte don Melillo, I-84084 Fisciano (SA), Italy
adicrescenzo@unisa.it

[2] Dipartimento di Matematica e Applicazioni, Università di Napoli Federico II
Via Cintia, I-80126 Napoli, Italy
maria.longobardi@unina.it

Abstract. The cumulative entropy is a new measure of information, alternative to the classical differential entropy. It has been recently proposed in analogy with the cumulative residual entropy studied by Wang *et al.* (2003a) and (2003b). After recalling its main properties, including a connection to reliability theory, we discuss estimates of random lifetimes based on the empirical cumulative entropy, which is suitably expressed in terms of the dual normalized sample spacings.

1 Introduction

The description of the behavior of biological and engineering systems often requires use of concepts of information theory, and in particular of entropy. As is well known, the entropy was proposed by Shannon (1948) and Wiener (1948) as an essential tool to describe information coding and transmission. More recently it has been also used in the context of theoretical neurobiology (see, for instance, Johnson and Glantz (2004)).

We recall that for a non-negative absolutely continuous random variable X with probability density function $f(x)$, the differential entropy is defined by

$$H(X) = -\mathrm{E}[\log f(X)] = -\int_0^{+\infty} f(x)\, \log f(x)\, \mathrm{d}x, \qquad (1)$$

where 'log' means natural logarithm. We point out that $H(X)$ is location-free. Indeed, X possesses the same differential entropy as $X+b$, for any $b>0$. The differential entropy (1) presents various drawbacks when used as a continuous counterpart of the entropy for discrete random variables. Various efforts have been made by some authors to define alternative information measures. One example is given in Schroeder (2004), where a measure is proposed that is always positive and invariant with respect to linear coordinate transformations. Another measure is studied in Di Crescenzo and Longobardi (2006), where a 'length

* Work performed under partial support by G.N.C.S.-INdAM and Regione Campania.

J. Mira et al. (Eds.): IWINAC 2009, Part I, LNCS 5601, pp. 132–141, 2009.

biased' shift-dependent weighted entropy that assigns larger weights to larger values of X is discussed.

Moreover, the 'cumulative residual entropy' (CRE) has been defined recently as (see Rao *et al.* (2004))

$$\mathcal{E}(X) = -\int_0^{+\infty} \overline{F}(x) \log \overline{F}(x) \, \mathrm{d}x, \tag{2}$$

with $\overline{F}(x) = P(X > x)$ denoting the cumulative residual distribution, or survival function, of X. Differently from (1), this entropy allows one to deal with discrete and continuous random variables in a unified setting. It has been applied to image alignment and to measurements of similarity between images by Wang and Vemuri (2007) and by Wang *et al.* (2003a), (2003b); see also Rao (2005) for its properties. In addition, the cumulative residual entropy (2) is suitable to describe information in problems of reliability theory involving the mean residual life function. Generalized versions of it have been discussed in Drissi *et al.* (2008), and in Zografos and Nadarajah (2005).

A new information measure similar to $\mathcal{E}(X)$ has been proposed recently by Di Crescenzo and Longobardi (2009). It is named 'cumulative entropy', and is defined as

$$\mathcal{CE}(X) = -\int_0^{+\infty} F(x) \log F(x) \, \mathrm{d}x, \tag{3}$$

where $F(x) = P(X \le x)$ is the distribution function of X. Similarly to (2), $\mathcal{CE}(X)$ is non-negative, the argument of the logarithm being a probability. It is useful to measure information on the inactivity time of a system, which is the time elapsing between the failure of a system and the time when it is found to be 'down'. In other terms, $\mathcal{CE}(X)$ is appropriate to deal with systems whose uncertainty is related to the past. To this concern, we recall that the role of entropy-based information measures to describe past lifetimes has been outlined in Di Crescenzo and Longobardi (2002).

Here we aim to present some basic results on the problem of estimating (3) from a random sample. Preliminarly, in Section 2 we discuss some properties of the cumulative entropy, including a connection to reliability theory, and determine various upper and lower bounds. Section 3 is devoted to the estimate of $\mathcal{CE}(X)$. Starting from the empirical distribution $\hat{F}_n$ we define the empirical cumulative entropy $\mathcal{CE}(\hat{F}_n)$, and express it in terms of the dual normalized spacings. Then we show that $\mathcal{CE}(\hat{F}_n)$ converges a.s. to the cumulative entropy as the size of the sample tends to $+\infty$. Finally, two examples are given in which the empirical cumulative entropy is used to gain information from real lifetime datasets.

2 Cumulative Entropy

In analogy with (2), Di Crescenzo and Longobardi (2009) defined the cumulative entropy of a non-negative random variable X as specified in (3). Due to Eq. (3)

Table 1. Cumulative entropy of various random variables

$F(x)$	$\mathcal{CE}(X)$
$1-\mathrm{e}^{-\lambda x},\ x\ge 0;\ \lambda>0$	$\frac{1}{\lambda}\left(\frac{\pi^2}{6}-1\right)$
$\frac{x-a}{b-a},\ a\le x\le b;\ b>a\ge 0$	$\frac{b-a}{4}$
$\exp\{-c\,x^{-\gamma}\},\ x>0;\ c>0,\ \gamma>1$	$\frac{c^{1/\gamma}}{\gamma}\,\Gamma\left(1-\frac{1}{\gamma}\right)$
$\exp\left\{\frac{-c}{e^x-1}\right\},\ x>0;\ c>0$	$c\,\mathrm{e}^c\,\Gamma(0,c)$

we have $0\le\mathcal{CE}(X)\le+\infty$, with $\mathcal{CE}(X)=0$ if and only if X is a degenerate random variable. Some examples of cumulative entropies are given in Table 1.

We remark that $\mathcal{CE}(X)$ is a shift-independent measure. Indeed, if $a>0$ and $b\ge 0$, then $\mathcal{CE}(a\,X+b)=a\,\mathcal{CE}(X)$. Note that the cumulative residual entropy shares the same property. Indeed, it is not hard to see from (2) that $\mathcal{E}(a\,X+b)=a\,\mathcal{E}(X)$.

Let us now point out a connection between the cumulative entropy and a notion of reliability theory. If X is the lifetime of a system, then the inactivity time of the system is denoted by $[t-X\,|\,X\le t]$, $t>0$, where $[Z\,|\,B]$ is a random variable whose distribution is identical to that of Z conditional on B. The inactivity time is thus the duration of the time occurring between an inspection time t and the failure time X, given that at time t the system has been found to be down. For all $t\ge 0$ such that $F(t)>0$, the 'mean inactivity time' is given by

$$\tilde{\mu}(t)=\mathrm{E}[t-X\,|\,X\le t]=\frac{1}{F(t)}\int_0^t F(x)\,\mathrm{d}x. \tag{4}$$

This function has been used in various contexts of reliability theory involving stochastic orders and characterizations of random lifetimes (see, for instance, Ahmad and Kayid 2005, Ahmad *et al.* 2005, Kayid and Ahmad 2004, Li and Lu 2003, Misra *et al.* 2008 and Nanda *et al.* 2003). The inactivity time is a reliability concept dual to the residual life $[X-t\,|\,X>t]$, which is the time elapsing between the failure time X and an inspection time t, given that at time t the system has been found working. Consequently, (4) is dual to the mean residual life

$$\mathrm{mrl}(t)=\mathrm{E}[X-t\,|\,X>t]=\frac{1}{\overline{F}(t)}\int_t^{+\infty}\overline{F}(x)\,\mathrm{d}x, \qquad \forall t\ge 0:\ \overline{F}(t)>0.$$

The forthcoming theorem, whose proof is given in Di Crescenzo and Longobardi (2009), is analogous to Theorem 2.1 of Asadi and Zohrevand (2007), where it is shown that the cumulative residual entropy (3) can be expressed as

$$\mathcal{E}(X)=\mathrm{E}[\mathrm{mrl}(X)].$$

Theorem 1. *For a non-negative random variable X with mean inactivity time $\tilde{\mu}(t)$ and cumulative entropy $\mathcal{CE}(X) < +\infty$ we have*

$$\mathcal{CE}(X) = \mathrm{E}[\tilde{\mu}(X)]. \tag{5}$$

Hereafter we shall focus on certain lower and upper bounds for the cumulative entropy.

Proposition 1. *Let X be a non-negative random variable. Then,*[1]
(i) $\mathcal{CE}(X) \geq C\,\mathrm{e}^{H(X)}$, *where* $C = \exp\left\{\int_0^1 \log(x\,|\log x|)\,\mathrm{d}x\right\} = 0.2065$;
(ii) $\mathcal{CE}(X) \geq \int_0^{+\infty} F(x)\,\overline{F}(x)\,\mathrm{d}x$;
(iii) $\mathcal{CE}(X) \geq -\int_{\mu}^{+\infty} \log F(z)\,\mathrm{d}z$;
(iv) $\mathcal{CE}(X) \leq \mathrm{E}(X)$;
(v) $\mathcal{CE}(X) \leq \mathrm{e}^{-1}\,b$;
(vi) $\mathcal{CE}(X) \leq [b - \mathrm{E}(X)]\,\left|\log\left(1 - \frac{\mathrm{E}(X)}{b}\right)\right|$,
bounds (v) and (vi) holding if X takes values in $[0, b]$, with b finite.

The proof of Proposition 1 can be found in Di Crescenzo and Longobardi (2009).

Example 1. Let X have distribution function

$$F(x) = \left(\frac{x-a}{b-a}\right)^{\theta}, \qquad a \leq x \leq b, \tag{6}$$

with $b > a \geq 0$ and $\theta > 0$. This distribution satisfies the proportional reversed hazards model (see Di Crescenzo 2000, Gupta and Gupta 2007 and references therein). Indeed, $F(x)$ is the θ-th power of the distribution function of a random variable uniformly distributed in $[a, b]$. From (3) and (6) we have

$$\mathcal{CE}_{\theta}(X) = \frac{\theta}{(1+\theta)^2}(b-a). \tag{7}$$

We point out that this function satisfies the following symmetry property:

$$\mathcal{CE}_{\theta}(X) = \mathcal{CE}_{1/\theta}(X), \qquad \forall \theta > 0.$$

For instance, when $a = 0$ and $b = 1$, if $\theta = 2$ then Eq. (7) yields

$$\mathcal{CE}_2(X) = \mathcal{CE}_{1/2}(X) = \frac{2}{9} = 0.2222.$$

Table 2 shows the corresponding bounds provided by Proposition 1.

[1] In the following, all numerical values are approximated to the fourth decimal digit.

Table 2. Bounds to the cumulative entropy (7), with $a = 0$ and $b = 1$

θ	(*i*)	(*ii*)	(*iii*)	(*iv*)	(*v*)	(*vi*)
2	0.1702	0.1333	0.1260	0.6667	0.3679	0.3662
1/2	0.1519	0.1667	0.1502	0.3333	0.3679	0.2703

3 Empirical Cumulative Entropy

In this section estimates of the cumulative entropy are constructed by means of the empirical cumulative entropy.

Let $X_1, X_2, \dots, X_n$ be non-negative, absolutely continuous, independent and identically distributed random variables, that form a random sample drawn from a population having distribution function $F(x)$. As is customary, we denote the empirical distribution of the sample by

$$\hat{F}_n(x) = \frac{1}{n} \sum_{i=1}^{n} I_{\{X_i \le x\}}, \qquad x \in \mathbb{R}.$$

According to (2), we define the 'empirical cumulative entropy' as

$$\mathcal{CE}(\hat{F}_n) = -\int_0^{+\infty} \hat{F}_n(x) \log \hat{F}_n(x) \, \mathrm{d}x.$$

Hence, denoting by $(0 = X_{(0)} \le)\, X_{(1)} \le X_{(2)} \le \ldots \le X_{(n)}$ the order statistics of the random sample, we immediately have

$$\mathcal{CE}(\hat{F}_n) = -\sum_{j=1}^{n-1} \int_{X_{(j)}}^{X_{(j+1)}} \hat{F}_n(x) \log \hat{F}_n(x) \, \mathrm{d}x. \tag{8}$$

Recalling that

$$\hat{F}_n(x) = \frac{j}{n} \qquad \text{if } X_{(j)} \le x < X_{(j+1)}, \quad j = 1, 2, \dots, n-1,$$

from (8) we thus obtain

$$\mathcal{CE}(\hat{F}_n) = -\sum_{j=1}^{n-1} U_{j+1} \frac{j}{n} \log \frac{j}{n}, \tag{9}$$

where use of the sample spacings

$$U_i = X_{(i)} - X_{(i-1)}, \qquad i = 1, 2, \dots, n$$

has been made. Moreover, recalling that the dual normalized spacings are defined as (cf. Bartoszewicz 2005, or Hu and Wei 2001)

$$\tilde{C}_i = (i-1)\, U_i, \qquad i = 1, 2, \dots, n,$$

from (9) we have

$$\mathcal{CE}(\hat{F}_n) = -\frac{1}{n}\sum_{j=1}^{n-1} \tilde{C}_{j+1} \log \frac{j}{n}. \tag{10}$$

This expression can be given an alternative form that involves the sample mean $\bar{X}$. Indeed, noting that

$$\begin{aligned}\sum_{j=1}^{n-1} \tilde{C}_{j+1} &= \sum_{j=1}^{n-1}\sum_{k=1}^{j} \left[X_{(j+1)} - X_{(j)}\right] = \sum_{k=1}^{n-1}\sum_{j=k}^{n-1} \left[X_{(j+1)} - X_{(j)}\right] \\ &= \sum_{k=1}^{n-1} \left[X_{(n)} - X_{(k)}\right] = n\left[X_{(n)} - \bar{X}\right],\end{aligned}$$

Eq. (10) finally becomes:

$$\mathcal{CE}(\hat{F}_n) = \left[X_{(n)} - \bar{X}\right] \log n - \frac{1}{n}\sum_{j=1}^{n-1} \tilde{C}_{j+1} \log j. \tag{11}$$

Let us now face the problem of assessing the convergence of $\mathcal{CE}(\hat{F}_n)$ as n tends to $+\infty$. We recall that Glivenko-Cantelli theorem asserts that

$$\sup_x |F_n(x) - F(x)| \to 0 \quad \text{a.s. as } n \to +\infty.$$

Using this result we now prove the following theorem, which discloses an asymptotic property of the empirical cumulative entropy.

Proposition 2. Let X be a non-negative random variable in L^p for some $p > 1$; the empirical cumulative entropy converges to the cumulative entropy of X, i.e.

$$\mathcal{CE}(\hat{F}_n) \to \mathcal{CE}(X) \quad \text{a.s. as } n \to +\infty.$$

Proof. It goes similarly to the proof of Theorem 9 of Rao *et al.* (2004). By the dominated convergence theorem the integral of $\hat{F}_n(x) \log \hat{F}_n(x)$ converges to that of $F(x) \log F(x)$ on any finite interval. Hence, we only have to show that

$$\left| \int_1^{+\infty} \hat{F}_n(x) \log \hat{F}_n(x)\,\mathrm{d}x - \int_1^{+\infty} F(x) \log F(x)\,\mathrm{d}x \right| \to 0 \quad \text{a.s. as } n \to +\infty.$$

Denoting by Z_n the random variable uniformly distributed over the sample points $X_1, X_2, \ldots, X_n$, by the strong law of large numbers we have

$$\mathrm{E}(Z_n^p) = \frac{1}{n}\sum_{i=1}^{n} {X_i}^p \to \mathrm{E}({X_1}^p) \quad \text{a.s. as } n \to +\infty.$$

In particular,

$$\alpha := \sup_n \mathrm{E}(Z_n^p) < \infty \quad \text{a.s.}$$

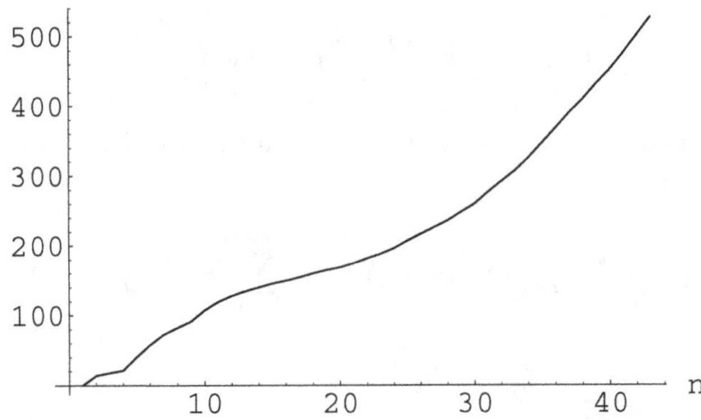

Fig. 1. Empirical cumulative entropy for the sample data of Example 2

Recalling that $|u \log u| \leq 1-u$, $0 \leq u \leq 1$, for $x \geq 1$ we obtain

$$|\hat{F}_n(x) \log \hat{F}_n(x)| \leq 1 - \hat{F}_n(x) = P(Z_n > x) \leq \frac{1}{x^p}\, \mathrm{E}(Z_n^p),$$

and thus

$$\int_1^{+\infty} \left|\hat{F}_n(x) \log \hat{F}_n(x)\right| \mathrm{d}x \leq \alpha \int_1^{+\infty} \frac{1}{x^p}\, \mathrm{d}x = \frac{\alpha}{p-1},$$

with $p > 1$. The proof finally follows by applying the dominated convergence theorem and by virtue of the Glivenko-Cantelli theorem.

Hereafter we make use of the empirical cumulative entropy in order to perform some lifetime estimations.

Example 2. We consider the following set of 43 sample lifetime data, taken from Bryson and Siddiqui (1969):

$$\{7, 47, 58, 74, 177, 232, 273, 285, 317, 429, 440, 445, 455, 468, 495, 497, 532, 571, 579, 581, 650, 702, 715, 779, 881, 900, 930, 968, 1077, 1109, 1314, 1334, 1367, 1534, 1712, 1784, 1877, 1886, 2045, 2056, 2260, 2429, 2509\}.$$

From (11), the following estimate of the cumulative entropy holds: $\mathcal{CE}(\hat{F}_{43}) = 527.3$. In Figure 1 we show the empirical cumulative entropy of the first n data, for $n = 1, 2, \ldots, 43$, by assuming that the sample data are observed in increasing order.

We remark that when the random sample containes repeated values, then Eq. (9) can be rewritten as

$$\mathcal{CE}(\hat{F}_k) = -\sum_{j=1}^{k-1} \left[Y_{(j+1)} - Y_{(j)}\right] \frac{n_j}{n} \log \frac{n_j}{n}, \tag{12}$$

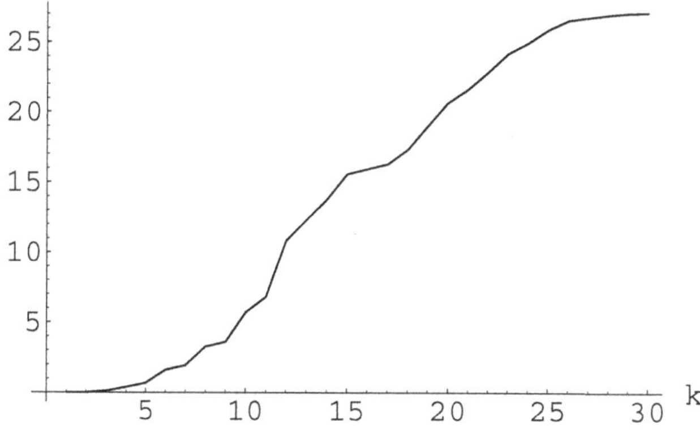

Fig. 2. Same as Fig. 1, for the data of Example 3

where $Y_{(1)} < Y_{(2)} < \ldots < Y_{(k)}$ denote the ordered different modalities of a n-sized random sample, and n_j is the multiplicity of the j-th value, so that $Y_{(j)}$ appears n_j times in the sample, with $n_1 + n_2 + \ldots + n_k = n$.

Example 3. Let us consider the following set of 50 sample lifetime data given in Lai *et al.* (2003), and originally presented in Aarset (1987):

$$\{0.1, 0.2, 1, 1, 1, 1, 1, 2, 3, 6, 7, 11, 12, 18, 18, 18, 18, 18,\\ 21, 32, 36, 40, 45, 46, 47, 50, 55, 60, 63, 63, 67, 67, 67, 67,\\ 72, 75, 79, 82, 82, 83, 84, 84, 84, 85, 85, 85, 85, 85, 86, 86\}.$$

Due to (12), the cumulative entropy is estimated as $\mathcal{CE}(\hat{F}_{30}) = 27.1$. Figure 2 shows the empirical cumulative entropy of the first k modalities, with $k = 1, 2, \ldots, 30$, if one assumes that the sample data are observed in an increasing order.

Acknowledgements

The authors thank their colleagues and friends A. Buonocore, B. Martinucci, E. Pirozzi and L.M. Ricciardi for useful remarks.

References

1. Aarset, M.V.: How to identify bathtub hazard rate. IEEE Trans. Rel. R-36, 106–108 (1987)
2. Ahmad, I.A., Kayid, M.: Characterizations of the RHR and MIT orderings and the DRHR and IMIT classes of life distributions. Probab. Eng. Inform. Sci. 19, 447–461 (2005)

3. Ahmad, I.A., Kayid, M., Pellerey, F.: Further results involving the MIT order and IMIT class. Probab. Eng. Inform. Sci. 19, 377–395 (2005)
4. Asadi, M., Zohrevand, Y.: On the dynamic cumulative residual entropy. J. Stat. Plann. Infer. 137, 1931–1941 (2007)
5. Bartoszewicz, J.: Dispersive ordering between order statistics and spacings from an IRFR distribution (preprint, 2005)
6. Bryson, M.C., Siddiqui, M.M.: Some criteria for aging. J. Amer. Stat. Assoc. 64, 1472–1483 (1969)
7. Di Crescenzo, A.: Some results on the proportional reversed hazards model. Stat. Prob. Lett. 50, 313–321 (2000)
8. Di Crescenzo, A., Longobardi, M.: Entropy-based measure of uncertainty in past lifetime distributions. J. Appl. Prob. 39, 434–440 (2002)
9. Di Crescenzo, A., Longobardi, M.: On weighted residual and past entropies. Scientiae Math. Japon. 64, 255–266 (2006)
10. Di Crescenzo, A., Longobardi, M.: On cumulative entropies (submitted, 2009)
11. Drissi, N., Chonavel, T., Boucher, J.M.: Generalized cumulative residual entropy for distributions with unrestricted supports. Res. Lett. Sign. Proc., 5 pages, Article ID 790607 (2008)
12. Gupta, R.C., Gupta, R.D.: Proportional reversed hazard rate model and its applications. J. Stat. Plann. Infer. 137, 3525–3536 (2007)
13. Hu, T., Wei, Y.: Stochastic comparisons of spacings from restricted families of distributions. Stat. Prob. Lett. 53, 91–99 (2001)
14. Johnson, D.H., Glantz, R.M.: When does interval coding occur? Neurocomputing 59-60, 13–18 (2004)
15. Kayid, M., Ahmad, I.A.: On the mean inactivity time ordering with reliability applications. Probab. Eng. Inform. Sci. 18, 395–409 (2004)
16. Lai, C.D., Xie, M., Murthy, D.N.P.: A Modified Weibull Distribution. IEEE Trans. Rel. 52, 33–37 (2003)
17. Li, X., Lu, J.: Stochastic comparisons on residual life and inactivity time of series and parallel systems. Probab. Eng. Inform. Sci. 17, 267–275 (2003)
18. Misra, N., Gupta, N., Dhariyal, I.D.: Stochastic properties of residual life and inactivity time at a random time. Stoch. Models 24, 89–102 (2008)
19. Nanda, A.K., Singh, H., Misra, N., Paul, P.: Reliability properties of reversed residual lifetime. Comm. Statist. Theory Methods 32, 2031–2042 (2003); With correction in Comm. Statist. Theory Methods 33, 991–992 (2004)
20. Rao, M.: More on a new concept of entropy and information. J. Theor. Probab. 18, 967–981 (2005)
21. Rao, M., Chen, Y., Vemuri, B.C., Wang, F.: Cumulative residual entropy: a new measure of information. IEEE Trans. Inf. Theory 50, 1220–1228 (2004)
22. Schroeder, M.J.: An alternative to entropy in the measurement of information. Entropy 6, 388–412 (2004)
23. Shannon, C.E.: A mathematical theory of communication. Bell System Techn. J. 27, 279–423 (1948)
24. Wang, F., Vemuri, B.C.: Non-rigid multi-modal image registration using cross-cumulative residual entropy. Intern. J. Computer Vision 74, 201–215 (2007)
25. Wang, F., Vemuri, B.C., Rao, M., Chen, Y.: A new & robust information theoretic measure and its application to image alignment. In: Taylor, C.J., Noble, J.A. (eds.) IPMI 2003. LNCS, vol. 2732, pp. 388–400. Springer, Heidelberg (2003a)

26. Wang, F., Vemuri, B.C., Rao, M., Chen, Y.: Cumulative residual entropy, a new measure of information & its application to image alignment. In: Proceedings on the Ninth IEEE International Conference on Computer Vision (ICCV 2003), vol. 1, pp. 548–553. IEEE Computer Society, Los Alamitos (2003b)
27. Wiener, N.: Cybernetics. MIT Press and Wiley, New York (1948) (2nd edn. 1961)
28. Zografos, K., Nadarajah, S.: Survival exponential entropies. IEEE Trans. Inf. Theory 51, 1239–1246 (2005)

Activity Modulation in Human Neuroblastoma Cultured Cells: Towards a Biological Neuroprocessor

J.M. Ferrández-Vicente[1,2], M. Bongard[1], V. Lorente[2], J. Abarca[3], R. Villa[4,5], and E. Fernández[1,5]

[1] Instituto de Bioingeniería, Universidad Miguel Hernández, Alicante
[2] Dpto. Electrónica, Tecnología de Computadoras, Univ. Politécnica de Cartagena
[3] Servicio de Neurocirugía, Hospital General Universitario de Alicante
[4] Instituto de Microelectrónica de Barcelona (IMB-CNM-CSIC), Bellaterra
[5] CIBER-BBN
jm.ferrandez@upct.es

Abstract. The main objective of this work is to analyze the computing capabilities of human neuroblastoma cultured cells and to define stimulation patterns able to modulate the neural activity in response to external stimuli. Multielectrode Arrays Setups have been designed for direct culturing neural cells over silicon or glass substrates, providing the capability to stimulate and record simultaneously populations of neural cells. This paper describes the process of growing human neuroblastoma cells over MEA substrates and tries to modulate the natural physiologic responses of these cells by tetanic stimulation of the culture. If we are able to modify the selective responses of some cells with a external pattern stimuli over different time scales, the neuroblastoma-cultured structure could be trained to process pre-programmed spatio-temporal patterns. We show that the large neuroblastoma networks developed in cultured MEAs are capable of learning: stablishing numerous and dynamic connections, with modifiability induced by external stimuli

1 Introduction

Using biological nervous systems as conventional computer elements is a fascinating problem that permits the hybridation between Neuroscience and Computer Science. This synergic approach can provide a deeper understanding of natural perception and may be used for the design of new computing devices based on natural computational paradigms. Classical computational paradigms consist in serial and supervised processing computations with high-frequency clocks silicon processors, with moderate power consumption, and fixed circuits structure. However the brain uses millions of biological processors, with dynamic structure, slow commutations compared with silicon circuits, low power consumption and unsupervised learning. This kind of computation is more related to perceptual recognition, due to the natural variance of the perceptive patterns and the a priori lack of knowledge about the perceptual domain.

J. Mira et al. (Eds.): IWINAC 2009, Part I, LNCS 5601, pp. 142–154, 2009.

There exist many research approaches based on mimicking this bioinspired parallel processing, not only from the algorithm perspective [1], but also from the silicon circuits design [2]. These bioinspired approaches are useful for pattern recognition applications, like computer vision or robotics, however they are implemented over a serial and artificial silicon processors with fixed and static structure. A real biological processor with millions of biological neurons and a huge number of interconnections, Would provide much more computational power instead of their low transition rates due to high number of computing elements and the extraordinary network capability of adaptation and reconfiguration to unknown environments. This extraordinary capability is related with natural unsupervised learning.

Learning is a natural process that needs the creation and modulation of sets of associations between stimuli and responses. For understanding the process of learning, is necessary to define the physiological mechanisms that support the creation and modulation of associations and determine the relation that modulate the configuration between stimuli and responses associations. These mechanisms and relation have been studied by many neurophysiological studies at different levels mainly in single cell experimentation.

Since the christening of the neurophysiology, neuroscientists have been recording and stimulating individual aisle neural cells to study their role in the functioning of nervous systems. A considerable number of studies have focused on single cell responses [3], [4]. Traditionally, the spiking rate, or even the spontaneous firing rate has been used as information carrier due to the close correlation with the stimulus intensity in all sensory systems, however single neurons produce a few spikes in response to different presentations and they must code a huge spectrum in their firings, and also the concept of rate implies some temporal averaging, and the decoding must be fast enough and unequivocal, so the system is able to response in a few milliseconds. Temporal dimension, the exact temporal sequence of action potentials may be coding the stimulus main features [5] as occurs in some systems (e.g. auditory coding [6]), providing accuracy, fast decoding and a wide dimension for representing all kind of stimuli. A combination between the spatio-temporal network strategies would provide robustness to the system.

There are, however some drawbacks on using single cells activity. First the response of a single cell cannot unequivocally describe the stimulus. The response from a single cell to repetitions of the same stimuli has a considerable variability for different presentations. The timing sequence differs not only in the time events but also in the number of spikes, producing uncertainty in the decoding. And second, the same response, the same sequence of neural events may be obtained by providing different stimulus, introducing ambiguity in the neural response. So, it is a complex task to "understand" the neural response just by analizing a single ganglion cell response. In this way the brain requires the combined activity of millions of interconnected neurons and glial cells associated with them.

New register techniques, and the emergence of new technologies, allow simultaneous recordings from populations of retinal ganglion cells. At this time, recordings of the order of a hundred simultaneous spike trains may be obtained. This technology permit to analyse the interactions between populations of neural cells, and to describe the emergent properties, result of the neural population computations.

Our learning experiments were performed in neural cultures containing 120.000 human neuroblastoma SY-5Y, under the assumption that this kind of cells are able to respond electrically to external stimuli and modulate their neural firing by changing the stimulation parameters. Such cultured neuroblastoma networks showed that they have dynamical configurations, and that they are able to develop and adapt functionally and maybe morphologically in response to external stimuli over a broad range of configuration patterns. We are especially interested in analizing if populations of neuroblastoma cells are able to process and store information, and if learning can be implemented over this biological structure. Neural cultures usually have substantial limitations because of the fact that they are aisled from a brain structure creating in some cases artificial behaviour, like spontaneous bursting over the culture. Learning over this neural culture will have to deal with these processes.

The main objective of this work is to analyze the computing capabilities of human neuroblastoma cultured cells and to define stimulation patterns able to modulate the neural activity in response to external stimuli. Multielectrode Arrays Setups have been designed for direct culturing neural cells over silicon or glass substrates, providing the capability to stimulate and record simultaneously populations of neural cells . This paper describes the process of growing human neuroblastoma cells over MEA substrates and tries to change the natural physiologic responses of these cells by external stimulation of the culture. Modifying the global responses of some cells with a external pattern stimuli means adjusting the biological network behaviour due to changes in synaptic efficiency or long-term potentiation (LTP). Therefore, the neuroblastoma-cultured structure could be trained to process pre-programmed spatio-temporal patterns. In what follows, we show that the large neuroblastoma networks developed in cultured MEAs are capable of learning: stablishing numerous and dynamic connections, with modifiability induced by external stimuli.

2 Human Neuroblastoma SY-5Y

The physiological function of neural cells are modulated by the underlying mechanisms of adaptation and reconfiguration in response to neural activity. Hebbian learning describes a basic mechanism for synaptic plasticity wherein an increase in synaptic efficacy arises from the presynaptic cell's repeated and persistent stimulation of the postsynaptic cell. The theory is commonly evoked to explain some types of associative learning in which simultaneous activation of cells leads to pronounced increases in synaptic strength. The N-methyl-D-aspartate (NMDA) receptor, a subtype of the glutamate receptor, has been implicated as playing a key role in synaptic plasticity in the CNS [7], where as dopamine receptors are involved in the regulation of motor and cognitive behaviors. For most synaptic ion channels, activation (opening) requires only the binding of neurotransmitters. However, activation of the NMDA channel requires two events: binding of glutamate (a neurotransmitter) and relief of Mg2+ block. NMDA channels are located at the postsynaptic membrane. When the membrane potential is at rest, the NMDA channels are blocked by the Mg2+ ions. If the membrane potential is depolarized due to excitation of the postsynaptic neuron, the outward depolarizing field may repel Mg2+ out of the channel pore. On the other hand, binding of glutamate may open the gate of NMDA channels (the gating mechanisms of most ion channels are not known). In the normal physiological process, glutamate is released from the presynaptic terminal when the presynaptic neuron is excited. Relief of Mg2+ block is due to excitation of the postsynaptic neuron. Therefore, excitation of both presynaptic and postsynaptic neurons may open the NMDA channels.

Another important feature of the NMDA channel is that it conducts mainly the Ca2+ ion which may activate various enzymes for synaptic modification, even nictric oxide has been identified as a relevant element in synaptic regulation. The enhancement of synaptic transmission is called the long-term potentiation (LTP), which involves two parts: the induction and the maintenance. The induction refers to the process which opens NMDA channels for the entry of Ca2+ ions into the postsynaptic neuron. The subsequent synaptic modification by Ca2+ ions is referred to as the maintenance of LTP.

A human neuroblastoma SY5Y cell line, that express clonal specific human dopamine receptors, and also NMDA receptors, will be the biological platform for studying learning in cultured cells.

Neuroblastoma SH-SY5Y cells are known to be dopaminergc, acetylcholinergic, glutamatergic and adenosinergic, so in this line they respond to different neurotransmitters. The cells have very different growth phases, as it can be seen in Figure 1. The cells both propagate via mitosis and differentiate by extending neurites to the surrounding area. The dividing cells can form clusters of cells which are reminders of their cancerous nature, but chemicals can force the cells to dendrify and differentiate, in some kind of neuritic growth.The loss of neuronal

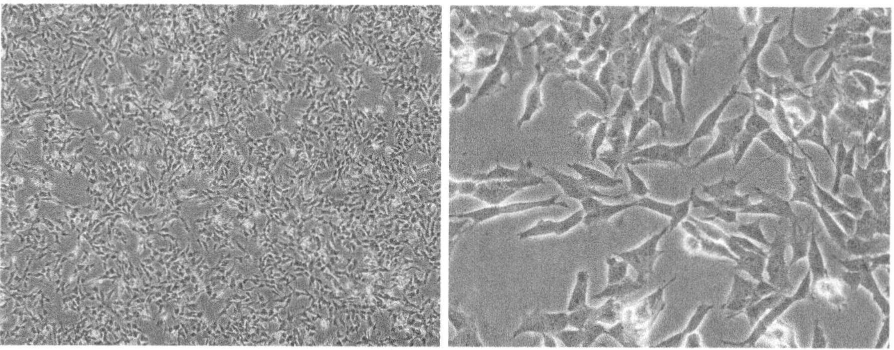

Fig. 1. Human Neuroblastoma SY-5Y cultures

characteristics has been described with increasing passage numbers. Therefore it is recommended not to be used after passage 20.

As conclusion, neuroblastoma culture cells show electrophysiological responses similar to standard neurons, as potential actions generation sensible to tetrodotoxin (TTX) and acetylcholyn. They have neurotransmitters synthesis process and are able to neuritic growth in culture medium.

3 Experimental Setup

The neuro-physiology setup provides a complete solution for stimulation, heating, recording, and data acquisition from 64 channels. The MEA (microelectrode array) system is intended for extracellular electrophysiological recordings in vitro of different applications that include acute brain, heart, and retina slices; cultured slices; and dissociated neuronal cell cultures.

The basic components of the MEA Systems are shown in Figure 2. These components are:

- A microelectrode array is an arrangement of 60 electrodes that allows the simultaneous targeting of several sites for extracellular stimulation and recording. Cell lines or tissue slices are placed directly on the MEA and can be cultivated for up to several months. Almost all excitable or spontaneously active cells and tissues can be used.
- The temperature controller regulates the MEA temperature.
- Raw data from the MEA electrodes are amplified by MCS filter amplifiers with custom bandwidth and gain, which are built very small and compact using SMD (Surface Mounted Devices) technology. The small-sized amplifier combines the interface to the MEA probe with the signal filtering and the amplification of the signal. The compact design reduces line pick up and

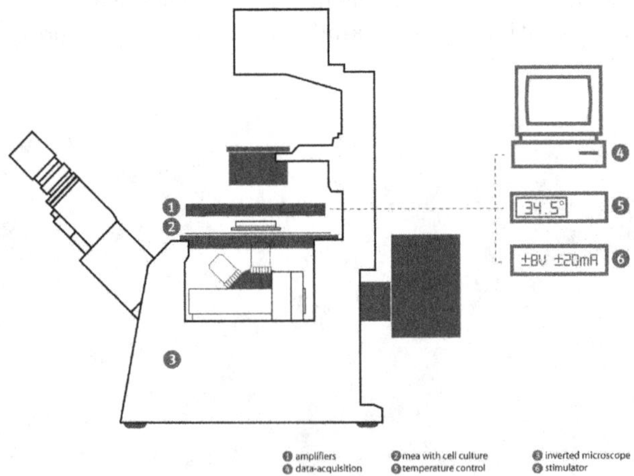

Fig. 2. Experimental Setup

keeps the noise level down. The amplifiers are mounted over an inverted microscopes.
- The analog input signals are then acquired and digitized by the MC-Card that is preinstalled on the data acquisition computer, that supplies the power for the amplifiers.

4 Methods

4.1 Neuroblastoma Cultures

Human neuroblastoma cultures were produced using the commercial line SH/ SY5Y . Neural cells were then plated on Micro-Electrode Arrays -MEAs (MultiChannel Systems, Reutlingen, Germany). Initially the nitrogen frozen cells, was immersed in a 37 degree bath, and centrifuged at 1000 rpm during 5 minutes, When cells have grown in a uniform mono-layer process, they are washed three time with buffer Phosphate-buffered saline (PBS) for keeping the pH approximately constant. 0,5 per cent trypsin was added to the solution in order to re-suspend cells adherent to the cell culture dish wall during the process of harvesting cells. The cells were kept in the incubator for 5 minutes and passed through a 40 microm cell strainer (Falcon, Bedford, MA) to remove large debris. Finally the cells are transferred to a specific medium in order to inactivate trypsin, and centrifugated again during 5 minutes at 1000 rpm.

For seeding the plate cells are stained with trypan blue, (because cells that loose their permeability get colorred with this solution) and counted with a Neubauer chamber. Finally, 80.000 or 120.000 total neuroblastoma cells has been placed over the MEA substrate.

Maintaining cells in culture is essential for studying their physiological properties. Cell culturing is dependent on the growth surfaces and cells must adhere to the electrode substrate in order to establish the best connection with the electrodes material. For most cultures coated tissue culture plates are prerequisite for seeding. The most commonly used coatings are positively charged polymers. In this work, the insulation layer (silicon nitride) of some of the plates was pretreated with polyethyleneimine (PEI), which provides adequate adhesive properties for cultured cell lines, forming a good surface for network development. The advantage of coating surfaces with PEI-based will be compared with no covered plates.

The neuroblastoma cultures are maintained in a 37 degree humidified incubator with 5 per cent CO2 and 95 per cent O2 with serum-free Neurobasal medium. Under the aforementioned conditions we were able to record stable electrophysiological signals over different days in vitro (Div). The medium was replaced one-half of the medium every 5 days.

4.2 Electrophysiological Methods

The substrate-embedded multielectrode array technology was used. It includes arrays of 60 Ti/Au/TiN electrodes, 30 microm in diameter, and spaced 200 microm from each other [MultiChannelSystems (MCS), Reutlingen, Germany].

A commercial 60 channel amplifier (B-MEA-1060; MCS) with frequency limits of 10-3000 Hz and a gain of 1024X was used. Stimulation through the MEA is performed using a dedicated two channel stimulus generator (MCS) able of time multiplexing the signals for stimulating the 60 electrodes in a nearly parallel process. The microincubation environment was arranged to support long-term recordings from MEA dishes. by electrically heating the MEA platform to 37C. Data were digitized using MCS analog-to-digital boards. Each channel is sampled at a frequency of 24,000 samples/sec and prepared for analysis using the MC-RACK softeware provide also by Multichannel Systems.

Neural spike events were detected by comparing the instantaneous electrode signals to level thresholds set for each data channel using standard procedures described elsewhere [8]. Standard deviation of -3 was used for identifying the signals above the average level of the recordings. When a supra-threshold event occurs, the signal window surrounding the event was time-stamped and stored together with the state of the visual stimulus for later, offline analysis. For spike sorting we used a free program, NEV2kit, which has been recently developed by our group [9] and runs under Windows, MacOSX and Linux (source code and documentation is freely available at: http://nev2lkit.sourceforge.net/). NEV2kit loads multielectrode data files in various formats (ASCII based formats, LabView formats, Neural Event Files, etc) and is able to sort extracted spikes from large sets of data. The sorting was done using principal component analysis (PCA) and performed simultaneously on many records from the same experiment.

5 Results

5.1 Neuroblastoma Cultures over Multielectrode Array

The cultured neuroblastoma cells establish synaptic connections. In Figure 3-left it can be seen differentiated and non-differentiated neuroblastoma cell bodies growing around the whole electrode population. The dendritic arborescence is more evident in the magnification Figure 3-right where differentiated neural cells surround the four electrodes while the rest of the cells are in their growing process. This Figure corresponds to 80.000 neuroblastoma cells seeded in a no-PEI MEA at 2nd day in vitro (div). At 10 div, the cells decrease their adhesion to the plate, so more of the population lost their contact with the metal elements. So, the same neuroblastoma cells where seeded in polyethyleneimine (PEI) covered MEAs.

The cultured neuroblastoma also cells establish numerous synaptic connections. This is evident in Figure 4-left, where many neuroblastoma cells forma large number of dendritic structures around some of the electrodes. This Figure correspond to 80.000 neuroblastoma cells seeded in a no-PEI MEA at 5th day in vitro (div). There exist few neuronal cluster around some electrodes with clear denditric connections. In Figure 4-right it is shown a magnification of four electrodes, where differentiated and non-differentiated neuroblastoma can be seen around the metal elements, and some of them, this usually happens after 15 div, get an embryonic configuration, upper-right electrode. The main drawback of

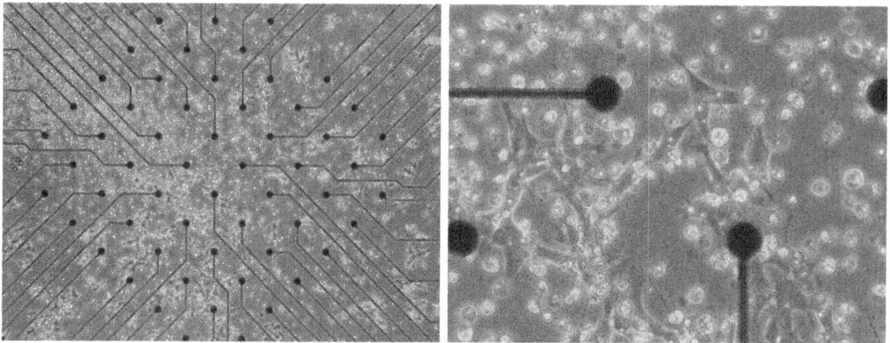

Fig. 3. Human Neuroblastoma Culture over Multielectrode Array (80.000 cells)

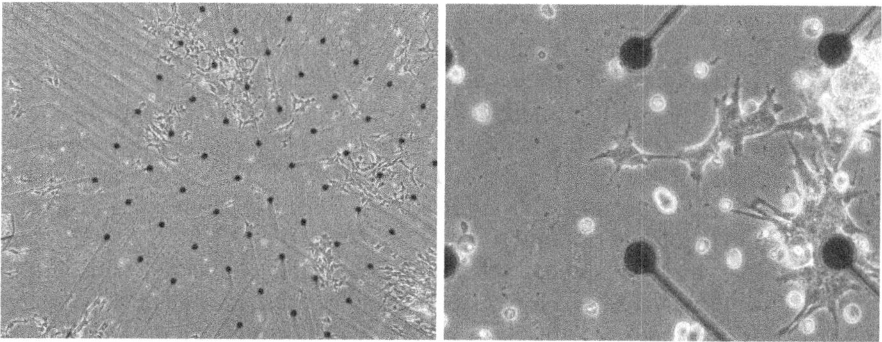

Fig. 4. Human Neuroblastoma Culture over Multielectrode Array with PEI covering

this covering is that only few electrodes are covered with cells, so we will get very limited responses from the whole electrode matrix.

We finally decide to use no-PEI MEAS and increase the number of seeded cells to 120.000 in order to cover the full array. Figure 5-left show the whole array with differentiated neuroblastoma cells over the whole structure with high-density connections between them. This can be clearly observed in the magnified image, Figure 5-right. Both images corresponds to 120.000 neuroblastoma cells seeded in a no-PEI MEA at 5th day in vitro (div).

5.2 Electrophysiological Recordings

The electrophysiological properties of the neuroblastoma cultures were analized by recording the spontaneous activity of the network. Time course of experiments was over 15 days, recordings were done using two MCS-Meas with two neuroblastoma cell cultures (but only in one the cells survived till day 15). In vitro neuroblastoma networks show spontaneously firing. This firing rates change

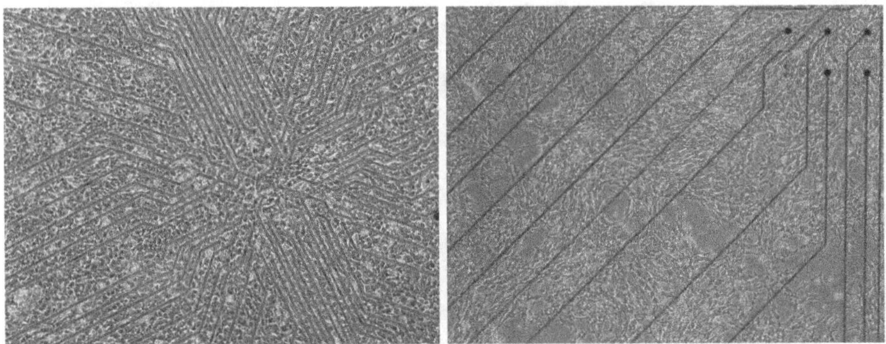

Fig. 5. Human Neuroblastoma Culture over Multielectrode Array (120.000 cells)

during the culture development with marked day differences and the global rate is closely related to the age of the network. The physiological recordings correspond to neuroblastoma cultures in the range of 1-7 div. They show bursting and spiking activity (Fig.6), with usually negative depolarisations. Figure 4 shows the spontaneous activity registered in an electrode center in the negative depolarisation peak, and show the signal recorder 1 ms before and 2 ms after the detected spike. The spikes magnitude is around -30 mV, showing a typical signal:noise ration around 3:1.

Figure 7-left show the spiking activity of the neural population with an automatic detection level for each electrode. This is very convenient if you have multiple channels for extracting spikes. The standard deviation of each data trace is used to estimate its spike threshold. A time interval of 500 ms is used to calculate the standard deviation. By fixing the factor, by which the standard deviation is multiplied, the sign of the factor determines whether the spike detection level is positive or negative, only values above this will be extracted as spiking activity. A value between -1 and -4 is appropriate for most applications the threshold was fixed at standard deviation equal to -3 with respect to the electrode activity in order to identify spikes embedded in the noisy signals. Figure 7-right show the culture activity with a grey map in order to analyse the spatial distribution. The brighter colours correspond to the most active electrode while the darker ones correspond to few spiking. Black electrodes correspond to the electrodes that are not covered by the culture or the stimulating electrode. During the neuroblastoma development, a wide range of population bursting or synchronized activity has been observed, according to some studies in neural cultures preparations [10]. The burst usually contains a large number of spikes at many channels, with variable duration, from milliseconds to seconds.

5.3 Tetanic Stimulation

Spontaneous activity was recorded for intervals of 3 minutes before stimulation (PRE-data), and the total number of spikes extracted was counted. The

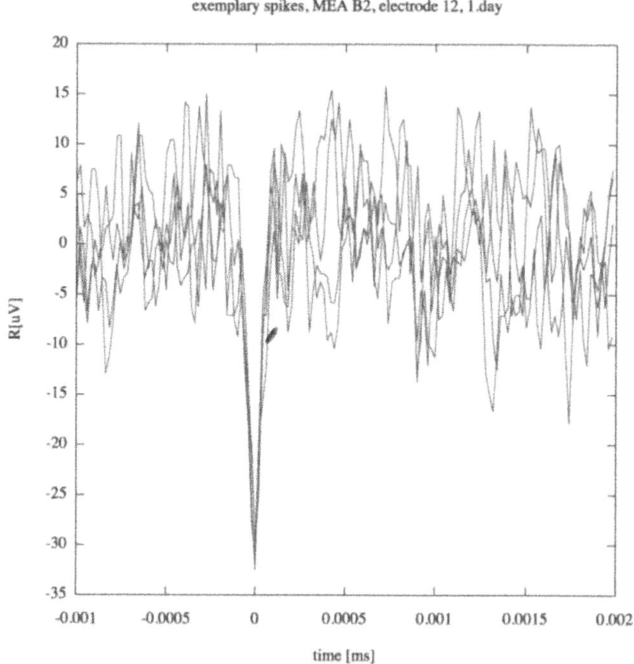

Fig. 6. Spontaneous activity register at a single electrode

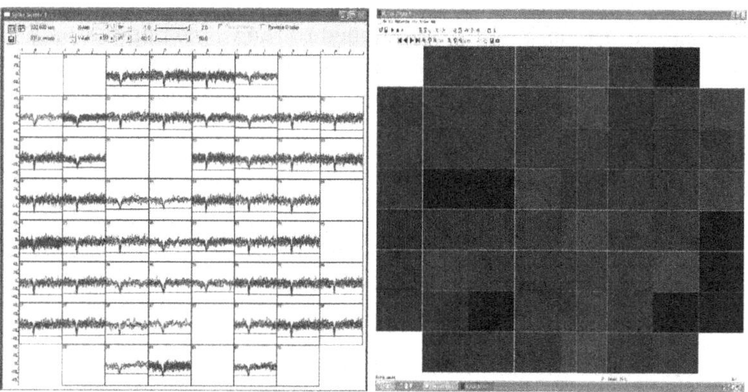

Fig. 7. A: Spontaneous activity register at the multielectrode array. B: Spiking activity grey map

biphasic stimulus consists in a 10 trains of a 100 anodic-first waveform with 1 Volt amplitude delivered to all 60 electrodes in order to propagate a tetanization stimulus to the neuroblastoma culture. The stimulation waveform is shown in Figure 8.

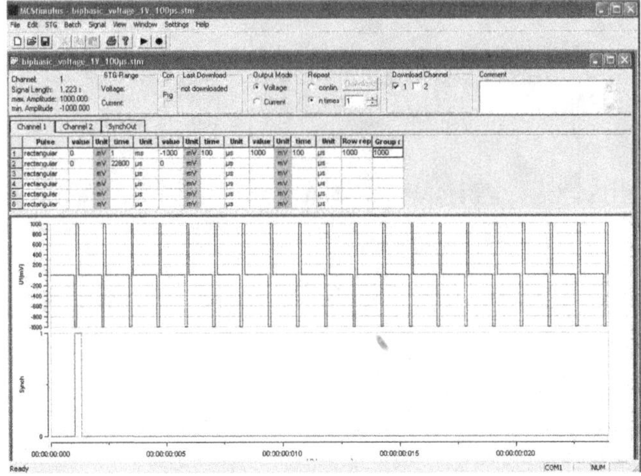

Fig. 8. Tetanization stimuli

In neurobiology, a tetanic stimulation consists of a high-frequency sequence of individual stimulations of a neuron. It is associated with long-term potentiation, the objective of this work. High-frequency stimulation causes an increase in transmitter release called post-tetanic potentiation [11]. This presynaptic event is caused by calcium influx. Calcium-protein interactions then produce a change in vesicle exocytosis. Some studies [12] use repetitive stimulation for training neural cultures, achieving activity potentiatiation or depresion.

Once the tetanization stimulus was applied to the whole population 5 minutes after the stimulation a 3 minutes interval was recorded (POST-data). Only neuronal signals which had at least a 2:1 signal:noise ration were valued as "spikes".

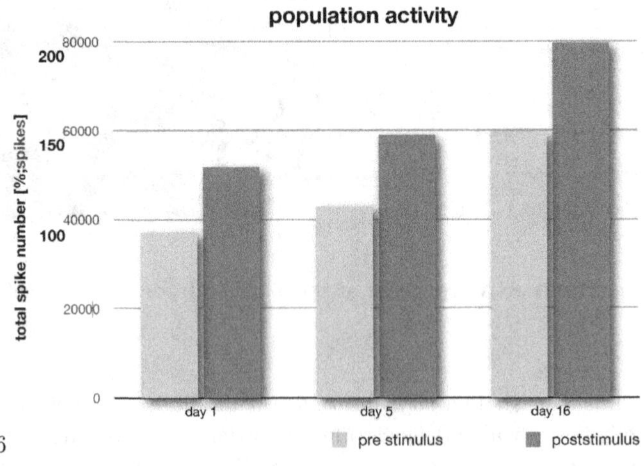

Fig. 9. Modulated activity evoked by tetanization stimuli

Again, the total number of spikes extracted was counted. This process was made for cultures at 1 day in vitro (div), 5 div and 16 div. Figure 9 represents the counted spikes with bar charts for the different recordings. The conclusion from this Figure is: 1) While the neuroblastoma culture is growing new connections are created, and the number of spikes increases as the culture expands over the MEA. 2) After a tetanic stimulation the cells continue with their increased spiking rate, providing a persistent change in the culture behaviour. When this change in the network response lasts, these changes can be called learning. In all the experimentation performed, tetanic stimulation was applied as training method, and the electrophysiological properties of the neuroblastoma culture change, getting a potentiation effect on the spontaneous firing, modulating in this way the culture neural activity.

6 Discussion

Learning in cultured neuroblastoma networks by a stimulation process, without the involvement of a natural adaptation process to the environment requires identifying the correct stimuli to provide to the neurons maintained ex vivo. These neuroblastoma networks form a large culture covering the whole electrode array and generating a rich dendritic configuration. The connectivity can be modulated by external stimulation as has been described in many studies, but also the activity of the network can be modulated with the appropriate stimulation scheme.

Tetanization consists in high-frequency stimulation to the culture, in order to cause an increase in transmitter release called post-tetanic potentiation. The results illustrate the existence of qualitatively different responses to stimulation, ranging from one day in vitro to sixteen days in vitro. Our results indicate the existence of a clear facilitation mechanism in response to the tetanization stimuli at different stages of cell development. Since this kind of stimulation has been used in attempts to induce plasticity in neuroblastoma, refining some crucial aspects of the stimulation is still indispensable.

It is very important to adjust the frequency of the train pulses of the stimulation for suppressing bursting in the culture. While in vivo networks suppress bursting naturally with the tissue development and sensory inputs, ex-vivo cultures need to reduce this synchronized activity by adjusting the stimulation parameters. Also, for superimposing a desired behaviour on the biological networks it is necessary to stimulate locally some part of the culture in order to facilitate some parts of the networks, or achieve some kind of electrical stimulation that depress the local activity of a restricted location. With this local potentiation-inhibition scheme the culture global behaviour could be controlled.

A detailed protocol for causing plasticity implies determining particular patterns of stimulation, which evoke the desired response in the neural cultured platform. The results of this paper allow inducing that there exists a kind of stimulation, tetanic, that elicits activity-induced plasticity at different stages of neuroblastoma development. The results presented imply that neuroblastoma networks display general properties expected from neural systems capable of learning: stablishing numerous and dynamic connections, with modifiability induced by external stimuli.

Future work consists in determine the optimal spatio-temporal stimulation parameters that induce a permanent change in culture behaviour, quantify the time evolution of this changes, and analyse the spatial resolution of the local plasticity. These aspects will then constitute the basis for inducing stable goal-directed plasticity, and hence for designing new biological neuroprocessors.

Acknowledgements

This work was supported by the Spanish Government through grants TIN2008-06893-C03, TEC2006-14186-C02-02 and SAF2008-03694, Cátedra Bidons Egara, Fundación Séneca 08788/PI/08, CIBER-BBN and by the European Comission through the project "'NEUROPROBES"' IST-027017.

References

1. Anderson, J.A., Rosenfeld, E.: Neurocomputing: Foundations of research. MIT Press, Cambridge (1988)
2. The FACETS Project, `http://facets.kip.uni-heidelberg.de/`
3. Ammermuller, J., Kolb, H.: The organization of the turtle inner retina. I. ON- and OFF-center pathways. J. Comp. Neurol. 358(1), 1–34 (1995)
4. Ammermuller, J., Weiler, R., Perlman, I.: Short-term effects of dopamine on photoreceptors, luminosity- and chromaticity-horizontal cells in the turtle retina. Vis. Neurosci. 12(3), 403–412 (1995)
5. Berry, M.J., Warland, D.K., Meister, M.: The Structure and Precision of Retinal Spike Trains. Proc. Natl. Acad. Sci. USA 94(10), 5411–5416 (1997)
6. Secker, H., Searle, C.: Time Domain Analysis of Auditory-Nerve Fibers Firing Rates. J. Acoust. Soc. Am. 88(3), 1427–1436 (1990)
7. Bading, H., Greenberg, M.E.: Stimulation of protein tyrosine phosphorylation by NMDA receptor activation. Science 253(5022), 912–914 (1991)
8. Bongard, M., Micol, D., Fernández, E.: Nev2lkit: a tool for handling neuronal event files, `http://nev2lkit.sourceforge.net/`
9. Bonomini, M.P., Ferrández, J.M., Bolea, J.A., Fernández, E.: RDATA-MEANS: An open source tool for the classification and management of neural ensemble recordings. J. Neurosci. Meth. 148, 137–146 (2005)
10. Wagenaar, D.A., Pine, J., Potter, S.M.: An extremely rich repertoire of bursting patterns during the development of cortical cultures. BMC Neuroscience 7, 11 (2006)
11. Antonov, I., Antonova, I., Kandel, E.R.: Activity-Dependent Presynaptic Facilitation and Hebbian LTP Are Both Required and Interact during Classical Conditioning in Aplysia. Neuron 37(1), 135–147 (2003)
12. Jimbo, Y., Robinson, H.P., Kawana, A.: Strengthening of synchronized activity by tetanic stimulation in cortical cultures: application of planar electrode arrays. IEEE Transactions on Biomedical Engineering 45(11), 1297–1304 (1998)

A Network of Coupled Pyramidal Neurons Behaves as a Coincidence Detector

Santi Chillemi, Michele Barbi, and Angelo Di Garbo

Istituto di Biofisica CNR, Sezione di Pisa,
Via G. Moruzzi 1, 56124 Pisa, Italy
{santi.chillemi,michele.barbi,angelo.digarbo}@pi.ibf.cnr.it
http://www.pi.ibf.cnr.it

Abstract. The transmission of excitatory inputs by a network of coupled pyramidal cells is investigated by means of numerical simulations. The pyramidal cells models are coupled by excitatory synapses and each one receives an excitatory pulse at a random time extracted from a Gaussian distribution. Moreover, each cell model is injected with a noisy current. It was found that the excitatory coupling promotes the transmission of the the synaptic inputs on a time scale of a few msec.

1 Introduction

Let us consider some external sensory event triggering input excitatory signals to some population of pyramidal cells. Then some cells of the population will fire and then the processing of the incoming information starts. In this context an interesting problem is to understand the relationship between the timing of the input signals with that of the action potentials generated by the population of pyramidal cells. In particular - dues to the relevance of synchronization phenomena as a possible mechanism of information coding - it is interesting to study the relationship between the temporal jitter of the action potentials with that of the input excitatory signals. Such problem was addressed by several authors in the case of a single pyramidal cell[1,2,3]. For instance in [1] was shown that under physiological condition the temporal dispersion (or jitter) of the output spikes is linearly related to the input jitter of the input signals. Moreover it was found that the constant of proportionality is less than one. In [3] the same problem was investigated by asking whether cortical neurons behave as integrators or coincidence detectors. It was found that the neuron model is able to operate as an integrator or coincidence detector depending on the degree of synchrony of the synaptic inputs. Moreover, it was shown, in agreement with the findings of [1], that the output jitter is smaller than that of the synaptic inputs. In addition it was found that the presence of synaptic background activity strongly affects the operating mode of the neural model. In the present paper the previous works are extended to the case of a population of coupled pyramidal cells by using a biophysical modeling approach. In particular it is investigated how the transmission properties of the excitatory inputs depend on the parameters describing the network of coupled cells.

J. Mira et al. (Eds.): IWINAC 2009, Part I, LNCS 5601, pp. 155–163, 2009.

2 Methods

2.1 Model Description

It is adopted a single compartment biophysical model of a pyramidal neuron defined in [4]. The $j-th$ cell model of a population of N coupled units reads:

$$C\frac{dV_j}{dt} = I_{E,j} - g_{Na}m_j^3h_j(V_j - V_{Na}) - g_K n_j^4(V_j - V_K) - g_M w_j((V_j - V_M))$$
$$-g_L(V_j - V_L) + g_{exc}\sum_{k=1}^{n_p} P_j(t - t_k) + I_{Sy,j} + \eta\xi_j(t) \tag{1}$$

$$\frac{dm_j}{dt} = \alpha_{m,j}(1 - m_j) - \beta_{m,j}m_j \tag{2}$$

$$\frac{dh_j}{dt} = \alpha_{h,j}(1 - h_j) - \beta_{h,j}h_j, \tag{3}$$

$$\frac{dn_j}{dt} = \alpha_{n,j}(1 - n_j) - \beta_{n,j}n_j, \tag{4}$$

$$\frac{dw_j}{dt} = \frac{w_{j,\infty} - w_j}{\tau_{j,w}}, \tag{5}$$

where $C = 1\ \mu F/cm^2$, $I_{E,j} = I_E (j = 1, 2, ..N)$ is the external stimulation current. The maximal specific conductances and the reversal potentials adopted in this paper are respectively: $g_{Na} = 100\ mS/cm^2$, $g_K = 80\ mS/cm^2$, $g_M = 1\ mS/cm^2$, $g_L = 0.15\ mS/cm^2$ and $V_{Na} = 50\ mV$, V_K = -100 mV, V_M = -100 mV, V_L = - 72 mV. The term $P_j(t - t^*)$ represents an excitatory pulses starting at time t^* and it is defined by: $P_j(t-t^*) = H(t-t^*)\{C_N[e^{-(t-t^*)/\tau_D} - e^{-(t-t^*)/\tau_R}]\}$ where, C_N is a normalization constant ($| P |\leq 1$), $\tau_D = 2ms$ and $\tau_R = 0.5ms$ are, respectively, realistic values of the decay and rise time constants of the excitatory pulse and $H(t - t^*)$ is the Heaviside function. n_p represent the number of excitatory pulses that each pyramidal neuron receive. The rate variables describing the currents are defined by: $\alpha_{m,j}(V_j) = 0.32(V_j + 54)/[1 - exp((V_j + 54)/4)]$,$\beta_{m,j}(V_j) = 0.28(V_j + 27)/[exp((V_j + 27)/5) - 1]$, $\alpha_{h,j}(V_j) = 0.128exp(-(V_j + 50)/18)$, $\beta_{h,j}(V_j) = 4/[1 + exp(-(V_j + 27)/5)$, $\alpha_{n,j}(V_j) = 0.032(V_j + 52)/[1 - exp(-(V_j + 52)/5)]$, $\beta_{n,j}(V_j) = 0.5exp(-(V_j + 57)/40)$, $w_{j,\infty} = 1/[1+exp(-(V_j+35)/10)]$, $\tau_{j,w} = 400/[3.3exp((V_j+35)/20)+exp(-(V_j+35)/20)]$. In this model the onset of periodic firing occurs through an Hopf bifurcation for $I_E \cong 3.25\ \mu A/cm^2$ with a well defined frequency ($\nu \cong 5Hz$).

The starting times $\{t_k : k = 1, 2, ..., Nn_p\}$ of the excitatory currents are extracted from a Gaussian distribution of mean value T_{in} and standard deviation σ_{in}. The value of the parameter σ_{in} quantifies the synchrony level - or equivalently the time dispersion - of the excitatory synaptic inputs. $I_{Sy,j}$ represents the excitatory coupling current that will be defined in the next section. To reproduce the membrane potential fluctuations each $j-th$ cell model is injected with the noisy current $\eta\xi_j(t)$, ξ_j being an uncorrelated Gaussian random variable of zero mean and unit standard deviation $< \xi_i, \xi_j >= \delta_{ij}, i \neq j = 1, 2, 3, , N)$. The values of the parameters $I_{E,j} = I_E(j = 1, 2, ..N)$ and η are chosen so that no firing occurs in absence of the excitatory pulse. Lastly, to get a good statistics the stimulation protocol is repeated $N_{Trials} = 400$ times by using independent realizations of the applied noisy current. For each trial the set of times $\{t_k : k = 1, 2, ..., Nn_p\}$ is updated by using a new realization of the Gaussian distribution from which they are extracted.

2.2 Synaptic Coupling and Pulse Timing

The excitatory synaptic coupling between pyramidal cells is assumed to be all-to-all. Therefore the excitatory synaptic current acting on the $j-th$ cell is defined by

$$I_{Sy,j} = -\frac{1}{N-1} \sum_{k \neq j} g_e s_k(t)(V_j - V_{Rev}) \tag{6}$$

where $g_e = 0.025 mS/cm^2$ represents the maximal amplitude of the excitatory coupling, the function $s_k(t)$ describes the time evolution of the postsynaptic current and V_{Rev} is the corresponding reversal potential. According to [4] the time evolution of $s_k(t)$ is described by

$$\frac{ds_k(t)}{dt} = T(V_k)(1 - s_k) - s_k \tau_s \tag{7}$$

where $T(V_k) = 5(1 + tanh(V_k/4)$ and τ_s is the decay time constant. The value of this last parameter will be assumed to be equal to τ_D (see 2.1).

2.3 Quantification of the Network Response

Let be $\{t_F(j) : j = 1, 2, ...M\}$ the set of all firing times of the network collected for the $N_{Trials} = 400$ trials and corresponding to a given σ_{in} value. Then the corresponding mean value

$$T_{out} = \frac{1}{M} \sum_{k=1}^{M} t_F(k) \tag{8}$$

and corresponding standard deviation

$$\sigma_{out} = \frac{1}{\sqrt{M-1}} \sqrt{\sum_{k=1}^{M} (t_F(k) - T_{out})^2} \tag{9}$$

were estimated. Analogously to σ_{in}, the quantity σ_{out} measures the temporal dispersion of the network response. The network transmission of the information on the timing of the excitatory synaptic inputs is quantified by the relationship between σ_{out} and σ_{in}. In particular the transmission of the timing promotes (does not promotes) synchrony whether the inequality $\sigma_{out} < \sigma_{in}$ ($\sigma_{out} > \sigma_{in}$) holds. When it is $\sigma_{out} > \sigma_{in}$ the transmission of the synaptic inputs introduces more and more jitter, compromising the precision of the response; when $\sigma_{out} < \sigma_{in}$ the temporal dispersion of the response is lower to that of the excitatory inputs and synchronous rhythms are promoted. Lastly, in the following output (input) jitter will be used as a synonymous of $\sigma_{out}(\sigma_{in})$.

3 Results

3.1 Transmission Properties of a Single Pyramidal Cell Model

Let us begin by studying the transmission properties of a single pyramidal cell model receiving excitatory synaptic inputs. To mimic the experimental data the amplitude of each excitatory current is chosen so that the firing probability of

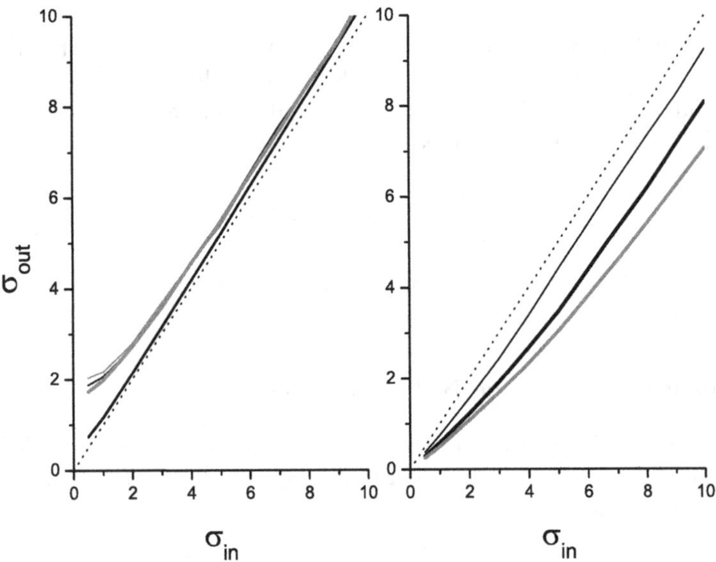

Fig. 1. Output jitter values (σ_{out}) against σ_{in} for a single pyramidal neuron.(**Left panel**) Black thin line: $n_p = 1$, $\eta = 0.2\mu A/cm^2$,$g_{exc} = 2.8\mu A/cm^2$,$I_E = 1.8\mu A/cm^2$,$\tau_D = 2msec$,$\tau_R = 0.5msec$;black thick line: as for the black thin line case, but with $g_{exc} = 3.5\mu A/cm^2$;gray thin line: as for the black thin line case, but with $\eta = 0.4\mu A/cm^2$;gray thick line: as for the black thin line case, but with $I_E = 1.88\mu A/cm^2$. (**Right panel**) For all plots it is $\eta = 0.2\mu A/cm^2$,$g_{exc} = 2.8\mu A/cm^2$,$I_E = 1.8\mu A/cm^2$,$\tau_D = 2msec$,$\tau_R = 0.5msec$ and $n_p = 2$ for the black thin line, $n_p = 3$ for the black thick line, $n_p = 4$ for the gray thick line. For both panels the dotted curves represent the identity line. The results reported in the figure were obtained by using $N_{Trials} = 2000$.

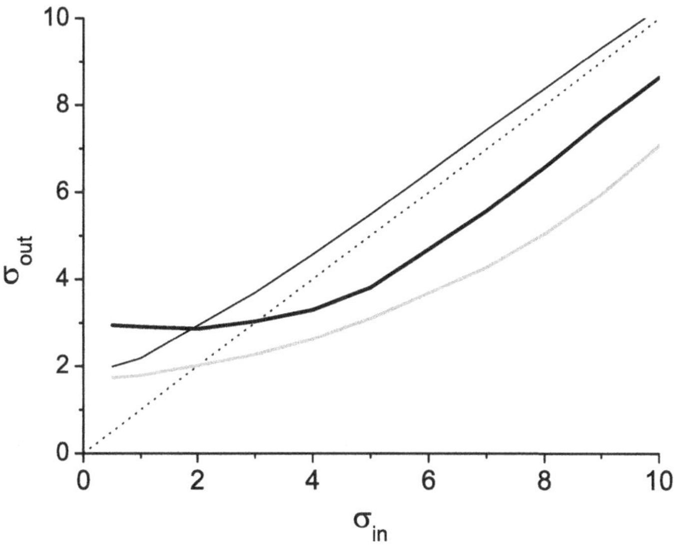

Fig. 2. Output jitter values (σ_{out}) against σ_{in} for a network of $N = 25$ coupled pyramidal neurons models. Black thin line: $n_p = 1$, $\eta = 0.2\mu A/cm^2$,$g_{exc} = 2.8\mu A/cm^2$, $g_e = 0mS/cm^2$, $I_E = 1.8\mu A/cm^2$,$\tau_D = 2msec$,$\tau_R = 0.5msec$; black thick line: as for the black thin line case, but with $g_e = 0.025mS/cm^2$; gray thick line: as for the black thin line case, but with $g_e = 0.05mS/cm^2$. The dotted curve represents the identity line. The results were obtained by using $N_{Trials} = 400$.

the cell model is about $P_F \cong 0.4$. The results obtained in the case $n_p = 1$ are reported in the left panel of figure 1.

To improve the statistic of all simulation with a single neuron model, the number of trial was increased to $N_{Trial} = 2000$. The σ_{out} values are all above the identity line and the corresponding distance decreases for σ_{in} value within 0 and about 5 msec. From general consideration it is expected that for large σ_{in} values the corresponding output jitter value should approach the identity line. However this does not occur for the corresponding data in figure 1. Which is the explanation for this discrepancy? A similar behavior was already found in the case of a single interneuron model in [5] and the corresponding explanation can be found in the cited paper. The increase of the noise amplitude deteriorate the transmission properties of the input jitter, while the opposite effect occurs as the stimulation current is increased ($P_F \cong 0.8$). Instead the increase of g_{exc} strongly impacts the transmission properties of the pyramidal cells. Additional simulations with increasing g_{exc} and I_E values (with the constraint $p_F < 1$) showed that the output jitter values were above the identity line. The stronger effect on the transmission properties of the cell model arises when the number of excitatory pulse (n_p) is increased. The corresponding results are reported on the right panel of figure 1. Now it is $\sigma_{out} < \sigma_{in}$ and therefore a more synchronous

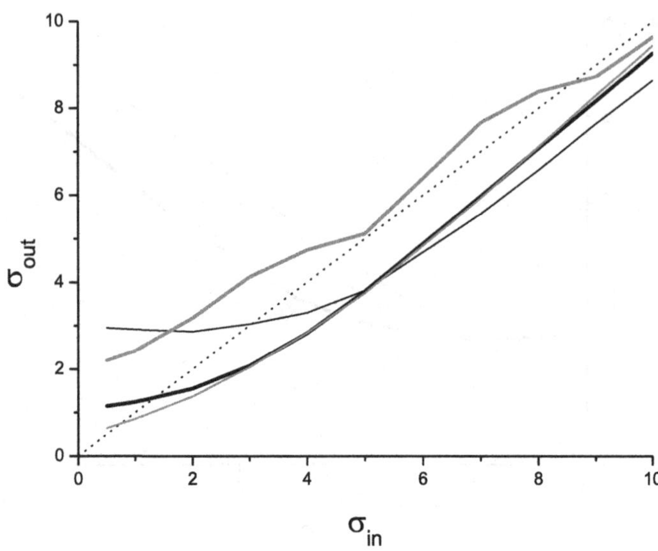

Fig. 3. Output jitter values (σ_{out}) against σ_{in} for a network of $N = 25$ coupled pyramidal neurons models. Black thin line: $n_p = 1$, $\eta = 0.2\mu A/cm^2$,$g_{exc} = 2.8\mu A/cm^2$, $g_e = 0.025mS/cm^2$, $I_E = 1.8\mu A/cm^2$,$\tau_D = 2msec$,$\tau_R = 0.5msec$; black thick line: as for the black thin line case, but with $g_{exc} = 3\mu A/cm^2$; gray thin line: as for the black thin line case, but with $\tau_D = 3msec$; gray thick line: as for the black thin line case, but with $\tau_R = 0.25msec$. The dotted curve represents the identity line. The results were obtained by using $N_{Trials} = 400$.

firing is promoted. Moreover, in agreement with [1] it was found that σ_{in} varies linearly for increasing σ_{in} values.

3.2 Transmission Properties of a Network of Pyramidal Cells Models

Let us now consider a network of $N = 25$ coupled pyramidal cell models. As will be shown later, the simulations with N values ranging from 50 to 100 coupled units give results qualitatively similar to those obtained with $= 25$. Thus, this choice is a compromise between two opposite requirements: an acceptable statistics and computational advantages. In figure 2 are reported the corresponding results for several values of the coupling amplitude g_e. When the excitatory coupling between pyramidal neuron is absent the dependence of the output jitter on σ_{in} look like that obtained in the case of a single cell. But a dramatic change occurs as the value of the coupling amplitude is increased to $g_e = 0.025mS/cm^2$: now there is a range of σ_{in} values where the inequality $\sigma_{out} < \sigma_{in}$ holds.

Therefore, in such cases, the nonlinear network processing of its synaptic inputs leads to a firing activity characterized by a synchrony level higher than that of its excitatory inputs. It is important to remark that this behaviour is

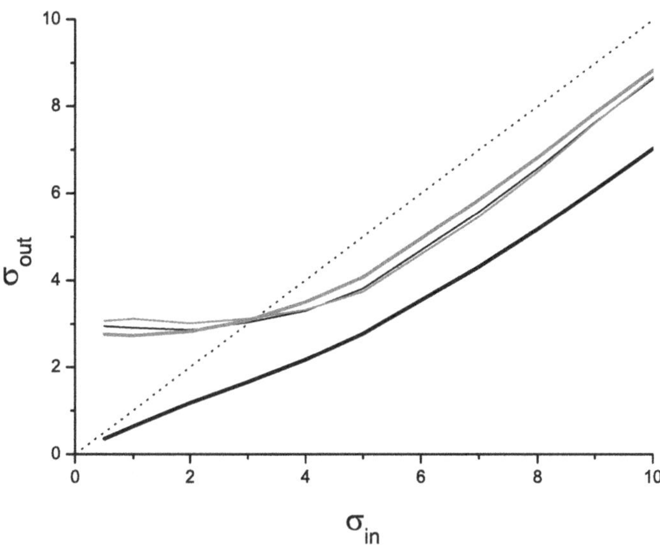

Fig. 4. Output jitter values (σ_{out}) against σ_{in} for a network of $N = 25$ coupled pyramidal neurons models. Black thin line: $n_p = 1$, $\eta = 0.2\mu A/cm^2$,$g_{exc} = 2.8\mu A/cm^2$, $g_e = 0.025mS/cm^2$, $I_E = 1.8\mu A/cm^2$,$\tau_D = 2msec$,$\tau_R = 0.5msec$; black thick line: as for the black thin line case, but with $n_p = 2$; gray thin line: as for the black thin line case, but with $\eta = 0.1\mu A/cm^2$; gray thick line: as for the black thin line case, but with $\eta = 0.4\mu A/cm^2$. The dotted curve represents the identity line. The results were obtained by using $N_{Trials} = 400$.

never observed in the case of a single cell with $n_p = 1$. Thus the coupling between the pyramidal cells models confers to the network the capability of promoting synchrony. How change the results when the amplitude or the time course of the excitatory current are varied? The corresponding simulation results are reported in figure 3. When the amplitude of the excitatory synaptic input is increased to $g_{exc} = 3\mu A/cm^2$ and $\sigma_{in} \in [0, 5ms]$ the output jitter gets values lower than those obtained for $g_{exc} = 2.8\mu A/cm^2$. For $\sigma_{in} < 5ms$ it is $\sigma_{in|(g_{exc}=3\mu A/cm^2)} < \sigma_{in|(g_{exc}=2.8\mu A/cm^2)}$.

Thus, the increase of the amplitude of the pulse promotes the synchronization in the region of small σ_{in} values. The data in figure 3 also shows that the increase of the decay time constant of the excitatory currents strongly impacts the transmission properties of the network by promoting synchronization. A similar effect arises when the value of the rise time constant of the current is increased. As suggested by the results obtained in the case of a single cell [1], an important parameter affecting the network transmission properties should be n_p (i.e. the number of excitatory pulses for each cell). The simulations with a network of coupled cells confirm that and the corresponding findings are reported in figure 4. The comparison of the plots shows that a dramatic change occurs in the

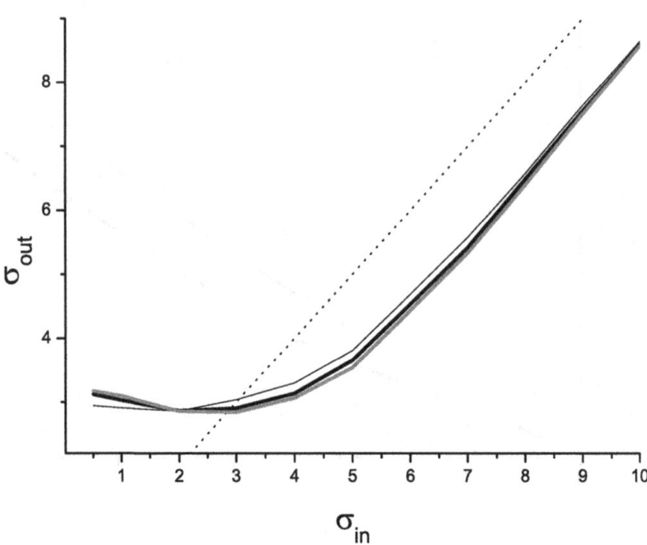

Fig. 5. Output jitter values (σ_{out}) against σ_{in} for a network of coupled pyramidal neurons models. Black thin line: $n_p = 1$, $\eta = 0.2\mu A/cm^2$,$g_{exc} = 2.8\mu A/cm^2$, $g_e = 0.025mS/cm^2$, $I_E = 1.8\mu A/cm^2$,$\tau_D = 2msec$,$\tau_R = 0.5msec$,$N = 25$; black thick line: as for the black thin line case, but with $N = 50$; gray thin line: as for the black thin line case, but with $N = 100$. The dotted curve represents the identity line. The results were obtained by using $N_{Trials} = 400$.

case $n_p = 2$. In figure 4 are also reported the results obtained by using different noise amplitudes: they exhibit a weak dependence on the η values. Additional simulations where performed with $N = 50$ and $N = 100$ to verify whether the previous findings depend of the network size of coupled pyramidal cells. The corresponding results are reported in figure 5 and clearly show that no appreciable changes occurs with respect to the case $N = 25$. The results obtained for $N = 50$ and $N = 100$ by changing the others network parameters were similar to those obtained with $N = 25$ (data not shown).

4 Conclusions

The synchronous firing activity of pyramidal cells seems to be associated to the processing of sensory information and coincidence detection appears as the key mechanism underlying this phenomenon [6]. However, the details and the exact mechanisms that govern these phenomena are not yet fully understood. The topic investigated in this paper concerned the study of how a population of coupled pyramidal cells models transmit its excitatory synaptic inputs. This problem was addressed, both experimentally and theoretically in several papers, but mainly at the single pyramidal neuron level [1,2,3]. Here the investigation was carried out numerically by using a biophysical modeling approach, in which the pyramidal cells models are coupled by excitatory synapses. It was found that the presence of the excitatory synapses plays a key role for the transmission properties of the excitatory inputs: the increase of the amplitude of the corresponding conductance promotes synchronous firing (or an enhanced coincidence detection network capability). Similarly, it was shown that both the increase of the amplitude of the excitatory input current and its time course strongly impact the network transmission properties. On the contrary, it was found that the above results exhibit a weak dependence on the noise amplitude. Instead, the increase of the number of excitatory synaptic inputs that each pyramidal cell model receives has a strong effect on the network transmission properties. Lastly, it was shown that the increase of network size ($N = 50, 100$) leads to results qualitatively and quantitatively similar to those obtained for $N = 25$. Summarizing, the above findings suggest that a network of coupled pyramidal cells behaves as a coincidence detector.

References

1. Marsalek, P., Koch, C., Maunsell, J.: On the relationship between +synaptic input and spike output jitter in individual neurons. Proc. Natl. Acad. Sci. 94, 735–740 (1997)
2. Kisley, M.A., Gerstein, G.L.: The continuum of operating modes for a passive model neuron. Neural Computation 11, 1139–1154 (1999)
3. Rudolph, M., Destexhe, A.: Tuning neocortical pyramidal neurons between integrators and coincidence detectors. J. Comp. Neurosci. 14, 239–251 (2003)
4. Olufsen, M., Whittington, M., Camperi, M., Kopell, N.: New roles for the gamma rhythm: population tuning and preprocessing for the beta rhythm. J. Comp. Neurosci. 14, 33–54 (2003)
5. Di Garbo, A.: The electrical coupling confers to a network of interneurons the ability of transmitting excitatory inputs with high temporal precision. Brain Research 1225, 47–56 (2008)
6. Fries, P., Nikolik, D., Singer, W.: The gamma cycle. Trends Neurosci. 30, 309–316 (2007)

Characterisation of Multiple Patterns of Activity in Networks of Relaxation Oscillators with Inhibitory and Electrical Coupling

Tiaza Bem[1,2] and John Hallam[1]

[1] Mærsk Mc-Kinney Møller Institute, University of Southern Denmark, Odense
[2] Institute of Biocybernetics and Biomedical Engineering, Polish Academy of Sciences, Warsaw

Abstract. Fully-connected neural networks of non-linear half-center oscillators coupled both electrically and synaptically may exhibit a variety of modes of oscillation despite fixed topology and parameters. In suitable circumstances it is possible to switch between these modes in a controlled fashion. Previous work has investigated this phenomenon the simplest possible 2 cell network. In this paper we show that the 4 cell network, like the 2 cell, exhibits a variety of symmetrical and asymmetrical behaviours. In general, with increasing electrical coupling the number of possible behaviours is reduced until finally the only expressed behaviour becomes in-phase oscillation of all neurons. Our analysis enables us to predict general rules governing behaviour of more numerous networks, for instance types of co-existing activity patterns and a subspace of parameters where they emerge.

1 Introduction

The highly interconnected networks of the mammalian brain can generate a wide variety of synchronised[1] activities. Widespread synchrony can emerge in neuronal populations involved in specific functions, depending on the state of the animal [1,5,4] it can also underlie epileptic seizure, which often appears as a transformation of otherwise normal brain rhythms [2].

Networks of inhibitory neurones are often involved in generation of coherent neural activity and they have been extensively studied. Although they have generally complex architecture and a diversity of synaptic interaction, important insights into their underlying mechanisms have been obtain with reduced, idealized models, both experimentally and theoretically. For example, the neural drive for alternately activating cycles of antagonistic muscle groups in locomotion, respiration, etc., is provided by central pattern generator (CPG) networks [20].

[1] *synchrony* has two meanings: it signifies rhythmic activity expressed with the same frequency, all neurons therefore must be phase-locked with some phase (for example equal to 0 — in-phase behaviour — or 0.5 of the cycle out-of-phase behaviour) or it signifies a phase locking equal 0 (i. e. in-phase behaviour). We will use here the term *synchrony* in its former, more general, meaning.

J. Mira et al. (Eds.): IWINAC 2009, Part I, LNCS 5601, pp. 164–173, 2009.

In isolated CPGs and in models for them the half-center oscillator, composed of reciprocally inhibitory pairs of pacemaker neurons plays a critical role [18,19]. Furthermore, they have been suggested to synchronise cells in the γ frequency in hippocampal and neocortical circuits [5,4,3]. Moreover, the dynamics of inhibitory neurones in the reticular nucleus were shown to determine whether the thalamus displays spindle or delta rhythms [6,7] and whether inhibitory neurons burst synchronously [22].

Recently, it has been shown that chemically mediated synaptic inhibition often coexists with electrical coupling mediated by gap junctions in many neuronal systems in the CNS [8,9]. A common view holds that whereas electrical coupling phase-locks the cells with the same phase, synaptic inhibition provides anti-phase co-ordination. However, computational studies suggest that these two coupling forces are not necessarily antagonistic, but on the contrary may act in synergy to enlarge and stabilise various activity patterns [10,11,12].

Indeed varying the strength of electrical coupling between two reciprocally-inhibitory neural oscillator models of short duty cycle (i.e., active for a brief portion of the cycle) leads to a sequence of different rhythmic states [12]. Of special interest, the anti-phase (AP) and in-phase (IP) patterns coexist in the model network. Each can be activated, by different transient, precisely timed stimuli, for the same intrinsic parameter values, thus demonstrating bistability. These theoretical predictions were confirmed by using the dynamic clamp system to introduce and control both electrical and inhibitory synapses between two biological neurones [14,17]. Striking multistability may also occur in large scale networks. Indeed, a 50-cell fully connected model network showed switching between IP and AP patterns in some range of coupling parameters which may be evoked by different and selective transient stimuli [21]. However robustness of this phenomenon was hard to assess because of the variety and complexity of co-existing stable activity patterns. We therefore decided to carry out a detailed study of the dynamics of a fully connected 4 cell network and to compare it with the 2 cell network behaviour. The aim of this study was to understand how multistable activity patterns emerge as a function of coupling parameters, i.e. electrical and inhibitory synaptic strength, as well as of network size.

We show here that 4 cell network, like 2 cell network expresses stable IP as well as AP behaviour which coexist in a range of moderate coupling. Moreover there is a number of asymmetrical states in which individuals or pairs of neurons oscillate in a more complex pattern. In general, with increasing electrical coupling the number of states is reduced until finally the IP state becomes the only expressed behaviour. Our analysis enables us to predict general rules governing behaviour of more numerous networks, for instance types of coexisting activity patterns and a subspace of parameters where they emerge.

2 The Neuron Model

Neurons in the network are modelled as a set of first order differential equations, each neuron contributing two state variables to the set. These state variables model the instantaneous membrane potential (V_i) and a slow recovery current (W_i) dependent on membrane potential. They have the (non-dimensionalised) dynamics defined by equation 1.

$$\frac{d}{dt}\begin{bmatrix} V_i \\ W_i \end{bmatrix} = -\begin{bmatrix} \frac{W_i+V_i-\tanh g^{fast}V_i+I_i^{syn}+I_i^{gap}+I_i^{in}(t)}{\tau_V} \\ \frac{\left(W_i-g^{slow}V_i\right)}{\tau_W(V_i)} \end{bmatrix} \quad (1)$$

In that equation, the parameter g^{fast} models a fast-acting membrane current, g^{slow} models the slow membrane current dynamics, while τ_V and $\tau_W(V)$ determine the time constants of the respective potentials, the latter depending on the instantaneous membrane potential.

The three I_i^x terms represent trans-membrane currents induced by inter-neuron connections, and an externally injected input current $I_i^{in}(t)$. I_i^{gap} models the electrical coupling between neurons via so-called Gap Junctions, which are modelled as constant electrical conductances g^{gap} thus:

$$I_i^{gap} = \sum_{j=1}^{j=N} g^{gap}(V_j - V_i) . \quad (2)$$

Similarly, I_i^{syn} represents the membrane current due to the activity of the inhibitory synaptic junctions,

$$I_i^{syn} = \sum_{j=1}^{j=N} g^{syn}\sigma\left(\frac{V_i - \Theta^{syn}}{k^{syn}}\right)\cdot[V_j - E^{syn}] \quad (3)$$

where the synaptic properties are defined by the basic conductance g^{syn}, the synaptic reversal potential E^{syn}, the synaptic threshold potential Θ^{syn} and the steepness k^{syn} of the synaptic activation curve. The function $\sigma(x) = \frac{1}{1+e^{-x}}$.

Finally, the variable time constant of the slow current process is

$$\tau_W(V) = \tau_2 + (\tau_1 - \tau_2)\cdot\sigma\left(\frac{V}{k^{tw}}\right) \quad (4)$$

with τ_1 and τ_2 specifying the minimum and maximum time constants — and thereby determining the durations of the active and silent phases of the oscillator — and k^{tw} quantifying the rate of voltage dependence.

The complete model defined by equations 1–4 has, therefore, three time constants, two membrane and two junction conductances, synaptic threshold and reversal potentials and two rate constants (the k parameters). In principle, the junction parameters can vary per junction while the neuron parameters may vary per neuron; for tractability in the current work these parameters are identical for all junctions and neurons respectively, with the values given in Table 1.

The parameters have been chosen to model neurons with relatively fast synaptic onset and short duty cycle, i.e., the fraction of the cycle when the cell is depolarised above threshold (its active phase) and may exert synaptic action.

Table 1. Parameter Values Used in the Simulations

Param.	Value	Param.	Value	Param.	Value	Param.	Value
g^{syn}	varies	g^{gap}	varies	g^{fast}	2	g^{slow}	2
E^{syn}	-4	Θ^{syn}	0	k^{syn}	0.02	k^{tw}	0.2
τ_v	0.16	τ_1	5	τ_2	50		

With this choice of parameters, equations 1–4 may be considered a model of spiking neurons. In the study presented here, the only parameters varied from the defaults given in Table 1 are the conductances g^{gap} and g^{syn} of the gap and synaptic junctions.

The model has been implemented as a set of Matlab functions which compute the quantities defined by the four equations above and integrate the set of ordinary differential equations using Matlab's standard `ode45` solver. The implementation has been compared to an independent realisation using the `xpp` tool and found to give identical results. The external input currents $I_i^{in}(t)$ are assumed to be piecewise constant and the equations are integrated over each piece of the input function in turn.

3 Analysis Methods

The investigation of the oscillatory behaviours generated by networks of the type under study is a time-consuming business and has been automated as much as possible. For a given choice of conductance parameters, the network dynamics are integrated from a suitable starting state. The first 30% of the simulation is discarded to mitigate the effects of transients and the membrane potentials are computed at time points with an interval of 0.2 units. Given these values, an attempt is made to estimate a period of regular oscillation for the network by computing the positions of peaks in the autocorrelation of the signals and looking for recurring inter-peak periods. If this calculation fails, the network is simulated further and the calculation repeated. If no period can be found with simulations up to 3000 units in duration, the signals are reported to be unanalysable. (The typical period of oscillation for the parameters used is 20–25 time units.)

Once a period has been determined, the network signals are analysed. Samples for a single period are generated and the traces of the individual cells compared, to group cells into classes executing the same behaviour with possibly differing phases. The networks studied here exhibit a single class of behaviour within the resolution of this test, which does not distinguish the small differences in dynamics between a neuron that spontaneously spikes and one that is triggered by a spike from one of its peers.

Once the cell behaviours have been grouped, the classification of oscillatory modes is performed. The analysis distinguishes the behaviours illustrated in Figure 1:

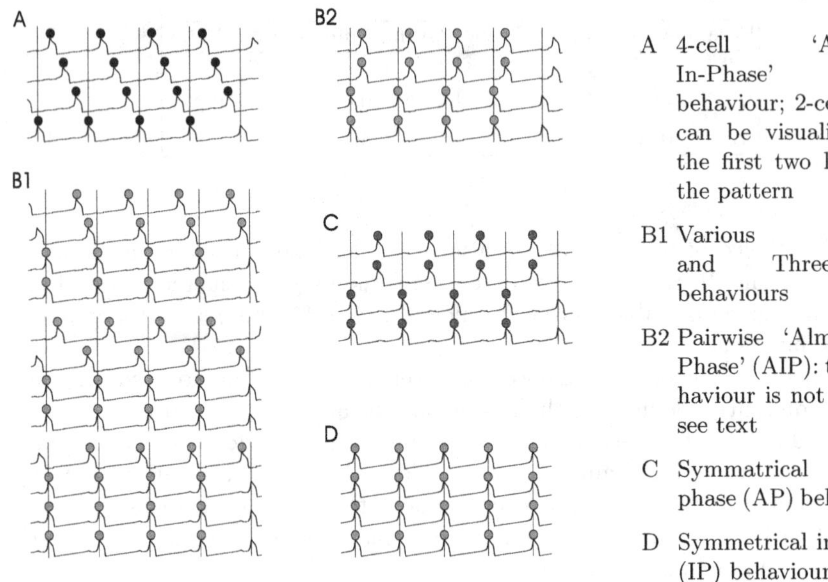

Figure Legend:

A 4-cell 'Almost-In-Phase' (AIP) behaviour; 2-cell AIP can be visualised as the first two lines of the pattern

B1 Various Two- and Three-Phase behaviours

B2 Pairwise 'Almost-In-Phase' (AIP): this behaviour is not stable, see text

C Symmatrical anti-phase (AP) behaviour

D Symmetrical in-phase (IP) behaviour

Fig. 1. Illustrating the Various Oscillatory Behaviours of a 4 Cell Network

In-phase (**IP**) behaviour, in which all cells have identical membrane potentials at all times [green];

Anti-phase (**AP**) behaviour in which there are two groups of cells whose membrane potential traces match but differ in phase by 180 degrees [red];

Two-phase behaviour apart from anti-phase: either two equal groups with less than 180 degree phase difference, or unequal groups of 1 and 3 cells [light-blue];

Three-phase behaviours (4-cell network only) with phase groups of cells of size 1, 1 and 2 [light-blue];

Four-phase behaviour (4-cell network only) with four phase 'groups' each containing a single cell [dark-blue].

The presence of oscillatory behaviours of these types is indicated in the main results figures below by a triangle of the colour mentioned above. Signals that cannot be analysed (because no period can be computed) are indicated by a [yellow] triangle.

3.1 Probing the Behaviours

For a given choice of parameters, the network exhibits a number of oscillatory behaviours. In this study, we vary the two principal conductance parameters g^{gap} and g^{syn}, over the range in which interesting behaviours occur, for a network comprising 2 cells and one comprising 4 cells. The behaviours found for each tested parameter combination are reported in Figures 2 and 3.

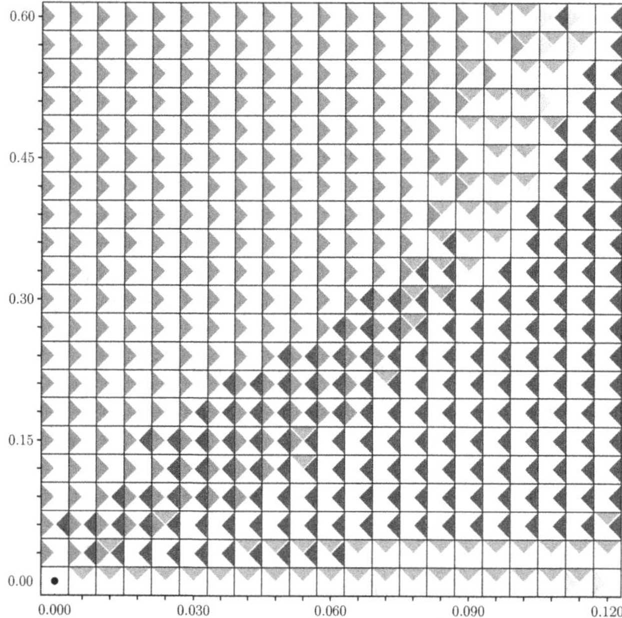

Figure Legend

Fig. 2. Occurrence of 2 Cell Oscillatory Behaviours for Differing Conductance Values

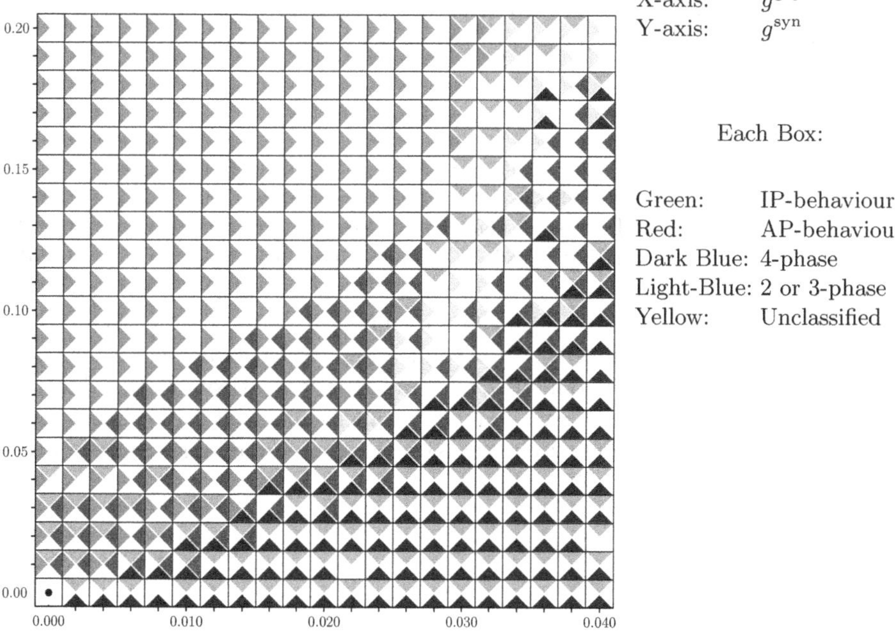

Fig. 3. Occurrence of 4 Cell Oscillatory Behaviours for Differing Conductance Values

The reported results generated as follows. For each pair of parameter values investigated

1. a set of 8 random initial states for the network are generated, using a zero mean Gaussian distribution with 0.025 standard deviation. The model is then integrated and the behaviour classified.
2. the model is integrated from a zero initial state, and the generation of IP behaviour is checked. If IP is generated, an attempt is made to switch the network to AP: half the cells receive a 1 unit positive current injection and half a 1 unit negative current injection for 0.2 time units, applied at varying times close to the peak of the active phase of the cells. The resulting behaviour is classified.
3. the model is integrated from the zero intial state as before, and the switching stimulus is applied mid-way through the cycle of oscillation. Again, the resulting behaviour is classified.

The reported set of behaviours is the union of the results of the three steps above.

This procedure, in our opinion, represents a reasonable compromise between computational effort and completeness of the results. Note, however, that it is *not* complete: the presence of any particular type of behaviour other than AP and IP is only detected if a suitable initial state is chosen (one that lies within the basin of attraction of that behaviour) and this cannot be guaranteed. The two special tests for the stability of AP, steps 2 and 3 above, are used because the initial states chosen in step 1 are close to the symmetric zero state and, when electrical coupling is large, all tend to fall in the basin of attraction of IP.

4 Results

As mentioned above, the results are presented in Figures 2 and 3. The general trend apparent in both figures is that fewer, simpler and more symmetrical, stable states exist as electrical coupling strength increases. A second key result is that in both networks a moderately large region in which only IP and AP behaviours are stable can be seen. The coupling strength parameters in the figures are scaled so that cells' total connection strengths are equivalent for corrsponding boxes in the two figures, allowing easy comparison. Notice that the region of AP occurrence in 4 cells is much smaller compared to 2 cell network. The reason is instability of a 2 cluster AIP solution containing 2 cells in each cluster. See below for discussion of this.

5 Discussion

For dominating inhibitory coupling, the 4 cell network expresses variety of activity patterns which are asymmetrical. The patterns consist of 2, 3 or 4 phases in which activity is expressed by individuals or by groups of 2 or 3 cells. It is important to note here that the 2 cell network also expresses only an asymmetrical solution (i.e. AIP) unless electrical coupling is involved. Indeed, when coordinated in the AIP pattern the two cells have different trajectories in the phase-space and different mechanisms for the transition from silent to active phase: the leading cell behaves as a free cell whereas the following cell is released from inhibition [12]. We believe that none of these mechanisms can provide a stable IP solution. Therefore, in the 4 cell network when 2 to or 3 cells are grouped together (i.e. coordinated in-phase), a transition from silent to active phase may occur through an escape from inhibition which can stabilize the IP behaviour of the group. This however requires a leading cell which provides the inhibition for this group. Interestingly, and in accordance with this hypothesis, stable asymmetrical patterns in 4 cell network always contain a single leader, i. e. a group is never leading.

As in the 2 cell network, IP and AP patterns can be achieved only if electrical coupling is present. However the emergence of the AP is different: whereas in the 2 cell network adding electrical coupling provided a transition from AIP to AP solution, here the two phasic AIP behaviour (consisting of 2 groups of 2 cells) is not stable for the reasons described above. Instead, 4 phasic AIP is present and this solution is transformed into AP with increasing electrical coupling. Since AP (consisting of 2 groups of 2 cells) is symmetrical, its stability cannot be provided by escape from inhibition as in the asymmetrical case of leader and follower described above. Only electrical coupling can stabilize a 2 cell cluster and thereby the AP solution. On the other hand, stability of 2 cell cluster was not necessary for the existence of the AP solution in the 2 cell network. Therefore emergence of AP in the 4 cell network requires a stronger electrical coupling then in the 2 cell network – compare Figures 2 and 3.

Although we do not analyse here the switching between different patterns (this will be described in the next paper) it is interesting to consider putative functions of such switching from the perspective of basic neuroscience and engineering.

Multistable Networks in Neuroscience: The occurrence of abnormal dynamics in physiological systems can become manifest as a sudden change in the behavior of characteristic physiological variables. This was proposed as a model of what happens to the brain with regard to epilepsy [13]. If the neural network involved in epilepsy possesses multistable dynamics it may display several dynamic states and sudden transitions between normal and abnormal behaviour are possible. Our findings suggest that such multistability may indeed occur in large-scale networks due to co-existence of electrical coupling and synaptic inhibition.

Multistable Networks in Engineering: Oscillating neural networks are well known in Nature and in Robotics as appropriate controllers for gaits such as walking, running and swimming, where the complex dynamical properties of a soft body in motion can entrain to oscillations of the controlling network. Study of such networks offers insight into animal locomotion [15] as well as interesting possibilities for novel controllers of engineering systems such as wave-power converters [16] or for biologically-inspired control of robot walking and running (for example over rough terrain).

The networks studied here are expected to be able to express a variety of sequences of activity, which could be used for control and switching of the order of events as well as for synchronisation with external event sequences. Such systems could be useful for sequence control, or as substrates for connectionist theories of how the brain deals with logical knowledge and inference using synchronised sequential firing of neurons.

One of the problems with neural network circuits in silicon is pin out: a network with n neurons has $O(n^2)$ connections whose properties have to be controllable from outside. Depending on the implementation, this need can result in routing problems inside the chip and interface problems with external electrical systems. Multistable networks have the useful property that they can express multiple stable states with *no* changes to topology or parameters. We believe that the number of stable states may grow factorially with the number of cells simulated, making such networks efficient in silicon implementations.

6 Conclusions

An analysis of the expression of different oscillatory behaviours in a fully-connected 4-cell neural network incorporating both electrical and synaptic inhibitory coupling has been presented. The 4-cell network, like its 2-cell counterpart, exhibits a moderate range of coupling parameters within which both in-phase and symmetrical anti-phase behaviours are stable. When inhibitory coupling dominates over electrical coupling, more complex multiphase patterns can be found. Although not shown here, it is easy to demonstrate fast switching of network behaviour between stable states, given suitably timed stimuli: thus a fixed topology fixed parameter network can express a controllable variety of oscillatory behaviours and may be suitable for modelling neural phenomena such as epilepsy or for control of movement sequences.

References

1. Riehle, A., Grun, S., Diesmann, M., Aersten, A.: Spike synchronization and rate modulation differently involved in motor cortical functions. Science 278, 1950–1953 (1997)
2. McCormick, D.A., Contreras, D.: On the cellular and network basis of epileptic seizures. Annu. Rev. Physiol. 63, 815–846 (2001)
3. Traub, R.D., Jeffreys, J.G.R., Whittington, M.A.: Simulations of gamma rhythms in networks of interneurons and pyramidal cells. J. Comp. Neurosci. 4, 141–150 (1997)
4. Wang, X.J., Buzsaki, G.: Alternating and synchronous rhythms in reciprocally inhibitory model networks. Neural Comp. 4, 84–97 (1992)
5. Whittington, M.A., Traub, R.D., Jeffreys, J.G.R.: Synchronized oscillations in interneuron networks driven by metabotropic glutamate receptor activation. Nature 373, 612–615 (1995)
6. Terman, D., Bose, A., Kopell, N.: Functional reorganization in thalamocortical networks: Transitions between spindling and delta sleep rhythms. Proc. Natl. Acad. Sci. USA 93, 15417–15422 (1996)

7. Rubin, J., Terman, D.: Geometric analysis of population rhythms in synaptically coupled neuronal networks. Neural Comp. 12, 597–645 (2000)
8. Galaretta, M., Hestrin, S.: A network of fast-spiking cells in the neocortex connected by electrical synapses. Nature 402, 72–75 (1999)
9. Gibson, J.R., Belerlein, M., Connors, B.W.: Two networks of electrically coupled inhibitory neurons in neocortex. Nature 402, 75–79 (1999)
10. Traub, R.D., Kopell, N., Bibbig, A., Buhl, E.H., Le Beau, F.E.N., Wittington, M.A.: Gap junctions between interneuron dendrites can enhance synchrony of gamma oscillations in distributed networks. J. Comp. Neurosci. 21, 9478–9486 (2001)
11. Lewis, T., Rinzel, J.: Dynamics of spiking neurons connected by both inhibitory and electrical coupling. J. Comp. Neurosci. 14, 283–309 (2003)
12. Bem, T., Rinzel, J.: Short duty cycle destabilizes half-center oscillator but gap junctions can restabilize the anti-phase pattern. J. Neurophysiol. 91, 693–703 (2004)
13. Lopez da Silva, F.H., Blanes, W., Kalitzin, S.N., Parra, J., Suffczynski, P., Velis, D.N.: Epilepesis as dynamic deaseas of brain systems: basic models of the transitions between normal and epileptic activity. Epilepsia 44, 72–83 (2003)
14. Bem, T., Le Feuvre, Y., Rinzel, J., Meyrand, P.: Electrical coupling induces bistability of rhythms in networks of inhibitory spiking neurons. European J. Neurosci. 22, 2661–2668 (2004)
15. Hallam, J., Ijspeert, A.J.: Using evolutionary methods to parameterize neural models: a study of the lamprey central pattern generator. In: Duro, R.J., Santos, J., Grana, M. (eds.) Biological Inspired Robot Behavior Engineering. Studies in Fuzziness and Soft Computing, vol. 109. Springer, Heidelberg (2002)
16. Mundon, T., Murray, A.F., Hallam, J., Patel, L.N.: Causal neural control of a latched ocean wave point absorber. In: Duch, W., Kacprzyk, J., Oja, E., Zadrożny, S. (eds.) ICANN 2005. LNCS, vol. 3697, pp. 423–429. Springer, Heidelberg (2005)
17. Merriam, E.B., Netoff, T.I., Banks, M.I.: Bistable Network behaviour of Layer-I Interneurons in Auditory Cortex. J. Neurosci. 25(26), 6175–6186 (2005)
18. Selverston, A.I., Moulin, M.: Oscillatory neural networks. Ann. Rev. Physiol. 47, 29–48 (1985)
19. Rowat, P., Selverston, A.I.: Modeling the gastric mill central pattern generator of the lobster with a relaxation-oscillator network. J. Neurophysiol. 70, 1030–1053 (1993)
20. Cohen, A.H., Rossignol, S., Grillner, S. (eds.): Neural control of rhythmic movements of vertebrates. Wiley, New York (1988)
21. Bem, T., Hallam, J., Meyrand, P., Rinzel, J.: Electrical coupling and bistability in inhibitory neural networks. Biocybernetics and Biomedical Engineering 26, 3–14 (2006)
22. Goloub, D., Wang, X.J., Rinzel, J.: Synchronization properties of spindle oscillations in a thalamic reticular nucleus model. J. Neurophysiol. 72, 1109–1126 (1994)

Scaling Power Laws in the Restoration of Perception with Increasing Stimulus in Deficitary Natural Neural Network

Isabel Gonzalo-Fonrodona[1] and Miguel A. Porras[2]

[1] Departamento de Óptica, Facultad de Ciencias Físicas, Universidad Complutense de Madrid, Ciudad Universitaria s/n, 28040-Madrid, Spain
igonzalo@fis.ucm.es

[2] Departamento de Física Aplicada, ETSIM, Universidad Politécnica de Madrid, Rios Rosas 21, 28003-Madrid, Spain

Abstract. Measurements of the restoration of visual and tactile perceptions with increasing stimulus, carried out by Justo Gonzalo (1910-1986) in patients with lesions in the cerebral cortex, constitute exceptional examples of quantification of perception. From an analysis of the data for different types of stimulus, we find that, at high enough intensity of stimulus, perception follows scaling power laws with dominance of quarter exponents, which are similar to the scaling laws found in the improvement of perception by multisensory facilitation, reflecting general mechanisms in the respective neural networks of the cortex. The analysis supports the idea that the integrative cerebral process, initiated in the projection path, reaches regions of the cortex of less specificity.

1 Introduction

Some lesions in the cerebral cortex, far from the projection paths, originate a deficit in the cerebral excitation that results in an incomplete integration in the cerebral processing. This is the case of the so-called central syndrome (or symmetric multi-sensory syndrome) as a result of an unilateral parieto-occipital lesion equidistant from the visual, tactile and auditory projection areas. This syndrome is featured by [1,2,3] (a) multisensory affection with symmetric bilaterality, (b) functional decomposition of perception, in the sense that sensory qualities are gradually lost in a well-defined order as the stimulus diminishes, or as the magnitude of the lesion grows, and (c) capability to improve the perception by increasing the intensity of the stimulus, or by multisensory facilitation, as for instance, strong muscular stress. The syndrome was interpreted [1,2,3] as a deficit of cerebral integration due to a deficit of cerebral nervous excitation caused by the loss of a rather unspecific (or multi-specific) neural mass. This interpretation arises from a model of functional gradients, where sensory functional densities for each sensory system are distributed in gradation through the cortex [3,4,5,6]. This accounts for multisensory interactions, which is a requirement formulated recently by some authors (e.g., [7,8]). Some works are devoted

J. Mira et al. (Eds.): IWINAC 2009, Part I, LNCS 5601, pp. 174–183, 2009.

to this syndrome and the model [9,10,11,12,13,14], or are closely related to it (e.g., [15,16,17]).

The interest of this syndrome is that the cerebral organization remains similar to that of a normal man except for a deficit of nervous excitability (scale reduction). Normal perceptions of all or nothing in the normal man, present instead intermediate stages of perception according to the intensity of the stimulus, which reveals a series of sensory levels in the sensory organization. This is the case of the striking phenomenon of inverted or tilted perception disorder (interpreted as an stage of incomplete cerebral integration). These intermediate stages were observed, could be quantified [1,3], and some of them are analyzed in this work.

In another context and for a normal man, Stevens [18] formulated his well-known physiological relation between sensation or perception P and the physical intensity of a stimulus S, expressed mathematically as a power law of the type

$$P = pS^m \,, \tag{1}$$

where p is a constant and the exponent m depends on the nature of the stimulus. This law is regarded as more accurate than the logarithmic Fechner's law ($P \propto \log S$), but is not exempt from criticism.

The stimulus induces nervous impulses which originate a cerebral excitation, which in essence can be assimilated to the activation of a number of neurons. As exposed in previous works, a reasonable assumption is that the activated neural mass follows a power law with the stimulus intensity with exponent close to unity [19,14], i.e., an approximate linear relation. In this way, the perception becomes described by a power law with respect to the activated neural mass, and then Stevens' law a manifestation of the scaling power laws of biological neural networks.

In all biological systems, the allometric scaling power laws are supposed to arise from universal mechanisms, as the optimization to regulate the activity of its subunits, as cells, leading to hierarchical fractal-like branching networks [21], or other type of network structure [22]. Examples are the animal circulatory, respiratory, renal and neural systems, the plant vascular system, etc. Most of exponents in power laws of biological variables versus the mass of the organism are surprisingly found to be multiples of the power $1/4$, as the metabolic rate ($n \simeq 3/4$), lifespan ($1/4$), growth rate ($-1/4$), height of trees ($1/4$), cerebral gray matter ($5/4$), among many others (see [23,24] and references therein, and also [25,26]). The quarter-power allometric laws can be theoretically derived from simple dimensionality reasonings that derive from the geometrical constraints inherent to the networks.

In previous works, data of restoration of perception in patients with central syndrome by facilitation by another stimulus (e.g., muscular stress), were found to fit accurately to a power law (exponents $1/4$ and $1/8$ in different cases) [14,5]. However, the data of perception of a stimulus as a function of the intensity of *that stimulus* were observed to follow a more complex dependence than a single power law. The significance of this fact is discussed in this work.

Here we analyze data of perception of a stimulus versus its intensity. For different types of visual and tactile stimulus, we find that for sufficiently high intensity of the stimulus, the data fit to power laws with dominance of quarter exponents, as in the case of improvement of perception by multisensory facilitation. However, for low stimuli, the data deviates significantly from this power law, fitting better to another power law. The analysis supports the idea that a very weak stimulus would activate only specific neurons which, being deficitary, produce an incomplete perception. Higher stimuli appear to be able to activate neurons of less specificity so as to improve the perception in the same way as facilitation does.

2 Improvement of Perception by Increasing Stimulus in a Deficitary Cerebral System

We analyze data from the measurements made by Gonzalo [1] in the acute case (called M) of central syndrome described above, where the deficit of cerebral excitability is due to the loss of rather unspecific neural mass. Most of data refer to the peculiar phenomena of inverted or tilted perception in visual and tactile systems. In all experiences the stimulus was acting until a stationary perception was reached.

The values perception-stimulus are represented in log–log scale, so that a power law is represented by a straight line with slope equal to the exponent. Note that the exponent, and hence the slope, is independent of the units used and other proportionality factors.

Contrary to the improvement of perception by facilitation, the improvement with increasing stimulus is seen to be well-described by two power laws, represented by two straight lines, in the limits of very low and very high stimulus (see Figs. 1,2,4 and 5) These two branches can also be appreciated, in general, in a representation of perception versus logarithm of the stimulus (Fechner type). We choose, however, the log–log representation because we consider that scaling power laws have more physical and biological background in relation to the laws of biological growth and the dynamics of neural networks.

In order to account mathematically for two limiting power laws in a single expression we use a fitting function of the type

$$P = \left[1 - \exp\left(-\frac{p_1}{p_2} S^{m_1 - m_2}\right)\right] p_2 S^{m_2} , \tag{2}$$

which represents a fast growth of perception (high m_1) at small stimuli, saturating into a slower growth (small m_2) at large stimuli. For very low intensities Eq. (2) reduces to the power law

$$P = p_1 S^{m_1} , \tag{3}$$

and for higher values, to the power law

$$P = p_2 S^{m_2} . \tag{4}$$

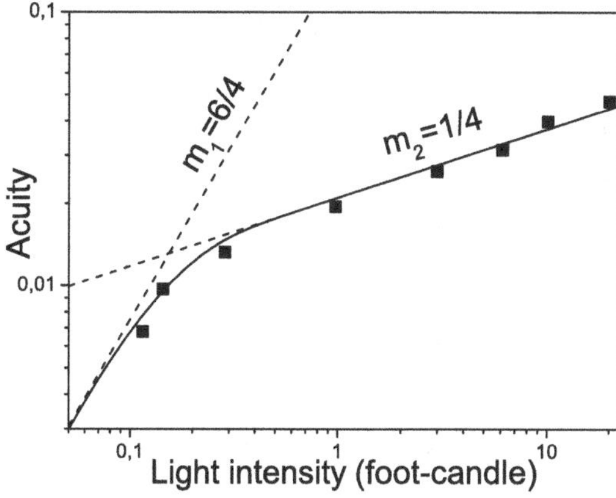

Fig. 1. Acuity of right eye of M (facilitated by a constant muscular stress), versus light intensity . Data from [1].

First we focus on the visual system. For visual acuity we found in previous works [14,5] that the data of acuity in central vision versus light intensity, fitted to a power law with exponents 1/4 in two cases (called M and T) and 1/8 in normal man. However, when data of acuity at very low light intensity are available, as in case M (facilitated by some muscular stress), fitting with a single power law fails in low intensity range, as seen in Fig. 1. Instead, the two power laws of Eq. (2) fit much better (see Fig. 1), with exponents $m_1 = 6/4$ for low intensity and $m_2 = 1/4$ for high intensity.

The next example deals with the perception of a vertical upright white arrow by M in central vision and being in inactive state (free of sensory facilitation). As was explained [1,3,6], with sufficiently high illumination, the perception of a vertical upright white arrow was upright, well-defined and with a slight green tinge. As lighting was reduced, the arrow was perceived to be more and more rotated in the frontal plane, at the same time that became smaller and loss its qualities of form and color in a well-defined physiological order. The phenomenon was reversible by increasing the stimulus intensity. The tilt was measured by rotating the arrow in the opposite direction until it was seen upright (180°). The direction perceived by M as a function of the light intensity that illuminates the arrow is shown in Fig. 2. For the case of M inactive, the fitting with Eq. (2) gives $m_1 = 1$ for low light intensity and $m_2 = 1/8$ for high intensity. For M facilitated by strong muscular stress a good fitting is possible by a single straight line with slope $m = 1/8$.

If the test object was situated in one side of the visual field, the object was seen to rotate with centripetal deviation, coming to rest, inverted and constricted, in contralateral position quite close to the center of the visual field.

In the tactile system, similar inversion mechanism to that in vision were described [1,3,6]. For a mechanical pressure stimulus on one hand in case M, five phases in the dissociated perception were distinguished successively as the

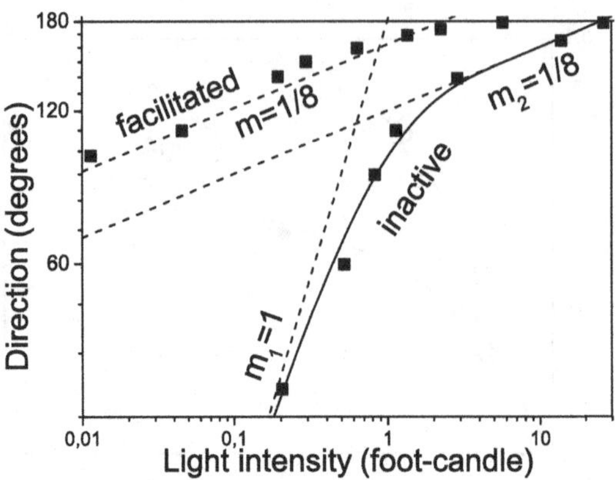

Fig. 2. Perceived direction by patient M (right eye, central vision) of a vertical white 10 cm size test arrow, versus light intensity illuminating the arrow on a black background, at 40 cm patient-arrow distance; for inactive state and facilitated state by strong muscular stress. Data from [1].

energy of the stimulus was increased (Fig. 3): I, primitive tactile sensation without localization; II, perception in the middle of the body with irradiation (spatial diffusion, similar to chromatic irradiation); III, inversion phase (contralateral localization) but closer to the middle line of the body than the stimulus (sensory field constriction); IV, homolateral phase; V, normal localization, which required intense stimulus, or moderate stimulus and facilitation by muscular stress, for example. This phenomenon is similar to visual inversion when the test object was situated to one side of the visual field, as mentioned above, and illustrates the spiral development of the tactile field [2,3].

We analyze first the deviation of the perception towards the middle line of the body. As shown in Fig 3, the perception in phase II is highly delocalized and centered at the middle line of the body (completely reduced tactile field). The perception in phase III is contralateral and at a certain distance from the middle line (a more developed tactile field). In phase IV, the perception becomes homolateral and still farther from the middle line, until it reaches the normal localization on the hand in phase V. Measurements of the deviation versus intensity of the stimulus were made with the arms separated from the body and perpendicular to the middle line. They are seen in Fig. 4 to follow a similar trend as in Fig. 2, i.e., a double-power-law. For M inactive (free from facilitation), fitting with Eq. (2) gives $m_1 = 4$ for low intensity of the stimulus, and $m_2 = 1/4$ for higher intensity. For M facilitated by muscular stress, the asymptotic slope for high stimulus is instead $m_2 = 1/8$.

Next we consider a similar experience but now the stimulus is not punctual, but a 6 cm long line. In this case, the perception has a new component to be added to the previous description, namely, the tilted perception of the orientation of the line stimulus, depending on the stimulus intensity (pressure of the line stimulus). We consider, as in inverted vision, that a tactile direction of 180°

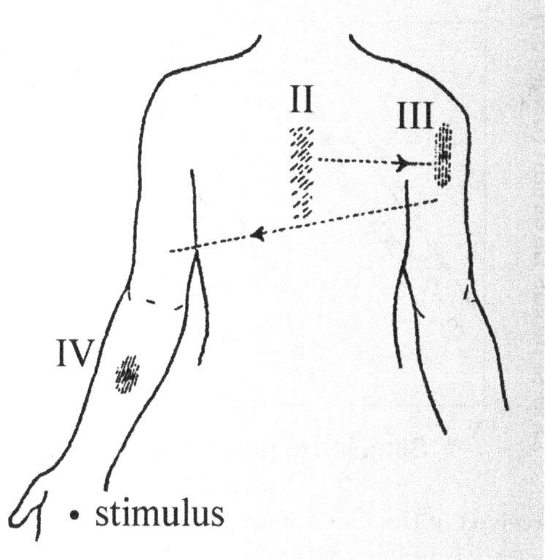

Fig. 3. Phases of tactile localization of a punctual stimulus on the hand of case M inactive (see the text) according to the intensity of the stimulus. The irradiation extension is shown by a dashed pattern [1]. Adapted from figure 2 (a) of [10], with permission of Springer Science and Business Media.

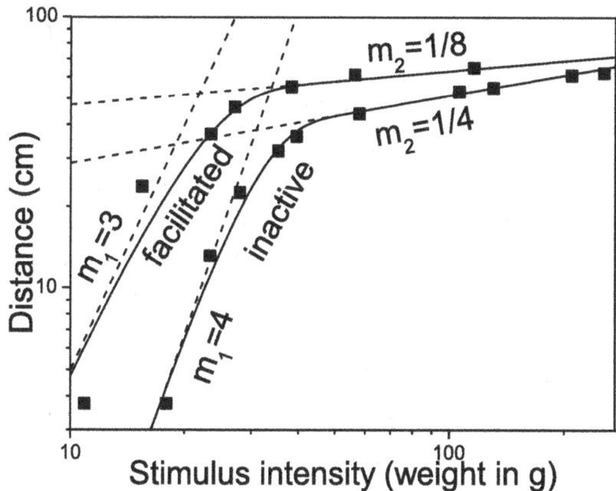

Fig. 4. Distance in cm from the the middle line of the body to the localization of perception, as a function of the intensity (weight in g) of a punctual stimulus on the hand of case M, in inactive and facilitated states. Data from [1].

means a correct (restored) perception of the line direction. The experimental data, showing a similar trend as before, and their fitting with Eq. (2) are plotted in log–log scale in Fig. 5. For case M inactive we obtained $m_1 = 3$ for low stimulus, and $m_2 = 1/4$ for high stimulus. For case M facilitated by muscular stress we obtained $m_1 = 2$ and and $m_2 = 1/8$.

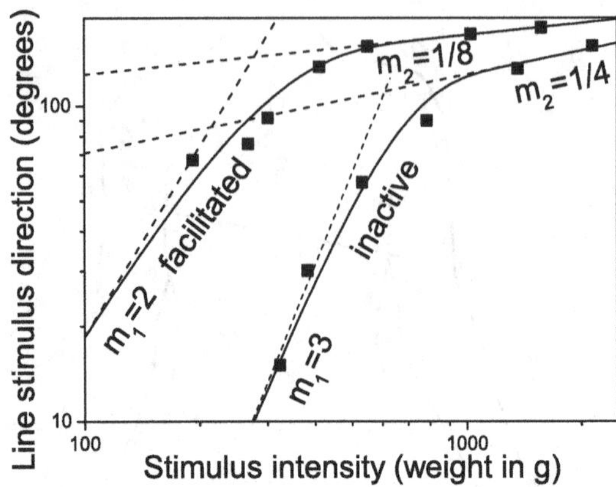

Fig. 5. Direction perceived of the line versus the intensity (weight in g) on a line 6 cm long on the hand of case M, in inactive and facilitated states. Data from [1].

Noticeable facts are the dominant exponent 1/4 in the power laws for high stimuli, and that this exponent holds irrespective of the sensory system (visual or tactile). The same exponent was also reported in [14,5] for visual acuity in case M inactive and in case T versus light intensity, as well as in the width of the visual field versus light intensity in case M inactive and in case M facilitated by muscular stress [14,5]. In some curves the exponent of the power law is found to be 1/8, as also found for visual acuity versus light intensity in a normal man and for the visual field width versus light intensity in case T [14,5]. A still more remarkable, and probably related fact, is that the exponents 1/4 and 1/8 also appear in the power laws that relate the perception to the intensity of a multisensory facilitating stimulus as, for example, the recovery of the upright direction in the perception of an upright arrow versus muscular stress (1/4) and versus light intensity on the eye which was not observing the test object (1/8) in case M [14,5].

Concerning facilitation, we recall [1,3,6,5], that it would afford a non-specific neural excitation that supplies in part the deficit of excitation due to the loss of neural mass. In the functional gradients scheme proposed in [3], the specificity is maximum in the respective projection path, decreases gradually towards more "central" zones and beyond, reaching in its final decline other specific areas and even the primary zones. Accordingly, the action of the region with maximum specificity would not suffice for the perception to be normal, and the whole specific functional density in gradation through the cortex should be involved in the integrative process. A sensory signal in the projection area is only an inverted and constricted outline that must be magnified, re-inverted (i.e., integrated) over the whole region where the functional specific gradient is extended [3]. In this way, it can be understood that a lesion where the functional specificities decline (as in the cases analyzed), originates an integration deficit and hence an incomplete perception. A very weak stimulus, as in the left part of Figs. 1, 2, 4 and 5, would activate only specific neurons, whose deficit does not permit a complete perception. However, a higher stimulus would be able to activate neurons with less specificity or of other specificities, in the same way as multisensory facilitation does, supplying in part the deficit and improving almost completely the perception. Within this interpretation, it is then not strange that the power laws found for higher stimuli (right parts of Figs. 1, 2, 4 and 5) are the same as those found only with facilitation, where the facilitating stimulus (different from that to be perceived) activates rather unspecific neurons or of different specificity.

3 Conclusion

We have analyzed data on visual and tactile perception in the so-called central syndrome, characterized by a deficit of cerebral excitability in a cerebral organization similar to that of a normal man.

In addition to the improvement of visual acuity with intensity of light, we have analyzed the recovery of upright direction in inverted or tilted perception with the intensity of the stimulus, either in the visual system or the tactile system. The experimental data have been seen to be adequately described by a function that has two limiting power laws, a steeper one for low intensity, and a smoother one for high intensity.

The power laws at high intensities are repeatedly characterized by exponents $1/4$ and $1/8$, even if the stimuli belong to different classes and sensory systems. These exponents are moreover the same as those found previously for the improvement of perception by increasing the intensity of a facilitating stimulus. In the limit of low stimulus intensity, however, the exponent of the power law is quite different in each case.

These results support the idea [1,3] that a very weak stimulus would activate only specific neurons, which being in deficit produce only an incomplete perception, while a higher stimulus seems to be able to activate unspecific neurons, or specific neurons of another sensory system, improving the perception in the same way as facilitation does. In a general way (including normal man), these results would be a reflect of the integrative process from the projection path to regions of the cortex of less specificity.

Under the assumption of an approximate linear relationship between the intensity of the stimulus and activated neural mass, the dominance of power laws with quarter exponents would reflect the characteristic behavior of the scaling power law of biological neural networks.

References

1. Gonzalo, J.: Investigaciones sobre la nueva dinámica cerebral. La actividad cerebral en función de las condiciones dinámicas de la excitabilidad nerviosa. Publicaciones del Consejo Superior de Investigaciones Científicas, Inst. S. Ramón y Cajal, Madrid, vol. I (1945), II (1950) (available in: Instituto Cajal, CSIC, Madrid)
2. Gonzalo, J.: La cerebración sensorial y el desarrollo en espiral. Cruzamientos, magnificación, morfogénesis. Trab. Inst. Cajal Invest. Biol. 43, 209–260 (1951)

3. Gonzalo, J.: Las funciones cerebrales humanas según nuevos datos y bases fisiológicas: Una introducción a los estudios de Dinámica Cerebral. Trab. Inst. Cajal Invest. Biol. 44, 95–157 (1952)
4. Gonzalo, I., Gonzalo, A.: Functional gradients in cerebral dynamics: The J. Gonzalo theories of the sensorial cortex. In: Moreno-Díaz, R., Mira, J. (eds.) Brain Processes, Theories and Models. An Int. Conf. in honor of W.S. McCulloch 25 years after his death, pp. 78–87. MIT Press, Massachusetts (1996)
5. Gonzalo-Fonrodona, I.: Functional grsdients through the cortex, multisensory integration and scaling laws in brain dynamics. Neurocomputing 72, 831–838 (2009)
6. Gonzalo-Fonrodona, I.: Inverted or tilted perception disorder. Revista de Neurología 44, 157–165 (2007)
7. Martuzzi, R., Murray, M.M., Michel, C.M., et al.: Multisensory interactions within human primary cortices revealed by BOLD dynamics. Cereb. Cortex 17, 1672–1679 (2007)
8. Wallace, M.T., Ramachandran, R., Stein, B.E.: A revised view of sensory cortical parcellation. Proc. Natl. Acad. Sci. USA 101, 2167–2172 (2004)
9. Gonzalo, I.: Allometry in the J. Gonzalo's model of the sensorial cortex. In: McCune, W. (ed.) CADE 1997. LNCS, vol. 1249, pp. 169–177. Springer, Heidelberg (1997)
10. Gonzalo, I.: Spatial Inversion and Facilitation in the J. Gonzalo's Research of the Sensorial Cortex. Integrative Aspects. In: Mira, J. (ed.) IWANN 1999. LNCS, vol. 1606, pp. 94–103. Springer, Heidelberg (1999)
11. Gonzalo, I., Porras, M.A.: Time-dispersive effects in the J. Gonzalo's research on cerebral dynamics. In: Mira, J., Prieto, A.G. (eds.) IWANN 2001. LNCS, vol. 2084, pp. 150–157. Springer, Heidelberg (2001)
12. Gonzalo, I., Porras, M.A.: Intersensorial summation as a nonlinear contribution to cerebral excitation. In: Mira, J., Álvarez, J.R. (eds.) IWANN 2003. LNCS, vol. 2686, pp. 94–101. Springer, Heidelberg (2003)
13. Arias, M., Gonzalo, I.: La obra neurocientífica de Justo Gonzalo (1910-1986): El síndrome central y la metamorfopsia invertida. Neurología 19, 429–433 (2004)
14. Gonzalo-Fonrodona, I., Porras, M.A.: Physiological Laws of Sensory Visual System in Relation to Scaling Power Laws in Biological Neural Networks. In: Mira, J., Álvarez, J.R. (eds.) IWINAC 2007. LNCS, vol. 4527, pp. 96–102. Springer, Heidelberg (2007)
15. Delgado, A.E.: Modelos Neurocibernéticos de Dinámica Cerebral. Ph.D.Thesis. E.T.S. de Ingenieros de Telecomunicación, Univ. Politécnica, Madrid (1978)
16. Mira, J., Delgado, A.E., Moreno-Díaz, R.: The fuzzy paradigm for knowledge representation in cerebral dynamics. Fuzzy Sets and Systems 23, 315–330 (1987)
17. Mira, J., Manjarrés, A., Ros, S., Delgado, A.E., Álvarez, J.R.: Cooperative Organization of Connectivity Patterns and Receptive Fields in the Visual Pathway: Application to Adaptive Thresholdig. In: Sandoval, F., Mira, J. (eds.) IWANN 1995. LNCS, vol. 930, pp. 15–23. Springer, Heidelberg (1995)
18. Stevens, S.S.: On the psychophysical law. Psychol. Rev. 64, 153–181 (1957)
19. Arthurs, O.J., Stephenson, C.M.E., Rice, K., Lupson, V.C., Spiegelhalter, D.J., Boniface, S.J., Bullmore, E.T.: Dopaminergic effects on electrophysiological and functional MRI measures of human cortical stimulus-response power laws. NeuroImage 21, 540–546 (2004)
20. Nieder, A., Miller, E.K.: Coding of cognitive magnitude. Compressed scaling of numerical information in the primate prefrontal cortex. Neuron 37, 149–157 (2003)
21. West, G.B., Brown, J.H.: A general model for the origin of allometric scalling laws in biology. Science 276, 122–126 (1997)

22. Banavar, J.R., Maritan, A., Rinaldo, A.: Size and form in efficient transportation networks. Nature 399, 130–132 (1999)
23. West, G.B., Brown, J.H.: Life's Universal Scaling Laws. Phys. Today, 36–42 (September 2004)
24. West, G.B., Brown, J.H.: The origin of allometric scaling laws in biology from genomes to ecosystems: towards a quantitative unifying theory of biological structure and organization. J. Exper. Biol. 208, 1575–1592 (2005)
25. Anderson, R.B.: The power law as an emergent property. Mem. Cogn. 29, 1061–1068 (2001)
26. Gisiger, T.: Scale invariance in biology: coincidence or footprint of a universal mechanisms? Biol. Rev. 76, 161–209 (2001)

Neuron-Less Neural-Like Networks with Exponential Association Capacity at *Tabula Rasa*

Demian Battaglia

Department of Nonlinear Dynamics, Max-Planck-Institute for Dynamics and Self-Organization, D37073 Göttingen, Germany
Bernstein Center for Computational Neuroscience, D37073 Göttingen, Germany
demian@nld.ds.mpg.de

Abstract. Artificial neural networks have been used as models of associative memory but their storage capacity is severely limited. Alternative machine-learning approaches perform better in classification tasks but require long learning sessions to build an optimized representational space. Here we present a radically new approach to the problem of classification based on the fact that networks associated to random hard constraint satisfaction problems display naturally an exponentially large number of attractor clusters. We introduce a warning propagation dynamics that allows selective mapping of arbitrary input vector onto these well-separated clusters of states, without need of training. The potential for such networks with exponential capacity to handle inputs with a combinatorially complex structure is finally explored with a toy-example.

1 Introduction

Classification of inputs sampled from high dimensional spaces involves two fundamental tasks: 1) description of the data which is typically achieved by building a codebook associated to a suitable quantization of the input space; and 2) assigning a particular input to one of the available internal categories which the system can represent. Recurrent artificial neural networks (ANNs) with multiple attractors provide a viable and neurally plausible paradigm for fast parallel association of an external stimulus to internal representations [1,2]. Standard associative attractor neural network models suffer however of restricted storage capacity (growing linearly as $\alpha_c N$, with N number of neurons and $\alpha_c \sim 1.144$ for the Hopfield model ad small distortion retrieval [2]). Classification in high dimensional spaces is challenging for other machine learning algorithms too, requiring the generation of representational spaces tailored to the stimuli.

In this paper, we adopt a radically novel approach to the problem of classification. We consider a simple warning propagation dynamics over the graphical models associated to instances of hard Constraint Satisfaction Problems (CSPs) [3]. The main idea of the paper is to capitalize on the fact that, in certain

J. Mira et al. (Eds.): IWINAC 2009, Part I, LNCS 5601, pp. 184–194, 2009.

phases, these network dynamics exhibit an exponentially large number of selectively addressable clusters of attracting states. This feature allows these states to be used for associating an exponentially large set of arbitrary new inputs to clusters of attractors in a reliable way and to perform classification without the need for learning a codebook. The associations performed by these "neuron-less" networks (the nodes will be *e.g.* boolean variables and boolean gates) are as well "neural-like" since: 1) they are achieved with a dynamical exchange of binary information spikes; 2) similar inputs are categorized into a same attractor; 3) incomplete knowledge of one attractor allows its associative restoration. But, in contrast with ANNs, the typical distortion between the presented input and the associated attractors will be significantly larger. In a Hopfield network approach, associations are performed by learning a small number of attractors highly adapted to the stimulus statistics. In a CSP-based network approach, associations will be performed using fair quality unlearned attractors whose number grows exponentially with N. The fine quantization grid provided by these exponentially many states will ensure universal classification capability, since a *tabula rasa* state of the network.

The paper is organized as follows. We start by briefly describing in Section 2 under which conditions CSP networks exhibit clustering and by reviewing how the Survey Propagation algorithm (see *e.g.* [4,5]) can be modified to probe selectively individual clusters [6,7]. We introduce then in Section 3 a simplified algorithm (SPike), which is based on the exchange of binary messages combined with an adaptive "short-term plasticity" rule. We show how SPike can be used to associate input vectors to clusters of attractors. We present, finally, a toy-application to demonstrate how one can take advantage of an exponential association capacity to handle inputs with a combinatorially complex structure.

2 Clustering of Solutions in Constraint Satisfaction Problems

2.1 Phase Diagram

An instance of a CSP can be described by a energy function. Given N variables s_i, $i = 1, \ldots N$ (each one assuming one of Q possible states $S = 1, \ldots Q$) and a set of M constraints $C_a = C_a(s_{a_1}, \ldots, s_{a_K}, \ldots)$, $a = 1 \ldots M$ (each one being function of the states assumed by a limited number of variables), it is possible to write an Hamitonian $\mathcal{H}(s_1, \ldots, s_N) = \sum_{a=1}^{M} C_a$. Resolution of the instance corresponds to determining one of the possibly many ground states of $\mathcal{H}$. Every instance of a CSP can be represented as a bipartite network (*factor graph*), with variable nodes and constraint nodes. Each variable node s_i is connected to the nodes corresponding to the costraints C_a in which it is is involved.

Statistical mechanics techniques can be used to derive the phase diagram of CSP networks [3,4,8]. Briefly, the relevant parameter for a broad class of random CSPs is the ratio $\alpha = M/N$. When $\alpha > \alpha_c$, the ground state energy is strictly positive and the instance is over-constrained (*SAT/UNSAT transition*).

In addition, several other transitions have been characterized in these networks. In a somewhat oversimplified picture, for $\alpha_G < \alpha < \alpha_c$ (*hard phase*), the space of solutions breaks into an exponential number of well separated clusters, each one with its own internal entropy. We will refer to them as solution clusters (denoted here *S-clusters*). The number of S-clusters is given by $\mathcal{N}_{cl} = \exp(\Sigma_0 N)$, where Σ_0 is the so-called *complexity* functional [4].

In addition, suboptimal assignments with a sufficiently small energy are also organized into clusters, even if their separation can be less sharp [9]. These suboptimal clusters are exponentially more numerous than the S-clusters, as quantified by the evaluation of the complexity functional Σ_E at different energies [4]. We will call them threshold clusters (*T-clusters*). We will demonstrate in Section 3 how T-clusters (and not only S-clusters) can be exploited as a computational resource to perform robust associations. Networks with positive Σ_0 and Σ_E can therefore achieve exponential representation capacity, whenever a suitable dynamics able to address selectively the available clusters is devised.

The specific CSP that we will consider in the rest of the paper is the *random balanced 3-SAT* problem. Variables s_i are boolean and can assume only two states $S = \pm 1$. The constraints are logical "ORs" of $K = 3$ randomly picked variables. When a variable appears in more than a single constraint, approximately half of the times it is negated and the other half it is not. This balancing ensures that S-clusters are homogeneously dispersed over the space of configurations. The Hamiltonian associated to an instance can be written as:

$$\mathcal{H} = \sum_{a=1}^{M} \frac{1 - J_a^{a_1} S_{a_1}}{2} \frac{1 - J_a^{a_2} S_{a_2}}{2} \frac{1 - J_a^{a_3} S_{a_3}}{2} \qquad (1)$$

where the couplings $J_a^i = \mp 1$ show whether the variable s_i appears negated or not in clause a. For random instances of the balanced 3-SAT problem, the critical values are estimated as $\alpha_G = 3.09, \alpha_c = 3.41$ [6]. All following simulations will use a network representing the factor graph of a large $N = 10^4$ instance of random balanced 3-SAT with $\alpha = 3.10$, without loss of generality. The ratio α is taken only slightly larger than α_G in order to maximize the number of available well-separated S-clusters and T-clusters, analogously to [6]. For that network $\Sigma_0 = 0.03$, yielding an astronomic number of S-clusters of the order of 10^{130}, and exponentially larger numbers of T-clusters...

2.2 Message-Passing Algorithms for the Clustered Phase

Local search strategies fail to obtain efficiently solutions of CSPs in the hard phase, because of the proliferation of T-clusters. Also bayesian message-passing strategies, like Belief Propagation (BP), which try to infer the probabilities that given variables assume specific states in a satisfying assignment, fail from converging in the hard-phase of random 3-SAT [5,8]. Survey Propagation (SP) generalizes BP to remedy this shortcoming. SP integrates the assumption that the space of ground states is clustered, and it can converge for 3SAT and many other CSPs instances well beyond α_G and almost up to α_c [4,5] .

SP can be described by defining two kinds of *warnings* traveling along the edges of the factor graph: *cavity biases* $u_{a \to i}$ traveling from a constraint node C_a to a neighboring variable node s_i to bias it toward some state; and *cavity fields* $h_{i \to a}$, traveling from a variable node s_i to a neighboring constraint node C_a to inform it about the biases applied to it by its neighboring clauses other than C_a. A key feature of SP is that the possibility of a constraint leaving a neighboring variable *unconstrained* is explicitly taken into account. Specifically, a cavity bias is fired by a constraint node C_a only when it is critically required for the target variable to assume a specific state in order to satisfy C_a itself. In addition to cavity biases and fields, *local fields* H_i are also defined in SP, and they measure the global bias acting on a variable as an effect of all the constraints. In the case of 3-SAT instances, described by a Hamiltonian (1), the expressions for the SP warnings are:

$$u_{a \to i} = \hat{u}(\{h_{j \to a}\}) = \prod_{j \in V(a) \setminus i} \theta(-J_a^j h_{j \to a}) \tag{2}$$

$$h_{i \to a} = \hat{h}(\{u_{b \to i}\}) = \sum_{b \in V(i) \setminus a} J_b^i u_{b \to i}, \qquad H_i = \hat{H}(\{u_{a \to i}\}) = \sum_{a \in V(i)} J_a^i u_{a \to i} \tag{3}$$

where $V(i)$ denotes the set of the indices of the constraints neighboring to s_i, $V(a)$ the set of the indices of the variables involved in constraint C_a and the symbol $A \setminus a$ denotes substraction of element a from set A. The cavity biases $u_{a \to i}$ can assume two possible values, $u_{a \to i} = 1$, or $u_{a \to i} = 0$, which can be interpreted as a null warning not imposing any bias to the target variable [5].

SP (and BP) are probabilistic algorithms. Different self-consistent sets of local fields, cavity fields and cavity biases are associated with different S-clusters, reflecting the fact that distinct S-clusters have distinct patterns of variables frozen to specific states. The SP algorithm evaluates iteratively the probability distribution functions (pdfs) $Q_{a \to i}(u)$, $P_{i \to a}(h)$ and $P_i(H)$ of cavity biases, cavity and local fields respectively over the different S-clusters. This probabilistic information can be used to determine satisfying assignments, for example by means of a decimation procedure [5] or of a reinforcement strategy, suitable for parallel and hardware implementations. The interested reader is invited to refer to the original literature for more details about SP and its variants [4,5,6,7,10] .

The introduction of additional forcing cavity biases (playing the role of an external driving input to the network) endows SP with the capacity to find solutions only in S-clusters whose overlap with the input direction is large. This is a crucial step for the implementation of associative dynamic CSP networks. In the resulting SP-ext algorithm [6] a free parameter π can be tuned to assign increasingly larger weights to the S-clusters close to the external reference forcing vector, thereby biasing the message-passing dynamics toward them. This allows to associate distinct S-clusters selectively to different input patterns. The cluster association and retrieval features of SP-ext have been used to implement lossy compression protocols reminiscent of vector quantization [6].

The selection of S-clusters in SP-ext requires to exchange between network nodes sets of real-valued messages, parameterizing the pdfs $Q_{a \to i}(u)$ and $P_{i \to a}(h)$.

We will show in the next section that a much simpler and faster network dynamics based on the exchange of spikes of binary information is able to achieve meaningful associations between inputs and T-clusters.

3 Associating Clusters to Input Vectors via Dynamic Exchange of Binary Warnings

3.1 The SPike Algorithm

SP and SP-ext compute directly the equilibrium pdfs of warnings in a probabilistic space. The actual exchange of warnings between nodes of the factor graph is not simulated. In contrast, we implement here a genuine warning propagation dynamics over a given factor graph, and use it to perform associative classification and storage of input vectors. This new algorithm, termed SPike, can be described as follows:

- Equations (2) and (3) are directly used as *update rules* in a warning exchange dynamics over the factor graph. Cavity biases are initialized randomly and new cavity and local fields are computed accordingly using the update rule (3). New cavity biases are then computed using the update rule (2). Note that only the sign of the incoming cavity fields is relevant for the computation of cavity biases, the information exchanged between nodes is therefore limited to mere one-bit spikes (hence the name of the algorithm).
- At any time t, an istantaneous configuration $\{S_i(t)\}$ —and the associate energy given by the Hamiltonian (1)— can be defined by setting the variable s_i to a state $S_i = \text{sign}\,\overline{H_i}$, where $\overline{H_i}$ is the average of the local field over a short integration window (*e.g.* $\tau_s = 8$ preceding iterations). Instantaneous configurations can be compared between them and with the input forcings to compute relative overlaps.
- As in the case of SP-ext, external forcing can be applied to drive the dynamics toward a cluster close to an input vector $f = (f_1, \ldots, f_N)$. The update rule (3) is modified in the following way during the application of the input forcing:

$$\texttt{INPUT FORCING:}\quad \text{add to } h_{i\to a} \text{ or } H_i \text{ a term } + g_{in} f_i \text{ with prob. } p_{in} \tag{4}$$

where g_{in} is the strength of the coupling with the input forcing source (*e.g.*, $g_{in} = 1$) and p_{in} is the rate of this source (*e.g.*, $p_{in} = 0.9$).

Unfortunately, the simple dynamics described by update rules (2) and (3) in conjunction with (4), converges to an attractor cluster, generally a T-cluster, only if the applied input vector f is sufficiently close to an exact solution s belonging to some S-cluster (*e.g.*, overlap between f and s of an unacceptable order of the 60%). In such cases, the SPike dynamics lead to a very fast decrease of the network energy, while the input forcing is applied (Fig. 1). However, when the forcing is removed, SPike behaves differently depending on the energy level E of the reached configuration. If E is sufficiently small (smaller than a certain

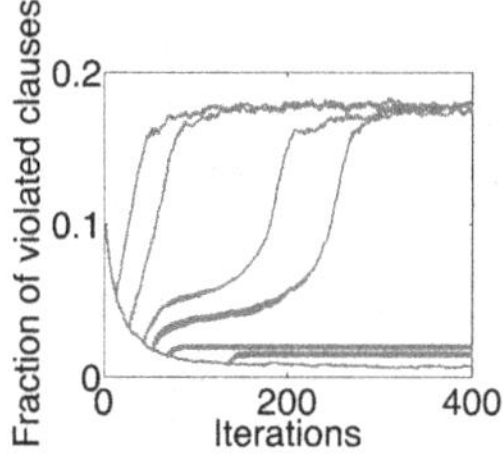

Fig. 1. Pure SPike dynamics without threshold-shifts for an input forcing f sufficiently close to some solution s; trajectories after removal of the forcing escape or are trapped depending on the reached energy level

threshold $E_{th} = E_{th}(\alpha, s)$), the dynamics gets trapped into a T-cluster. Such a stable dynamical state, characteristic of the *evoked activity*, provides a valid internal representation of the input, since the selected T-cluster has a significant overlap to the input f (see Fig. 2 left panel). However, if the energy is too large, $E > E_{th}$, the reached configuration is not stable and the SPike dynamics escapes toward dramatically higher energy states, which are characteristic of the *spontaneous activity* in absence of external forcing.

3.2 The SPike Algorithm with Activity-Dependent Threshold-Shifts

In the spirit of the reinforcement technique introduced in [10], we enhance the convergence properties of SPike introducing intrinsically generated activity-dependent consolidating fields. The update rule (3) is slightly modified by introducing a unitary threshold-shift which is updated on a slower time-scale than the warning dynamics ($\tau_{shift} > 1$, *e.g.* $\tau_{shift} = 4$). This shift follows the direction of the mean field $\overline{H_i}$, averaged over the preceding τ_{shift} iterations:

$$\texttt{THRESHOLD SHIFT:} \text{ add to } h_{i\to a}(\text{or } H_i) \ + \text{sign } \overline{H_i} \ \text{ with prob. } p_{shift} \quad (5)$$

For each variable node, such a shift increase of one unit the number of incoming warning biases required to revert the state of the network node. In a sense, this

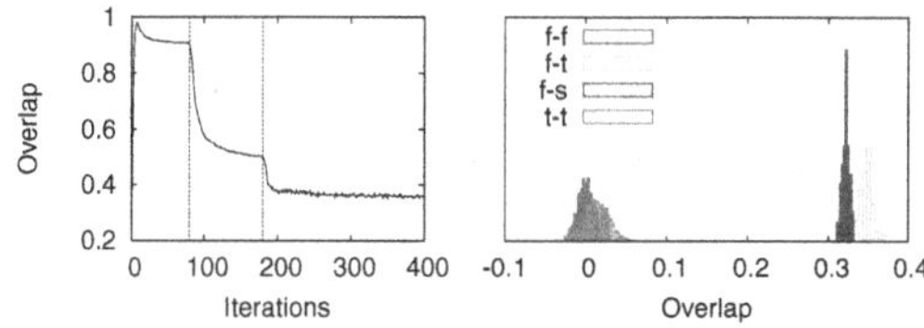

Fig. 2. *On the left*: convergence of SPike with reinforcement toward a T-cluster (vertical lines separate, from left to right, imprinting, correction and trapped phase). *On the right*: distance relations between input forcings and associated S-clusters and T-clusters.

is analogous to *short-term plasticity*, which consolidates the state of a variable unit, by depressing the effect of spikes which try to modify it.

To associate an internal cluster to a random input, the input forcing is applied intensely ($g_{in} = 1 \div 3$) for a short period of time (*e.g.* $\tau_{input} = 80$) and then it is removed. The threshold-shifts given by (5) are kept active for a slightly longer time T_{shift} (*e.g.* $T_{input} = 180$). When the input is removed, a large number of threshold-shifts dynamically occur and support a swift decrease of the energy of the istantaneous configuration (*correcting phase*). When the energy becomes $E < E_{th}$, the dynamics of the network is kept trapped within a single T-cluster, even when all the threshold-shifts are removed.

As shown by an example in Fig. 2 (right panel), the selected T-clusters in response to an input are non trivially correlated with the associated input vectors (overlap between the input vector and the associated T-cluster ~ 0.36). During and shortly after the application of the forcing (*imprinting phase*), the reinforcement field relaxes toward a "corrected" version of the input forcing pattern f, more compatible with the constraints coded by the network architecture (*correction phase*). If the forcing pattern pushes a variable to a direction which is not compatible with any of the available attracting (T or S) clusters of the network, the internal recurrent dynamics of the network, boosted by the threshold-shifts, will generate a large number of warning biases, in order to develop a large local field $\overline{H_i}$ opposite to the local direction of the forcing f_i. It is important to see that the short-term plastic threshold-shifts used in SPike are an enhancement of the dynamics and *not* a form of learning. The T-cluster attractor which is selected by the dynamics is *unlearned* since it was already present in the original network and can trap the dynamics even when all the thresholds-shifts have been cleaned off.

We cannot prove rigorously the convergence of the SPike algorithm to a T-cluster. However SPike was successful in associating a T-cluster to the input vector in all the attempted numerical experiments, without failures. To test the performance, we generated randomly $150000 = 15N$ different input vectors f, approximately uncorrelated between them (peak f-f in Fig. 2 right panel at an overlap of ~ 0). After 400 iterations SPike always converged to a T-cluster, whose average overlap with the presented pattern was of ~ 0.34 (peak f-t in Fig. 2 right panel). The representation provided by the associated T-clusters was distorted. However the input-output association was still meaningful, because the overlap between T-clusters associated to different input is almost vanishing (peak t-t in Fig. 2 right panel at an overlap of ~ 0.03). In the figure, we report also the overlaps between the input and the S-clusters associated to them by the full bayesian algorithm SP-ext [6] (peak f-s in Fig. 2 right panel at an overlap of ~ 0.32). Interestingly, the S-cluster and the T-clusters associated respectively by SP-ext and by SPike to a same input vector are significantly correlated between them (overlap ~ 0.35), hinting to the fact that SPike provide an approximated emergent "spiking" computation of the attractor retrieved by SP-ext with a bayesian probabilistic computation. Remark also that the number of tested successful associations is very large (150000 for a network size of $N =$

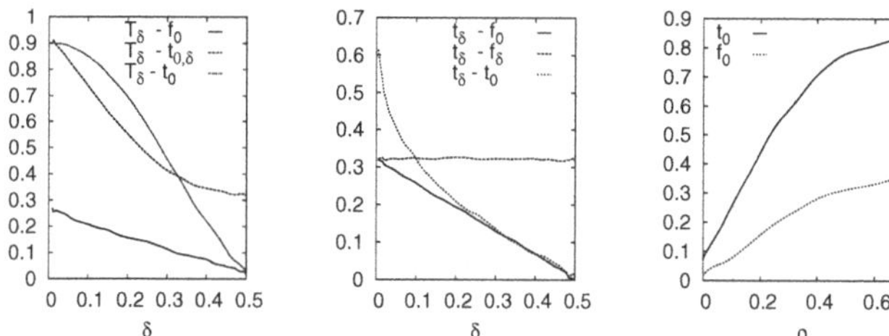

Fig. 3. *On the left*: stability of a T-cluster. *On the middle*: basin of attraction of a T-cluster. *On the right*: retrieval of a T-cluster.

10000 variable nodes), and definitely beyond the capacity bound for a Hopfield network[1].

3.3 Properties of Association and Retrieval

In order to study the properties of the association, we performed three experiments. First, we wanted to study the stability of T-clusters. Let consider a reference input forcing f_0. SPike associates to it a T-cluster. Let take a reference configuration t_0 in this T-cluster. Let generate a family of forcing $t_{0,\delta}$ by flipping randomly δ bits of the T-cluster configuration t_0. Let present as inputs to the network these vectors $t_{0,\delta}$, SPike will associate to them new configurations T_δ. Where are they? The left panel of Fig. 3 shows the overlap between T_δ and the original reference input f_0 (solid line), between T_δ and the perturbed T-cluster configuration $t_{0,\delta}$ (dashed line) and between T_δ and the T-cluster "center" t_0 (dotted line). The T-cluster is a stable attractor: for perturbations up to $\delta \sim 0.34$, the overlap between T_δ and t_0 remains significative and is larger than the overlap with the perturbed input $t_{0,\delta}$, indicating that the network dynamics tries to compensate for the perturbation in the input restoring it to the original t_0 state (*pattern restoration*).

Second, we generate a family of input forcings f_δ by flipping randomly δ bits of the reference input f_0. SPike associates to these inputs f_δ new configurations t_δ. Where are they? The middle panel of Fig. 3 shows the overlap between t_δ and the original reference input f_0 (solid line), between t_δ and the perturbed reference input f_δ (dashed line) and between t_δ and the T-cluster "center" t_0 (dotted line). The T-cluster t_0 has a basin of attraction of non-vanishing size (*i.e.* the T-clusters provide a fine *quantization* of the input space) since for perturbations up to $\delta \sim 0.34$, the overlap between t_δ and t_0 remains significative and is larger than the overlap with the perturbed input f_δ (*pattern association*). Remark that the overlap with f_0 is decreasing linearly. This happens because the T-cluster is... a cluster with internal degrees of freedom and not a point-like attractor.

[1] This comparison is not completely fair because we consider here highly distorted internal representations.

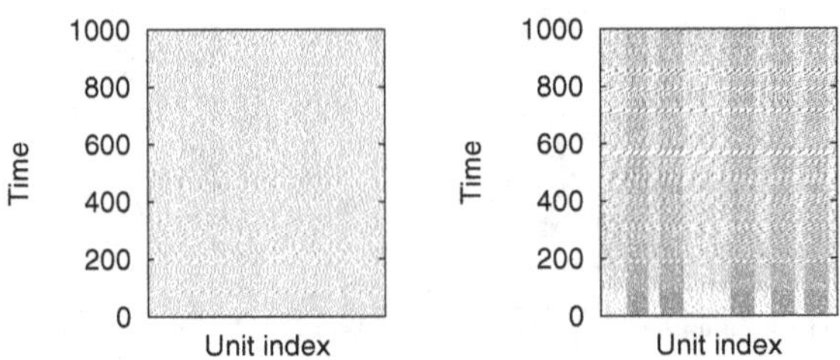

Fig. 4. (*Left*): a raster plot of a dynamics trapped into a random T-cluster. (*Right*): since there are exponentially many T-clusters, some of them have very peculiar structure and allow unstructured networks to behave as a multi-item working memory.

The SPike dynamics selects therefore, within the cluster whose center is t_0, the configuration which is the closest to the last presented input f_δ.

Third, we address retrieval. We select randomly ρN bits of the T-cluster center t_0. We apply a strong input along these ρN bits. It is worth to remind that, as usual, the input is presented just for τ_{input} iterations and then removed, letting the network relax to a T-cluster under the SPike dynamics. Where is this T-cluster? The right panel of Fig. 3 shows the overlap of the retrieved state with the T-cluster attractor t_0 (solid line) and with the associated input f_0 (dotted). Since t_0 belongs to a T-cluster of the network, the dynamics can retrieve it with a very large overlap after the presentation of a quite small number of bits (*pattern completion*).

3.4 Combinatorial Working Memory

A brief application of any input vector drove the network dynamics to enter into a T-cluster attractor which is highly correlated with the input itself. This behavior can be interpreted as an implementation of *working memory*, since the activity persists in the associated T-cluster after the application of the stimulus. Among the exponentially many dynamical attractors naturally available in the network at *tabula rasa* there will always be a T-cluster sufficiently close to any arbitrary f providing a distorted but meaningful internal representation of f. Fig. 4 shows two example raster plots of working memory activity. In the raster plots, the horizontal axis correspond to "space" (different variable units in an arbitrary progressive order) and the vertical axis to time. The left panel shows the persistent activity associated to a typical random T-cluster. Since the network is spatially unstructured most of its attractors will not have any clear spatial structure as well.

Spatially structured inputs are just special cases of arbitrary inputs. The right panel of Fig. 4 shows an example of persistent activity containing five inhomogeneously displaced "bumps". This example illustrates that our untrained random network is embedded with versatile memory capabilities able to mimick multi-item working memory. It is important to remark that such combinations of bumps at arbitrary positions are perceived by the network just as a special single item. The number of such storable "compound single items" is itself exponentially growing with the network size. This allows to classify objects of combinatorially growing complexity without a parallel combinatorial growth in the network size.

4 Conclusions

We developed a new framework for pattern associations based on the exponentially abundant number of clusters of attractors in random hard CSP networks. Unlike more classical ANNs, this new approach can handle exponentially large number of input patterns, and it implements a randomized representational structure reminiscent of the ones used in current classification systems but without need for learning. Furthermore, our approach has a number of features typical for biological memory systems, such as the capability of classifying inputs never seen before and having bias-dependent spontaneous and evoked activities [11]. Hypothetical neural networks with an exponential capacity analogous to the CSP networks here presented might be useful to implement superior cognitive functions relying on a workspace network [12]. Although in present form the framework is not suitable to make the link with neuroscience explicit, CSP-like networks have a potential to be implemented in biological neural systems, bridging the gap from "neuron-less neural-like" up to "truly neural".

The author is indebted to Riccardo Zecchina for inspiring this line of research. He is also grateful to Joël Chavas and more particularly to Jozsef Fiser for long stimulating discussions opening exciting development perspectives.

References

1. Hopfield, J.J.: Neural Networks and Physical Systems with Emergent Collective Computational Abilities. Proc. Nat. Ac. Sci. 79, 2554–2558 (1982)
2. Amit, D.: Modeling Brain Function. Cambridge University Press, NY (1989)
3. Dubois, O., Monasson, R., Selman, B., Zecchina, R. (eds.): Theoretical Computer Science 265(1-2) (2001); Special issue on NP-hardness and phase transitions
4. Mézard, M., Zecchina, R.: Random K-Satisfiability: from an analytic solution to an efficient algorithm. Physical Review E 66, 056126 (2002)
5. Braunstein, A., Mézard, M., Zecchina, R.: Survey Propagation: an algorithm for satisfiability. Random Structures and Algorithms 27, 201–226 (2005)
6. Battaglia, D., Braunstein, A., Chavas, J., Zecchina, R.: Source coding by efficient probing of ground state clusters. Physical Review E 72, 015103 (2005)
7. Chavas, J., Battaglia, D., Cicutin, A., Zecchina, R.: Construction and VHDL Implementation of a Fully Local Network with Good Reconstruction Properties of the Inputs. In: Mira, J., Alvarez, J.R. (eds.) IWINAC 2005. LNCS, vol. 3562, pp. 385–394. Springer, Heidelberg (2005)
8. Krzakala, F., et al.: Gibbs states and the set of solutions of random constraint satisfaction problems. Proc. Nat. Ac. Sci. 104(25), 10318 (2007)

9. Montanari, A., Parisi, G., Ricci-Tersenghi, F.: Instability of one-step replica-simmetry-broken phase in satisfiability problems. Journal of Physics A 37, 2073 (2004)
10. Chavas, J., Furtlehner, C., Mézard, M., Zecchina, R.: Survey-propagation decimation through distributed local computations. Journ. Stat. Mech., P11016 (2005)
11. Fiser, J., Chiu, C., Weliki, M.: Small modulation of ongoing cortical dynamics by sensory input during natural vision. Nature 431, 573–578 (2004)
12. Dehaene, S., Naccache, L.: Towards a cognitive neuroscience of consciousness: basic evidence and a workspace framework. Cognition 79, 1–37 (2001)

Brain Complexity: Analysis, Models and Limits of Understanding

Andreas Schierwagen

Institute for Computer Science, Intelligent Systems Department
University of Leipzig
Leipzig, Germany
schierwa@informatik.uni-leipzig.de
http://www.informatik.uni-leipzig.de/~schierwa

Abstract. Manifold initiatives try to utilize the operational principles of organisms and brains to develop alternative, biologically inspired computing paradigms. This paper reviews key features of the standard method applied to complexity in the cognitive and brain sciences, i.e. decompositional analysis. Projects investigating the nature of computations by cortical columns are discussed which exemplify the application of this standard method. New findings are mentioned indicating that the concept of the basic uniformity of the cortex is untenable. The claim is discussed that non-decomposability is not an intrinsic property of complex, integrated systems but is only in our eyes, due to insufficient mathematical techniques. Using Rosen's modeling relation, the scientific analysis method itself is made a subject of discussion. It is concluded that the fundamental assumption of cognitive science, i.e., cognitive and other complex systems are decomposable, must be abandoned.

1 Introduction

During the last decade, the idea has gained popularity that time is ripe to build new computing systems based on information processing principles derived from the working of the brain. Thus, corresponding research programs have been initiated by leading research organizations (see [1], and references therein).

Obviously, these research initiatives take for granted that the operational principles of the brain as a complexly organized system are sufficiently known to us, and that at least a qualitative concept is available which only needs to be implemented into an operational, quantitative model. Tuning the model then could be achieved since lots of empirical data are available, due to the ever-improving experimental techniques of neuroscience.

Trying to put this idea into practice, however, has generally produced disenchantment after high initial hopes and hype. If one rhetorically ask "What is going wrong?", possible answers are: (1) The parameters of our models are wrong; (2) We are below some complexity threshold; (3) We lack computing

J. Mira et al. (Eds.): IWINAC 2009, Part I, LNCS 5601, pp. 195–204, 2009.

power; (4) We are missing something fundamental and unimagined (see [2] for related problems in robotics). In most cases, only answers (1)-(3) are considered by computer engineers and allied neuroscientists, and appropriate conclusions are drawn. If answer (1) is considered true, still better experimental methodologies are demanded to gather the right data, preferably at the molecular genetic level [3]. Answers (2) and (3) often induce claims for concerted, intensified efforts relating phenomena and data at many levels of brain organization [4].

Together, any of answers (1)-(3) would mean that there is nothing in principle that we do not understand about brain organization. All the concepts and components are present, and need only to be put into the model. This view is widely taken; it represents the belief in the efficiency of the *scientific method*, and it leads one to assume that our understanding of the brain will major advance as soon as the 'obstacles' are cleared away.

As I will show in this paper, there is, however, substantial evidence in favour of answer (4). I will argue that, by following the standard scientific method, we are in fact ignoring something fundamental, namely that biological and engineered systems are basically different in nature.

2 The Standard Approach to Brain Complexity

There is general agreement that brains, even those of simple animals, are enormously complex structures. At the first moment, it seems almost impossible to cope with this complexity. Which methods and approaches should be used? Brains are said to have fortunately - miraculously? - a property that allows us to study them scientifically: they are organized in such a way that the specific tasks they perform are largely constrained to different sub-regions. These regions can be further subdivided in areas that perform sub-tasks [4,5,6,7].

A well-known exponent of this concept is Marr [8] who formulated much of these ideas. In order to explain the human capacity of vision, he discussed detection of contours, edges, surface textures and contrasts as sub-tasks. Their results, he suggested, are combined to synthesize images, 2 1/2-D sketches and the representation of form. This kind of approach has been also employed in other areas of cognition such as language and motor control. Common assumption is that human behavior and cognition can be partitioned into different functions, each of which can be understood independently and with algorithms specific to the area of study. Obviously, this strategy illustrates the standard method used in science for explaining the properties and capacities of complex systems. It consists in applying a decompositional analysis, i.e. an analysis of the system in terms of its components or subsystems.

Since Simon's *The Sciences of the Artificial* [9], decomposability of cognitive and other complex systems has been accepted as fundamental for the Cognitive and Computational Neuroscience (CCN). We call this the fundamental assumption for CCN, for short: FACC. Simon [9], Wimsatt [10] and Bechtel and Richardson [11] have spent much work to elaborate this concept. They consider decomposability a continously varying system property, and state, roughly, that systems fall on a continuum from aggregate (full decomposable) to integrated (non-decomposable). The FACCN states that real systems are non-ideal aggregate systems; the capacities of the components are internally realized (strong intra-component interactions), and interactions between components do not appreciably contribute to the capacities; they are much weaker than the intra-component interactions. Hence, the description of the complex system as a set of weakly interacting components seems to be a good approximation. This property of complex systems, which should have evolved through natural selection, was called *near-decomposability* [9]. Simon characterizes near-decomposability as follows: "(1) In a nearly decomposable system, the short-run behaviour of each of the component subsystems is approximately independent of the short-run behaviour of the other components; (2) in the long run the behaviour of any one of the components depends in only an aggregate way on the behaviour of the other components" [9, p.100].

Thus, if the capacities of a near-decomposable system are to be explained, to some approximation its components can be studied in isolation, and based on their known interactions, their capacities eventually combined to generate the system's behavior. In CCN (and in other areas of science), the components of near-decomposable systems are called *modules.* This term originates from Engineering; it points at the assembly of a product from a set of building blocks with standardized interfaces. Thus, modularization denotes the process of decomposing a product into building blocks (modules) with specified interfaces, driven by the designers interests and intended functions of the product. Modularized systems are linear in the sense that they obey an analog of the *superposition principle* of Linear System Theory in Engineering [13]. This principle represents the counterpart of the decomposition analysis method which therefore is often denoted as *reverse engineering method.* A corresponding class of systems is characterized in Mathematics by a theorem stating that for homogeneous linear differential equations, the sum of any two solutions is itself a solution. The terms *linear* and *nonlinear* are often used in this sense: linear systems are decomposable into independent modules with negligible interactions while nonlinear systems are not [13,14].

Applying this concept to the systems at the other end of the complexity scale, the integrated systems are basically non-decomposable, due to the nonlinear interactions involved. Thus, past or present states or actions of any or most subsystems always affect the state or action of any or most other subsystems. In practice, analyses of integrated systems nevertheless try to apply the methodology for decomposable systems, in particular if there is some hope that the interactions can be linearized. Such linearizable systems were denoted above as nearly decomposable. However, in the case of strong nonlinear interactions, we must accept that decompositional analysis is not applicable to integrated systems. Their capacities depend in non-negligible way on the interaction between components, and it is not possible to identify component functions contributing to the system capacity under study. The question then arises, should we care about integrated systems, given the FACCN that all relevant systems are

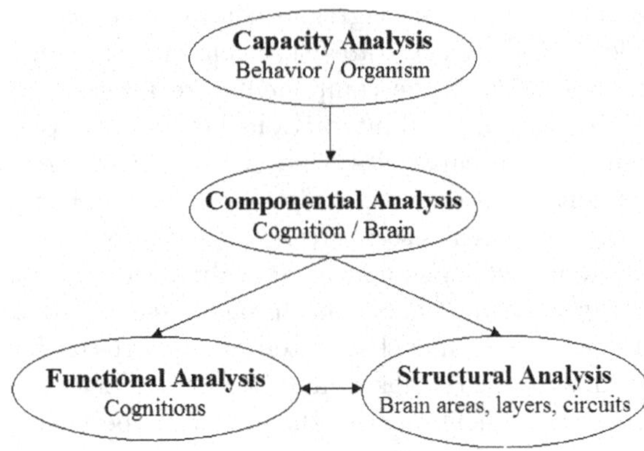

Fig. 1. View on decompositional analysis of brain and behavior. See text for details.

nearly decomposable? Non-decomposability then would be only in our eyes, and not an intrinsic property of strongly nonlinear systems, and - as many cognitive and computer scientists believe - scientific progress will provide us with the new mathematical techniques required to deal with nonlinear systems. We will return to this problem in Section 4.

In CCN, two types of componential analysis must be differentiated, i.e. functional and structural decomposition (see [12] for a clear, intelligible exposition of these matters). If one attempts to identify a set of functions performed by some (as yet unspecified) structural components of the system, a functional analysis is undertaken. Structural analysis involves to attempt to identify the structural, material components of the system. Functional analysis and structural analysis must be clearly differentiated, although in practice, there is a close interplay between them (as indicated by the arrows in Figure 1). Functional analysis should also be differentiated from capacity analysis. The former is concerned with the functions performed by components of the whole system which enable this whole system to have certain capacities and properties. The latter is concerned with the dispositions or abilities of the whole system, whereas functional and structural analysis is concerned with the functional and structural bases of those dispositions or abilities.

Especially important in the present context is this caveat: There is no reason to assume that functional and structural components match up one-to-one! Of course, it might be the case that some functional components map properly onto individual structural components - the dream of any cognitive scientist working as *reverse engineer.* It is rather probable, however, for a certain functional component to be implemented by non-localized, spatially distributed material components. Conversely, a given structural component may implement more than one distinct function. According to Dennett [15, p. 273]: "In a system as complex as the brain, there is likely to be much 'multiple, superimposed functionality' ". With other words, we cannot expect specific functions to be mapped to structurally bounded neuronal structures, and vice versa.

3 Decompositional Brain Analysis

A guiding idea about the organization of the brain is the hypothesis of the columnar organization of the cerebral cortex. It was developed mainly by Mountcastle, Hubel and Wiesel, and Szenthágothai (e.g. [16,17,18]), in the spirit of the highly influential paper " The basic uniformity in structure of the neocortex" published in 1980 by Rockel, Hiorns, and Powell [19]. According to this hypothesis (which has been taken more or less as fact by many experimental as well as theoretical neuroscientists), the neocortex is composed of *building blocks* of repetitive structures, the *columns* or *neocortical microcircuits*, and it is characterized by a basic canonical pattern of connectivity. In this scheme all areas of neocortex would perform identical or similar computational operations with their inputs.

Referring to and based on these works, several projects started recently, among them the *Blue Brain Project*. It is considered to be "the first comprehensive attempt to reverse-engineer the mammalian brain, in order to understand brain function and dysfunction through detailed simulations" [20]. The central role in this project play cortical microcircuits. As Maas and Markram [21] formulate, it is a "tempting hypothesis regarding the computational role of cortical microcircuits ... that there exist genetically programmed stereotypical microcircuits that compute certain basis functions." Their paper well illustrates the modular approach fostered, e.g. by [4,22,23]. The tenet is that there exist fundamental correspondences among the anatomical structure of neuronal networks, their functions, and the dynamic patterning of their active states.

Starting point is the 'uniform cortex' with the cortical microcircuit or column as the structural component. The question for the functional component is answered by assuming that there exists a one-to-one relationship between the structural and the functional component (see Section 2). Experimental results confirming these assumptions are cited, but also some with contrary evidence. Altogether the modularity hypothesis of the brain is considered to be both structurally and functionally well justified. As quoted above, the goal is to substantiate the hypothesis "that there exist genetically programmed stereotypical microcircuits that compute certain basis functions".

Let us consider the general structure of the decompositional analysis of the cortex performed from computational point of view. In the modular approach, the problem of the capacity to be analyzed often is not discussed explicitly. Founding assumption of Cognitive Science is that "cognition is computation", i.e. the brain produces the cognitive capacities by computing functions. We know from mathematical analysis and approximation theory that a continuous function $f : R \to R$ can be expressed by composition or superposition of basis functions. This leads to the functional decomposition as follows: The basic functions are computed by the structural components (cortical microcircuits), and the composition rules are contained implicitly in the interconnection pattern of the circuits.

Obviously, this type of approach simplifies the analysis very much. The question is, however, Are the assumptions and hypotheses made appropriate, or must they considered as too unrealistic?

In fact, most of the underlying hypotheses have been questioned only recently. To start with the assumptions about the structural and functional components of the cortex, the notion of a basic uniformity in the cortex, with respect to the density and types of neurons per column for all species, turned out to be untenable (e.g. [24,25,26]). It has been impossible to find the cortical microcircuit that computes specific basis functions. No genetic mechanism has been deciphered that designates how to construct a column. It seems that the column structures encountered in many species (but not in all) represent spandrels (structures that arise non adaptively, i.e. as an epiphenomenon) in various stages of evolution.

If we evaluate the modular approach as discussed above, it is obvious that the caveat expressed in Section 2 has been largely ignored. There is evidence, however, for a certain functional component to be implemented by spatially distributed networks and, vice versa, for a given structural component to implement more than one distinct function. With other words, it is not feasible for specific functions to be mapped to structurally bounded neuronal structures [24,25,26]. This means, although the column is an attractive idea both from neurobiological and computational point of view, it cannot be used as an unifying principle for understanding cortical function. Thus, it has been concluded that the concept of the cortex as a large network of identical units should be replaced with the idea that the cortex consists of large networks of diverse elements whose cellular and synaptic diversity are important for computation [27,28,29]. It is worth to notice that the reported claims for changes of the research concept belong to the category of answers (1)-(3) to the question "What is going wrong?" (Section 1). A more fundamental point of criticism is formulated in the spirit of answer (4); it concerns the method of decompositional analysis itself and will be discussed in the next section.

4 Salient Features of Complex, Integrated Systems

In Section 2, we concluded that integrated systems are basically not decomposable, thus resisting the standard analysis method. We raised the question, Should we at all care about integrated systems, given the FACCN that all relevant systems are nearly decomposable? According to the prevalent viewpoint in CCN, non-decomposability is not an intrinsic property of complex, integrated systems but is only in our eyes, due to insufficient mathematical techniques (e.g. [5,30]). Bechtel and Richardson, instead, warn that the assumption according to which nature is decomposable and hierarchical might be false: "There are clearly risks in assuming complex natural systems are hierarchical and decomposable" [11, p. 27].

Rosen [31,32] has argued that understanding complex, integrated systems requires that the scientific analysis method itself is made a subject of discussion. A powerful method of understanding and exploring the nature of the scientific method provides Rosen's modeling relation. It is this relation by which scientists bring "entailment structures into congruence" [31, p. 152]. What does this mean?

The modeling relation is the set of mappings shown in Figure 2 [33,34]. It relates two systems, a natural system N and a formal system F, by a set of

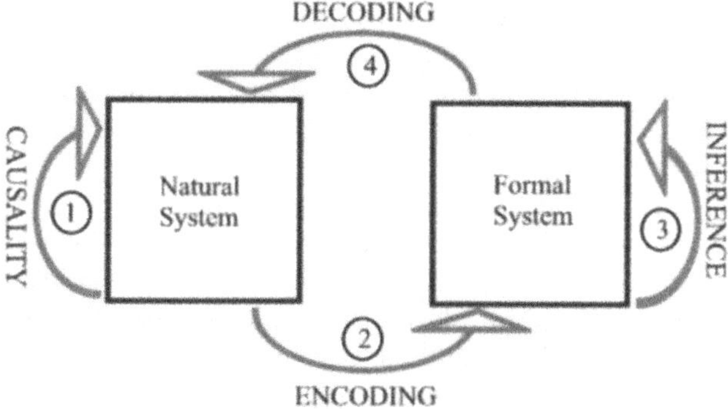

Fig. 2. Rosen's Modeling Relation. A natural system N is modeled by a formal system F. Each system has its own internal entailment structures (arrows 1 and 3), and the two systems are connected by the encoding and decoding processes (arrows 2 and 4). From `http://www.panmere.com`.

arrows depicting processes and/or mappings. The assumption is that this diagram represents the various processes which we are carrying out when we perceive the world. N is a part of the physical world that we wish to understand (in our case: organism, brain), in which things happen according to rules of causality (arrow 1). On the right, F represents symbolically the parts of the natural system (observables) which we are interested in, along with formal rules of inference (arrow 3) that essentially constitute our working hypotheses about the way things work in N, i.e. the way in which we manipulate the formal system to try to mimic causal events observed or hypothesized in the natural system on the left. Arrow 2 represents the encoding of the parts of N under study into the formal system F, i.e. a mapping that establishes the correspondence between observables of N and symbols defined in F. Predictions about the behavior in F, according to Fs rules of inference, are compared to observables in N through a decoding represented by arrow 4. When the predictions match the observations on N, we say that F is a successful model for N.

It is important to note that the encoding and decoding mappings are independent of the formal and/or natural systems. In other words, there is no way to arrive at them from within the formal system or natural system. That is, the act of modeling is really the act of relating two systems in a subjective way. That relation is at the level of observables; specifically, observables which are selected by the modeler as worthy of study or interest. Given the modeling relation and the detailed structural correspondence between our percepts and the formal systems into which we encode them, it is possible to make a dichotomous classification of systems into those that are *simple* or *predicative* and those that are *complex* or *impredicative.* This classification can refer to formal inferential systems such as mathematics or logic, as well as to physical systems. As Rosen showed [33], a simple system is one that is definable completely by algorithmic method: all the models of such a system are Turing-computable or simulable. When a single dynamical description is capable of successfully modeling a system, then the behaviors of that system will, by definition, always be correctly predicted. Hence, such a system will be *predicative* in the sense, that there will exist no unexpected or unanticipated behavior. Simple systems are decomposable sensu Simon [9](Section 2). This is the basis for the classical scientific method, the compositional analysis.

A complex system is thus by exclusion not a member of the syntactic, algorithmic class of systems. Its main characteristics are as follows. A complex system possesses non-computable models; it has inherent impredicative loops in it. This means, it requires multiple partial dynamical descriptions - no one of which, or combination of which, suffices to successfully describe the system. It is not a purely syntactic system, it necessarily includes semantic elements, and is not formalizable. Complex systems also differ from simple ones in that complex systems are not simply summations of parts - they are non-decomposable. This means, when a complex system is decomposed, its essential nature is broken by breaking its impredicative loops. This has several effects. Decompositional analysis is inherently destructive to what makes the system complex - such a system is not decomposable without losing the essential nature of the complexity of the original system! In addition, by being not decomposable, complex systems no longer have analysis and synthesis as simple inverses of each other. How you build a complex system is not simply the inverse of any analytic process of decomposition into parts. In other words, reverse engineering a cognitive system (which is a complex, integrative and thus non-decomposable system) will not enable its full undertanding!

5 Conclusions

Given the characteristics of complex systems - being non-decomposable, non-formalizable, non-computable - can such systems be studied by the scientific method at all? Indeed, they can, provided we acknowledge the inherent limitations of the compositional analysis if questions on the scale of the complex whole are to be answered. In the present context, this means that the fundamental assumption for CCN (cognitive and other complex systems are decomposable) must be abandoned.

Instead, we must consider the set of simple, predicative models of the organism, its behavior and brain *in the limit*, i.e. the infinite set of models, each providing partial dynamical descriptions. Thus, we cannot expect any ultimate model but a multitude of models corresponding to the infinite possible aspects of analysis.

In the case of complex biological systems, Rosen argued in favor of an approach oriented to study them at the level of the organizational structure of the system. This approach of *Relational Biology* - originally created by Rashevsky - involves composing descriptions of organisms at the functional level, thereby retaining the impredicative complexity. Of course, this approach (as well as, e.g., the related concept of autopoietic systems [35]) is not compatible with the standard engineering approach which is oriented to gain control over systems, be it natural or artificial. We must learn, however, to take into account the impredicativities as essential characteristics of complex, integrative systems. This will avoid exaggerated expectations und pitfalls in projects investigating the brain in order to derive operational principles which can be used for unconventional computing models.

References

1. Schierwagen, A.: Brain Organization and Computation. In: Mira, J., Álvarez, J.R. (eds.) IWINAC 2007. LNCS, vol. 4527, pp. 31–40. Springer, Heidelberg (2007)
2. Brooks, R.: The relationship between matter and life. Nature 409, 409–410 (2001)
3. Le Novere, N.: The long journey to a Systems Biology of neuronal function. BMC Syst. Biol., 1–28 (2007)
4. Grillner, S., Markram, H., De Schutter, E., Silberberg, G., LeBeau, F.E.N.: Microcircuits in action - from CPGs to neocortex. Trends in Neurosciences 28, 525–533 (2005)
5. van Vreeswijk, C.: What is the neural code? In: van Hemmen, J.L., Sejnowski Jr., T. (eds.) 23 Problems in System neuroscience, pp. 143–159. Oxford University Press, Oxford (2006)
6. Furber, S., Temple, S.: Neural systems engineering. J. Roy. Soc. Interface 4, 193–206 (2007)
7. Anderson, J.A.: A brain-like computer for cognitive software applications: the Ersatz Brain project. In: Fourth IEEE International Conference on Cognitive Informatics, pp. 27–36 (2005)
8. Marr, D.: Vision. W. H. Freeman & Co., New York (1982)
9. Simon, H.: The Sciences of the Artificial. MIT Press, Cambridge (1969)
10. Wimsatt, W.: Forms of aggregativity. In: Donagan, A., Perovich, A.N., Wedin, M.V. (eds.) Human Nature and Natural Knowledge, pp. 259–291. D. Reidel, Dordrecht (1986)
11. Bechtel, W., Richardson, R.C.: Discovering complexity: Decomposition and localization as strategies in scientific research. Princeton University Press, Princeton (1993)
12. Atkinson, A.P.: Wholes and their parts in cognitive psychology: Systems, subsystems, and persons (1998), `http://www.soc.unitn.it/dsrs/IMC/IMC.htm`
13. Schierwagen, A.: Real neurons and their circuitry: Implications for brain theory. iir-reporte, Akademie der Wissenschaften der DDR, Institut für Informatik und Rechentechnik, Seminar "Neuroinformatik", Eberswalde, 17–20 (1989)
14. Forrest, S.: Emergent Computation: Self-Organizing, Collective, and Cooperative Phenomena in Natural and Artificial Computing Networks. Physica D 42, 1–11 (1990)
15. Dennett, D.C.: Consciousness explained, p. 273. Little, Brown & Co., Boston (1991)
16. Hubel, D.H., Wiesel, T.N.: Shape and arrangement of columns in cat's striate cortex. J. Physiol. 165, 559–568 (1963)
17. Mountcastle, V.B.: The columnar organization of the neocortex. Brain 120, 701–722 (1997)
18. Szentágothai, J.: The modular architectonic principle of neural centers. Rev. Physiol. Biochem. Pharmacol. 98, 11–61 (1983)
19. Rockel, A.J., Hiorns, R.W., Powell, T.P.S.: The basic uniformity in structure of the neocortex. Brain 103, 221–244 (1980)

20. Markram, H.: The Blue Brain Project. Nature Rev. Neurosci. 7, 153–160 (2006)
21. Maass, W., Markram, H.: Theory of the computational function of microcircuit dynamics. In: Grillner, S., Graybiel, A.M. (eds.) The Interface between Neurons and Global Brain Function, Dahlem Workshop Report 93, pp. 371–390. MIT Press, Cambridge (2006)
22. Arbib, M., Érdi, P., Szentágothai, J.: Neural Organization: Structure, Function and Dynamics. MIT Press, Cambridge (1997)
23. Bressler, S.L., Tognoli, E.: Operational principles of neurocognitive networks. Intern. J. Psychophysiol. 60, 139–148 (2006)
24. Rakic, P.: Confusing cortical columns. PNAS 105, 12099–12100 (2008)
25. Horton, J.C., Adams, D.L.: The cortical column: a structure without a function. Phil. Trans. R. Soc. B 360, 386–462 (2005)
26. Herculano-Housel, S., Collins, C.E., Wang, P., Kaas, J.: The basic nonuniformity of the cerebral cortex. Proc. Natl. Acad. Sci. USA 105, 12593–12598 (2008)
27. Destexhe, A., Marder, E.: Plasticity in single neuron and circuit computations. Nature 431, 789–795 (2004)
28. Bullmore, E., Sporns, O.: Complex brain networks: graph theoretical analysis of structural and functional systems. Nature Rev. Neurosci. 10, 186–198 (2009)
29. Frégnac, Y., et al.: Ups and downs in the genesis of cortical computation. In: Grillner, S., Graybiel, A.M. (eds.) Microcircuits: The Interface between Neurons and Global Brain Function, Dahlem Workshop Report 93, pp. 397–437. MIT Press, Cambridge (2006)
30. Poirier, P.: Be There, or Be Square! On the importance of being there. Semiotica 130, 151–176 (2000)
31. Rosen, R.: Life Itself: A Comprehensive Inquiry into the Nature, Origin, and Fabrication of Life. Columbia University Press, New York (1991)
32. Rosen, R.: Essays on Life Itself. Columbia University Press, New York (2000)
33. Rosen, R.: Anticipatory Systems: Philosophical, Mathematical and Methodological Foundations. Pergamon Press, Oxford (1985)
34. Mikulecky, D.C.: Robert Rosen: the well posed question and its answer–Why are organisms different from machines? Syst. Res. 17, 419–432 (2000)
35. Maturana, H.R., Varela, F.J.: Autopoieses and Cognition: The Realization of the Living. D. Reidel, Dordrecht (1980); 49, pp. 27–29 (2006)

Classifying a New Descriptor Based on Marr's Visual Theory

J.M. Pérez-Lorenzo[1], S. García Galán[1], A. Bandera[2], R. Vázquez-Martín[2], and R. Marfil[2]

[1] Dept. Ing. Telecomunicación, Universidad de Jaén, Linares-23700, Spain
jmperez@ujaen.es
[2] Dept. Tecnología Electrónica, Universidad de Málaga, Málaga-29071, Spain

Abstract. Descriptors are a powerful tool in digital image analysis. Performance of tasks such as image matching and object recognition is strongly dependent on the visual descriptors that are used. The dimension of the descriptor has a direct impact on the time the analysis take, and less dimensions are desirable for fast matching. In this paper we use a type of region called curvilinear region. This approach is based on Marr's visual theory. Marr supposed that every object can be divided in its constituent parts, being this parts cylinders. So, we suppose also that in every image there must be curvilinear regions that are easy to detect. We propose a very short descriptor to use with these curvilinear regions in order to classify these regions for higher visual tasks.

1 Introduction

Image matching is a fundamental aspect of many problems in computer vision. Many approaches have been presented in the literature for achieving this goal [1] [3] [5] [8] . In a general way, an image matching method can be divided into three steps. In a first step a technique for searching visual regions with a high repeatability is employed. In a second one these regions are characterized with descriptors, and in the third step the descriptor vectors are matched between different images.

In [9] an approach to visual scenes matching was presented. In biological object recognition no much work has solved the question of specifity, that is, which properties of an object are exactly encoded in the neural representations used to recognize that object. Also the problem of feature selection is a fundamental question [10], although these features seem to be very sensitive to particular combinations of local shape, colour, and texture properties. The approach proposed in [9] was based in Marrs "Generalizad Cilinders" visual theory [7]. Marrs suggests that objects can be formed with an alphabet of simple geometrical shapes, being these shapes cylinders along the main axes of the object. By this way, objects can be represented in terms of their constituent parts.

So the scheme of image matching proposed in [9] defends that a particular type of regions, called curvilinear regions, can be easily detected in digital images, and those regions could be compared in their complexity to those features

J. Mira et al. (Eds.): IWINAC 2009, Part I, LNCS 5601, pp. 205–212, 2009.

analysed by the IT cortex for achieving objects recognition. The algorithm can be divided in several steps. Firstly the curvilinear regions are obtained thanks to a colour segmentation and the verification of some previously established geometrical properties, called curvilinear properties. As the result of this analysis step the method obtains the set of detected curvilinear regions. In a second step the normalized regions are employed as the input of a shape descriptor, which is a contour-based approach to object representation that uses a curvature function [2]. In the third step the recognition stage matches the obtained individual features to a database of features from known scenes using a nearest-neighbour algorithm. This algorithm uses the curvature description, colour description and position of the region in the image. The length of the descriptor used in the matching step was of 260 parameters.

In a general way, the dimension of the descriptor has a direct impact on the time the matches take, and less dimensions are desirable for fast interest point matching [1]. However, lower dimensional feature vectors are in general less distinctive than their high-dimensional counterparts. So, a good goal is to develop both a detector and descriptor that are fast to compute while not sacrificing performance [1].

In this paper a new descriptor is employed. The descriptor uses the colour parameters of the segmentation detection and the parameters that are computed for each region to decide whether it is a curvilinear region or not. So, this descriptor is a vector formed by colour and geometric properties. In the matching step a classification based on standard statistical pattern recognition has been employed. Experiments show that some visual objects can be correctly matched by using a descriptor with just a few parameters. The main difference of the work presented in this paper with the work in [9] is that the descriptor and matching steps have been replaced by new steps that use a much shorter descriptor, resulting in a much more efficient performance of the system. Although only a few experiments have been achieved we believe that they are promising for a further and deeper research in this visual matching method.

The rest of the paper is organized as follows: Section 2 briefly defines a curvilinear region and its properties. Section 3 describes the implementation of the curvilinear region detector. Section 4 presents the descriptor we use in this work for characterizing the curvilinear regions. Section 5 presents the methods we have used for the classification stage. Experimental results for the whole strategy is presented in Section 6. The paper concludes along with conclusions and future work in Section 7.

2 Curvilinear Regions Definition

We define a curvilinear region as a parameter vector $\{a(l), w_l(l), w_r(l)\}_{l=0..L}$, being L the length of the region, $a(l)$ a vector defining the axis between the right and left borders ($b_r(l)$ and $b_l(l)$), and $w_r(l)$ and $w_l(l)$ the widths of the curvilinear region (see Fig. 1).The curvilinear conditions to be satisfied by these regions are:

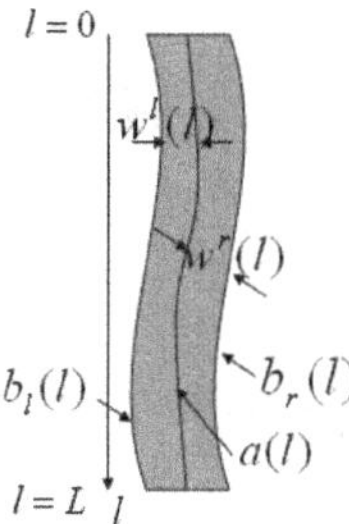

Fig. 1. Definition of curvilinear region. $b_l(l)$,$b_r(l)$: left and right borders, $a(l)$: medial axis, $w_l(l)$,$w_r(l)$: left and right widths.

i) Geometric similarity around the region axis
ii) The ratio between its average width and its total length must be less than a predefined threshold
iii) Left and right borders must be locally parallel
iv) Colour along this axis should be homogeneous.

3 Curvilinear Regions Detector

The implementation of the curvilinear region detector uses in a first step a segmentation in the HSI colour space, based on the Bounded Irregular Pyramid (BIP) [6]. With this segmentation the method ensures that the obtained regions comply with the homogeneous colour property of the curvilinear regions. Next, properties i), ii) and iii) are checked based on some geometrical parameters of the regions. For the estimation of the parameters, the skeleton of the region is extracted with the algorithm based on the d8-distance described in [4], which can approximate the distance transform inside the region in only two steps. The skeletons are generally not connected, so a post-process is needed before estimating further parameters. The skeleton obtained is defined as the set of connected pixels $p_s = (i_s, j_s), 0 \leq s \leq N-1$, being N the number of pixels being evaluated of the skeleton. In Fig. 2.a, 2.b and 2.c an original image, the segmentation image and the estimated skeletons are represented.

i) Symmetry around the skeleton. The method looks for those pixels which comply with the requirement of symmetry around the axis. The normal vector is calculated for each pixel ps in the skeleton, and the cross-points between the normal and the left and right borders of the region are estimated. If we define p_s^l and p_s^r as these cross-points, then we obtain the triplets $(p_s, p_s^l, p_s^r), 0 \leq s \leq N-1$. The symmetry condition can be defined as:

$$E_{\Delta w} \leq U_{\Delta w}(1 - e^{-\frac{N^2}{2\sigma^2_{\Delta w}}}) \tag{1}$$

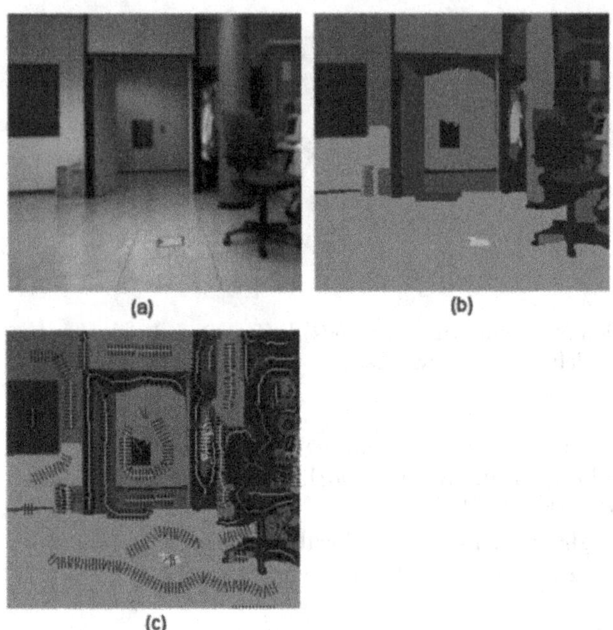

Fig. 2. Estimated skeletons over the segmented image. a) Original image. b) Segmented image. c) Estimated skeletons with the estimated normal vectors.

with

$$E_{\Delta w} = \frac{1}{N} \sum_{s=0}^{N-1} (\Delta w_s - \overline{\Delta w})^2 \tag{2}$$

$$\Delta w_s = |w_s^l - w_s^r| \tag{3}$$

$$\overline{\Delta w} = \frac{1}{N} \sum_{s=0}^{N-1} \Delta w_s \tag{4}$$

being w_s^l the Euclidean distance between pixels p_s and p_s^l and w_s^r the Euclidean distance between pixels p_s and p_s^r. $U_{\Delta w}$ is a parameter of the method.

ii) Ratio. Given a position s in the skeleton, the width w_s of the region is estimated as the Euclidean distance between pixels p_s^l and p_s^r. The following condition must be satisfied:

$$L_{max} \geq U_w \frac{1}{N} \sum_{s=0}^{N-1} w_s \tag{5}$$

being L_{max} the maximum length that the curvilinear region could have, which is estimated with all the connected pixels of the skeleton. U_w is a parameter of the method.

iii) Borders Parallelism. To check the borders parallelism requirement we estimate the tangential vectors on the borders at pixels p_s^l and p_s^r, and then we calculate the angle between those vectors and the normal vector given a position s, obtaining α_s^l and α_s^r. The following condition must be satisfied:

$$\overline{\Delta\alpha} \leq U_{\Delta\alpha} \tag{6}$$

being

$$\overline{\Delta\alpha} = \frac{1}{N} \sum_{s=0}^{N-1} |\alpha_s^l - \alpha_s^r| \tag{7}$$

and $U_{\Delta\alpha}$ is a parameter of the method.

The detector tries to join as many pixels as possible to form a curvilinear skeleton. For doing that, the algorithm starts in an endpoint of the skeleton and it looks for adding the connected pixels checking if Ec. 1, Ec. 5 and Ec. 6 are true with the new added pixel. When all the pixels have been evaluated inside a region, the curvilinear skeletons with close endpoints are linked. Those parts of the objects with a skeleton evaluated as a curvilinear skeleton are considered curvilinear region by the detector. In our experiments, we demand that these regions must have a minimum length of 10 pixels.

4 Curvilinear Regions Descriptor

The purpose of this work is trying to use a descriptor as shorter as possible and being able to match visual regions in a proper way. In [9] we decided to use a 260-dimensional descriptor, but in this new approach we suggest to use some of the previously estimated parameters as part of the regions descriptor.

In [9] we used a curvature function to characterize the region boundaries. In this new approach we apply this curvature function to the skeleton of the region and calculate:

$$C_T = \sum_{s=0}^{N-1} fc_s \tag{8}$$

$$\overline{C_T} = \frac{C_T}{N} \tag{9}$$

being fc_s the curvature function of the object shape. An angle-based curvature estimator is used, where the curve orientation is estimated at each point with respect to a reference direction. The approach is based on a k-slope algorithm which estimates the curvature using a k value which is adaptively changed [2]. C_T and $\overline{C_T}$ are part of the region descriptor.

The same geometrical parameters that are used to decide if the region complies with the curvilinear properties, plus some of the colour values of the segmented region and the curvature function mean values, form the region descriptor. Specifically, the descriptor has got the next parameters:

i) H (Hue value of the colour of the region),
ii) S (Saturation value of the colour of the region),
iii) $E_{\Delta w}$ (Estimated value in Ec. 2),
iv) $\overline{\Delta w}$ (Estimated value in Ec. 4),
v) $\overline{\Delta \alpha}$ (Estimated value in Ec. 7),
vi) C_T (Estimated value in Ec. 8) and
vii) $\overline{C_T}$ (Estimated value in Ec. 9)

5 Classification Stage: Gaussian Mixture Model-Based Classifier

For classifications purposes, a number of standard statistical pattern recognition (SPR) exists. The basic idea behind SPR is to estimate the probability density function (pdf) for the feature vectors of each class. In the Gaussian Mixture Model (GMM) classifier, each class pdf is assumed to consist of a mixture of a specific number K of multidimensional Gaussian distributions. The GMM classifier only needs to store the set of estimated parameters for each class. The Expectation-Maximization (EM) algorithm is used to estimate the parameters of each Gaussian component and the mixture weights.

We have used a three-component GMM classifier with diagonal covariance matrices because it has shown a slightly better performance than other SPR classifiers. The performance of the system does not improve when using a higher number of components in the GMM classifier. The GMM classifier is initialized using the K-means algorithm with multiple random starting points.

6 Experimental Results

For the experiment results an indoor environment has been captured in 30 images. Four different curvilinear regions were detected in each of the 30 images. In Fig. 3, we can see the four regions of interest detected in 4 of the 30 images. So, the experiments were focused to match those regions and see the classification success.

The classifications results were calculated using a ten-fold cross-validation evaluation where the dataset to be evaluated was randomly partitioned, so that 10% was used for testing and 90% was used for training. The process was iterated with different random partitions and the results were averaged.

The algorithm to classify one region uses a nearest neighbour-based matching strategy. Once the classifier has been trained, the algorithm estimates the membership of the region to be tested with every region class. If the probability index is above a threshold, the region is considered to be a member of the class. If the region to be tested belongs to more than one class, then the algorithm decides that the correct class is the one with the highest probability.

In our experiment the average ratio of success of the ten-fold cross-validation evaluation has been 94.46%.

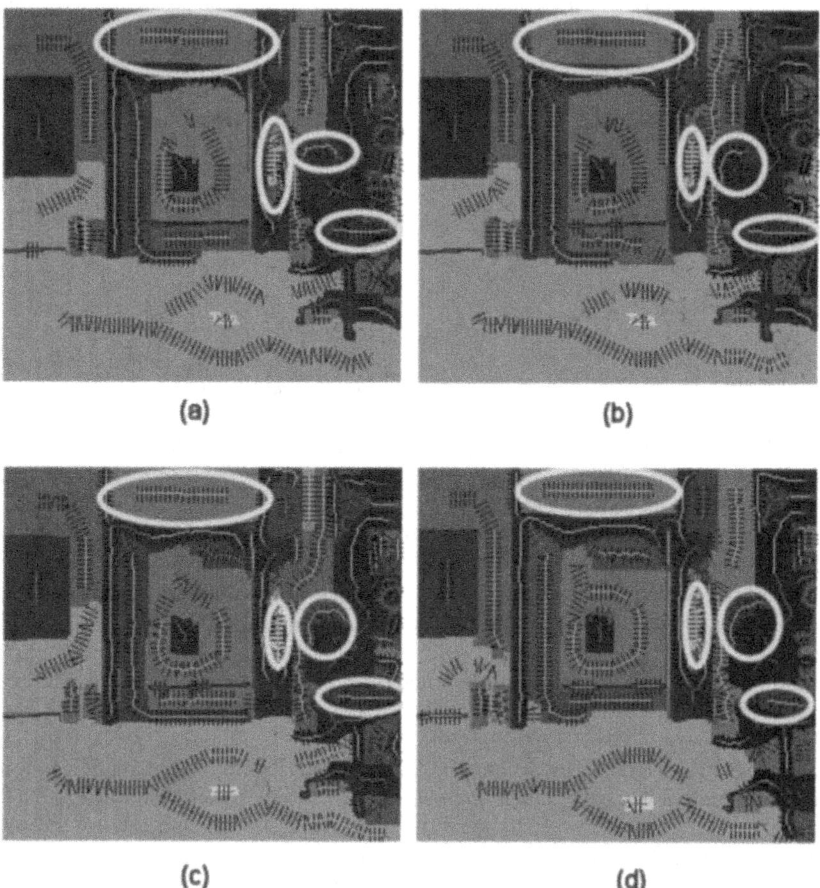

Fig. 3. Images captured from the same scene with slightly different viewpoints. The four curvilinear regions were detected in every image. These four regions were used to test the classifier.

7 Conclusions and Future Work

We have continued the work presented in [9]. Our work propose a scheme of image matching based on the detection of curvilinear regions in the images. In this paper we have used a very short curvilinear region descriptor in order to get a better performance in the matching stage. This descriptor is formed by the values of a colour segmentation and a few geometrical features of the regions. We have tested this descriptor in an indoor environment, where several regions have been matched with a high success percentage in a set of 30 images.

However, the results presented still belong to a preliminary stage of the experiments. More experiments should be taken in more diverse scenes in order to test the repeatability of the detection and the performance of the classifier, and a comparative study with other techniques would be also desirable. So, these issues should be our future work.

Acknowledgements

This work has been partially granted by the Spanish Junta de Andalucía project P07-TIC-02713 and P07-TIC-03106.

References

1. Bay, H., Tuytelaars, H., Van Gool, L.: SURF: Speeded up robust features. In: Leonardis, A., Bischof, H., Pinz, A. (eds.) ECCV 2006. LNCS, vol. 3951, pp. 404–417. Springer, Heidelberg (2006)
2. Bandera, A., Urdiales, C., Arrebola, F., Sandoval, F.: Corner detection by means of adaptively estimated curvature function. Electronics Letters 36(2), 124–126 (2000)
3. Ke, Y., Sukthankar, R.: PCA-SIFT: A more distinctive representation for local image descriptors. In: 2004 IEEE Computer Society Conference on Computer Vision and Pattern Recognition (CVPR 2004), vol. 2, pp. 506–513 (2004)
4. Klette, G.: A comparative discussion of distance transformation and simple deformations in digital image processing. Machine Graphics & Vision 12(2), 235–256 (2003)
5. Lowe, D.: Towards a computational model for object recognition in IT Cortex. In: First IEEE International Workshop on Biologically Motivated Computer Vision, Seoul, Korea, pp. 20–31 (2000)
6. Marfil, R., Rodriguez, J.A., Bandera, A., Sandoval, F.: Bounded irregular pyramid: a new structure for color image segmentation. Pattern Recognition 37(3), 623–626 (2004)
7. Marr, D., Nishihara, H.K.: Representation and recognition of the spatial organization of three-dimensional shapes. Proceedings of the Royal Society of London B: Biological Sciences 200, 269–294 (1978)
8. Matas, J., Chum, O., Urban, M., Pajdla, T.: Robust wide baseline stereo from maximally stable extremal regions. In: Proceedings of the British Machine Vision Conference, vol. 1, pp. 384–393 (2002)
9. Pérez-Lorenzo, J.M., Bandera, A., Reche-López, P., Marfil, R., Vázquez-Martín, R.: An approach to visual scenes matching with curvilinear regions. In: Mira, J., Álvarez, J.R. (eds.) IWINAC 2007. LNCS, vol. 4528, pp. 409–418. Springer, Heidelberg (2007)
10. Ullman, S., Vidal-Naquet, M., Sali, E.: Visual features of intermediate complexity and their use in classification. Nature Neuroscience 5, 682–687 (2002)

Solving the Independent Set Problem by Using Tissue-Like P Systems with Cell Division

Daniel Díaz-Pernil[1], Miguel A. Gutiérrez-Naranjo[2], Mario J. Pérez-Jiménez[2], and Agustín Riscos-Núñez[2]

[1] Research Group on Computational Topology and Applied Mathematics
sbdani@us.es
[2] Research Group on Natural Computing
magutier@us.es, ariscosn@us.es, marper@us.es
University of Sevilla
Avda. Reina Mercedes s/n, 41012, Sevilla, Spain

Abstract. Tissue-like P systems with cell division is a computing model in the framework of Membrane Computing inspired by the intercellular communication and neuronal synaptics. It considers the cells as unit processors and the computation is performed by the parallel application of given rules. Division rules allow an increase of the number of cells during the computation. We present a polynomial-time solution for the Independent Set problem via a uniform family of such systems.

1 Introduction

In the last years, Membrane Computing is becoming one of the pillars of Natural Computing. In the similar way as other research fields in Natural Computing, as *Genetic Algorithms*, *Neural Networks*, or *DNA Computing*, it takes advantage from the way in which Nature computes.

Membrane Computing is a theoretical model of computation inspired by the structure and functioning of cells as living organisms able to process and generate information. The computational devices in Membrane Computing are called *P systems*. Roughly speaking, a P system consists of a membrane structure, in the compartments of which one places multisets of objects which evolve according to given rules. In the most extended model, the rules are applied in a synchronous non-deterministic maximally parallel manner, but some other semantics are being explored (see [14] for details).

Since the seminal paper [10], different models of P systems have been studied. According to their architecture, these models can be split into two sets: cell-like P systems and tissue-like P systems. This paper is devoted to the second approach: tissue-like P systems. According to the architecture, the main difference with cell-like P systems is that the structure of membranes is defined by a general graph instead of a tree-like graph. This kind of models was first presented by Martín–Vide *et al.* in [7] and it has two biological inspirations (see [8]): intercellular communication and cooperation between neurons. The communication among cells is based on symport/antiport rules [12]. Symport rules move objects across a

J. Mira et al. (Eds.): IWINAC 2009, Part I, LNCS 5601, pp. 213–222, 2009.

membrane together in one direction, whereas antiport rules move objects across a membrane in opposite directions.

The search of the abilities of tissue-like P systems as computational model has leaded many authors to consider new models with small differences from the original one (see, for example, [4,5,6]) . One of the most interesting variants of tissue P systems was presented in [13] (and it was studied in depth in [1]). In that paper, tissue P systems are endowed with the ability of getting new cells based on cellular division, yielding *tissue-like P systems with cell division.*

This is a further step in the analogy between a theoretical computational model and living tissues. In a living tissue, new cells are obtained via cellular division or *mitosis.* This ability allows the tissue to grow till reaching its maturity or repairing damages. Cells divide themselves till reaching a number of copies big enough to perform its activity successfully. Following such analogy, tissue-like P systems with cell division are endowed with the ability of obtaining a number of cells greater than the original by means of a rule which implement the idea of mitosis. The computational meaning is that we can create an exponential number of cells in linear time. Each individual cell can be seen as a processor and the cells work in parallel.

The ability of obtaining an exponential amount of workspace in polynomial-time has been the basis to open a new landscape for the study of a key problem in the theory of computational complexity: **P** vs. **NP**. Finding biologically inspired differences among the different P system models that may determine different borderlines between tractability and intractability is the key problem of the Complexity Theory in P systems. Some **NP**-complete problems have been efficiently solved with tissue-like P systems with cell division: SAT [13], 3–coloring [2] or Subset Sum [3]. In this paper, we extend our study by presenting a polynomial–time solution to the Independent Set problem.

The paper is organized as follows: in Section 2 we recall the definition of Tissue-like P systems with cell division. A linear–time solution to the Independent Set problem with the necessary resources and the main results are presented in the following section. Section 4 includes a short overview of the computations. Finally, some conclusions and new open research lines are presented.

2 Tissue-Like P Systems with Cell Division

In the first definition of the model of tissue P systems [7,8] the membrane structure did not change along the computation. Based on the cell-like model of P systems with active membranes, Gh. Păun et al. presented in [13] a new model of tissue P systems *with cell division.* The biological inspiration is clear: alive tissues are not *static* network of cells, since cells are duplicated via mitosis in a natural way.

The main features of this model, from the computational point of view, are that cells have not polarizations (the contrary holds in the cell-like model of P systems with active membranes, see [11]); the cells obtained by division have the same labels as the original cell; and if a cell is divided, its interaction with other cells or with the environment is blocked during the mitosis process.

Formally, a *tissue-like P system with cell division* of degree $q \geq 1$ is a tuple of the form $\Pi = (\Gamma, \mathcal{E}, w_1, \ldots, w_q, \mathcal{R}, i_0)$, where:

1. Γ is a finite *alphabet* (a set of symbols that will be called *objects*).
2. $\mathcal{E} \subseteq \Gamma$ (the objects in the environment).
3. $w_1, \ldots, w_q \in \Gamma^*$ are strings over Γ representing the multisets of objects associated with the cells at the initial configuration.
4. $\mathcal{R}$ is a finite set of rules of the following form:
   (a) *Communication rules*: $(i, u/v, j)$, for $i, j \in \{0, 1, \ldots, q\}, i \neq j$, $u, v \in \Gamma^*$.
   (b) *Division rules*: $[a]_i \to [b]_i[c]_i$, where $i \in \{1, 2, \ldots, q\}$ and $a, b, c \in \Gamma$.
5. $i_0 \in \{0, 1, 2, \ldots, q\}$.

A tissue-like P system with cell division of degree $q \geq 1$ can be seen as a set of q cells (each one consisting of an elementary membrane) labelled by $1, 2, \ldots, q$. We will use 0 to refer to the label of the environment, and i_0 denotes the output region (which can be the region inside a cell or the environment).

The communication rules determine a virtual graph, where the nodes are the cells and the edges indicate if it is possible for pairs of cells to communicate directly. This is a dynamical graph, because new nodes can be produced by the application of division rules.

The strings $w_1, \ldots, w_q$ describe the multisets of objects placed in the q cells of the system. We interpret that $\mathcal{E} \subseteq \Gamma$ is the set of objects placed in the environment, each one of them available in an arbitrary large amount of copies.

The communication rule $(i, u/v, j)$ can be applied over two cells labelled by i and j such that u is contained in cell i and v is contained in cell j. The application of this rule means that the objects of the multisets represented by u and v are interchanged between the two cells. Note that if either $i = 0$ or $j = 0$ then the objects are interchanged between a cell and the environment.

The division rule $[a]_i \to [b]_i[c]_i$ is applied over a cell i containing object a. The application of this rule divides this cell into two new cells with the same label. All the objects in the original cell are replicated and copied in each of the new cells, with the exception of the object a, which is replaced by the object b in the first one and by c in the other one.

Rules are used as usual in the framework of membrane computing, that is, in a maximally parallel way (a universal clock is considered). In one step, each object in a membrane can only be used for one rule (non-deterministically chosen when there are several possibilities), but any object which can participate in a rule of any form must do it, i.e, in each step we apply a maximal set of rules. This way of applying rules has only one restriction: when a cell is divided, the division rule is the only one which is applied for that cell in that step.

2.1 Recognizer Tissue-Like P Systems with Cell Division

NP-completeness has been usually studied in the framework of *decision problems*. Let us recall that a decision problem is a pair (I_X, θ_X) where I_X is a language over a finite alphabet (whose elements are called *instances*) and θ_X is a total boolean function over I_X.

In order to study the computing efficiency for solving **NP**-complete decision problems, a special class of tissue P systems with cell division is introduced in [13]: *recognizer tissue-like P systems.* The key idea of such recognizer systems is the same one as from recognizer cell-like P systems.

Recognizer (cell-like) P systems were introduced in [9] and they provide a natural framework to study and solve decision problems within Membrane Computing, since deciding whether an instance of a given problem has an affirmative or negative answer is equivalent to deciding whether a string belongs or not to the language associated with the problem.

In the literature, recognizer (cell-like) P systems are associated with P systems with *input* in a natural way. The data encoding an instance of the decision problem has to be provided to the P system in order to compute the appropriate answer. This is done by codifying each instance as a multiset to be placed in an *input membrane.* The output of the computation (`yes` or `no`) is sent to the environment, and in the last step of the computation.

A recognizer tissue-like P system with cell division of degree $q \geq 1$ is a tuple $\Pi = (\Gamma, \Sigma, \mathcal{E}, w_1, \ldots, w_q, \mathcal{R}, i_{in}, i_0)$, where

- $(\Gamma, \mathcal{E}, w_1, \ldots, w_q, \mathcal{R}, i_0)$ is a tissue-like P system with cell division of degree $q \geq 1$ (as defined in the previous section), $i_0 = env$ and $w_1, \ldots, w_q$ strings over $\Gamma \setminus \Sigma$.
- The working alphabet Γ has two distinguished objects `yes` and `no`, present in some initial multiset w_i, but not present in $\mathcal{E}$.
- Σ is an (input) alphabet strictly contained in Γ.
- $i_{in} \in \{1, \ldots, q\}$ is the input cell.
- All computations halt.
- If $\mathcal{C}$ is a computation of Π, then either the object `yes` or the object `no` (but not both) must have been released into the environment, and only in the last step of the computation.

The computations of the system Π with input $w \in \Sigma^*$ start from a configuration of the form $(w_1, w_2, \ldots, w_{i_{in}} w, \ldots, w_q; \mathcal{E})$, that is, after adding the multiset w to the contents of the input cell i_{in}. We say that $\mathcal{C}$ is an accepting computation (respectively, rejecting computation) if the object `yes` (respectively, `no`) appears in the environment associated to the corresponding halting configuration of $\mathcal{C}$.

Definition 1. *We say that a decision problem $X = (I_X, \theta_X)$ is solvable in polynomial time by a family $\mathbf{\Pi} = \{\Pi(n) : n \in \mathbb{N}\}$ of recognizer tissue-like P systems with cell division if the following holds:*

- *The family $\mathbf{\Pi}$ is* polynomially uniform *by Turing machines, that is, there exists a deterministic Turing machine working in polynomial time which constructs the system $\Pi(n)$ from $n \in \mathbb{N}$.*
- *There exists a pair (cod, s) of polynomial-time computable functions over I_X (called a polynomial encoding from I_X in $\mathbf{\Pi}$) such that:*
  - *for each instance $u \in I_X$, $s(u)$ is a natural number and $cod(u)$ is an input multiset of the system $\Pi(s(u))$;*

- *the family* $\mathbf{\Pi}$ *is* polynomially bounded *with regard to* (X, cod, s), *that is, there exists a polynomial function* p, *such that for each* $u \in I_X$ *every computation of* $\Pi(s(u))$ *with input* $cod(u)$ *performs at most* $p(|u|)$ *steps;*
- *the family* $\mathbf{\Pi}$ *is* sound *with regard to* (X, cod, s), *that is, for each* $u \in I_X$, *if there exists an accepting computation of* $\Pi(s(u))$ *with input* $cod(u)$, *then* $\theta_X(u) = 1$;
- *the family* $\mathbf{\Pi}$ *is* complete *with regard to* (X, cod, s), *that is, for each* $u \in I_X$, *if* $\theta_X(u) = 1$, *then every computation of* $\Pi(s(u))$ *with input* $cod(u)$ *is an accepting one.*

In the above definition we have imposed to every system $\Pi(n)$ to be *confluent*, in the following sense: every computation of $\Pi(n)$ with the *same* input multiset must always give the *same* answer.

We denote by $\mathbf{PMC}_{TD}$ the set of all decision problems which can be solved by means of recognizer tissue-like P systems with cell division in polynomial time. This class is closed under polynomial reduction and under complement.

3 A Solution for the Independent Set Problem

Let us recall that an *independent set* of a non-directed graph is a subset of its vertices such that it does not contain any pair of vertices adjacent between them. The number of nodes in the subset is called the size of the independent set.

The `Independent Set Problem (IS)` can be settled as follows: given a non-directed graph, $G = (V, E)$, and a natural number $k \leq |V|$, to determine whether or not G has an independent set of size k.

Next, we will prove that `IS` can be solved in linear time (in the number of nodes and edges of the graph) by a family of recognizer tissue-like P systems with cell division. To this aim, let us construct a family $\mathbf{\Pi} = \{\Pi(n, m, k) : n, m, k \in \mathbb{N}\}$ where each system of the family will process every instance u of the problem given by a graph $G = (V, E)$ with n vertices and m edges, and by a size k of the independent set. More formally, we define $s(u) = \langle n, m, k\rangle = \langle n, \langle m, k\rangle\rangle$, where $\langle x, y\rangle = (x + y)(x + y + 1)/2 + x$ is the Gödel mapping. In order to provide a suitable encoding of this instances into the systems, we will use the objects A_{ij}, with $1 \leq i < j \leq n$, to represent the edges of the graph, and we will provide $cod(u)$ as the initial multiset for the system, where $cod(u)$ is the multiset (A, f) with $A = \{A_{ij} : 1 \leq i < j \leq n\}$ and $f : A \to \mathbb{N}$ such that $f(A_{ij}) = 1$ if $\{v_i, v_j\} \in E$ and $f(A_{ij}) = 0$ if $\{v_i, v_j\} \notin E$. It is easy to check that (cod, s) is a polynomial encoding from I_{IS} in $\mathbf{\Pi}$.

Then, given an instance u of the `IS` problem, the system $\Pi(s(u))$ with input $cod(u)$ decides that instance by a brute force algorithm, implemented in the following four stages:

- *generation stage*: all possible subsets of vertices are generated by applying division rules;
- *pre-checking stage*: only those subsets of size k are selected;

- *checking stage*: we check for each selected subset if they do not contain a pair of adjacent vertices;
- *output stage*: an affirmative or negative answer to the problem is given, according to the results of the previous stage.

The family $\mathbf{\Pi} = \{\Pi(n,m,k) : n,m,k \in \mathbb{N}\}$ of recognizer tissue-like P systems with cell division of degree 2 is defined as follows: for each $n,m,k \in \mathbb{N}, \Pi(n,m,k) = (\Gamma, \Sigma, \mathcal{E}, w_1, w_2, \mathcal{R}, i_{in})$, where

- $\Gamma = \{A_i, B_i, \overline{B}_i, B'_i, P_i : 1 \le i \le n\} \cup \{a_i : 1 \le i \le 3n+m+\lceil \lg n \rceil + 14\} \cup$
  $\{c_i, d_i : 1 \le i \le n+1\} \cup \{e_i : 1 \le i \le 2n+1\} \cup$
  $\{f_i : 0 \le i \le n-1\} \cup \{g_i : 1 \le i \le \lceil \lg n \rceil + 1\} \cup$
  $\{l_i : 1 \le i \le m + \lceil \lg n \rceil + 7\} \cup \{h_i : 1 \le i \le \lceil \lg m \rceil + 1\} \cup$
  $\{B_{ij} : 1 \le i \le n \wedge 1 \le j \le m\} \cup \{A_{ij}, P_{ij} : 1 \le i < j \le n\} \cup$
  $\{D_{ij} : 1 \le i,j \le n\} \cup \{b, D, F_1, F_2, p, T, S, N, \alpha, \beta, \mathtt{yes}, \mathtt{no}\}$.
- $\Sigma = \{A_{ij} : 1 \le i < j \le n\}$.
- $\mathcal{E} = \Gamma \setminus \{a_1, b, c_1, \mathtt{yes}, \mathtt{no}, D, A_1, \ldots, A_n\}$.
- $w_1 = a_1\, b\, c_1\, \mathtt{yes}\, \mathtt{no}$ and $w_2 = D\, A_1 \ldots A_n$.
- $\mathcal{R}$ is the following rules set:
  1. *Division rules:*
     $r_{1,i} \equiv [A_i]_2 \rightarrow [B_i]_2[\alpha]_2$, for $i = 1, \ldots, n$
  2. *Communication rules:*
     $r_{2,i} \equiv (1, a_i/a_{i+1}, 0)$, for $i = 1, \ldots, 3n + m + \lceil \lg m \rceil + 13$
     $r_{3,i} \equiv (1, c_i/c_{i+1}^2, 0)$, for $i = 1, \ldots, n$
     $r_4 \equiv (1, c_{n+1}/D, 2)$
     $r_5 \equiv (2, c_{n+1}/d_1 e_1, 0)$
     $r_{6,i} \equiv (2, e_i/e_{i+1}, 0)$, for $i = 1, \ldots, 2n$
     $r_{7,ij} \equiv (2, d_j B_i/D_{ij}, 0)$, for $1 \le i,j \le n$
     $r_{8,ij} \equiv (2, D_{ij}/\overline{B}_i d_{j+1}, 0)$, for $1 \le i,j \le n$
     $r_{9,j} \equiv (2, e_{2n+1} d_j/f_{j-1}, 0)$, for $j = 1, \ldots, n$
     $r_{10} \equiv (2, f_k/l_1 F_1, 0)$
     $r_{11,i} \equiv (2, l_i/l_{i+1}, 0)$, for $i = 1, \ldots, m + \lceil \lg n \rceil + 6$
     $r_{12} \equiv (2, F_1/pF_2, 0)$
     $r_{13} \equiv (2, F_2/g_1 h_1, 0)$
     $r_{14,i} \equiv (2, g_i/g_{i+1}^2, 0)$, for $i = 1, \ldots, \lceil \lg n \rceil$
     $r_{15,i} \equiv (2, h_i/h_{i+1}^2, 0)$, for $i = 1, \ldots, \lceil \lg m \rceil$
     $r_{16,ij} \equiv (2, \overline{A}_{ij} h_{\lceil \lg m \rceil + 1}/P_{ij}, 0)$, for $1 \le i < j \le n$
     $r_{17,i} \equiv (2, g_{\lceil \lg n \rceil + 1} \overline{B}_i/B_{i1}, 0)$, for $i = 1, \ldots, n$
     $r_{18,ij} \equiv (2, B_{ij}/B_{ij+1} B'_i, 0)$, for $= 1, \ldots, n$ and $j = 1, \ldots, m$
     $r_{19,ij} \equiv (2, P_{ij} B'_i/P_j, 0)$, for $1 \le i < j \le n$
     $r_{20,j} \equiv (2, P_j B'_j/\beta, 0)$, for $j = 1, \ldots, n$
     $r_{21} \equiv (2, p\,\beta/\alpha, 0)$
     $r_{22} \equiv (2, l_{m+\lceil \lg n \rceil + 7} p/T, 0)$
     $r_{23} \equiv (2, T/\alpha, 1)$
     $r_{24} \equiv (1, bT/S, 0)$
     $r_{25} \equiv (1, S\mathtt{yes}/\alpha, 0)$

$r_{26} \equiv (1, a_{2n+m+\lceil \lg n \rceil+14} b \,/N, 0)$
$r_{27} \equiv (1, \mathtt{no}N/\alpha, 0)$

– $i_{in} = 2$, is the input cell.

In order to establish that the family $\mathbf{\Pi}$ is polynomially uniform by deterministic Turing machines we firstly note that the sets of rules associated with the systems $\Pi(n, m, k)$ are recursively defined. Hence, it suffices to justify that the amount of necessary resources for defining the systems is polynomial in $\max\{n, m, \lceil \lg k \rceil\}$. In fact, it is polynomial in $\max\{n, m\}$, since those resources are the following:

1. Size of the alphabet: $n(5n-1)/2 + n \cdot m + 8n + 2m + 3\lceil \lg n \rceil + \lceil \lg m \rceil + 36 \in \Theta(n^2 + m)$.
2. Initial number of cells: $2 \in \Theta(1)$.
3. Initial number of objects: $n + 6 \in \Theta(n)$.
4. Number of rules: $3n^2 + n \cdot m + 9n + 2\lceil \lg n \rceil + \lceil \lg m \rceil + 31 \in \Theta(n^2 + nm)$.
5. Upper bound for the length of the rules: $3 \in \Theta(1)$

As we will see in the following section, the family $\mathbf{\Pi}$ is also polynomially (in fact, linearly) bounded, sound and complete with regard to $(\mathtt{IS}, cod, s)$. So, we have the main result of the paper.

Theorem 1. *IS* $\in \mathbf{PMC}_{TD}$

Taking into account that IS is an **NP**–complete problem, and that the class $\mathbf{PMC}_{TD}$ is closed under complement, the following is deduced.

Corollary 1. $\mathbf{NP} \cup \mathbf{co} - \mathbf{NP} \subseteq \mathbf{PMC}_{TD}$

4 An Overview of the Computations

Next, we describe in detail the steps followed by the system $\Pi(s(u))$ when the input multiset $cod(u)$ is supplied, for an arbitrary instance u of IS. Let us note that the system starts with only two cells, one labelled by 1 and the other labelled by 2, and that division rules are only applied to cells labelled by 2. This means that along the computations there will always be a unique cell labelled by 1 (which we will call the 1-cell), but that new cells labelled by 2 (which we will call the 2-cells) will be produced.

In order for the system $\Pi(s(u))$ to decide the instance of the IS problem encoded by $cod(u)$, it starts with the *generation stage*, where cells for all the possible subsets of nodes of the graph are generated. This is performed by the successive application of the division rules. These rules take the objects A_i in the 2-cells, which encode the vertices of the graph, and produce two new 2-cells, one of them with the object B_i, meaning that we include the vertex in the subset, and the other without it, meaning that we do not include the vertex in the subset. Of course, all the remaining objects contained in the original 2-cells are replicated into the new ones. This way, at the end of this stage, which spends n steps, the system will have 2^n 2-cells, each of them encoding one and only one subset of vertices of the graph, by means of the objects B_i.

To control the end of this stage, the objects c_i in the 1-cell of the system are used as counters. Initially object c_1 is interchanged for two objects c_2 from the environment using rule $r_{3,1}$; each of these objects are again interchanged for two objects c_3 using rule $r_{3,2}$; and so on. Thus, at the end of the generation stage the 1-cell will contain 2^{n+1} objects c_{n+1}. On the other hand, objects a_i in the 1-cell of the system are used as global counters of the computation by means of the rules $r_{2,i}$. Note that this generation stage is non-deterministic, but it is easy to check its confluence: independently of the way the division rules are applied, at the end of the stage the same configuration is always reached.

Once all the subsets of vertices of the graph are generated, the *pre-checking stage* selects only those of size k. This stage is activated by the rules r_4 and r_5, which interchange the object D of each 2-cell (recall that there are 2^n of them) with an object c_{n+1} of the 1-cell (recall that there are 2^{n+1} of them), and then each of the latter in each 2-cell with one object d_1 and one object e_1 of the environment (recall that there are infinitely many of them). From now on, the 1-cell will wait counting the number of steps of the computation by means the objects a_i and the rules $r_{2,i}$.

The objects d_1 and e_1 start two processes of counting in each 2-cell. The first one counts the number of steps of the stage that have been performed, and it is controlled by the objects e_i, which are repeatedly interchanged by objects e_{i+1} from the environment using the rules $r_{6,i}$.

The second process counts the number of vertices in the subset. It is performed using the rules $r_{7,ij}$ and $r_{8,ij}$, which interchange the objects B_i in the 2-cells by objects $\overline{B}_i$ (indicating this way that the corresponding vertex has been counted) and increase the counter d_j (the only purpose of the objects D_{ij} is to decrease the length of the rules). Note that this is a non-deterministic process, since the vertex "counted" in each step is chosen in a non-deterministic way. However, as the size of the subsets of vertices is upper bounded by n, after $2n$ steps the same configuration is always reached, so this stage is also confluent.

Note that for the counter d_j of a 2-cell to increase, it is necessary and sufficient that in that cell there exist objects B_i left. This means that at the end of the process explained in the previous paragraph, the only 2-cells that contain objects encoding subsets of size k are those containing the object d_{k+1}. At this moment, those cells also contain the counter e_{2n+1}, which then in two steps cause (using the rules $r_{9,j}$ and r_{10}, and the intermediate objects f_j for rules size reduction) the object d_{k+1} to be interchanged by objects l_1 and F_1 of the environment.

The *checking stage* starts now, but before we can check if any of the subsets of vertices of size k selected in the previous stage is an independent set of the graph, we need some preparation steps. First of all, the objects l_i will be used as a counter of the number of steps performed, controlled by rules $r_{11,i}$.

Simultaneously, the objects p and F_2 are traded against the object F_1 (by applying the rule r_{12}). In this way, the counters g and h appear (rule r_{13}) and they are duplicated (rules $r_{14,i}$ and $r_{15,i}$, respectively) till producing at least n copies of the object $g_{\lceil \lg n \rceil+1}$ and m copies of the object $h_{\lceil \lg m \rceil+1}$ respectively.

In the step $3n + \lceil \lg m \rceil + 7$, the rule $r_{16,ij}$ produces the trade of the objects $\overline{A}_{ij}$ and $h_{\lceil \lg m \rceil + 1}$ against the objects P_{ij}. On the other hand, in the step $3n + \lceil \lg n \rceil + 7$ the objects $\overline{B}_i$ and $g_{\lceil \lg n \rceil + 1}$ are traded against the objects B_{i1}. Along the following m steps, the rules $r_{18,ij}$ (i=1, …, n and $j = 1, \ldots, m$) are applied and the produce the occurrence of m objects B'_i (one in each step) in the cells with label 2 which remain active after the pre-checking stage.

In the next step, the rules $r_{19,ij}$ are applied in the cells with label 2. Such rules trade the objects P_{ij} and B'_i against the objects P_j, and, in the next step, the objects P_j and B'_j are trade against an object β by means of the rules $r_{20,j}$.

Since m copies of the object B'_i are obtained, then the last step in which an object β can appear in a cell 2 is the step $3n + m + \lceil \lg n \rceil + 9$. In this way, the following step is the last one in which an object p can be sent to the environment (by applying the rule r_{21}, and by finishing the checking stage.

The *answer stage* starts when an object $l_{m+\lceil \lg n \rceil + 7}$ appears in a cell with label 2. Two possibilities must be considered.

If there exists an object p in a cell labelled by 2 when the checking stage is finished, then the subset encodes an independent set of size k of the graph $G = (V, E)$, and, by applying the ruler_{22}, at least one object T appears in the cell. Then, the rules r_{23}, r_{24} and r_{25} are applied and an object yes is sent to the environment. This ends the computation.

If no object p remains in any cell with label 2 after the checking stage, then the subset does not encode an independent set of the graph $G = (V, E)$ of size k and the rule r_{22} cannot be applied. Then, in the step $3n + m + \lceil \lg n \rceil + 12$, no object T arrives to the cell with label 1 and the rules r_{24} and r_{25} cannot be applied. In this way, the counter a reaches the object $a_{3n+m+\lceil \lg n \rceil + 14}$ which is traded together the object b against an object N by means of the rule r_{26}. Such object N is sent together with an object no to the environment in the step $3n + m + \lceil \lg n \rceil + 15$, and this ends the computation.

5 Conclusion and Future Work

As pointed above, the study of tractability in Membrane Computing opens a new perspective of the classical problem **P** vs. **NP** from a biologically inspired point of view. This research field is only a few years old and much work needs to be done in this line, not only in the theoretical results about tractability, but also in the development of new skills on the design of cellular solutions.

Following the ideas used in [2] (where an efficient solution of the 3–coloring problem was presented), an schema for solving **NP**–complete problems of graph theory has been inferred. It is used in this paper for presenting a first (linear–time) solution to the Independent Set problem.

More open questions in the framework of Membrane Computing related to tractability can be considered. For example, it is interesting to investigate the possibility to address efficient solutions to **NP**-complete problems in different frameworks allowing to produce an exponential number of cells in linear time (e.g. cell creation or separation).

It is also worth investigating a lower bound on the length of the symport/antiport rules used by the system (in this paper, rules of length at most 3 are used). Another research problem is related to the number of objects in the environment. One could also consider a tissue-like P system where the environment always has a finite amount of objects.

Acknowledgment

The authors wish to acknowledge the support of the project TIN2006-13425 of the Ministerio de Educación y Ciencia of Spain, cofinanced by FEDER funds, and the support of the project of excellence TIC-581 of the Junta de Andalucía.

References

1. Díaz–Pernil, D.: Sistemas P de Tejido: Formalización y Eficiencia Computacional. PhD Thesis, University of Seville (2008)
2. Díaz–Pernil, D., Gutiérrez–Naranjo, M.A., Pérez–Jiménez, M.J., Riscos–Núñez, A.: A uniform family of tissue P system with cell division solving 3-COL in a linear time. Theoretical Computer Science 404, 76–87 (2008)
3. Díaz–Pernil, D., Gutiérrez–Naranjo, M.A., Pérez–Jiménez, M.J., Riscos–Núñez, A.: Solving Subset Sum in linear time by using tissue P systems with cell division. In: Mira, J., Álvarez, J.R. (eds.) IWINAC 2007. LNCS, vol. 4527, pp. 170–179. Springer, Heidelberg (2007)
4. Bernardini, F., Gheorghe, M.: Cell Communication in Tissue P Systems and Cell Division in Population P Systems. Soft Computing 9(9), 640–649 (2005)
5. Freund, R., Păun, G., Pérez-Jiménez, M.J.: Tissue P Systems with Channel States. Theoretical Computer Science 330, 101–116 (2005)
6. Krishna, S.N., Lakshmanan, K., Rama, R.: Tissue P Systems with Contextual and Rewriting Rules. In: Păun, G., Rozenberg, G., Salomaa, A., Zandron, C. (eds.) WMC 2002. LNCS, vol. 2597, pp. 339–351. Springer, Heidelberg (2003)
7. Martín Vide, C., Pazos, J., Păun, G., Rodríguez Patón, A.: A New Class of Symbolic Abstract Neural Nets: Tissue P Systems. In: H. Ibarra, O., Zhang, L. (eds.) COCOON 2002. LNCS, vol. 2387, pp. 290–299. Springer, Heidelberg (2002)
8. Martín Vide, C., Pazos, J., Păun, G., Rodríguez Patón, A.: Tissue P systems. Theoretical Computer Science 296, 295–326 (2003)
9. Pérez-Jiménez, M.J., Romero-Jiménez, A., Sancho-Caparrini, F.: A polynomial complexity class in P systems using membrane division. Journal of Automata, Languages and Combinatorics 11(4), 423–434 (2006)
10. Păun, G.: Computing with membranes. Journal of Computer and System Sciences 61(1), 108–143 (2000)
11. Păun, G.: Membrane Computing. An Introduction. Springer, Berlin (2002)
12. Păun, A., Păun, G.: The power of communication: P systems with symport/antiport. New Generation Computing 20(3), 295–305 (2002)
13. Păun, G., Pérez-Jiménez, M.J., Riscos-Núñez, A.: Tissue P System with cell division. International Journal of Computers, Communications & Control III (3), 295–303 (2008)
14. P systems web page, `http://ppage.psystems.eu/`

How to Do Recombination in Evolution Strategies: An Empirical Study

Juan Chen, Michael T.M. Emmerich*, Rui Li,
Joost Kok, and Thomas Bäck

Natural Computing Group, Leiden University,
P.O. Box 9500, 2333 CA Leiden, The Netherlands
{emmerich,ruili,joost,baeck}@liacs.nl

Abstract. In Evolution Strategies (ES) mutation is often considered to be the main variation operator and there has been relatively few attention on the choice of recombination operators. This study seeks to compare advanced recombination operators for ES, including multi-parent weighted recombination. Both the canonical $(\mu \overset{+}{,} \lambda)-$ES with mutative self-adaptation and the CMA-ES are considered. The results achieved on scalable (non-)separable test problem indicate that the right choice of recombination has an considerable impact on the performance of the ES. Moreover, the study will provide empirical evidence that weighted multi-parent recombination is a favorable choice for both ES variants.

1 Introduction

Evolution Strategies (ESs) are optimization meta-heuristics that especially aim for approximating minima or maxima of continuous optimization problems. The mutation is considered to be the most essential variation operator in ES. Like genetic algorithms (GA), ES employ a variation-selection scheme in order to gradually move a population of individuals towards the optimum of a given objective function. The mutation operator in ES has been intensively studied. Many alternative techniques for strategy parameter adaptations are implemented in ES, like cumulative step-size adaptation (CSA), and covariance matrix adaptation (CMA). However, only a few, simple ways of how to implement the recombination operator in ES have been studied, such as discrete recombination and intermediate recombination. As a promising new variant of recombination, weighted multi-parent recombination has been suggested by various authors ([1], [6], and [8]). Here we adopt the idea and study it for the first time in the context of canonical evolution strategy. Among other results, our empirical investigation suggests that weighted recombination is an interesting alternative to other recombination operators and deserves to be studied in more detail.

The remainder of this paper is organized as follows: In section 2 we will give further introduction of recombination in ES, especially the weighted recombination. Section 3 briefly presents the test functions and the experimental settings.

* Corresponding author.

J. Mira et al. (Eds.): IWINAC 2009, Part I, LNCS 5601, pp. 223–232, 2009.

Then a comparative review of recombination (mechanisms) in ESs is given in the fourth section. Conclusions and outlook form the last section.

2 Recombination in ES

Recombination is recommended to be used in ESs for the creation of all offspring individuals, when $\mu > 1$. There are two main theories: building block hypothesis (BBH) and genetic repair hypothesis (GRH) [3], attempting to explain the possible increase of performance by recombination.

In 1975, Holland suggested the BBH, which attempts to explain the working mechanism of crossover in GA: by combining the partial (relatively small) highly fit schemata, the so-called building blocks (BBs), the beneficial properties of the parents are aggregated in the offspring by crossover. BBH is widely believed in GA area, however, it seems not useful in the context of evolution strategies, as empirical results suggest that learning building blocks seems not to be very beneficial for the performance of ES [4]. A possible explanation for this negative result is, that in ESs a relatively high ratio λ/μ leads to a rapid loss of genetic diversity in the population ([9]).

The alternative hypothesis on the effect of recombination, GRH, aims to explain the effect of recombination on ES performance. It is believed that the usefulness of recombination is provided by its ability to conserve the beneficial parts of the mutations and dampening their harmful parts through genetic repair.

In Beyer's proof of GRH on the sphere model, the global intermediate recombination operator (multi-recombination) is investigated to show that since more parents take part in the recombination, more beneficial information is gathered and better components are preserved. Unlike the standard recombination operators in ES, which are all based on two parents, the global intermediate recombination operator averages the whole population to generate one offspring individual. In the global intermediate recombination, every individual in the population takes the same probability in generating offspring. But there is still some other available information for a better recombination: the fitness value. Weighted recombination makes effective use of this information based on the fitness rank and applies it to both object variables and strategy parameters in recombination.

Weighted recombination was firstly employed by Hansen and Ostermeier [6] in their implementation of covariance matrix adaptation evolution strategy. The use of weighted recombination in the CMA-ES was suggested by Rechenberg in a personal communication with Hansen [6]. However, to our knowledge, the choice of this particular recombination operator has not been justified by published results yet. Moreover, weighted recombination has not yet been tested in the canonical ES with mutative step-size adaptation.

In our experiments, we implemented weighted recombination in the standard (μ, λ)-ES. Here we give an example about how we implemented it, based on the standard (μ, λ)-ES with non-isotropic adaptation:

Table 1. Recombination Operators

Notation	Meaning	Notation	Meaning
N	No recombination	G	Global intermediate recombination
D	Discrete recombination	W	Weighted recombination
I	Intermediate recombination	1	Only the one with highest fitness is selected

$$\sigma_i' = (\frac{1}{\sum_{i=1}^{\mu} w_i} \sum_{i=1}^{\mu} w_i \sigma_i) \exp(\tau' \mathcal{N}(0,1) + \tau \mathcal{N}_i(0,1)) \tag{1}$$

$$x_i' = (\frac{1}{\sum_{i=1}^{\mu} w_i} \sum_{i=1}^{\mu} w_i x_i) + \sigma_i' \mathcal{N}_i(0,1) \tag{2}$$

In the above equations, the weights for weighted recombination are set as:

$$w_i = \log(\mu + 1) - \log(i); \forall i \in \{1, ..., \mu\} \tag{3}$$

following [6]. Here i denotes the rank of the individual according to its fitness (highest i corresponds to best individual).

The optimal recombination weights depend on the function to be optimized, and it remains an open question how to choose the weights for different kinds of problems in order to achieve a better performance. About the choice of weights, Rudolph [8] made analytical investigation of the consequences of different weights for optimization on the 3-dimensional sphere model, on the basis of ES with uniform mutation on a spherical surface. He found that negative weights could yield a beneficial effect for the progress rate of the weighted strategy. Arnold [1] studied the optimal weights of a so-called $(\lambda)_{opt}$-ES with cumulative step length adaptation on the sphere model both experimentally and analytically (for infinite population size) [1]. A study by Rudolph deals with 3-D cases with a modified mutation operator that places the offspring randomly at the boundary of a sphere around the parent [8]. However, an analysis of weighted recombination for the ES variants used in practise, such as the CMA and the canonical ES with mutative step-size adaptation, is still missing.

In order to see the usefulness of recombination, both ESs with and without recombination will be investigated in this paper. Because recombination is also applied to strategy parameters (standard deviations and rotation angles), the notation of recombination operators are defined as r_{XXX} (Following the representation of recombination operators in [2]), in which r means recombination operator and X denotes the recombination type (see table 1). The three indices refer to object variables, standard deviations, and rotation angles, respectively. This is an example for correlated ES (CorrES), in which two sets of strategy parameters exist. For isotropic ES (IES) and non-isotropic ES (NIES), only two indices are required because they only have standard deviations as their strategy parameters. For different recombination, please refer to table 1 for the corresponding notations.

Table 2. N-dimensions test functions(minimization)

Function	Bounds
$f_{Cigar}(y) = y_1^2 + \sum_{i=2}^{n}(1000y_i)^2$	$[-100, 100]$
$f_{Tablet}(y) = (1000y_1)^2 + \sum_{i=2}^{n} y_i^2$	$[-100, 100]$
$f_{Elliptic}(y) = \sum_{i=1}^{n}(1000^{(i-1)/(n-1)}y_i)^2$	$[-100, 100]$
$f_{Schwefel}(y) = \sum_{i=1}^{n}(\sum_{j=1}^{i} y_j)^2)$	$[-100, 100]$
$f_{Rosenbrock}(y) = \sum_{i=1}^{n-1}(100(y_i^2 - y_{i+1})^2) + (y_i - 1)^2)$	$[-5.12, 5.12]$

3 Experimental Setup

Three types of *test functions* are investigated in this paper: separable problems, non-separable problems, and semi-separable problems (SSP). They are all real-valued functions with a scalable number of variables and without noise. Because this paper mainly compares the performance of different recombination operators by testing their local convergence speed, the experiments focus on unimodal problems, though with the Rosenbrock's function also one multimodal problem has been included. Furthermore, test functions with only one non-separable problem were also tested. These non-separable problems are suited to see how ESs deal with tight linkage.

In addition to Schwefel's function and step-up function, Rosenbrock's function is also added to this test category. Rosenbrock's function is a multi-modal, non-separable and scalable problem, and its global optimum is $y^* = 1$, $f(y^*) = 0$. In [10], Rosenbrock's function is classified as a multi-modal function. It has a very narrow valley from local optimum to global optimum. A detailed analysis of Rosenbrock's function can be found in [11].

It is reasonable to test separable problems like the sphere function, because they are widely used in experimental and theoretical investigations. Since the sphere function is an equally scaled problem, some other unequally scaled problems (i.e. Elliptic function) are also used to test the performance of weighted recombination. These functions are selected from the benchmarks of CMA-ES [6].

There are no general *parameter settings* in ESs. Rather, optimal settings vary with the problem characteristics. However, in [2] and [6] some default settings, which can be followed for most of the situations, are mentioned. The parameter settings, especially for strategy parameters, are chosen according to these suggestions. Table 3 provides the settings implemented in the test cases whenever no extra mention about the parameterizations is made. In comparisons between ESs and CMA, the settings of (μ_W, λ)-CMA-ES follows the default settings in [6] (the Table 1 in p.19), and $\mu = \lfloor \lambda/2 \rfloor$.

Most of the performance assessment criteria follow the setting given in [10]. The termination conditions of the evaluation are decided by two stopping rules: 1. Maximum number of function evaluations (FES); 2. Fitness values drop below threshold. The default maximum number of function evaluations is set to $10000 \cdot n$. The threshold in the second criterion is set to an absolute error of 10^{-10} (or 10^{-6}). A run would be terminated if the error in the function value reaches 10^{-10} (or 10^{-6}), or else it would be stopped as the number of function evaluations (FES)

Table 3. Parameter Settings

Parameters	Notation	Value
Initial standard deviation	$\sigma^{(0)}$	1
Initial rotation angles	$\alpha^{(0)}$	0
Initial object variables	x_0	random generated within the given bounds
Offspring population size	μ	15
Parent population size	λ	100
Stop fitness	f_{stop}	10^{-10} or 10^{-6}
Max number of function evaluations	MaxFES	$10000 \cdot n$
Success rate	p_s	(# of successful runs)/(total # of runs)

Table 4. Mutation settings

S	Mutation Settings
Isotropic ES	$\tau_0 = 1/\sqrt{n}$
Non-Isotropic ES	$\tau = 1/\sqrt{2\sqrt{n}}$; $\tau' = 1/\sqrt{2n}$
Correlated ES	$\tau = 1/\sqrt{2\sqrt{n}}$;$\tau' = 1/\sqrt{2n}$;$\beta = 5°$

reaches the maximum value. A run is successful if it reaches the stopping criterion based on fitness before the maximum number of function evaluations is exceeded. All tests run 25 times for each algorithm, and figures are drawn according to the median of the 25 runs. We also provided a technical report with more detailed results. Whenever a parameter of these default settings reported in this section is changed, it will be explicitly mentioned in the corresponding figure.

4 Results and Analysis

As mentioned in the third section, mainly three different types of problems are evaluated: SSP, separable problems, and non-separable problems. Different kinds of crossover are implemented in (μ,λ)-ESs. Since weighted recombination shows unexpected good performance, this part mainly focuses on the performance of weighted recombination on the above functions. At the same time, the other recombination operators (i.e. discrete recombination, global intermediate recombination) are also investigated to make a comparison. The type of ES (i.e. IES, NIES, and CorrES) applied is chosen according to the test functions.

The elliptic function, cigar function and tablet function are all badly scaled *separable functions.* As suggested in [5], NIES "is capable of learning to self-adapt to problems where variables are unequally scaled in the objective function". Therefore unlike the sphere function, which performs best with IES, the solution of these problems demands for NIES.

Figure 1 shows the changing trend of the performance of ESs w.r.t. problem dimension and recombination mechanism. The studied multi-recombination operators show a high performance for separable problems in ESs. In these experiments, n varies from 10 to 80, except for the sphere function, global intermediate recombination stays close to weighted recombination. When the dimension

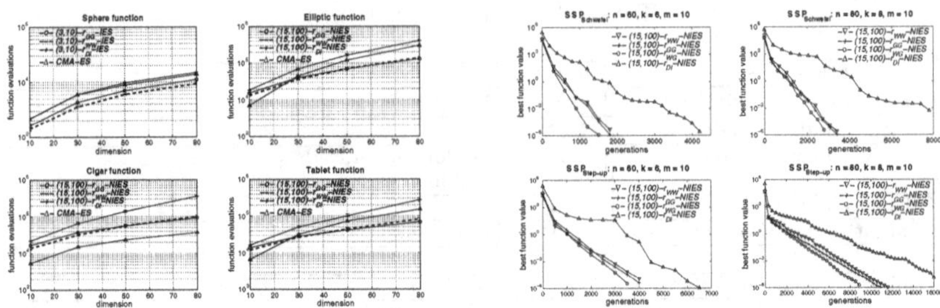

Fig. 1. Separable problems with IES/NIES

Fig. 2. SSP function (sub-problems: Schwefel function) with (15,100)-NIES

is large ($n > 50$), it overtakes weighted recombination w.r.t. convergence speed. Discrete recombination performs roughly 1.3 to 4 times slower on these separable problems, and the factor increases with the increase of the problem dimension.

Since CMA-ES [6] is often suggested as the state-of-the-art algorithm in real-valued evolutionary algorithms and ESs with weighted recombination show outstanding performance, the comparison between these two is also investigated on these separable problems ($n = 10; 30; 50; 80$), as shown in figure 1. On the sphere function, the relative difference between them is constant with the growth of problem dimension. In the experiment, the initial step-size is set to 1 on the sphere functions, for both CMA-ES and r_{WW}-IES. CMA-ES converges approximately 1.5 times slower than r_{WW}-IES. r_{WW}-IES with optimal step size could perform better. Like the sphere function, the relative difference between CMA-ES and r_{WW}-IES on the Cigar function stays constant with different dimension. The only difference is that in this case, r_{WW}-IES performs around 2 times slower than CMA-ES. On the Elliptic function and Tablet function, CMA-ES outperforms r_{WW}-NIES when $n < 30$, the speed-up factor is up to 2 (when n is small). Since the function evaluations cost here is not too much and the dimension is relatively small, there is not too much difference between these two algorithms. However, CMA-ES is 2 times slower when $n > 50$. Meanwhile, because of expensive matrix operations, the CPU-time cost of CMA-ES here is much more than r_{WW}-NIES, it growth superlinearly as the dimension increases [6].

Like in case of the completely separable problems, weighted recombination makes a remarkable acceleration of the convergence speed on these SSP. According to figure 2, we can figure out that multi-recombination is good for SSP. When n is large, weighted recombination could achieve a speed-up factor of two to three compared with the other non-global recombination. There are three multi-recombination operators used in the experiments, other than r_{WW} and r_{GG}, r_{WG}. r_{WW} and r_{GG} have a similar performance. r_{WG} shows better performance, especially for $SSP_{step-up}$ function. This implies that the performance of ESs could benefit from rank-based weights applied to strategy parameters, and equal weights applied to the object variables. Note, that a variant of discrete crossover that respects the building blocks of the semi-separable problems,

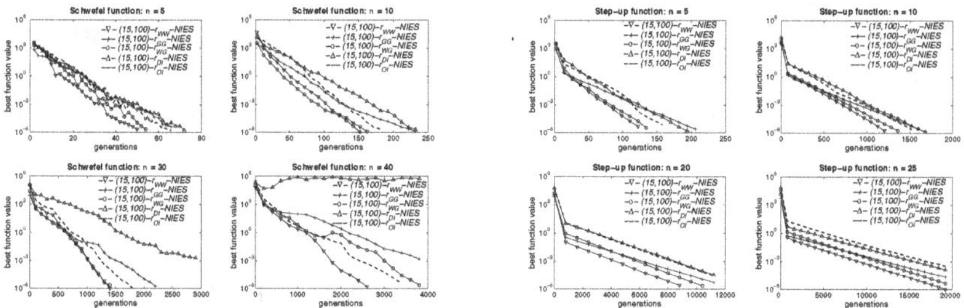

Fig. 3. Schwefel function with (15,100)-NIES

Fig. 4. Step-up function with (15,100)-NIES

could not find as good results as these variants, though it outperformed discrete crossover.

Comparing the results on *non-separable problems* in figure 3 and 4, weighted recombination performs best compared with the other recombination with (15,100)-NIES. Still, discrete crossover is the worst one in the comparison list, even worse, it cannot converge when n is large (see $n = 40$ in figure 3).

Since weighted recombination allows better individuals to contribute more when generating offsprings, and it performs well in all the experiments, does this mean that the fittest individual is the most essential for its success and maybe should we use the fittest one as the only parent for every generation? However, according to the tests the performance of the (1,λ)-IES/NIES, varies with test functions, sometimes clearly better and sometimes clearly worse, w.r.t. weighted recombination. In the next section we will present the experiment results and analyze them.

In figure 5 and 6, a new recombination operator r_{1W} is added aiming to identify the main effect of weighted recombination facilitating the performance of ESs. From these figures, it is clear to see that r_{1W} appears similar with r_{WW} on Schwefel's function, and faster than r_{WW} on Rosenbrock's function and step-up function, when n is smaller than 30. This could be explained by the effect of the

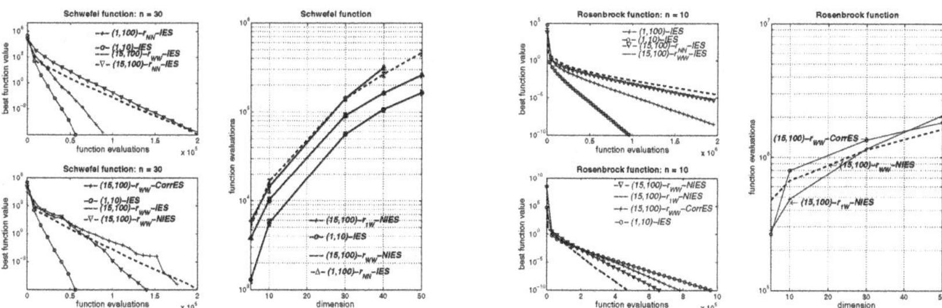

Fig. 5. Schwefel function with IES/NIES/CorrES. Stop fitness: 10^{-6}; maxFES: 10000·n.

Fig. 6. Rosenbrock function with IES/NIES/CorrES. Stop fitness: 10^{-10}; maxFES: 200000·n.

fittest individual and the weighted recombination of strategy parameters. It is not hard to imagine that the fittest individual contains more desired information about the linkage. Object variables and strategy parameters with higher fitness should participate more in the generation of offspring. With guidance by the best fitness, mutation could follow a better direction and improve the local search performance of ESs. However, it is not enough to just take the best one as the parent for the next generation. (1,λ)-ESs will be studied in the next section.

Until now, the ESs analyzed still remain in IES and NIES. This section concentrates on correlated problems, such that it is reasonable to think about the usage of CorrESs. Rosenbrock's function is a standard function for introducing the importance of mutation according to the covariance of variables. Therefore, it is chosen to be the test function for comparison between NIES and CorrES. It also explains why we chose NIES in the experiments here.

In the right part of figure 6, for Rosenbrock's function, when n is small (around 2 to 5), (15,100)-r_{WW}-CorrES performs best, but as n is increased, (15,100)-r_{WW}-NIES performs better than CorrES. Meanwhile, because NIES is simpler than CorrES in the mutation part in every generation is less than CorrES, and the total computation time would be much less, especially when n is large. Considering how large the number of mutation angles gets as dimension increases, the cost of CorrES in terms of CPU time can be very unreasonable. Its computational complexity is $o(n^2)$. This time difference between NIES and CorrES would be enlarged dramatically when n is larger than 40. Therefore, (15,100)-r_{WW}-NIES would be preferable as function evaluations are computationally cheap for Rosenbrock's function in ESs.

As figure 5 shows, for Schwefel function, (1,10)-IES only need half of the function evaluations needed by (1,100)-IES, possibly because the (1,10)-IES updates the search point more often than (1,100)-ES. However, (1,10)-IES is not suitable for global search. In this section, we mainly show the performance of (1,10)-IES with unimodal problems.

As there is no recombination in (1,λ)-IES, meanwhile, with only one identical step-size, it is already the simplest ES. Consequently, for Schwefel and step-up related functions, (1,λ)-IES not only performs well for a smaller number of generations, but also much faster in CPU time for each function evaluation.

Although the (1,10)-IES performs significantly better for $SSP_{Schwefel}$, $SSP_{Step-up}$, Schwefel's function, and step-up function, the other IES with local (or sexual) or global recombination still performs worse than the NIES with corresponding recombination. Ostermeier et al. [7] explains that (1,10)-IES outperforms (15/2,100)-CorrES "because of the rotation of the ellipsoid axes, the initialization with identical individual step-sizes is optimal or nearly optimal", but only (1,λ)-IES outperforms NIES and CorrES, the other IES do not perform well, which also mutate with identical individual step-sizes.

Usually, selective pressure is not recommended to be too strong. But in (1,λ)-IES for Schwefel's function, step-up function, $SSP_{Schwefel}$ and $SSP_{step-up}$, the advantage of best fitness is more important. But there remains a need for further

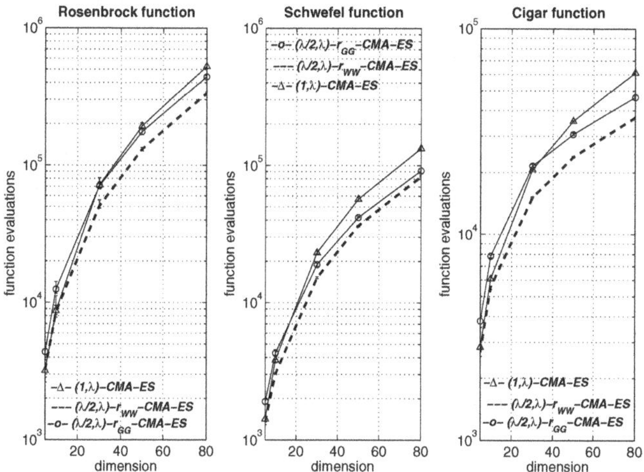

Fig. 7. Rosenbrock's function, Schwefel's function and Cigar function with CMA-ES. Stop fitness: 10^{-10}; maxFES: 10000·n; $\lambda = 4 + \lfloor 3 * \log(n) \rfloor$; $\sigma_0 = 1$.

effort to figure out the underlying reason, and to assess the robustness on noisy and multimodal problems.

Also for the *CMA* the weighted recombination scheme seems to achieve much better results, as indicated by the results on three different problems (cf. Fig. 7). It is remarkable that on the one hand, a strategy with no recombination, and on the other hand a strategy with equally weighted recombination partners achieve worse results. However, it remains a problem for future research if there are even better weighting schemes than the logarithmic one presented here.

Supplementary results (standard deviations, etc.) and the matlab implementation of the ES are provided as supplementary material in the technical report (`http://www.liacs.nl/$^\sim$emmerich/IWINAC09supp.pdf`).

5 Conclusions and Outlook

As observed in the experiments, weighted recombination provides an efficient mechanism that remarkably accelerates the local search performance of ESs, without significantly increasing computational costs per function evaluation. According to the test results in separable problems, non-separable problems and semi-separable problems, weighted recombination always outperforms discrete recombination significantly. Speed-up factors range from 1.3 to 4, increasing with the problem dimension.

In conclusion, based on the experimental results in this paper, weighted recombination should be preferred in (μ, λ)-ESs, at least if the precision of convergence to local optimum needed to be increased. Moreover, the comparison with CMA suggests that the canonical ES, using weighted recombination, can be a favorable choice for separable problems and for high-dimensional non-separable problems, for which the CPU time of CMA can be prohibitive high.

This paper mainly focused on experimental evaluation of different recombination operators applied in (μ,λ)-ESs, with a focus on weighted recombination. Comparing with all the other recombination operators investigated, weighted recombination turns out to be the best choice for ESs. However, theoretical investigations of weighted recombination in (μ,λ)-ESs and the optimal choice weights is still an open question. Based on our results it will be worthwhile to theoretically investigate the effect of weighted recombination.

References

1. Arnold, D.V.: Optimal Weighted Recombination. In: Wright, A.H., Vose, M.D., De Jong, K.A., Schmitt, L.M. (eds.) FOGA 2005. LNCS, vol. 3469, pp. 215–237. Springer, Heidelberg (2005)
2. Bäck, T.: Evolutionary algorithms in theory and practice: Evolution Strategies, Evolutionary Programming, Genetic Algorithms. Oxford University Press, New York (1996)
3. Beyer, H.-G.: Toward a Theory of Evolution Strategies: On the Benefit of Sex-the (μ/μ, λ)-Theory. Evolutionary Computation 3(1), 81–111 (1995)
4. Chen, J.: On Recombination in (μ,λ) Evolution Strategies. Technical Report, LIACS, Leiden (2006)
5. Deb, K., Beyer, H.-G.: Self-Adaptive Genetic Algorithms with Simulated Binary Crossover. Evolutionary Computation 9(2), 197–221 (2001)
6. Hansen, N., Ostermeier, A.: Completely Derandomized Self-Adaptation in Evolution Strategies. Evolutionary Computation 9(2), 159–195 (2001)
7. Ostermeier, A., Gawelczyk, A., Hansen, N.: A derandomized approach to self adaptation of evolution strategies. Evolutionary Computation 2(4), 369–380 (1995)
8. Rudolph, G.: Convergence Properties of Evolutionary Algorithms. Verlag Dr. Kovac, Hamburg (1997)
9. Schönemann, L., Emmerich, M., Preuss, M.: On the extinction of evolutionary algorithm subpopulations on multimodal landscapes. Informatica (Slovenia) 28(4), 345–351 (2004)
10. Suganthan, P.N., Hansen, N., Liang, J.J., Deb, K., Chen, Y.-P., Auger, A., Tiwari, S.: Problem Definitions and Evaluation Criteria for the CEC 2005 Special Session on Real-Parameter Optimization. Technical Report AND KanGAL Report 2005005, IIT Kanpur, India (2005)
11. Shang, Y.-W., Qiu, Y.-H.: A Note on the Extended Rosenbrock Function. Evolutionary Computation 14(1), 119–126 (2006)

Serial Evolution

V. Fischer[1,3], A.M. Tomé[2], and E.W. Lang[1]

[1] CIMLG, Institute of Biophysics, University of Regensburg, D-93040 Regensburg, Germany
elmar.lang@biologie.uni-regensburg.de

[2] DETI/IEETA, Universidade de Aveiro, 3810-193 Aveiro, Portugal
ana@ieeta.pt

[3] Experimental Psychology, University of Regensburg, D-93040 Regensburg, Germany
volker.fischer@psychologie.uni-regensburg.de

Abstract. Genetic algorithms (GA) represent an algorithmic optimization technique inspired by biological evolution. A major strength of this meta-heuristic is its ability to explore the search space in independent parallel search routes rendering the algorithm highly efficient if implemented on a parallel architecture. Sequential simulations of GAs frequently result in enormous computational costs. To alleviate this problem, we propose a serial evolution strategy which results in a much smaller number of necessary fitness function evaluations thereby speeding up the computation considerably. If implemented on a parallel architecture the savings in computational costs are even more pronounced. We present the algorithm in full mathematical detail and proof the corresponding schema theorem for a simple case without cross-over operations. A toy example illustrates the operation of serial evolution and the performance improvement over a canonical genetic algorithm.

1 Introduction

The idea of evolutionary computing was introduced in the 1950s by N. A. Barricelli [3] and shortly afterwards applied to solve large scale technical problems by I. Rechenberg [9]. Genetic Algorithms (GA) [4], [5], [6], [10] belong to the group of evolutionary algorithms and represent a search and optimization method inspired by the evolution of biological systems. A genetic algorithm is generally any population based model that uses selection and recombination operators to generate new sample points in a search space. The first GAs were developed by mathematician J. H. Holland [7] and his coworkers in the 1970s and have since been applied to many real world optimization problems. While being slower than other optimization techniques like, e.g., gradient descent, GAs have the considerable advantage of being applicable to discrete or discontinuous problems or to object functions with many local minima. We propose a serial evolution strategy which results in a much smaller number of necessary fitness function evaluations thereby speeding up the computation considerably. If implemented on a parallel architecture the savings in computational costs are even more pronounced. We present the algorithm in full mathematical detail and proof the corresponding schema theorem for a simple case without cross-over operations. A toy example illustrates the operation of serial evolution and the performance improvement over a canonical genetic algorithm.

J. Mira et al. (Eds.): IWINAC 2009, Part I, LNCS 5601, pp. 233–244, 2009.

2 Canonical Genetic Algorithm

The core of a GA consists of a population of individuals as well as of a set of mathematical methods which are used to mimic biological reproduction. Each population $P(t)$ comprises N_{ind} individuals (chromosomes, bit strings) $\mathbf{x}^i(t)$. Every individual in the population will be a binary string of length L which corresponds to the problem encoding. Such a string is called a "chromosome" which is usually initiated randomly. It represents an attempt to solve the optimization problem under investigation. After creating an initial population each string is then evaluated and assigned a fitness value. Afterwards these individuals are allowed to propagate by mating, a process, in which the information stored in two of the original individuals (also called parent individuals) is combined to produce an offspring. In this process, parent individuals which exhibit a better performance with respect to the optimization problem are allowed to produce more offsprings than those which perform poorly. Hence, the majority of offsprings inherits critical information from already well performing individuals and is thus assumed to solve the optimization problem even better than the parents.

After mating, small fragments of the information stored in the offsprings are altered randomly in a process called mutation. This is done in order to enable offsprings to explore areas of the search space which were not covered by any parent individuals yet. Finally, the offsprings are inserted into the overall population and the cycle of selection, mating, mutation and reinsertion is repeated until a suitable solution is found, all the individuals represent the same solution or the search is abandoned.

Usually there are only two main components of most genetic algorithms that are problem dependent: the problem encoding and the evaluation function f. The latter is related with the fitness function $\tilde{f}_i(t) = f_i(t)/ < f > (t)$ where $f_i(t) = f(\mathbf{x}^i(t))$ is the evaluation associated with string $\mathbf{x}^i(t)$ and $< f > (t) = N_{ind}^{-1} \sum_i f_i(t)$ is the average evaluation of all strings in the population.

A schematic metaprogram illustrates the main steps of a canonical genetic algorithm:

procedure evolution
begin
  $t \leftarrow 0$
  initialize population $P(t)$
  evaluate all individuals $\mathbf{x}^i(t) \in P(t)$
  while (**not** termination-condition) **do**
    $t \leftarrow t+1$
    select $P(t)$ from $P(t-1)$
    alter $P(t)$
    evaluate $P(t)$
  end
end

In the following, the individual steps, i.e. selection followed by cross-over and mutation, carried out during one generation of a canonical genetic algorithm will be elucidated considering as example the optimization of a fitness function $\tilde{f}(\mathbf{x})$.

2.1 Population Representation and Initialization

The parameter set $\mathbf{x}^{max}(t) = [x_1^{max}(t), x_2^{max}(t), ..., x_U^{max}(t)], x_u^{max}(t) \in \{0,1\}$ is sought for which the evaluation function $f(x)$ reaches its optimum. For this purpose, a population consisting of N_{ind} individuals is needed which explores the search space and which is assumed to eventually converge to the global optimum of $f(\mathbf{x}^{max})$. Each of the individuals represents a certain realization of the parameters of $f(\mathbf{x})$, i.e. the i-th individual, $i = 1, 2, ..., N_{ind}$, consists for every generation t of a binary vector $\mathbf{x}^i(t)$ of size U. At the beginning of the genetic algorithm, these individuals are initialized at random in order to spread them unbiased over the search space.

2.2 Fitness Function and Selection Probability

During each iteration, also called generation, of a GA a certain number $N_{off} \leq N_{ind}$ of offsprings has to be produced by mating individuals of the current population. In how many mating processes an individual of the current population may participate depends on its fitness with respect to the optimization problem. This means that individuals which lead to small values of the evaluation function have a higher probability to participate in mating processes than those which lead to larger values. In order to determine these probabilities quantitatively the target function value $f(\mathbf{x}^i), \quad i = 1, ..., N_{ind}$ for each of the N_{ind} individuals has to be computed first. Principally, the mating probability $p^i(t)$ of the i-th individual could simply be set to

$$p^i(t) = \frac{f(\mathbf{x}^i(t))}{\sum_{j=1}^{N_{ind}} f(\mathbf{x}^j(t))} = N_{ind}\tilde{f}_i(t)$$

however, such an approach often leads to a premature convergence of the GA. The root cause for this problem is that individuals which are significantly fitter than the remaining ones in the beginning of the GA often dominate the population excessively and thus keep the population from exploring the search space. This problem can be circumvented if the reproductive range of each individual is limited such that it cannot generate an exaggerated number of offsprings. One approach to achieve such a limitation is to use performance ranks in lieu of raw evaluation function values in order to quantize the fitness the individuals for reproduction [2]. For this purpose the evaluation function value of each individual is computed and the individuals are ranked in the descending order of these values. This ranking can be used to build an appropriate fitness function [8] .

2.3 Selection

During selection N_{off} individuals are chosen from the current population which will be used to produce the desired N_{off} offsprings by mating. In this process individuals with a higher selection probability $p^i(t)$ have to be chosen more often than those with lower values. An often used scheme to select the N_{off} parent individuals is *stochastic universal sampling* (SUS) [2] which implements an unbiased remainder stochastic sampling scheme. In this procedure the mating probabilities $p^i(t)$ of the individuals are arranged along a line segment of length one. Furthermore, N_{off} equally spaced pointers with distances of $1/N_{off}$ are placed over this line whereas the position of the first pointer is given by a random number in the range $[0, 1/N_{off}]$. The number of times an individual is selected for mating then corresponds to the number of pointers lying in its corresponding line segment. In SUS the maximal number n_{max} of times an individual may be selected for reproduction is limited by $n_{max} = p_{max} N_{off}$ with $p_{max} = max\{p^i(t), i = 1, 2, ..., N_{ind}\}$, such that no individual can dominate the entire population in the early stages of the GA. A related but simpler selection scheme corresponds to roulette wheel selection employing a stochastic sampling with replacement.

2.4 Mating

During the selection step of the GA the number of offsprings an individual may produce is determined. Based on this information a mating population is generated in which each individual is represented as many times as it may produce offsprings [5]. From this population always two individuals, called parents, are chosen randomly. These parents then produce offsprings by applying cross-over operations with probability p_c. Cross-over is performed at a single randomly chosen recombination point usually. Afterwards, the parents are removed from the mating population and the process is repeated until the mating population is empty. The mating between two individuals $\mathbf{x}^i(t)$ and $\mathbf{x}^j(t)$ can be implemented as follows: First, a binary vector $\mathbf{v} = [v_1, v_2, ..., v_U]$ of size U is generated at random. Based on this vector two offsprings are created whereas the u-th entry of the first offspring consists of $x^i_u(t)$ and the corresponding entry of the second offspring consists of $x^j_u(t)$ for $v_u = 1$. The same procedure is carried out if $v_u = 0$, however, then $x^i_u(t)$ constitutes the u-th element of the second offspring and $x^j_u(t)$ becomes the u-the element of the first offspring.

2.5 Mutation

During the mutation step [5] of a GA, a small fraction of the offsprings obtained by mating are altered slightly. These alterations occur at random but only with a less than 1% probability such that the major information stored in the affected offsprings is preserved. The motivation for including mutations in GAs is to ensure a minimum level of diversity in the population even if the algorithm has already converged. Binary individuals are mutated by flipping a tiny fraction of their elements. In this process the probability that a particular gene will be affected by mutation is given by a user provided parameter p_{mut} which usually ranges between 0.001 and 0.01.

2.6 Reinsertion

Finally, the N_{off} offsprings obtained by mating and mutation have to be inserted into the population [5]. As the size of the population should not increase from generation to generation individuals of the current population have to be replaced by newly created offsprings. Usually the N_{off} least fit individuals of the current population are replaced by the offsprings. In this context the fractional difference g between the population size N_{ind} and the number of offsprings to be inserted is called generation gap. Furthermore, the n_{elit} fittest individuals of the current population can be protected from replacement, i.e. these individuals are guaranteed to propagate to the next generation. After the offsprings have been inserted into the population their evaluation function values are computed before a new generation of the GA begins. Hence, the process of evaluation, selection, recombination and mutation forms one generation in the execution of a genetic algorithm.

2.7 Subpopulations

Apart from the ranking based fitness assignment also dividing the overall population into smaller subpopulations may help to keep the GA from converging prematurely [1]. These subpopulations evolve independently of each other for T_{mig} generations whereupon they exchange a certain fraction of their individuals. In the so-called complete net structure migration scheme each subpopulation exchanges a certain number $N_{migrate}$ of individuals with each other subpopulation. In this process the fittest individuals of each subpopulation are used as emigrants and replace the least fit individuals in their new host population. The purpose of migration is the following: after T_{mig} generations the individual subpopulations are assumed to have partly converged such that the diversity in their individuals is already low. This diversity is increased if individuals of other populations, which may have converged to another point in search space, immigrate. Hence, after migration the individual subpopulations will explore larger areas of search space and may hence be able to escape from local minima.

2.8 Termination

Eventually a suitable stopping criterion is needed which determines when a GA should terminate. One possibility is to stop the GA if the fittest individual of the population does not change within T_{mig} generations. For a large T_{mig} value this normally means that the entire population has converged to a single optimum.

3 Schemata

Holland [7] developed several arguments to explain why a genetic algorithm will work and result in a complex and robust search by implicitly sampling hyperplane partitions of a search space. Key to understand this is the concept of schemata. The latter denote bit strings which contain wild card match symbols like $*$. Each schema S corresponds to a hyperplane in the search space. The order $o(H)$ of a hyperplane refers to the number of actual bit values that appear in its schema. Thus $S_i = **1*$ is order one, while $S_j = *11*****0**0$ would be order four. A bit string matches a particular schema if it can be constructed from it by replacing the wild card match symbols with the appropriate bit values. All bit strings that match a particular schema belong to the related hyperplane partition. Every binary encoding is a chromosome which corresponds to a corner in the hypercube and is a member of $2^L - 1$ different hyperplanes. Here L denotes the length of the binary encoding.

Crossover and recombination obviously influence the distribution of bit strings. Holland formulated the fundamental **schema theorem** which provides a lower bound on the change in the sampling rate for a single hyperplane from generation t to generation $t + 1$:

$$P(H,t+1) \geq P(H,t)\frac{f(H,t)}{f}\left[1 - p_c\frac{\Delta(H)}{L-1}(1 - P(H,t)\frac{f(H,t)}{f}\right](1-p_{mut})^{o(H)}$$

Here $P(H,t)$ denotes the proportional representation of hyperplane H obtained by dividing $M(H,t)$ by the population size, and $M(H,t)$ denotes the number of strings sampling hyperplane H at the current generation t in some population. Furthermore, $f(H,t)$ is the average evaluation of the sample of strings in partition H in the current population, and $\Delta(H)$ is the defining length of a schema representing a hyperplane H. It is based on the distance between the first and last bits in the schema with value either 0 or 1. Finally p_c and p_{mut} denote the crossover and mutation probabilities, respectively, and the function $o(H)$ returns the order of the schema, i.e. the hyperplane H it represents. The schema theorem clearly emphasizes the role of crossover and hyperplane sampling in genetic search.

4 Serial Evolution

4.1 Notations

We begin with a population $P(t)$ with N_{ind} individuals (chromosomes, bit strings) $\mathbf{x}^i(t)$. The individual fitness of each string at time t is $f(\mathbf{x}^i(t)) \equiv f_i(t) \equiv f(i,t)$. Now we can define the overall fitness $f(t)$ and the averaged overall fitness $< f > (t)$:

$$f(t) = \sum_{i=1}^{N} f(i,t), \qquad < f > (t) = \frac{f(t)}{N_{ind}}.$$

Without loss of generality we will assume the fitness function positive $f(i,t) \geq 0$ and we are interested in its maximum.

Let S be a schema of dimension U, then $P(S,t)$ denotes the subset of strings in $P(t)$ containing the schema S. We will call the number of such strings $N(S,t)$. For the schema S we define the schema-fitness $f(S,t)$ and the averaged schema-fitness $\langle f(S,t)\rangle$ as:

$$f(S,t) = \sum_{\mathbf{x}^i(t)\in P(S,t)} f(i,t), \qquad < f > (S,t) = \frac{f(S,t)}{N(S,t)}.$$

Usually we would now choose N_{ind} strings on the basis of their fitness for the next generation $P(t+1)$. But instead we will choose one *creation-string* and one *elimination-string* based on their fitness and replace the elimination-string by the creation-string. After this selection step we can apply various genetic mutation operators to the creation-string to yield the next generation of the population $P(t+1)$. This type of selection will henceforth be called **Serial Evolution** (SE).

Note that it is possible that a string is replaced by itself. Note also that in the canonical GA the fitness function must be evaluated per time step N_{ind} times, whereas in the serial evolution it has to be evaluated only once for the (mutated, etc.) creation-string. In the following, the number of strings containing a specific schema S will be important.

We denote this number $\mathscr{N}(S,t)$ for the canonical type of selection and $N(S,t)$ for the serial evolution.

Consider a schema S whose average schema-fitness $< f > (S,t)$ is above the overall average fitness $\langle f(t)\rangle$:

$$\langle f\rangle(S,t) = \langle f\rangle(t)(1+\varepsilon)$$

Within a canonical GA we then have the well known increase of strings matching the schema S over generations:

$$\mathscr{N}(S,t+1) = \mathscr{N}(S,t)\frac{\langle f\rangle(S,t)}{\langle f\rangle(t)} = \mathscr{N}(S,t)(1+\varepsilon).$$

We will now derive a similar rule for serial evolution. For this we refer to the event of identifying a string $\mathbf{x}^i(t)$ as being the creation-string as $C(i,t)$ and as being the elemination-string as $E(i,t)$. To these events we allocate the following probabilities:

$$P(C(i,t)) := \frac{f(i,t)}{f(t)}, \qquad P(E(i,t)) := \frac{1-P(C(i,t))}{N_{ind}-1}.$$

As one easily verifies the following closure relations hold:

$$\sum_{i=1}^{N_{ind}} P(C(i,t)) = 1, \qquad \sum_{i=1}^{N_{ind}} P(E(i,t)) = 1.$$

Analogously we define, for a schema S, the events $C(S,t)$ and $E(S,t)$ that a string matching the schema S is choosen to be the creation-string or the elemination-string, respectively. From the previous definition we get:

$$\begin{aligned} P(C(S,t)) &= \sum_{\mathbf{x}^i(t)\in P(S,t)} P(C(i,t)) = \sum_{x(i,t)\in P(S,t)} \frac{f(i,t)}{f(t)} = \frac{f(S,t)}{f(t)} \\ P(E(S,t)) &= \sum_{\mathbf{x}^i(t)\in P(S,t)} \frac{1-P(C(i,t))}{N_{ind}-1} = \frac{N(S,t)-P(C(S,t))}{N_{ind}-1}. \end{aligned} \tag{1}$$

For a better overview we will write shortly

$$e := P(E(S,t)), \quad c := P(C(S,t))$$

Note that $P(C(S,t))$ is not the probability for having one more string matching S in the next generation, because another (or the same) string matching the schema S might be choosen to be the elemination-string. Considering this, we get:

$$\begin{aligned} p_+(S,t) &:= P(N(S,t+1) = N(S,t)+1) = c(1-e) \\ p_0(S,t) &:= P(N(S,t+1) = N(S,t)+0) = ce + (1-c)(1-e) \\ p_-(S,t) &:= P(N(S,t+1) = N(S,t)-1) = (1-c)e \end{aligned} \tag{2}$$

Now $p_+(S,t)$ (resp. $p_0(S,t)$, $p_-(S,t)$) denotes the probability that going from $P(t) \rightarrow P(t+1)$ during one generation of serial evolution, the number of strings matching S is increased by 1 (resp. $0,\ -1$). One easily verifies the following completeness relation:

$$p_+(S,t) + p_0(S,t) + p_-(S,t) = 1.$$

Note that we neglected the fact that through mutation or other bit operators strings who match S can be created or eleminated. We will deal with this complication later.

A schematic metaprogram illustrates the main steps of a serial genetic algorithm:

procedure serial evolution
begin
  $t \leftarrow 0$
  initialize population $P(t)$
  evaluate all individuals $\mathbf{x}^i(t) \in P(t)$
  while (**not** termination-condition) **do**
    $t \leftarrow t+1$
    select individuals $\mathbf{x}^c, \mathbf{x}^e \in P(t-1)$
    alter $\mathbf{x}^c \Rightarrow \tilde{\mathbf{x}}^c$
    replacement of $\mathbf{x}^e$ by $\tilde{\mathbf{x}}^c$ gives $P(t)$
    evaluate $\tilde{\mathbf{x}}^c$
  end
end

4.2 Schema Theorem for Serial Evolution

Now we are ready to formulate the **schema theorem for serial evolution**.

Theorem. For a schema S with $\langle f\rangle(S,t) = \langle f\rangle(t)(1+\varepsilon)$ the following is true:

$$\begin{aligned}\langle N(S,t+1)\rangle &= N(S,t) + (p_+(S,t) - p_-(S,t)) \\ &= N(S,t) + (c-e) \\ &= N(S,t) + \frac{cN_{ind} - N(S,t)}{N_{ind}-1} \\ &= N(S,t)\left(1 + \frac{\varepsilon}{N_{ind}-1}\right) \qquad (3)\end{aligned}$$

where $\langle N(S,t+1)\rangle$ denotes the expectation operator for the number of strings matching S.

Proof. By definition the expected number of strings in $P(t+1)$ matching schema S is given by:

$$\langle N(S,t+1)\rangle = \sum_{i=0}^{N_{ind}} i \cdot P(N(S,t+1) = i)$$

Because $N(S,t)$ can only increase or decrease by $\Delta N(S,t) = \pm 1$, and if one substitutes $i - N(S,t)$ by k, then the following identity results:

$$\langle N(S,t+1)\rangle = \sum_{k=-1}^{1} (N(S,t)+k)P(N(S,t+1) = N(S,t)+k)$$

The latter equality can be reformulated using the definitions of $p_+(S,t)$, $p_0(S,t)$ and $p_-(S,t)$ to yield:

$$\langle N(S,t+1)\rangle = (N(S,t)-1)p_-(S,t)+(N(S,t)-0)p_0(S,t)+(N(S,t)+1)p_+(S,t)$$

Using the definitions of $p_\star(S,t)$ again we have

$$\langle N(S,t+1)\rangle = (N(S,t)-1)p_-(S,t) + N(S,t)p_0(S,t) + (N(S,t)+1)p_+(S,t) \quad (4)$$

Removing all annihilating terms in eqn. 4 yields the first line of eqn. 3

$$\langle N(S,t+1)\rangle = p_+(S,t) + N(S,t) - p_-(S,t)\rangle$$

Furthermore, considering the following identity

$$p_+(S,t) - p_-(S,t) = c(1-e) - (1-c)e = c - e$$

we easily obtain the second line of eqn. 3:

$$\langle N(S,t+1)\rangle = N(S,t) + (c-e)$$

Considering the definition of e it follows:

$$\begin{aligned}\langle N(S,t+1)\rangle &= N(S,t) + \frac{(N_{ind}-1)c}{(N_{ind}-1)} - \frac{N(S,t)-c}{N_{ind}-1}\\ &= N(S,t) + \frac{cN_{ind}-N(S,t)}{N_{ind}-1}\end{aligned} \quad (5)$$

which proofs the third equality in eqn. 3. And finally considering the definition of c one obtains the following identity

$$\begin{aligned}\langle N(S,t+1)\rangle &= N(S,t)\left(1 + \frac{\langle f\rangle(S,t)/\langle f\rangle(t) - 1}{N_{ind}-1}\right)\\ &= N(S,t)\left(1 + \frac{\varepsilon}{N_{ind}-1}\right).\end{aligned} \quad (6)$$

which completes the proof of the theorem.

4.3 Corollary

Let $N_{ind} \geq 2$, and be S a schema with $\langle f\rangle(S,t) = \langle f\rangle(t)(1+\varepsilon)$, $\varepsilon > 0$. Then the number of strings matching schema S increases faster using serial evolution than with the standard type of selection, i.e. there are much fewer evaluations of the fitness function needed with serial evolution.

Proof. Because the usual selection strategy uses N_{ind} evaluations of the fitness function we can apply serial evolution N_{ind} times, which results in an increase of:

$$N(S,t+N_{ind}) = N(S,t)\left(1 + \frac{\varepsilon}{N_{ind}-1}\right)^{N_{ind}}$$

Employing Bernoulli's inequality yields

$$\left(1 + \frac{\varepsilon}{N_{ind}-1}\right)^{N}_{ind} > 1 + \frac{N_{ind}}{N_{ind}-1}\varepsilon > 1 + \varepsilon$$

which completes the proof of the corollary.

4.4 Serial Evolution with Mutation

In the same way the canonical GA uses the mutation operator we can implement mutation into serial evolution. Again p_{mut} is the probability that a particular gene (bit) will be affected and therefore altered by mutation. For a schema S of order $o(S)$ we obtain the survival probability s of a string $\mathbf{x}$ matching S under mutation:

$$s = (1 - p_{mut})^{o(S)}.$$

and we obtain similar to SE without mutation:

$$\begin{aligned} p_{+,m}(S,t) &= sc(1-e) \\ p_{0,m}(S,t) &= sce + (1-c)(1-e) + (1-s)c(1-e) \\ p_{-,m}(S,t) &= (1-c)e + (1-s)ce \end{aligned}$$

Again $p_{+,m}(S,t)$ is the probability that from $P(t) \rightarrow P(t+1)$ the number of strings matching S is increased by 1. One also easily sees again:

$$p_{+,m}(S,t) + p_{0,m}(S,t) + p_{-,m}(S,t) = 1.$$

And we get the schema theorem for serial evolution with mutation:

Theorem. For a schema S with $\langle f\rangle(S,t) = \langle f\rangle(t)(1+\varepsilon)$ the following is true:

$$\begin{aligned} \langle N(S,t+1)\rangle &= N(S,t) + sc - e \\ &= N(S,t)\left(1 + \frac{\varepsilon}{N_{ind}-1} - (1-s)\frac{1+\varepsilon}{N_{ind}}\right) \\ &= N(S,t)\left(1 + \frac{\varepsilon}{N_{ind}-1}\right)(1-(1-s)\xi(N_{ind},\varepsilon)) \end{aligned} \quad (7)$$

and for $N_{ind} \geq 2$:

$$0 \leq \xi(N_{ind},\varepsilon) \leq 1 \qquad \lim_{N_{ind}\rightarrow\infty} \xi(N_{ind},\varepsilon) = 0$$

5 Toy Example

To test Serial Evolution we compared it with a standard genetic algorithm. For simplicity we used a very simple fitness function, which just counts the number of ones in a string $\mathbf{x}$. For both types of evolution we used a population size of $N_{ind} = 10$, $U = 30$ bit-dimensions and $t = 1\ldots250$ generations, which means that for each evolution type we had 2500 evaluations of the fitness function. Varying the mutation parameter p_{mut} within the limits $[0.005, 1.0]$ showed that for both genetic algorithms their performance dependence on p_{mut} in the same way Fig. 1. For better comparability we computed for each p_{mut} both algorithms 200 times and averaged the results. In addition, the elemination of the fittest string was permitted in both evolution types to achieve a smoother convergence.

In Fig. 1 we show the results of this comparison. As claimed, one can see that serial evolution needs significantly less individuals to reach a particular fitness than the canonical genetic algorithm for the averaged (over p_{mut}) case as well as for the optimal case.

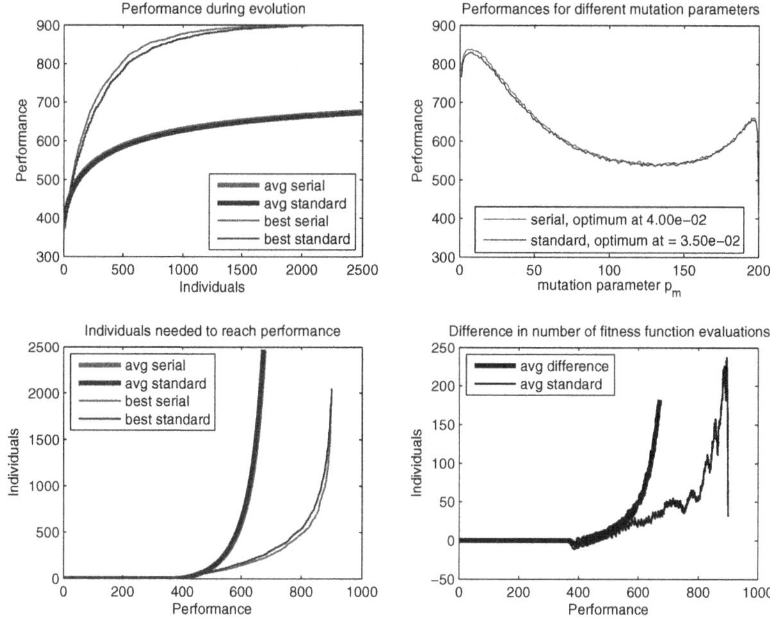

Fig. 1. Serial evolution (red) versus canonical evolution (blue)

6 Conclusion

We propose a new evolutionary algorithm called *serial evolution*. Compared to canonical genetic algorithms, serial evolution needs substantially less evaluations of the fitness function. This results in a substantial reduction of the computational load. This is achieved by taking into account the result from previous function evaluations thus avoiding unnecessary evaluations of the target function, rendering serial evolution much more elegant than classical canocial evolutionary algorithms. Currently the algorithm encompasses only mutations but no cross-over. The inclusion of the latter is currently under investigation.

Acknowledgment

Support by the German Ministry of Science and Education (BMBF) is gratefully acknowledges.

References

1. Computer simulations of genetic adaptation: Parallel subcomponent interaction in a multilocus model. Technical report, PhD thesis, University of Michigan (1985)
2. Baker, J.E.: Reducing bias and inefficiency in the selection algorithm. In: Proc. 2nd Int. Conf. Genetic Algorithms, pp. 14–21 (1987)

3. Barricelli, N.A.: Esempi numerici di processi di evoluzione. Methodos, 45–68 (1954)
4. Cantú-Paz, E.: Efficient and accurate parallel genetic algorithms. In: Genetic Algorithms and Evolutionary Computation, vol. 1. Kluwer Academic Publishers Group, Norwell (2000)
5. Goldberg, D.E.: Genetic Algorithms in Search, Optimization, and Machine Learning. Addison-Wesley Professional, Boston (1989)
6. Goldberg, D.E.: The Design of Innovation: Lessons from and for Competent Genetic Algorithms. In: Genetic Algorithms and Evolutionary Computation, vol. 7. Kluwer Academic Publishers Group, Norwell (2002)
7. Holland, J.H.: Adaptation in Natural and Artificial Systems. MIT Press, Cambridge (1975)
8. Tomé, A.M., Puntonet, C.G., Górriz, J.M., Stadlthanner, K., Theis, F.J., Lang, E.W.: Hybridizing sparse component analysis with genetic algorithms for microarray analysis. Neurocomputing 71, 2356–2376 (2008)
9. Rechenberg, I.: Evolutionsstrategie: Optimierung technischer Systeme nach Prinzipien der biologischen Evolution. Frommann-Holzboog Verlag, Stuttgart (1973)
10. Whitely, D.: A genetic algorithm tutorial. Technical report, Colorado State University (1993)

A Sensitivity Clustering Method for Hybrid Evolutionary Algorithms

F. Fernández-Navarro[1], P.A. Gutiérrez[1], C. Hervás-Martínez[1], and J.C. Fernández[1]

Department of Computer Science and Numerical Analysis, University of Cordóba, Campus de Rabanales, C2 building,1004
i22fenaf@uco.es

Abstract. The machine learning community has traditionally used the correct classification rate or accuracy to measure the performance of a classifier, generally avoiding the presentation of the sensitivities (i.e. the classification level of each class) in the results, especially in problems with more than two classes. Evolutionary Algorithms (EAs) are powerful global optimization techniques but they are very poor in terms of convergence performance. Consequently, they are frequently combined with Local Search (LS) algorithms that can converge in a few iterations. This paper proposes a new method for hybridizing EAs and LS techniques in a classification context, based on a clustering method that applies the k-means algorithm in the sensitivity space, obtaining groups of individuals that perform similarly for the different classes. Then, a representative of each group is selected and it is improved by a LS procedure. This method is applied in specific stages of the evolutionary process and we consider the minimun sensitivity and the accuracy as the evaluation measures. The proposed method is found to obtain classifiers with a better accuracy for each class than the application of the LS over the best individual of the population.

1 Introduction

The machine learning community has traditionally used the correct classification rate or accuracy, C, to measure the performance of a classifier, generally avoiding the presentation of the sensitivities (i.e. the classification level of each class) in the results. However, the pitfalls of using accuracy have been pointed out by several authors [11]. Actually, it suffices to realize that accuracy cannot capture all the different behavioural aspects found in two different classifiers.

On the other hand, Evolutionary Algorithms (EAs) [2], generally require a great number of iterations and they converge slowly, especially in the neighbourhood of the global optimum. It thus makes sense to incorporate a faster Local Search (LS) algorithm into the EA in order to overcome this lack of efficiency while keeping advantages of both optimization methods. However the use of the LS in every generation and for every individual of the EA would result in a prohibitive computational cost. Clustering methods can help to select

J. Mira et al. (Eds.): IWINAC 2009, Part I, LNCS 5601, pp. 245–254, 2009.

the individuals of an EA to which the LS is applied. They are a class of global optimization methods of which an important part includes a cluster-analysis technique [4]. These methods create groups (clusters) of mutually close points that could correspond to relevant regions of attraction. The use of a clustering algorithm allows the selection of individuals representing similar behaviours in the classification problem. In this way, the optimized individuals are more likely to converge towards different local optima.

This paper proposes the combination of an EA, a sensitivity clustering process, and a LS procedure for the evolutionary design of Radial Basis Function Neural Networks (RBFNNs) for multi-classification problems. The sensitivity clustering process is applied based on the sensitivity of each RBFNN model for each class.

This paper is organized as follows: Section II presents the sensitivity measure; Section III describes the proposed sensitivity clustering; Section IV is dedicated to a short description of the RBFNN model; Section V states the most relevant aspects of the EA used for training the RBFNNs; Section VI describes the proposed hybridization in depth; Section VII explains the experiments carried out; and Section VIII summarizes the conclusions of the paper.

2 Sensitivity Measure

A classification problem with Q classes and N training or testing patterns is considered with g as a classifier obtaining a $Q \times Q$ contingency or confusion matrix $M(g)$:

$$M(g) = \left\{ n_{ij};\ \sum_{i,j=1}^{Q} n_{ij} = N \right\} \tag{1}$$

where n_{ij} represents the number of times the patterns are predicted by classifier g to be in class j when they really belong to class i. The diagonal corresponds to correctly classified patterns and the off-diagonal to mistakes in the classification task, as shown in Table 1.

Let us denote the number of patterns associated with class i by $f_i = \sum_{j=1}^{Q} n_{ij}$, $i = 1, \ldots, Q$. Different scalar measures can be defined in order to take the elements of the confusion matrix into consideration from different points of view. Let $S_i = n_{ii}/f_i$ be the number of patterns correctly predicted to be in class i with respect to the total number of patterns in class i (sensitivity for class i). Therefore, the sensitivity for class i estimates the probability of correctly predicting a pattern of class i.

Table 1. Confusion matrix of a classifier

Class	1	2	...	Q	Priors
1	n_{11}	n_{12}	...	n_{1Q}	f_1
2	n_{21}	n_{22}	...	n_{2Q}	f_2
...	...	...	...	...	...
Q	n_{Q1}	n_{Q2}	...	n_{QQ}	f_Q

3 Sensitivity Clustering

In this paper, a clustering process for obtaining groups of RBFNN models is proposed based on the performance of each RBFNN for each class. This method could be used for any type of classifier. In this way, given a classifier g, the following application to the space R_Q is considered:

$$\mathbf{s}_g = \{S_{g1}, S_{g2}, ..., S_{gQ}\} \tag{2}$$

where S_{gi} is the sensitivity of the classifier g for the i-th class. Then, we can define the distance between two classifiers g_1 and g_2 as the Euclidean distance between the associated vectors $\mathbf{s}_{g_1}$ and $\mathbf{s}_{g_2}$:

$$d(g_1, g_2) = ||\mathbf{s}_{g_1} - \mathbf{s}_{g_2}|| = \sqrt{\sum_{i=1}^{Q} (S_{g_1 i} - S_{g_2 i})^2} \tag{3}$$

Now considering a group of N classifiers $G = \{g_1, ..., g_N\}$, the associated vectors are derived $\{\mathbf{s}_{g_1}, ..., \mathbf{s}_{g_N}\}$. With this distance measure, we apply the standard k-means [4] algorithm for obtaining a partition $P = \{C_1, ..., C_k\}$. The obtained groups represent classifiers that have a similar behaviour for the different classes of the problem. The classifier that is closest to the centroid of the i-th cluster is represented by $\overline{\mathbf{s}_i}$ and the members of the set $\{\overline{\mathbf{s}_1}, ..., \overline{\mathbf{s}_k}\}$ are selected as the most representative classifiers of the group G. An example for a group of 40

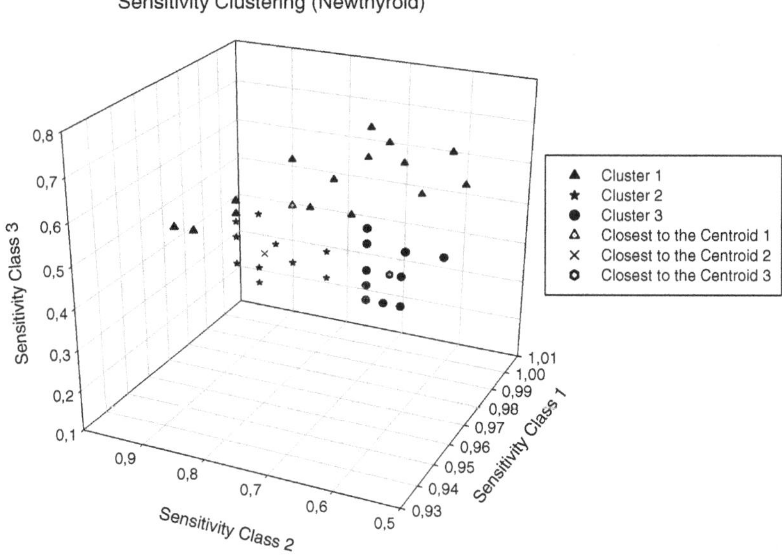

Fig. 1. Sensitivity Clustering: the example includes 40 RBFNNs trained for the Newthyroid dataset, and a k-means process with $k = 3$

classifiers for the Newthyroid dataset of 3 classes is given in Fig. 1, where we apply k-means with $k = 3$ clusters. The selected representatives for each cluster are also included in the figure.

4 Architecture of the RBFNN

We consider RBFNNs [3,5,9] as the base classification model. It can overcome the longer training time and the difficulty in determining hidden layer units of backpropagation network to a large extent. An RBFNN is a three-layer feed-forward neural network. Let the number of nodes of the input layer, of the hidden layer and of the output layer be p, m and n respectively. For any sample $\mathbf{x} = [x_1, x_2, \ldots, x_p]$, the output of the RBFNN is $\mathbf{f}(\mathbf{x}) = [f_1(\mathbf{x}), f_2(\mathbf{x}), \ldots, f_n(\mathbf{x})]$. The model of a RBFNN can be described with the following equation:

$$f_j(\mathbf{x}) = \beta_{0j} + \sum_{i=1}^{m} \beta_{ij} \cdot \phi_i(\|\mathbf{x} - \mathbf{c_i}\|),\ j = 1, 2, \ldots, n, \tag{4}$$

where $\phi_i(\mathbf{x})$ is a non-linear mapping from the input layer to the hidden layer, while the mapping from hidden layer to output layer (i.e. $\phi_i(x)$ to $f_j(\mathbf{x})$) is linear. $\boldsymbol{\beta}_j = [\beta_{1j}, \beta_{2j}, \ldots, \beta_{mj}]^T, (j = 1, 2, ..., n)$ is the connection weight between the hidden layer and the output layer, β_{0j} is the bias value for class j, $\|.\|$ represents the Euclidean norm and $\phi(.)$ is the Gaussian function:

$$\phi(\|\mathbf{x} - \mathbf{c}_i\|) = e^{-\frac{\|\mathbf{x}-\mathbf{c}_i\|^2}{2\cdot\sigma_i^2}}, \tag{5}$$

where $\mathbf{c_i} = [c_i^1, c_i^2, \ldots, c_i^m] \in R_m, i = 1, 2, \ldots, m$ represents the i-th centre in the hidden layer and σ_i controls the attenuation speed of Gaussian function [13].

The activation function of each output node is the softmax function given by:

$$p_j(\mathbf{x}) = \frac{e^{f_j(\mathbf{x})}}{\sum_{i=1}^{Q} e^{f_i(\mathbf{x})}}; \quad j = 1, ..., Q \tag{6}$$

where $p_j(\mathbf{x})$ is the probability that pattern $\mathbf{x}$ belongs to class j.

Under this probabilistic interpretation of the model outputs, it is possible to evaluate the model using the cross-entropy error function, given by:

$$l(g) = -\frac{1}{n}\sum_{i=1}^{n}\sum_{j=1}^{J}(y_j(\mathbf{x_i})ln(p_j(\mathbf{x}_i))) \tag{7}$$

where $y_j(\mathbf{x_i})$ is the expected value for the class j and pattern i [$y_j(\mathbf{x_i})$ is 1 if $\mathbf{x_i}$ belongs to the class j and 0 otherwise] and g is the evaluated RBFNN.

5 Base Evolutionary Algorithm

The basic framework of the EA is the following: the search begins with an initial population of RBFNNs and, in each iteration, the population is updated using a population-update algorithm which evolves both its structure and weights. The population is subject to operations of replication, mutation and recombination.

The main characteristics of the algorithm are the following:

1. *Representation of the Individuals.* The algorithm evolves architectures and connection weights simultaneously, each individual being a fully specified RBFNN. The neural networks are represented using an object-oriented approach and the algorithm deals directly with the RBFNN phenotype. Each connection is specified by a binary value indicating if the connection exists, and the real value representing its weights.
2. *Error and Fitness Functions.* We consider $l(g)$ (see Eq. 7) as the error function of an individual g of the population. The fitness measure needed for evaluating the individuals is a strictly decreasing transformation of the error function $l(g)$ given by $A(g) = \frac{1}{1+l(g)}$, where $0 < A(g) \leq 1$.
3. *Initialization of the Population.* The initial population is generated trying to obtain RBFNNs with the maximun possible fitness. First, 5,000 random RBFNNs are generated. The number of connections between all RBFs of an individual and the input layer is a random value in the interval $[1, k]$, and all of them are connected with the same randomly chosen input variables. In this way, all the RBFs of each individual are initialized in the same random subspace of the input variables. A random value in the $[-I, I]$ interval is assigned for the weights between the input layer and the hidden layer and in the $[-O, O]$ interval for those between the hidden layer and the output layer. The obtained individuals are evaluated using the fitness function and the initial population is finally obtained by selecting the best 500 RBFNNs. In order to improve the randomly generated centres, the standard k-means clustering algorithm is applied using these random centres as the initial centroids for the algorithm and a maximun number of 100 iterations.
4. *Recombination.* Different crossover operators are applied [10]:

   - *Binary Crossover Operator.* The binary-operator needs two RBFNNs to be applied, although it only changes one of them. This operator takes an uniformly randomly chosen number of consecutive hidden neurons from the first network, and another random sequence from the second; then it replaces the first of these sequences by the second one, so that the second individual remains unchanged.
   - *Multipoint Crossover Operator.* This operator replaces with probability 0.2 every hidden neuron of the first RBFNN by a randomly chosen neuron coming from the second net.

5. *Structural Mutation.* Structural mutation implies a modification in the structure of the RBFNNs and allows the exploration of different regions in the search space, helping to keep the diversity of the population. There are four different structural mutations: hidden node addition, hidden node deletion, connection addition and connection deletion. These four mutations are applied sequentially to each network, each one with a specific probability.

6. *Parametric Mutation.* Different weight mutations are applied:
   - *Centre and Radii Mutation.* These parameters are modified in the following way:
     - Centre creep. It changes the values for the centres applying a Normal noise. For each dimension of the centre point, the Normal distribution is centred on the current value and it is as wide as the radius.
     - Radius creep. Changes the values for the radii applying another Normal noise. The Normal distribution is centred on the current value and is as wide as the range of each dimension.
     - Randomize centres. Changes the values of the centres of the hidden neurons to random values in the range allowed for each dimension of the input space.
     - Randomize radii. It changes radius values randomly, always with values in the corresponding range of each input space dimension.
   - *Output-to-Hidden Node Connection Mutations* [8]. These connections are modified by adding a Normal noise, $w(t+1) = w(t) + \xi(t)$, where $\xi(t) \in N(0, T(g))$ and $N(0, T(g))$ represents a one-dimensional normally distributed random variable with mean 0 and variance the network temperature ($T(g) = 1 - A(g)$).

6 Hybrid Evolutionary Algorithms

The proposed methodology is based on the combination of the EA, a clustering process, and a local improvement procedure. The local optimization algorithm used in our paper is the *iRprop+* [6] optimization method.

We have two different versions of the hybrid EA, depending on the stage when we carry out the local search and the clustering partitioning. The base Evolutionary Algorithm presented in Section 5 without the clustering process and local search is represented by EA. In the Hybrid EA (HEA), we run the EA and then apply the local optimization algorithm to the best solution obtained by the EA in the last generation. The HEA with the Sensitivity Clustering (HEACS) algorithm applies the clustering process presented on Section 3 on the complete set of the individuals of the last generation. After that, we apply *iRprop+* algorithm to the individual closest to the centroid obtained in each cluster and return the optimized individual with the best minimum training sensitivity value as the final solution. Finally, the algorithm named Dynamic HEA with Sensitivity Clustering (HEACSD) carries out both the clustering process and the *iRprop+* local search dynamically every G_0 generations, where G_0 must be fixed by the user. The final solution is the individual with the best minimum training sensitivity value among the local optima found during the evolutionary process. HEACS and HEACDS, are the ones that implement the ideas proposed in this paper. The other two method (EA, HEA) [7] are considered for comparison purposes. The proposed methodology is described in Fig. 2.

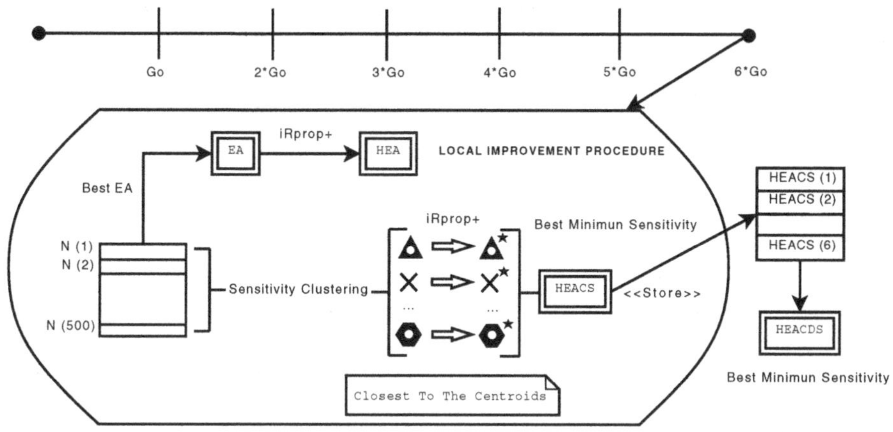

Fig. 2. Proposed Methodologies

7 Experiments

The proposed methodologies are applied to six datasets taken from the UCI repository [1], to test its overall performance when compared each other. The selected datasets include both binary problems (Sonar and Vote) and multi-class problems (Balance, Iris and Newthyroid with 3 classes and Zoo with 7 classes). The first subsection defines the experimental design and the next one shows the results obtained.

7.1 Experimental Design

The experimental design was conducted using a 10-fold cross validation, with 10 repetitions per each fold. The performance of each method has been evaluated using the correct classification rate (C) in the generalization set and the Minimun Sensitivity (MS) value for the generalization set, i.e. the accuracy for the class that is worst classified.

All the parameters used in the evolutionary algorithm except the maximun and minimun number of RBFs in the hidden layer have the same values in all problems analyzed below. We have done a simple linear rescaling of the input variables in the interval $[-2, 2]$, X_i^* being the transformed variables. The centres c_{ji} are initialized in this interval (i.e. $[-I, I] = [-2, 2]$), and the connection between hidden and output layer are initialized in the $[-5, 5]$ interval (i.e. $[-I, I] = [-5, 5]$). The initial value of the radii r_j is obtained in the interval $(0, d_{max}]$, where d_{max} is the maximun distance between two training input examples.

The size of the population is $N = 500$. For the structural mutation, the number of nodes that can be added or removed is within the $[1, 2]$ interval, and the number of connections to add or delete in the hidden and the output layer during structural mutations is within the $[1, 7]$ interval. The number of the clusters is $k = 6$ for the HEACS and HEACDS methodologies. The local improvement procedure is performed every 40 generations (i.e. $G_0 = 40$) and the local searches are applied 6 times. In this way, the algorithm stops when 240 generations are completed.

7.2 Comparison to EA and HEA

In Table 2, the mean and the standard deviation of the correct classification rate in the generalization set (C_G) is shown for each dataset and a total of 100 executions. From the analysis of the results, it can be concluded, from a purely descriptive point of view, that the HEACDS method obtained the best results for four datasets, and the EA and HEACS methods yield the higher performance for one dataset.

In Table 3, the mean and the standard deviation of the Minimun Sensitivity in the generalization set (MS_G) is shown for each dataset and a total of 100 executions. The HEACDS method obtained the best results for three datasets, HEACS got the higher performance for two datasets, and the EA method only for one dataset.

Table 2. Mean and Standard Deviation of the Accuracy Results (C_G) From 100 Executions of a 10-Fold Cross Validation

	Method(C_G(%))			
	EA	HEA	HEACS	HEACDS
Dataset	*Mean* ± *SD*	*Mean* ± *SD*	*Mean* ± *SD*	*Mean* ± *SD*
Balance	91.09 ± 1.13	91.97 ± 2.00	*92.32 ± 2.24*	**92.50 ± 2.42**
Iris	95.87 ± 6.05	96.07 ± 5.85	*96.33 ± 5.85*	**96.33 ± 5.39**
Newthyroid	95.56 ± 3.49	96.13 ± 2.66	*96.37 ± 2.69*	**96.63 ± 2.84**
Sonar	78.14 ± 9.86	80.95 ± 8.80	**81.66 ± 9.48**	*81.66 ± 9.93*
Vote	**95.84 ± 2,82**	95.56 ± 2,63	95.56 ± 2.81	*95,72 ± 2.79*
Zoo	89.85 ± 8.49	94.64 ± 7.01	*95.34 ± 6.73*	**96.52 ± 6.30**

The best result is in bold face and the second best result in italic

Table 3. Mean and Standard Deviation of Minimun Sensitivity Results From 100 Executions of a 10-Fold Cross Validation

	MS_G(%)			
	EA	HEA	HEACS	HEACDS
Dataset	*Mean* ± *SD*	*Mean* ± *SD*	*Mean* ± *SD*	*Mean* ± *SD*
Balance	1.00 ± 4.38	16.75 ± 19.49	*25.41 ± 23.68*	**27.56 ± 22.17**
Iris	93.40 ± 9.45	93.40 ± 9.45	**93.60 ± 9.37**	*93.60 ± 9.70*
Newthyroid	76.07 ± 21.19	83.15 ± 15.05	*83.73 ± 15.84*	**86.67 ± 14.68**
Sonar	68.45 ± 13.94	71,90 ± 12.08	**73.97 ± 11.82**	*73.85 ± 12.60*
Vote	**93.31 ± 4.50**	92.92 ± 3.92	92.92 ± 4.09	*93.23 ± 4.12*
Zoo	32.00 ± 46.02	59.25 ± 49.15	*64.25 ± 48.01*	**73.00 ± 45.05**

The best result is in bold face and the second best result in italic

Table 4. Asintotic signification of the Wilcoxon signed-rank test

Dataset	HEACDS Vs	C_{G} (p-value)	MS_{G} (p-value)	Dataset	HEACDS Vs	C_{G} (p-value)	MS_{G} (p-value)
Balance	EA	$0.000_{(*)}$	$0.000_{(*)}$	Sonar	EA	$0.000_{(*)}$	$0.000_{(*)}$
	HEA	$0.038_{(*)}$	$0.000_{(*)}$		HEA	0.218	0.131
	HEACS	0.434	0.232		HEACS	0.730	0.576
Iris	EA	$0.000_{(*)}$	$0.000_{(*)}$	Vote	EA	0.950	0.945
	HEA	$0.000_{(*)}$	$0.000_{(*)}$		HEA	0.516	0.304
	HEACS	$0.000_{(*)}$	$0.000_{(*)}$		HEACS	0.257	0.257
Newthyroid	EA	$0.001_{(*)}$	$0.000_{(*)}$	Zoo	EA	$0.000_{(*)}$	$0.000_{(*)}$
	HEA	0.295	$0.013_{(*)}$		HEA	$0.003_{(*)}$	$0.004_{(*)}$
	HEACS	0.748	$0.045_{(*)}$		HEA	$0.035_{(*)}$	$0.039_{(*)}$

(*): Statistically significant differences with $\alpha = 0.05$

To determine the statistical significance of the C_{G} and MS_{G} differences observed for each method in the different datasets, we have carried out a non-parametric Wilcoxon signed-rank test [12] with the ranking of C_{G} and MS_{G} of the best models as the test variable (since a previous evaluation of the C_{G} and MS_{G} values results in rejecting the normality and the equality of variances' hypothesis). The results of this test for $\alpha = 0.05$ can be seen in Table 4. As can be seen in these results, the differences in C_{G} and MS_{G} of the HEACDS method with respect to the EA method are significant for all datasets except Vote. When comparing HEACDS to HEA, the differences are significant in C_{G} and MS_{G} for three out of the six datasets considered and in S_{G} for one more dataset. The comparison of the proposed method to the HEACS method results in significant differences in C_{G} and MS_{G} for Iris and Zoo and in MS_{G} for Newthyroid.

8 Conclusions

In this paper, we have proposed a new approach to solve multi-classification problems. This approach is based on the combination of an EA, a new sensitivity clustering process, and a local-search procedure. The proposed clustering process allows the selection of individuals representing different regions in the search space. These selected individuals are the ones subject to local optimization. In this way, the optimized individuals are more likely to converge towards different local optima. We have proposed two different versions of the hybrid EA, depending on the stage when we carry out the local searches and the cluster partitioning.

The hybrid EA proposed was applied to six classification problems of the UCI repository [1]. The results show that the dynamic version obtains not only better results in the mean but also in the standard deviation. Moreover, the sensitivity clustering have shown not only to improve the generalization accuracy for many datasets but also to obtain more balanced classifiers, that result in a good performance for all classes of the datasets. Although the computational cost is only slightly higher, the differences in accuracy/sensitivity performance between the proposed method and the method that does not use clustering is significant. This result suggest that the use of a clustering algorithm to select just a few individuals to optimize, instead of optimizing many of them, provides a very good compromise between performance and computational cost.

References

1. Asuncion, A., Newman, D.J.: UCI machine learning repository (2007)
2. Back, T.: Evolutionary Algorithms in Theory and Practice. Oxford (1996)
3. Freeman, J.A.S., Saad, D.: Learning and generalization in radial basis function networks. Neural Computation 7(5), 1000–1020 (1995)
4. Fukunaga, K.: Introduction to Statistical Pattern Recognition, 2nd edn. Academic Press, London (1999)
5. Hwang, Y.S., Bang, S.Y.: An efficient method lo construct radial basis function neural network classifier. Neural Networks 10(8), 1495–1503 (1997)
6. Igel, C., Hüsken, M.: Empirical evaluation of the improved rprop learning algorithms. Neurocomputing 50(6), 105–123 (2003)
7. Martinez-Estudillo, A.C., Hervas-Martínez, C., Martinez-Estudillo, F.J., Garcia-Pedrajas, N.: Hybridization of evolutionary algorithms and local search by means of a clustering method. IEEE Transactions on Systems, Man and Cybernetics, Part B: Cybernetics 36(3), 534–545 (2006)
8. Martínez-Estudillo, F.J., Hervás-Martínez, C., Gutiérrez, P.A., Martínez-Estudillo, A.C.: Evolutionary product-unit neural networks classifiers. Neurocomputing 72(1-2), 548–561 (2008)
9. Orr, M.J.L.: Regularisation in the selection of radial basis function centres. Neural Computation 7(3), 606–623 (1995)
10. Parras-Gutierrez, E., del Jesus, M.J., Merelo, J.J., Rivas, V.M.: A symbiotic CHC co-evolutionary algorithm for automatic RBF neural networks design. In: Advances in Soft Computing, vol. 50, pp. 663–671 (2009)
11. Provost, F., Fawcett, T.: Robust classification system for imprecise environments. In: Proccedings of the Fithteenth National Conference on Artificial Intelligence, pp. 706–713 (1998)
12. Wilcoxon, F.: Individual comparisons by ranking methods. Biometrics 1, 80–83 (1945)
13. Yang, Z.R.: A novel radial basis function neural network for discriminant analysis. IEEE Transactions on Neural Networks 17(3), 604–612 (2006)

A Genetic Algorithm for the Open Shop Problem with Uncertain Durations

Juan José Palacios[1], Jorge Puente[1], Camino R. Vela[1], and Inés González-Rodríguez[2]

[1] A.I. Centre and Department of Computer Science, University of Oviedo, Spain
{UO165228,puente,crvela}@uniovi.es
http://www.aic.uniovi.es/Tc

[2] Department of Mathematics, Statistics and Computing, University of Cantabria, Spain
ines.gonzalez@unican.es

Abstract. We consider a variation of the open shop problem where task durations are allowed to be uncertain and where uncertainty is modelled using fuzzy numbers. We propose a genetic approach to minimise the expected makespan: we consider different possibilities for the genetic operators and analyse their performance, in order to obtain a competitive configuration. Finally, the performance of the proposed genetic algorithm is tested on several benchmark problems, modified so as to have fuzzy durations, compared with a greedy heuristic from the literature.

1 Introduction

The open shop scheduling problem is often regarded as a variation of the job shop scheduling problem, and traditionally it has received considerably less attention by researchers. However, its significantly larger solution space and the scarcity of specific methods to solve it make it an important problem in itself, with an increasing presence in the recent literature [1],[2],[3]. It is also a problem with clear applications. Consider for instance testing facilities, where units go through a series of diagnostic tests that need not be performed in a specified order and where different testing equipment is usually required for each test. To enhance the range of applications of scheduling, part of the research is devoted to model the uncertainty and vagueness pervading real-world situations. The approaches are diverse [4] and, among these, fuzzy sets have been used in a wide variety of approaches, ranging from representing incomplete or vague states of information to using fuzzy priority rules with linguistic qualifiers or preference modelling [5]. Incorporating uncertainty to scheduling usually requires a significant reformulation of the problem and solving methods, in order that the problem can be precisely stated and solved efficiently and effectively. Some attempts have been made to extend heuristic methods to shop scheduling problems where durations are modelled via fuzzy intervals, most commonly and successfully for the flow shop problem (for instance, in [6] and [7]) and also for the job shop [8], [9], [10] and [11].

J. Mira et al. (Eds.): IWINAC 2009, Part I, LNCS 5601, pp. 255–264, 2009.

2 Open Shop Scheduling with Uncertain Durations

The *open shop scheduling problem*, or *OSP* in short, consists in scheduling a set of n jobs $J_1, \ldots, J_n$ to be processed on a set of m physical resources or machines $M_1, \ldots, M_m$. Each job consists of m tasks or operations, each requiring the exclusive use of a different machine for its whole processing time without preemption, i.e. all operations must be processed without interruption. In total, there are mn operations, $\{O_k, 1 \leq k \leq mn\}$. A solution to this problem is a *schedule*–an allocation of starting times for all operations– which is *feasible*, in the sense that all constraints hold, and is also optimal according to some criterion. Here, the objective will be minimising the makespan C_{max}, that is, the time lag from the start of the first operation until the end of the last one, a problem often denoted $O||C_{max}$ in the literature.

2.1 Uncertain Durations

In real-life applications, it is often the case that it is not known in advance the exact time it will take to process one operation and only some uncertain knowledge is available, for instance, an interval of possible durations, or a most likely duration with a certain error margin. Such knowledge can be modelled using a *triangular fuzzy number* or TFN, given by an interval $[n^1, n^3]$ of possible values and a modal value n^2 in it. For a TFN N, denoted $N = (n^1, n^2, n^3)$, the membership function takes the following triangular shape:

$$\mu_N(x) = \begin{cases} \frac{x-n^1}{n^2-n^1} & : n^1 \leq x \leq n^2 \\ \frac{x-n^3}{n^2-n^3} & : n^2 < x \leq n^3 \\ 0 & : x < n^1 \text{ or } n^3 < x \end{cases} \quad (1)$$

In the open shop, we essentially need two operations on processing times (fuzzy numbers), the sum and the maximum. These are obtained by extending the corresponding operations on real numbers using the *Extension Principle*. However, computing the resulting expression is cumbersome, if not intractable. For the sake of simplicity and tractability of numerical calculations, we follow [8] and approximate the results of these operations, evaluating the operation only on the three defining points of each TFN. It turns out that for any pair of TFNs M and N, the approximated sum $M + N \approx (m^1 + n^1, m^2 + n^2, m^3 + n^3)$ coincides with the actual sum of TFNs; this is not necessarily so for the maximum $M \vee N \approx (m^1 \vee n^1, m^2 \vee n^2, m^3 \vee n^3)$, although they have identical support and modal value.

The membership function of a fuzzy number can be interpreted as a possibility distribution on the real numbers. This allows to define its expected value [13], given for a TFN N by $E[N] = \frac{1}{4}(n^1 + 2n^2 + n^3)$. It coincides with the neutral scalar substitute of a fuzzy interval and the centre of gravity of its mean value [5]. It induces a total ordering $\leq_E$ in the set of fuzzy intervals [8], where for any two fuzzy intervals M, N $M \leq_E N$ if and only if $E[M] \leq E[N]$.

Require: an instance of $FuzO||E[C_{max}]$, P
Ensure: a schedule for P
  1. generate and evaluate the initial population;
  while No termination criterion is satisfied **do**
    2. select chromosomes from the current population;
    3. apply recombination operators to the chromosomes selected at step 2.;
    4. evaluate the chromosomes generated at step 3;
    5. apply replacement using chromosomes from step 2. and step 3.;
  return the schedule from the best chromosome evaluated so far;

Algorithm 1. Genetic Algorithm

2.2 Fuzzy Open Shop Scheduling

If processing times of operations are allowed to be imprecise and such imprecision or uncertainty is modelled using TFNs, the resulting schedule is fuzzy in the sense that starting and completion times for each operation and hence the makespan are TFNs. Each TFN can be seen as a possibility distributions on the values that the time may take. Notice however that there is no uncertainty regarding the task processing ordering given by the schedule.

An important issue with fuzzy times is to decide on the meaning of "optimal makespan". It is not trivial to optimise a fuzzy makespan, since neither the maximum nor its approximation define a total ordering in the set of TFNs. Using ideas similar to stochastic scheduling, we follow the approach taken for the fuzzy job shop in [9] and use the total ordering provided by the expected value and consider that the objective is to minimise the expected makespan $E[C_{max}]$. The resulting problem may be denoted $FuzO||E[C_{max}]$.

3 Genetic Algorithms for the FOSP

The open shop scheduling problem is NP-complete for a number of machines $m \geq 3$ [12].This motivates the use of metaheuristic techniques to solve the general m-machine problem. For instance, [1] proposes two heuristic methods to obtain a list of operation priorities later used in a basic algorithm of list scheduling. Also, local search and genetic algorithms are used, by themselves or combined with each other and with dispatching rules: [14] presents an iterative improvement algorithm with a heuristic dispatching rule to generate the initial solutions; [15] proposes a tabu search algorithm, [2] introduces a genetic algorithm hybridised with local search, and a genetic algorithm using heuristic seeding is proposed in [16]. In [17], a local search with constraint propagation and conflict-based heuristics framework is applied to OSP, and [3] proposes a solution based on particle swarm optimisation. However, to our knowledge, none of these metaheuristic techniques have been adapted to the case where durations are fuzzy numbers.

From all heuristic strategies, genetic algorithms (GAs) have proved effective techniques to solve scheduling problems, specially when hybridised with other strategies, such as local search. The structure of a standard genetic algorithm is described in Algorithm 1. First, the initial population is generated and evaluated. Then the genetic algorithm iterates for a number of steps or generations and in each iteration, a new population is built from the previous one by applying the genetic operators such as selection, mutation, crossover, etc,. Clearly, the choice of chromosome codification and genetic operators is decisive in order to obtain a competitive GA.

3.1 Chromosome Codification and Evaluation

Following [2], we use operation-based chromosomes. This representation encodes a schedule as an ordered sequence of operations, with one gene per operation. Operations are listed in the order in which they are scheduled, so a chromosome is just a permutation of numbers from 1 to nm, where number i corresponds to operation O_i. Such a permutation expresses partial orders among operations in each job and each machine.

Decodification of a chromosome may be done in different ways. Here, we consider two approaches. The first one is to schedule every operation so its starting time is the maximum of the completion times of the previous operation in its machine and the previos operation in its job, according to the ordering provided by the chromosome. This strategy produces *semi-active* schedules, meaning that for any operation to start earlier, the relative ordering of at least two operations must be swapped.

The second approach consists in using an insertion strategy, scheduling operations directly from the order expressed in the chromosome: the starting time of each operation is obtained as the earliest possible time given the operations which are already scheduled (previous in the chromosome) both on the same machine and on the same job. It is easy to check that this strategy yields active schedules, where a schedule is *active* if one operation must be delayed for any other one to start earlier. Active schedules are good in average and, most importantly, the space of active schedules contains at least an optimal one [12].

To obtain the initial population, chromosomes are obtained as random permutations of (1 2 3 4 5 6 7 8 9). Notice that, given the codification schema, this initialisation method, albeit simple, always produces feasible individuals.

3.2 Genetic Operators

Selection. In order to select chromosomes from one population to be combined and generate the next population, we propose to group the chromosomes in pairs at random, so as to give every individual the opportunity to contribute to the next generation.

Crossover. Once we have a pair of chromosomes, we consider several crossover operators proposed in the literature (c.f. [18]):

First, we consider the *Partially-Mapped Crossover* or PMX, which builds an offspring by choosing a subsequence from one parent and preserving the order and position of as many operations as possible from the other parent. A subsequence is selected by choosing two random cut points, which serve as boundaries for swapping operations, for instance, we may have as parents $p_1 = (1\,2\,3\,|4\,5\,6\,7|\,8\,9)$, $p_2 = (4\,5\,2\,|1\,8\,7\,6|\,9\,3)$ with two cut points marked by $|$. The segment between cut points in the first parent is copied in the same positions onto the first offspring: $o_1 = (x\,x\,x\,|4\,5\,6\,7|\,x\,x)$ and similarly with the segment from the second parent onto the second offspring $o_2 = (x\,x\,x\,|1\,8\,7\,6|\,x\,x)$ (the symbol x may be interpreted as 'unknown at present'). The cut also defines a mapping between the operations in the same position in both parents: $1 \leftrightarrow 4$, $8 \leftrightarrow 5$, $7 \leftrightarrow 6$, $6 \leftrightarrow 7$. Next, we fill further operations in the first (resp. second) offspring from the second (resp. first) parent for which there is no conflict: $o_1 = (x\,x\,2\,|4\,5\,6\,7|\,9\,3)$ and $o_2 = (x\,2\,3\,|1\,8\,7\,6|\,x\,9)$. Finally, we resolve the conflicts by replacing the operation from the second (resp. first) parent which already appears between the cut points with the corresponding operation from the first (resp. second) parent, according to the obtained mapping: $o_1 = (1\,8\,2\,4\,5\,6\,7\,9\,3)$ and $o_2 = (4\,2\,3\,|1\,8\,7\,6|\,5\,9)$.

A second operator is *Linear-Order Crossover* or LOX in short, a modification of the well-known order crossover so as not to treat chromosomes as circular. This operator selects a subsequence of operations at random and copies this subsequence from the first (resp. second) parent onto the first (resp. second) offspring. It then completes each offspring by placing operations from left to right in the same order as they appear in the other parent. For instance, for the same parents as above, $p_1 = (1\,2\,3\,|4\,5\,6\,7|\,8\,9)$, $p_2 = (4\,5\,2\,|1\,8\,7\,6|\,9\,3)$ and the same selected subsequence, the offsprings would be $o_1 = (2\,1\,8\,4\,5\,6\,7\,9\,3)$ and $o_2 = (2\,3\,4\,|1\,8\,7\,6|\,5\,9)$. LOX is designed so as to preserve as much as possible both the relative positions between genes and the absolute positions relative to the extreme operations in the parents, which correspond to the high and low priority operations.

A modification of LOX is the PBX or *Position-Based Crossover* where, instead of selecting one subsequence of operations to be copied, several operations are selected at random for that purpose.

Mutation. Mutation is applied to a chromosome by randomly modifying its features, thus helping to preserve population diversity and providing a mechanism to escape local minima. Among the several mutation operators proposed for permutation-based codification [18], we consider insertion and swap mutations. The *insertion* operator selects an operation at random and then inserts it into another random position. The *swap* or *gene swapping* operator selects two positions at random and then swaps the operations in these positions.

Replacement. From each pair of parents we obtain offsprings by applying crossover and mutation. This forms a set of four solutions from which two will be accepted as members of the next generation using tournament: the two old chromosomes and the new ones are put together and the best two are selected to replace the parents in the next generation. Here, we have two possibilities, namely, accept two chromosomes with the same fitness (same expected makespan), or force the two accepted chromosomes to have different fitness. Notice that with this replacement scheme, inferior solutions are eliminated only through newborn superior solutions.

Therefore, both the best and worst makespans of the solutions in the population at each generation are non-increasing and there is an implicit elitism.

As we can see, by selecting one operator or another from those explained herein, we obtain a wide range of configurations for a GA. Our goal is to choose the best possible configuration based on thorough experimentation.

4 Experimental Results

For the experimental study, we follow [8] and generate a set of fuzzy problem instances from well-known benchmark problems. In particular, we consider a subset of the problems proposed by Taillard [19], consisting of four families of problems of sizes 4×4, 5×5, 7×7 and 10×10 and where each family contains 10 problem instances. From each of these crisp problem instances, we generate 10 fuzzy instances, so we have 400 fuzzy problem instances in total. To generate a fuzzy instance from a crisp one, we transform each crisp processing time x into a symmetric fuzzy processing time $p(x)$, so its modal value is $p^2 = x$ and p^1, p^3 are random values, symmetric w.r.t. p^2 and generated so the TFN's maximum range of fuzziness is 30% of p^2. By doing this, the optimal solution to the crisp problem provides a lower bound for the fuzzified version [8]. The obtained benchmarks for the fuzzy open shop are available at `http://www.aic.uniovi.es/tc/spanish/repository.htm`

4.1 Configuration and Parameter Setting

A first set of experiments is conducted to choose the GA's configuration, in an incremental approach, where the first step is to fix the population size and the number of generations, to ensure convergence. We outline the experimentation process and give a summary of results, but we omit a more detailed exposition, due to lack of space and the large number (400) of problem instances used.

The **convergence** study is performed using the following base configuration: fitness, active scheduling; crossover, LOX with probability $p_x = 0.70$; mutation, insertion with probability $p_m = 0.05$; selection, random pairs; replacement, tournament with repetition. For this base configuration, the GA has been run varying the number of maximum iterations depending on the problem sizes: 1000 (4×4), 5000 (5×5), 10000 (7×7 and 10×10). Tests have been repeated with two different population sizes: 100 and 200 and for each possibility we have considered the average performance across ten runs. The quality of a solution is assessed using the following makespan relative error w.r.t. a lower bound: $(E[C_{max}] - LB)/LB$ with $LB = \max(\max_j\{\sum_{i=1}^{n} p_{ij}\}, \max_i\{\sum_{j=1}^{m} p_{ij}\})$, where p_{ij} is the crisp processing time of the operation belonging to job J_i which has to be processed on machine M_j. Here, we have considered the relative makespan error of the best individual in each generation and have followed the error evolution along the generations for both population sizes, establishing as convergence point the point where the difference between the generation's error and the minimum error obtained with the maximum number of generations is less than 1%. With this criterion, we set the population size as 100 for all problems, with the number of

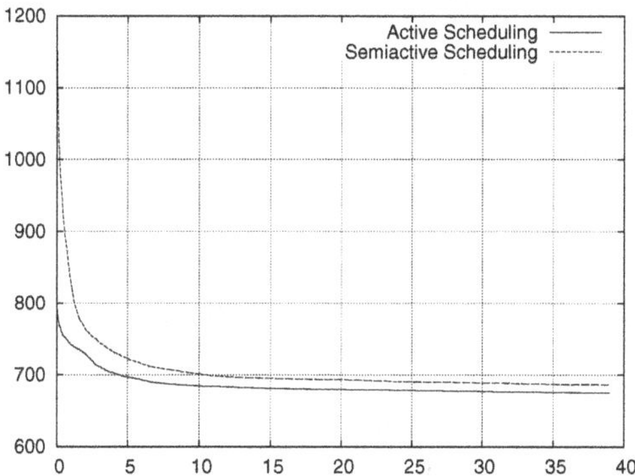

Fig. 1. Convergence of base configuration of GA for problem tai10_1_05

generations depends on the size as follows: $4 \times 4 : 40$, $5 \times 5 : 130$, $7 \times 7 : 1600$, $10 \times 10 : 3000$.

Having fixed the base configuration for population size and number of generations, we decide on the type of **chromosome evaluation**: active or semi-active scheduling. To compare both approaches, we extend the running time for the GA with semi-active scheduling so it is approximately the same as that of the GA with active scheduling. The conclusion is that, under equivalent conditions, the relative makespan error for semi-active scheduling is between 3.074% and 4.438% worse than for active scheduling. We therefore opt for the latter. Figure 1 illustrates the convergence of the GA with both active and semi-active scheduling for problem `tai10_1_05`, that is, the fifth fuzzy instance obtained from the first problem proposed by Taillard of size 10×10. The x axis represents the time taken in seconds and the y axis represents the expected makespan.

To perform **replacement**, we only need choose whether to filter those chromosomes with the same expected makespan or not. Again, with the base configuration and using active scheduling, the experimental results show that the cost of filtering individuals with identical makespan is insignificant w.r.t. the total running time of the algorithm and it always generates slightly better results.

Using the GA with active scheduling and replacement with no repetition, we analyse the performance of the three different **crossover** operators using as crossover probability values $0.7, 0.8, 0.9$. Again, the conclusions are drawn based on the relative makespan error w.r.t. LB and also on computational cost. The results show that there is no significant difference between PMX and LOX, both for the makespan error and running times, being 0.7 the best value for the crossover probability. PMX is always better, although the improvement w.r.t. LOX is less than 0.4%. PMX is also preferred to PBX, since the latter takes approximately 28.38% more running time and the makespan does not always improve. For the ten sets of fuzzy problems of size 10×10, PBX with $p_x = 0.9$ (the best of the three values) only improves the makespan obtained with PMX with $p_x = 0.7$ in three of the sets and this improvement is less than 0.9%. We conclude that the crossover for the final configuration will be PMX with $p_x = 0.7$ (although LOX is very similar).

Table 1. Average relative makespan error (in %) for all families of problems

Problem	GA		Random Pop.	DS/LRPT
	B	A		
4×4	3.002	4.020	4.774	12.473
5×5	3.432	6.263	9.840	16.100
7×7	1.480	4.123	11.698	10.938
10×10	1.382	3.624	13.723	7.069

As above, we try the two **mutation** operators and adjust the mutation probability choosing from 0.05, 0.10, 0.15. Here the best results for insertion are obtained with $p_m = 0.05$ and the best results for swap are obtained with $p_m = 0.10$; overall, insertion behaves better than swap.

After this experimental analysis, the chosen configuration for the GA will be: fitness, active scheduling; crossover, PMX with probability $p_x = 0.70$; mutation, insertion with probability $p_m = 0.05$; selection, random pairs; replacement, tournament without repetition.

4.2 Performance

To evaluate the performance of the proposed GA, we run the GA 30 times for each problem instance and consider the average value across these 30 runs of the relative makespan error for the best individual (B) and the average (A) in the last generation. Table 1 shows these values averaged across the 100 problem instances of each size, as a summary of the GA's behaviour, compared to the makespan relative error obtained by a dispatching rule DS/LRPT proposed in [14], adapted to fuzzy durations, and also compared to relative error of the best individual from a population of random permutations, with as many individuals as the total number of chromosomes evaluated by the GA. No other comparisons are made since, to our knowledge, there are no heuristic methods proposed in the literature for $FuzO||E[C_{max}]$.

Each row of Table 1 summarises the information relative to 100 problem instances: 10 fuzzy versions of each of the 10 crisp instances of one size. More detailed results are presented in Table 2, where each row corresponds to one set of 10 fuzzy versions of a crisp instance of size 10×10. As expected, the GA compares favourably with the random population. Concentrating on the second table, the makespan for the latter is between 9.259% and 10.904% worse (it also takes between 36.238% and 37.959% more running time). The GA also performs better than the priority rule DS/LRPT with, of course, smaller difference in makespan error (between 2.263% and 4.564% compared to the average performance of the GA). Notice as well that the relative errors for the best (B) and average (A) solution do not differ greatly, which means that the GA is quite stable. Also, relative errors are relatively small, showing that the GA either obtains an optimum solution or is quite close, even for the problems of greater size.

Table 2. Average relative makespan error (in %) for sets of problems of size 10 × 10

Problem	GA		Random Pop.	DS/LRPT
	B	A		
tai10_1	2.743	5.245	16.044	7.508
tai10_2	0.923	2.842	12.904	5.940
tai10_3	2.140	4.591	14.101	7.738
tai10_4	0.568	2.404	11.659	5.884
tai10_5	1.687	3.987	13.559	8.160
tai10_6	0.520	2.608	14.150	6.018
tai10_7	1.088	3.497	12.890	6.920
tai10_8	1.361	3.839	14.063	7.815
tai10_9	1.366	3.541	14.513	8.105
tai10_10	1.422	3.686	13.343	6.607

5 Conclusions and Future Work

We have considered an open shop problem with uncertain durations modelled as triangular fuzzy numbers, $FuzO||E[C_{max}]$, and have proposed a genetic algorithm to solve this problem. Using a permutation-based codification, we have considered several genetic operators and have conducted a thorough experimentation in order to select operators and set GA parameters to obtain a final competitive configuration. The performance of the GA has been assessed on a set of problems obtained from classical ones in what would constitute a first benchmark for $FuzO||E[C_{max}]$. The GA has obtained good results both in terms of relative makespan error and also in comparison to a priority rule and a population of random solutions. These promising results suggest directions for future work. First, the GA should be tested on more difficult problems, fuzzy versions of other benchmark problems from the literature. Also, the GA provides a solid basis for the development of more powerful hybrid methods, in combination with local search techniques, an already successful approach in classical shop problems [15],[2] and also in fuzzy open shop [7] and fuzzy job shop [9],[20].

Acknowledgements. This work is supported by MEC-FEDER Grant TIN2007-67466-C02-01.

References

1. Guéret, C., Prins, C.: Classical and new heuristics for the open-shop problem: A computational evaluation. European Journal of Operational Research 107, 306–314 (1998)
2. Liaw, C.F.: A hybrid genetic algorithm for the open shop scheduling problem. European Journal of Operational Research 124, 28–42 (2000)
3. Sha, D.Y., Cheng-Yu, H.: A new particle swarm optimization for the open shop scheduling problem. Computers & Operations Research 35, 3243–3261 (2008)

4. Herroelen, W., Leus, R.: Project scheduling under uncertainty: Survey and research potentials. European Journal of Operational Research 165, 289–306 (2005)
5. Dubois, D., Fargier, H., Fortemps, P.: Fuzzy scheduling: Modelling flexible constraints vs. coping with incomplete knowledge. European Journal of Operational Research 147, 231–252 (2003)
6. Celano, G., Costa, A., Fichera, S.: An evolutionary algorithm for pure fuzzy flowshop scheduling problems. International Journal of Uncertainty, Fuzziness and Knowledge-Based Systems 11, 655–669 (2003)
7. Ishibuchi, H., Murata, T.: A multi-objective genetic local search algorithm and its application to flowshop scheduling. IEEE Transactions on Systems, Man, and Cybernetics–Part C: Applications and Reviews 67(3), 392–403 (1998)
8. Fortemps, P.: Jobshop scheduling with imprecise durations: a fuzzy approach. IEEE Transactions of Fuzzy Systems 7, 557–569 (1997)
9. González Rodríguez, I., Vela, C.R., Puente, J.: A memetic approach to fuzzy job shop based on expectation model. In: Proc. of IEEE International Conference on Fuzzy Systems, London, pp. 692–697. IEEE, Los Alamitos (2007)
10. González Rodríguez, I., Puente, J., Vela, C.R., Varela, R.: Semantics of schedules for the fuzzy job shop problem. IEEE Transactions on Systems, Man and Cybernetics, Part A 38(3), 655–666 (2008)
11. Tavakkoli-Moghaddam, R., Safei, N., Kah, M.: Accessing feasible space in a generalized job shop scheduling problem with the fuzzy processing times: a fuzzy-neural approach. Journal of the Operational Research Society 59, 431–442 (2008)
12. Pinedo, M.L.: Scheduling, 3rd edn. Theory, Algorithms, and System. Springer, Heidelberg (2008)
13. Liu, B., Liu, Y.K.: Expected value of fuzzy variable and fuzzy expected value models. IEEE Transactions on Fuzzy Systems 10, 445–450 (2002)
14. Liaw, C.F.: An iterative improvement approach for the nonpreemptive open shop scheduling problem. European Journal of Operational Research 111, 509–517 (1998)
15. Liaw, C.F.: A tabu search algorithm for the open shop scheduling problem. Computers and Operations Research 26, 109–126 (1999)
16. Puente, J., Diez, H., Varela, R., Vela, C., Hidalgo, L.: Heuristic Rules and Genetic Algorithms for Open Shop Scheduling Problem. In: Conejo, R., Urretavizcaya, M., Pérez-de-la-Cruz, J.-L. (eds.) CAEPIA/TTIA 2003. LNCS(LNAI), vol. 3040, pp. 394–403. Springer, Heidelberg (2004)
17. Jussien, N., Lhomme, O.: Local search with constraint propagation and conflict-based heuristics. Artificial Intelligence 139, 21–45 (2002)
18. Michalewicz, Z.: Genetic Algorithms + Data Structures = Evolution Programs, 3rd revised and extended edn. Springer, Heidelberg (1996)
19. Taillard, E.: Benchmarks for basic scheduling problems. European Journal of Operational Research 64, 278–285 (1993)
20. González Rodríguez, I., Vela, C.R., Puente, J., Varela, R.: A new local search for the job shop problem with uncertain durations. In: Proc. of the Eighteenth International Conference on Automated Planning and Scheduling, pp. 124–131. AAAI Press, Menlo Park (2008)

Genetic Algorithm Combined with Tabu Search for the Job Shop Scheduling Problem with Setup Times

Miguel A. González, Camino R. Vela, and Ramiro Varela

A.I. Centre and Department of Computer Science,
University of Oviedo, Spain
raist@telecable.es, {crvela,ramiro}@uniovi.es
http://www.aic.uniovi.es/Tc

Abstract. We face the Job Shop Scheduling Problem with Sequence Dependent Setup Times and makespan minimization. To solve this problem we propose a new approach that combines a Genetic Algorithm with a Tabu Search method. We report results from an experimental study across conventional benchmark instances showing that this hybrid approach outperforms the current state-of-the-art methods.

1 Introduction

The Job Shop Scheduling Problem with Sequence Dependent Setup Times (SDST-JSP) is a generalization of the classical Job Shop Scheduling Problem (JSP) in which a setup operation on a machine is required when the machine switches between two jobs. In this way the SDST-JSP models many real situations better than the JSP, for example in semiconductor industry [1]. At the same time, the presence of setup times changes significatively the nature of this problem so as the well-known results and methods for the JSP are not directly applicable to the SDST-JSP.

In the last decades, scheduling problems with setup considerations have been subject to intensive research. Consequently, in the literature several approaches can be found that, in general, try to extend solutions that were successful for the classical JSP. For example, the branch and bound algorithm proposed in [2] is an extension of well-known algorithms proposed in [3] and [4]. Also, in [1] the authors extend the shifting bottleneck heuristic proposed in [5] for the JSP. In [6] and [7] two hybrid approaches are proposed that combine a genetic algorithm with local search procedures.

In this paper we propose a Tabu Search (TS) algorithm and combine it with a Genetic Algorithm (GA). The hybrid approach, termed GA+TS, uses the TS algorithm as local search procedure. Both metaheuristics, GA and TS, have not a theoretical performance guarantee but both of them have a solid track record of empirical performance in solving several scheduling problems [7], [6], [8], [9].

In the next section we formulate the SDST-JSP and introduce the notation used across the paper. In sections 3 and 4 we describe TS and GA algorithms,

J. Mira et al. (Eds.): IWINAC 2009, Part I, LNCS 5601, pp. 265–274, 2009.

and how they are used in combination. Section 5 reports results from the experimental study. Finally, in section 6 we summarize the main conclusions and propose some ideas for the future.

2 Description of the Problem

The SDST-JSP requires scheduling a set of N jobs $\{J_1, \ldots, J_N\}$ on a set of M physical resources or machines $\{R_1, \ldots, R_M\}$. Each job J_i consists of a set of tasks or operations $\{\theta_{i1}, \ldots, \theta_{iM}\}$ to be sequentially scheduled, each task θ_{ij} having a single resource requirement, a fixed duration $p_{\theta_{ij}}$ and a start time $st_{\theta_{ij}}$ whose value should be determined.

After an operation θ_{ij} leaves the machine and before an operation θ_{kl} enters the same machine, a setup operation is required with duration $S_{\theta_{ij}\theta_{kl}}$. $S_{0\theta_{ij}}$ is the setup time required before θ_{ij} if this operation is the first one scheduled on the machine, analogously $S_{\theta_{ij}0}$ is the cleaning time after operation θ_{ij} if this is the last operation scheduled on the machine.

The SDST-JSP has two binary constraints: precedence constraints and capacity constraints. Precedence constraints, defined by the sequential routings of the tasks within a job, translate into linear inequalities of the type: $st_{\theta_{ij}} + p_{\theta_{ij}} \leq st_{\theta_{i(j+1)}}$ (i.e. θ_{ij} before $\theta_{i(j+1)}$). Capacity constraints that restrict the use of each resource to only one task at a time translate into disjunctive constraints of the form: $st_{\theta_{ij}} + p_{\theta_{ij}} + S_{\theta_{ij}\theta_{kl}} \leq st_{\theta_{kl}} \vee st_{\theta_{kl}} + p_{\theta_{kl}} + S_{\theta_{kl}\theta_{ij}} \leq st_{\theta_{ij}}$, where θ_{ij} and θ_{kl} are operations requiring the same machine. The objective is to obtain a feasible schedule such that the completion time of all jobs, i.e. the *makespan*, denoted C_{max}, is minimized. This problem is denoted by $J|s_{ij}|C_{max}$ according to the $\alpha|\beta|\gamma$ notation used in the literature.

2.1 The Disjunctive Graph Model Representation

The Disjunctive Graph is a common representation model for scheduling problems. The definition of such graph depends on the particular problem. For the SDST-JSP, Disjunctive Graphs can be defined as follows. A problem instance may be represented by a directed graph $G = (V, A \cup E \cup I)$. Each node in the set V represents a task of the problem, with the exception of the dummy nodes *start* or 0 and *end* or $nm + 1$ which represent tasks with null processing times. For a task θ_{ij}, the label of the corresponding node will be $k = m(i-1)+j$. The arcs of set A are called *conjunctive arcs* and represent precedence constraints and the arcs of set E are called *disjunctive arcs* and represent capacity constraints. Set E is partitioned into subsets E_i, with $E = \cup_{i=1,\ldots,M} E_i$. E_i corresponds to resource R_i and includes an arc (v, w) for each pair of operations requiring that resource. Each arc (v, w) of A is weighted with the processing time of the operation at the source node, p_v, and each arc (v, w) of E is weighted with $p_v + S_{vw}$. The set I includes arcs of the form $(start, v)$ and (v, end) for each operation v of the problem. These arcs are weighted with S_{0v} and $p_v + S_{v0}$ respectively.

A feasible schedule is represented by an acyclic subgraph G_s of G, $G_s = (V, A \cup H \cup J)$, where $H = \cup_{i=1\ldots M} H_i$, H_i being a hamiltonian selection of E_i.

J includes arcs $(start, v_i)$ and (w_i, end) for all $i = 1 \ldots M$, v_i and w_i being the first and last operations of H_i respectively.

Therefore, finding a solution can be reduced to discovering compatible hamiltonian selections, i.e. processing orderings for the operations requiring the same resource, or partial schedules, that translate into a solution graph G_s without cycles. The makespan of the schedule is the cost of a *critical path*. A *critical path* is a longest path from node *start* to node *end*. Nodes and arcs in a critical path are termed *critical*. The critical path is naturally decomposed into subsequences $B_1, \ldots, B_r$ called *critical blocks*. A *critical block* is a maximal subsequence of operations of a critical path requiring the same machine.

Figure 1 shows a solution to a problem with 3 jobs and 3 machines. Dotted arcs represent the elements H and J, while arcs of A are represented by continuous arrows.

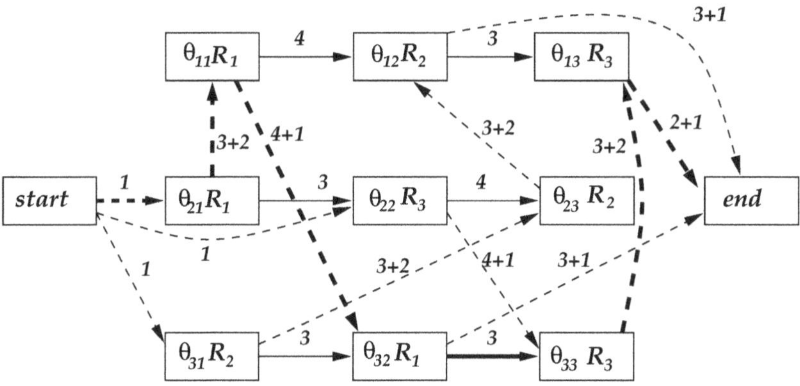

Fig. 1. A feasible schedule to a problem with 3 jobs and 3 machines. Bold face arcs show a critical path whose length, i.e. the makespan, is 22.

The concepts of critical path and critical block are of major importance for scheduling problems due to the fact that most of the formal properties and solution methods rely on them. For example, most of the neighborhood structures used in local search algorithms, such as that described in section 3, consist in reversing arcs in a critical path.

3 Tabu Search for the SDST-JSP

Tabu search is an advanced local search technique that can escape from local optima by selecting non improving neighbors temporarily. In order to avoid revisiting recently visited solutions and also to search new promising regions of the search space, a tabu search algorithm maintains a tabu list with a set of moves which are not allowed when generating the new neighborhood. In [10], [11] or [9] we can found examples of successful tabu search algorithms for the

```
Require: A scheduling problem instance P
Ensure: A schedule H for instance P
  Generate an initial solution s and calculate its makespan f(s);
  Set the current solution s* = s, the best solution s_B = s, and the best makespan
  f(s_B) = f(s);
  globalIter = 0, improveIter = 0;
  Empty the tabu list, and push s* onto the elite solution stack L (LIFO list);
  while globalIter < maxGlobalIter do
    if improveIter = maxImproveIter then
      if The solution stack L is empty then
        return The solution s_B and its makespan f(s_B),
      else
        Pop the solution s_L on top of the solution stack L;
        Set the current solution s* = s_L;
        Set improveIter = 0;
        Clear the tabu list and the cycle detection structure;
    Set globalIter = globalIter+1, set improveIter = improveIter+1;
    Generate neighbors of the current solution s* by means of the neighborhood struc-
    ture;
    Let s* be the best neighbor not being tabu or satisfying the aspiration criterion,
    update the tabu list and the cycle detection structure accordingly;
    if f(s*) < f(s_B) then
      Set s_B = s*, f(s_B) = f(s*), improveIter = 0, and push s* onto the elite solution
      stack L;
  return The solution s_B and its makespan f(s_B);
```

Algorithm 1. The Tabu Search Algorithm

JSP. In particular, in [9] the best solution known for a number of the largest JSP instances taken from conventional repositories are reported.

Algorithm 1 shows the tabu search algorithm we have considered here. This algorithm is quite similar to other tabu search algorithms described in the literature [12]. In the first step the initial solution is generated and evaluated. Then, it iterates over a number of steps. In each iteration, the neighborhood of the current solution is built and one of the neighbors is selected for the next iteration. After a number of iterations without improvement, the search is restarted from a previous solution taken from the elite solution stack. The tabu search finishes when the elite solution stack gets empty and returns the best solution reached so far.

Initial solution. We have chosen to start the search from a randomized active schedule. To generate active schedules we use the Serial Schedule Generation Scheme (SSGS), proposed in [13]. This schedule builder chooses one of the available operations at random and selects for it the earliest starting time that satisfies all constraints with respect to the previous scheduled operations. An operation is available when its predecessor in the job sequence has already been scheduled. It continues until all operations get scheduled. This algorithm always produces active schedules, provided that the triangular inequality for the setup times holds. Moreover any active schedule can be built by taking the appropriate sequence of selections. When TS is combined with GA, the initial solution cames from a chromosome generated by the GA.

The neighborhood structure. The neighborhood structure considered in this paper is that termed N^S in [7]. N^S extends some of the structures proposed in [14] and [15] that have given rise to some of the most outstanding methods for the JSP such as, for example, the algorithms proposed in [10], [11], [16], [9]. N^S is based on the following two results.

Theorem 1. *Given a critical block of the form* $(b'\ v\ b\ w\ b'')$, *where* b, b' *and* b'' *are sequences of operations, a sufficient condition for an alternative path from* v *to* w *not existing is that*

$$r_{PJ_w} < r_{SJ_u} + p_{SJ_u} + \min\{S_{kl}/(k,l) \in E, J_k = J_u\}, \forall u \in \{v\} \cup b \qquad (1)$$

Theorem 2. *Let* H *be a schedule and* $(b'\ v\ b\ w\ b'')$ *a critical block, where* b, b' *and* b'' *are sequences of operations of the form* $b = (u_1 \dots u_n)$, $b' = (u'_1 \dots u'_{n'})$ *and* $b'' = (u''_1 \dots u''_{n''})$. *Even if the schedule* H' *obtained from* H *by moving* w *just before* v *is feasible,* H' *does not improve* H *if the following condition holds*

$$S_{u_n v} + S_{u'_n w} + S_{w u''_1} \geq S_{u_n w} + S_{wv} + S_{u'_n u''_1}. \qquad (2)$$

If $n = 0$, u'_n *should be substituted by* v *in (2).*

Therefore, the neighborhood structure N^S is defined as follows.

Definition 1 (N^S)**.** *Let operation* v *be a member of a critical block* B. *In a neighboring solution* v *is moved to another position in* B, *provided that condition (2) doesn't hold and that the sufficient condition of feasibility (1) is preserved.*

Makespan estimation. As it is usual, for the sake of efficiency the selection rule is based on makespan estimations instead of computing the actual makespan of all neighbors. For this purpose, we use here the procedure *lpathS* given in [6] for the SDST-JSP that extends the *lpath* rule for the JSP. This procedure takes an input sequence of operations requiring the same machine of the form $(Q_1 \dots Q_q)$ after a move, which is a permutation of operations appearing as $(O_1 \dots O_q)$ before the move. For each $i = 1 \dots q$, *lpathS* computes the cost of the shortest path from node *start* to node *end* through Q_i. The maximum of these values is taken as the makespan estimation for the neighboring schedule, which is clearly a lower bound of its actual makespan. For N^S, if w is moved before v in a block of the form $(b'\ v\ b\ w\ b'')$, the input sequence is $(w\ v\ b)$.

The tabu list. The basic role of tabu list is to prevent the search process from turning back to solutions visited in previous steps. In order to reduce computational cost, we have chosen to store in the tabu list the attributes of moves rather than the attributes of solutions. So, the tabu list stores the arcs that have recently been reversed. These arcs can not be reversed, with the exception of those satisfying the aspiration criterion. When a neighbor is chosen for the next iteration, the tabu list is updated with all the arcs that have been reversed to generate this neighbor. Then, a new neighbor is marked as tabu if it requires reversing at least one arc of the tabu list.

The length of the tabu list is usually of critical importance. In this paper we have used a dynamic length for the tabu list, as it is proposed in [10]: it is decreased in one unit when the current solution is better than the previous one, otherwise it is increased in same amount. In general, the length is keep in an interval [*min,max*], being *min* and *max* parameters, but it is set to one when the current solution is better than the best solution reached so far. The rationale behind dynamic length is unrestrict the search when new promising areas are found and restrict the search only to new solutions over non promising areas in order to escape quickly towards more promising ones.

Cycle checking. The cycle prevention procedure is based on that proposed in [10] for the JSP. In a cyclic solution, a solution $s(i)$ after i iterations is equal to a solution $s(i+k)$ after $i+k$ iterations, for some positive k, and the arc selected to move from $s(i+k)$ to $s(i+k+1)$ is the same as the arc selected previously to move from $s(i)$ to $s(i+1)$. In order to detect this situation, a witness arc is chosen for each move and the value of the estimated makespan of the resulting neighbor is associated with it. Then, the makespan of each neighbor is compared with that associated to the witness arc of the corresponding move. If these values coincide for more than *Tcycle* consecutive iterations, the search is supposed to be in a cycle and so the neighbor is discarded, unless it satisfies the aspiration criterion given in the next section.

Move selection and aspiration criterion. The selection rule chooses the neighbor with the lowest estimated makespan, discarding suspect-of-cycle and tabu neighbors (unless they satisfy the aspiration criterion). A side effect of implementing a *partial attribute* tabu list is that it may lead to giving a tabu status to an unvisited solution or even an interesting solution. So the algorithm uses an aspiration criterion which allows to accept any move provided that its estimated makespan is lower than that of the current best solution found so far. If all possible moves are tabu or lead to a cycle, and none of them satisfy the aspiration criterion, a neighbor is chosen at random.

Recovery of elite solutions. Each time a new solution improves the current best so far, it is stored in a stack L. Then every *maxImproveIter* iterations without improvement, the current solution is replaced with one extracted from L (in the experiments we have chose $maxImproveIter = maxGlobalIter/Size(L)$). This is a technique commonly used to avoid the algorithm getting trapped in local optima. The size of L is restricted to a small value (20 in our experiments) in order to avoid bookkeeping bad solutions from the beginning of the search. When L is full, every new solution replaces the oldest one. When recovery to a previous solution from L, the state of the algorithm must be reconsidered. We have tried a number of possibilities with very similar results, so we have chosen to reset the tabu list, the bookkeeping structure and the number of iterations without improvement to their initial values. When TS is used in combination with GA, for the sake of efficiency the recovery solutions mechanism is not used. This simplification does not prevent the algorithm from search over the whole search space as TS is applied to every chromosome generated by GA.

Termination criterion. The algorithm stops after performing a given number of iterations *maxGlobalIter*, or when the elite solution stack has been exhausted.

4 Genetic Algorithm for the SDST-JSP

The GA used here is taken from [7]. The codification schema is based on permutations with repetition as it was proposed in [17]. A chromosome is a permutation of the set of operations, each one being represented by its job number. For example, the sequence (2 1 1 3 2 3 1 2 3) is a valid chromosome for a problem with 3 jobs and 3 machines. As it is demonstrated in [18], this codification has a number of interesting characteristics; for example, it tends to represent orders of operations as they appear in good solutions. For chromosome mating, the GA uses the Job Order Crossover (JOX) described in [17]. And to build schedules from chromosomes, the GA exploits the Sequential Schedule Generation Schema (SSGS) proposed by Artigues et al. in [13]. As it has commented above, when GA is combined with TS, this algorithm is applied to the schedule produced by SSGS. Then, the chromosome is rebuild from the improved schedule obtained by TS, so as the improvement obtained by TS can be transferred to subsequent offsprings.

5 Experimental Study

For experimental study we have used the set of problems proposed in [2] (the BT set) and also some new benchmark instances proposed in [6]. The BT set has a total of 15 instances named t2-ps01 to t2-ps15. Instances t2-ps01 to t2-ps05 have size 10×5 (10 jobs and 5 machines). Instances t2-ps06 to t2-ps10 have size 15×5. Almost all these instances are very easy to solve so we have not considered them in our experiments. Instances t2-ps11 to t2-ps15 are harder to solve and have size 20×5. Instances t2-pss12 and t2-pss13 differ only from t2-ps12 and t2-ps13 in the setup times. The second benchmark is a set of larger problem instances than those in the BT set. These instances are derived from the set of 10 selected problems identified in [19] as hard to solve for the classical JSP. The size of these problems are 15×10 for the smallest ones (La21 to La25), 20×10 for La27 and La29, 15×15 for La38 and La40, and 20×15 for the largest (ABZ instances). These instances are extended to the SDST-JSP by using the same criteria as in the BT instances, i.e. the setup times depend only on the jobs and are taken from one of two matrices (these matrices can be found, for example, in [13] and identify the type of setup times: 5 or 10). Instances with 15 jobs are of type 5 and instances with 20 jobs are type 10. Each instance is identified by the the name of the JSP instance followed by st.

We compare the proposed GA+TS algorithm with two previous approaches: The shifting bottleneck algorithm with guided local search (SB+GLS) proposed in [1] and another GA hybridized with local search (GA+LS) proposed in [7]. Both GA+LS and GA+TS were run 30 times for each instance. Then the best of the 30 solutions and their mean values are reported. SB+GLS is compared in [1] with a number of exact and non-exact algorithms and obtain the best results on the largest instances of the BT set (with only one exception in t2-ps14 instance where a solution with 1483 is known). To obtain similar run times the GA+TS parameters (/number of individuals/number of generations/number of iteration of TS/) were /40/60/200/ for the BT instances and from /40/60/200/ to /80/110/200/ for the selected instances depending on their size.

Table 1 summarizes the results of all three algorithms. As we can observe, GA+LS obtains better results than SB+GLS on BT instances. The results from GA+LS over the second benchmark are the first ones given on this set. GA+TS obtains better results than GA+LS. For instances t2-ps06 to t2-ps10 the best solution reached is the same for both algorithms, however the mean solution is better or at least equal in four of the five instances. For the remaining instances, both the best and mean solutions from GS+TS are better than those from GS+LS for most of the instances. Only in a few cases any of these values is equal.

It is important to be aware of the differences in the target machines. The SB+GLS was implemented in C language and run on a Sun Ultra 60 with

Table 1. Summary of previous and new results

Instance	SB+GLS		GA+LS		GA+TS		
	best	time	best	Avg.	best	Avg.	time
t2-ps11	1470	3033	1438	1439	1438	1441	34.92
t2-ps12	1305	2186	1269	1295	1269	1277	34.57
t2-ps13	1439	2506	1415	1416	1415	1416	34.57
t2-ps14	1485	2115	1452	1489	1452	1489	37.51
t2-ps15	1527	3029	1485	1505	1485	1496	36.02
t2-pss12	1290	2079	1269	1273	1258	1266	35.23
t2-pss13	1398	1864	1390	1390	1361	1379	36.81
La21_st			1277	1291	1268	1276	26.28
La24_st			1151	1156	1151	1153	23.24
La25_st			1183	1193	1183	1188	26.25
La27_st			1760	1775	1735	1754	73.37
La29_st			1664	1675	1659	1667	78.20
La38_st			1456	1469	1446	1452	33.22
La40_st			1492	1497	1482	1490	33.21
ABZ7_st			1239	1260	1231	1245	180.12
ABZ8_st			1266	1286	1257	1271	174.71
ABZ9_st			1220	1255	1217	1235	175.79

UltraSPARC-II processor at 360MHz. whilst GA+LS and GA+TS algorithms were coded in C++ and run in a Pentium IV (1.7GHz) and in a Intel Core 2 Duo (2.6GHz) machines respectively. Both algorithms GA+LS and GA+TS were parameterized so as to obtain similar run times as it is indicated in 1.

We have also experimented with TS and GA alone. TS was run 30 times starting from a random solution and leaving TS for a maximum of $5 * 10^5$ to $22 * 10^5$ iterations depending on the instance size. GA was parameterized from /500/600/ to /1000/1000/ to obtain similar run times. The results, not reported here, shown clearly that any of these algorithms alone is much worse than both of them in combination, when all algorithms run for similar amount of time. So, it is worth to mention that it is the synergy gained from their use in combination what produces a very efficient approach. For all the instances considered in the experimental study this approach obtains the best results known so far.

6 Conclusions

We have considered the job shop problem with sequence dependent setup times, where the objective is to minimize the makespan. We have used a hybrid algorithm that combines a simple tabu search with a genetic algorithm. We have reported results from an experimental study across the benchmarks proposed in [2] and [6], and have compared our algorithm with two state-of-the-art methods: the shifting bottleneck heuristic proposed in [1] and the hybrid genetic algorithm proposed in [7]. The results of this study show that our approach outperforms all the current state-of-the-art methods. As future work we plan to consider objective functions other than makespan, for example the total flow time, the tardiness or the lateness. We will consider single and multi-objective versions of all these problems.

Acknowledgements

This work has been supported by the Spanish Ministry of Science and Education under research project MEC-FEDER TIN2007-67466-C02-01 and by FICYT under grant BP07-109.

References

1. Balas, E., Simonetti, N., Vazacopoulos, A.: Job shop scheduling with setup times, deadlines and precedence constraints. Journal of Scheduling (2008), doi:10.1007s10951-008-0067-7
2. Brucker, P., Thiele, O.: A branch and bound method for the general-job shop problem with sequence-dependent setup times. Operations Research Spektrum 18, 145–161 (1996)
3. Brucker, P., Jurisch, B., Sievers, B.: A branch and bound algorithm for the job-shop scheduling problem. Discrete Applied Mathematics 49, 107–127 (1994)

4. Carlier, J., Pinson, E.: Adjustment of heads and tails for the job-shop problem. European Journal of Operational Research 78, 146–161 (1994)
5. Adams, J., Balas, E., Zawack, D.: The shifting bottleneck procedure for job shop scheduling. Managament Science 34, 391–401 (1988)
6. Vela, C.R., Varela, R., González, M.A.: Local search and genetic algorithm for the job shop scheduling problem with sequence dependent setup times. Journal of Heuristics (2009), doi:10.1007/s10732-008-9094-y
7. González, M.A., Vela, C.R., Varela, R.: A new hybrid genetic algorithm for the job shop scheduling problem with setup times. In: Proceedings of the Eighteenth International Conference on Automated Planning and Scheduling (ICAPS 2008), Sidney. AAAI Press, Menlo Park (2008)
8. González Rodríguez, I., Vela, C.R., Puente, J.: A memetic approach to fuzzy job shop based on expectation model. In: Proceedings of IEEE International Conference on Fuzzy Systems, London, pp. 692–697. IEEE, Los Alamitos (2007)
9. Zhang, C.Y., Li, P., Rao, Y., Guan, Z.: A very fast TS/SA algorithm for the job shop scheduling problem. Computers and Operations Research 35, 282–294 (2008)
10. Dell' Amico, M., Trubian, M.: Applying tabu search to the job-shop scheduling problem. Annals of Operational Research 41, 231–252 (1993)
11. Nowicki, E., Smutnicki, C.: An advanced tabu search algorithm for the job shop problem. Journal of Scheduling 8, 145–159 (2005)
12. Glover, F.: Tabu search–part I. ORSA Journal on Computing 1(3), 190–206 (1989)
13. Artigues, C., Lopez, P., Ayache, P.: Schedule generation schemes for the job shop problem with sequence-dependent setup times: Dominance properties and computational analysis. Annals of Operations Research 138, 21–52 (2005)
14. Matsuo, H., Suh, C., Sullivan, R.: A controlled search simulated annealing method for the general jobshop scheduling problem. Working paper 03-44-88, Graduate School of Business, University of Texas (1988)
15. Van Laarhoven, P., Aarts, E., Lenstra, K.: Job shop scheduling by simulated annealing. Operations Research 40, 113–125 (1992)
16. Balas, E., Vazacopoulos, A.: Guided local search with shifting bottleneck fo job shop scheduling. Management Science 44(2), 262–275 (1998)
17. Bierwirth, C.: A generalized permutation approach to jobshop scheduling with genetic algorithms. OR Spectrum 17, 87–92 (1995)
18. Varela, R., Serrano, D., Sierra, M.: New codification schemas for scheduling with genetic algorithms. In: Mira, J., Álvarez, J.R. (eds.) IWINAC 2005. LNCS, vol. 3562, pp. 11–20. Springer, Heidelberg (2005)
19. Applegate, D., Cook, W.: A computational study of the job-shop scheduling problem. ORSA Journal of Computing 3, 149–156 (1991)

Prediction and Inheritance of Phenotypes

Antonio Bahamonde, Jaime Alonso, Juan José del Coz, Jorge Díez, José Ramón Quevedo, and Oscar Luaces

Artificial Intelligence Center. University of Oviedo at Gijón, Asturias, Spain
www.aic.uniovi.es

Abstract. In the search for functional relationships between genotypes and phenotypes, there are two possible findings. A phenotype may be heritable when it depends on a reduced set of genetic markers. Or it may be predictable from a wide genomic description. The distinction between these two kinds of functional relationships is very important since the computational tools used to find them are quite different. In this paper we present a general framework to deal with phenotypes and genotypes, and we study the case of the height of barley plants: a predictable phenotype whose heritability is quite reduced.

1 Introduction

There is an increasing number of applications in which genomic information (genotypes) is functionally related to observable characteristics (phenotypes). The applications include the study of risks of human diseases and the improvement in food production.

When we look for such relationship there are two possible positive outputs. First, we can find that there is a small set (usually a singleton) of genetic markers (the smallest part of genotype descriptions) whose presence implies a disease or a useful property of food, in other words, a phenotype. In these cases, the phenotype is a hereditary characteristic. The piece of genome that contains the marker, according to Mendel's Laws, is transmitted from parent organisms to their children.

On the other hand, there is a second kind of genothype/phenotype relation. The phenotypes may be predicted from a genetic description of individuals. Predictions may have different degrees. For instance, heritable phenotypes are trivially predicted. But the opposite is not always true. In fact, if predictions are established from a high number of markers, the probability of transmitting a phenotype may be low: not all relevant markers are present in children if parents do not have two copies of all these markers (homozygotes).

The existence of predictable but not necessarily inheritable phenotypes is an interesting issue that has received little attention until a few years ago. From a biological point of view these phenotypes are complex traits driven by genetic markers which may be very distant one each other in the genome [1].

The utility of such predictions has been documented in [2]. Thus, prediction is a computable way to assess the genetic risk of a disease in healthy individuals

J. Mira et al. (Eds.): IWINAC 2009, Part I, LNCS 5601, pp. 275–284, 2009.

[3]. In livestock, it has been reported [4] that artificial selection on genetic values predicted from markers could substantially increase the rate of genetic gain in animals and plants. For instance, [5] details the benefits of artificial selection in beef cattle. The authors study the use of marker-assisted selection (MAS) by using estimates (predictions) of the effects of markers on commercial crossbred performance.

In this paper, we describe the difficulties to implement a prediction system in a context of food products. So, in the next section we review how to determine a phenotype in a food environment. In these circumstances, phenotype means assessment and usually this is done by human experts whose decisions are not always repeatable. Other times phenotypes are approximate estimations of a magnitude that may have environmental (not genetic) influences. In any case, the main characteristic of these assessments is the order. In other words, the absolute values of phenotypes are irrelevant, their utility is to compare different individuals.

In the last section, we present a set of experimental results conducted to illustrate the differences between predictions and inheritances. The experiments were carried out with a dataset of barley genotypes and phenotypes publicly available. In this section we shall propose a set of measurements to assess the goodness of predictions in this genetic context.

2 Phenotypes and Genotypes

Phenotypes are quantitative expressions of any genetic characteristic or trait. As was emphasized in the Introduction, these values have only a relative meaning. They are useful only if they can be employed to order samples according to a given trait. In this section we shall review two main kinds of phenotypes of food products: those that are given by assessments rated by experts or consumers, and those that are estimations inferred from a collection of measurements of some performance related to food quality or productivity.

On the other hand, genotypes are genetic descriptions of individuals. The final aim is to find a functional relationship between genotypes and phenotypes, if such relation exists.

2.1 Phenotypes as People's Assessments

Let us start with a typical assessment problem of food products. The paper [6] describes a methodology for establishing an objective measurement of *mushroom* quality. Four experts visually evaluated 300 mushrooms and graded them into three major and eight subclasses of commercial quality. Grader consistency was also assessed by repeated classification (four repetitions) of two 100-mushroom sets. Grader repeatability ranged from 6 to 15% misclassification.

To avoid these difficulties, a number of methods have been suggested [7]. The aim is to make as *objective* as possible the assessment. For this purpose, the best option is to rest on a computable formula that uses only a set of metric measurements. However, the complexity of the task of learning an assessment function stems from the low repeatability of human evaluations. Despite experts had been trained exhaustively and had accumulated a large and valuable body of knowledge.

Let us now recall how we dealt with the assessment of *beef cattle* as meat producers. This learning task was proposed by ASEAVA, the Association of Breeders of a beef breed of the North of Spain, *Asturiana de los Valles*. This is a specialized breed with many double-muscled individuals; their carcasses have dressing percentages over 60%, with muscle content over 75%, and with a low (8%) percentage of fat.

Even if the animals are not going to be slaughtered, the prediction of carcass value of a beef cattle is interesting since it is useful for breeders to select the progenitors of the next generation. The records of carcass value predictions can be used for the evaluation of programs of genetic selection. The growth of the scores over years of selection for specific goals can be seen as a measure of the success of the selection policy.

Traditionally, the assessment procedures were based on *visual* appreciations of well trained technicians that had to rank a number of morphological characteristics that include linear lengths of significant parts of animals' bodies. To test the reliability of these visual appreciations, we gathered (in [8]) the ASEAVA records of 2844 animals. These records include visual appreciations of the curvature of the round profile (see the curves of the buttocks in Figure 1). The ranks given by the experts of the Association of Breeders had only a poor correlation (0.70) with the EUROP ranks used to pay breeders for their carcasses. The so called *EUROP* assessment method is regulated by the European Union, and it is supposed to be correlated (for similar weights) with the curvature of the round profile.

The lack of coherence of people's assessments is not an unusual situation in beef cattle. Another application where we found assessment problems was studying *consumer tastes* about food products [9]. Consumers and experts tend

Fig. 1. The assessment of the round profile of a beef cattle is a measurement of the roundness of the the curves of the buttocks. The cows in the picture are representative of an extremely high (the bigger cow) and extremely low (the smaller) round profiles.

to rate their preferences in a relative way, comparing objects with other samples in the same batch or tasting session. There is a kind of *batch effect* that often biases the ratings. Thus, an object presented in a batch surrounded by worse objects will probably obtain a higher rating than if it were presented together with better objects.

Therefore consumer or experts ratings cannot be interpreted as absolute assessments. They are relative values to each assessment session. From a computational point of view, this fact is the reason to reject regression methods in order to learn automatic assessment functions defined from a set of objective metric measurements of food products. From a Machine Learning perspective, if E is a set of food samples, the usefulness of people assessments (consumers or experts) is not the set of pairs sample-rating, but a set of *preference judgments* given by

$$PJ = \{(\boldsymbol{u}, \boldsymbol{v}) : \boldsymbol{u}, \boldsymbol{v} \in E, r_i^j(\boldsymbol{u}) > r_i^j(\boldsymbol{v})\}. \qquad (1)$$

where $r_i^j(\boldsymbol{x})$ is the rating given by user i in session j for the product $\boldsymbol{x}$.

If phenotypes are going to be learned from people assessments, then the assessment function must be induced from datasets of preference judgments (Equation 1) using tools like those described in [10]. In [11,12] we proposed a new assessment method for beef cattle. On the other hand, in [13,14,15] we described a collection of methods to handle sensory data from consumer's opinions about food products.

2.2 Phenotypes as Approximate Measurements

There is a second kind of phenotypes, those that come from measurements of observable and quantifiable behaviors. In food products these phenotypes are related with production or performance. In the next section we are going to use a dataset of microarray descriptions of *barley plants*. The phenotypes considered are 9 traits, measured in up to 16 different environments. These traits are: α-amylase, diastatic power, heading date, plant height, lodging, malt extract, pubescent leaves, grain protein content, and yield. The number of individuals described in this dataset is quite limited, only 94 individuals.

In 8 times out of 9 the traits were represented by continuous values. The exception is the presence of *pubescent leaves*: a binary predicate. In the 8 continuos traits, to standardize the scores achieved by the learners used in these experiments, is frequent to use the percentiles of traits instead of their actual values. Therefore, all traits values range in the interval $[0, 100]$.

However, in addition to the standardization of trait measurements, the main difficulty to handle such values as phenotypes is the noise involved the estimations. This noise is due to environmental distortions. In fact, most traits, in addition to a genetic predisposition, can be increased or dismissed by factors like the season, feeding conditions, or the geological type of lands in plants.

Thus, to handle the estimations of trait measurements as phenotypes, we must have to consider such values as mere approximations. This means that no straightforward tools can be used in order to look for a computational relationship between genotypes and phenotypes.

2.3 Dealing with Genotypes

Once the phenotype of an individual i is fixed and represented by a real number (y_i), its genetic description is represented by a vector ($\boldsymbol{x}_i$) of feature values or genetic markers that are discrete values codifying the allelic composition, usually via SNP (Single Nucleotide Polymorphism). Notice that we do not consider any pedigree information.

Nowadays, microarray technology can be used to read hundreds of thousands to millions of SNPs in a single array experiment at a reasonable price. Thus, these kind of vectors are a reasonable representation of genotypes.

It is important to mention that when the genetic descriptions have physical locations in the genome sequence, the set of relevant attributes (i.e. positions in the genome sequence) somehow useful in relation to a quantitative (continuous valued) phenotype are called *QTL* (*quantitative trait loci*).

3 Prediction Tools for the Genetic Learning Task

Formally, the genetic learning task is given by a dataset

$$S = \{(\boldsymbol{x}_i, y_i) : i = 1, \ldots, n\}, \tag{2}$$

where $\boldsymbol{x}_i$ represents the genotype and y_i is a quantitative (i.e. a real number) phenotype of an individual i.

When phenotype values (y_i) are continuous (real) numbers, this learning task can be handled using regression tools. However, as was mentioned above, we must take into account that phenotypes are only approximate estimations. Thus, there are two possible approaches. The first identifies near values into qualitative labels. This approach rewrites S discretizing the range of phenotype values in a set of k bins of equal frequency. The aim is to redefine phenotypes in a scale of k ordered qualitative values. For instance, if we use $k = 5$, we can see this process as a translation into a ranking scale of 5 qualitative ranks: very low, low, medium, high, very high.

Rewritten in this way, S is a *classification* learning task. But in fact, it is an *ordinal regression* task since the classes are not only discrete, but also ordered. In [16], we have proposed the use of nondeterministic classifiers introduced in [8]. The aim of these classifiers is to build hypotheses that try to predict the true rank, but when the classification is uncertain the hypotheses predict an interval of ranks; a set of consecutive ranks, such that the set is as small as possible, while still containing the true rank. In symbols, a *nondeterministic hypothesis* is a function h from the space of genotype descriptions to the set of non-empty intervals (subsets of consecutive ranks)

$$h : \mathcal{X} \longrightarrow Intervals(\{1, 2, 3, 4, 5\}). \tag{3}$$

In the case of these classifiers, it is important to register not only the proportion of classifications that include the true rank, but also the average number of ranks of the predictions. We shall limit this number to 1 or 2 in the experiments reported in the next section.

To handle a learning task S defined as in Equation (2), to clean possible *noise*, it is usual to select the most meaningful components of vectors $\boldsymbol{x}_i$ in order to establish a functional relationship with the phenotypes y_i. These components are, actually, the QTLs from a biological perspective, and they are a *selection of features* for Machine Learning nomenclature.

The feature selection methods to find QTLs are different if we have a regression (continuous phenotypes) or a classification (integer phenotypes) learning task. However, at this stage the differences are not so important. We can discretize the continuous phenotypes to obtain a classification learning task, see the preceding section. Although there are some robust methods for selecting features in regression tasks, the noise level in the genetic datasets is frequently too high to hope good results. Thus, the advantage of dealing with classification tasks is that they are implicitly taking into account that phenotype values are only approximations. Additionally, classification entails a bundle of tools to filter data sets. We shall use the filter FCBF [17] to search for a reduced number of features. This filter has been frequently used for dealing with genetic data.

Let us recall that FCBF orders the features of a learning task according to their relevance to induce class values. For this purpose FCBF uses the so called *symmetrical uncertainty* (a normalized version of the mutual information) between feature and class values, which must be discrete, not necessarily ordered. Then the filter rejects those features that are somehow redundant with other features higher in the order of relevance.

To test the goodness of a selection of features we can use the proportion of successful classifications achieved by the hypothesis learned from different subsets of features. This is an *absolute* score which is not always meaningful in the genetic case. If the phenotype is discretized in k bins, using a binary feature, it is only possible to output 2 different predictions. Thus, the maximum proportion of successful classifications is $2/k$. Taking into account this remark, let us define the *relative* accuracy of a hypothesis learned with r features on a test set $S' = \{(\boldsymbol{x'}_i, y'_i) : i = 1, \ldots, n'\}$ as follows

$$\frac{\sum_{i=1}^{n'}(y' \in h(\boldsymbol{x'}))}{n' \cdot \min\{1, \frac{2^r}{k}\}}. \tag{4}$$

4 Experimental Results

The experiments reported in this section use publicly available datasets about barley plants whose source was [18]. We used the version described in [19] that is straightforward usable by Machine Learning algorithms. The genotypes considered are a subset of a collection of more than 1,000 RFLP, DArT and SSR markers. The genotypes of individuals of barley plants in [19] version are codified by 367 binary (0/1) features to express the presence/absence of parental allele

Table 1. List of QTLs identified using different approaches. The column labeled by *FCBF* reports the ordered list of markers returned by the filter FCBF. The next columns report the markers selected by Composite Interval Mapping (*CIM*), single Marker Regression (*MR*), and Statistical Machine Learning (*SML*). The second part of the Table lists markers selected by CIM and MR ordered by Chromosome. In all cases, markers are represented by centimorgans (cM): distance along a chromosome.

Chromosomes	FCBF	CIM	MR	SML
2H	118.9	120.7	118.9	117.8
2H	0.0		14.8	
3H	117.4	117.4	117.4	117.4
3H	74.5			
5H	65.9	68.1	71.5	68.1
4H	129.1		142.2	
1H	31.6		35.5	
7H	55.4	52.8	53.4	
1H	91.1		91.1	
5H	179.1		177.2	
3H	0.0			
7H	128.7			
6H	89.8			
2H		22.5	46.7	
2H		91.2		
4H		78.6		
4H		92.6		
6H		144.1	149.4	

calls. As was mentioned above, these collection of tasks report the phenotypic data for 9 traits (1 is binary, and the other 8 are continuous). The scores obtained with the 8 continuous phenotypes are described in [16]. The novelty of this paper is the discussion that follows about the genetic features relevant to predict plant heigh. All the scores reported here were obtained using a 10-fold cross validation.

Table 1 shows the list of 13 genetic markers relevant to the trait plant heigh according to FCBF and 3 other methods (see [19]). The first QTL, 2H 118.9, is also reported as relevant for this trait by the methods SML, CIM and MR [19]. This marker is inheritable, however, using this marker the quality of predictions are really very poor. The first row of Table 2 reports that only 39% of times it is possible to predict the true percentile interval using the value of 2H 118.9. But the relevance of this marker comes from its relative accuracy (98%). The biological interpretation is the following. With the exception of 1 sample, barleys with percentiles in $(80, 100]$ have value 0 in 2H 118.9; and all barleys with percentiles in barleys $[0, 20]$ have value 1 in 2H 118.9. In other words, this marker is a necessary but not sufficient condition to determine the heigh of a barley plant.

Table 2. Scores achieved for different number of relevant features (first column). The absolute and relative proportion of successful classifications are followed by the average number of bins predicted.

# features	absolute	relative	size pred.
1	0.39	0.98	1.00
2	0.50	0.63	1.23
3	0.52	0.52	1.60
4	0.65	0.65	1.57
5	0.70	0.70	1.57
6	0.69	0.69	1.54
7	0.70	0.70	1.45
8	0.71	0.71	1.44
9	0.75	0.75	1.36
10	0.68	0.68	1.22
13	0.77	0.77	1.44

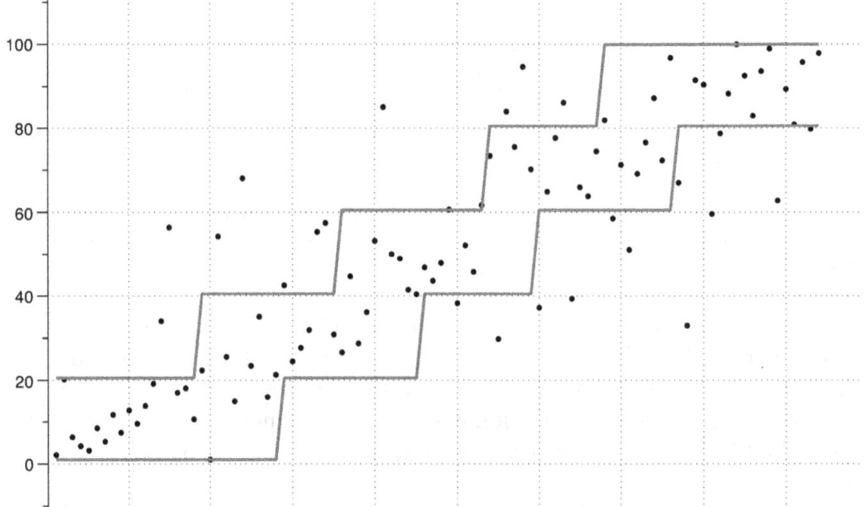

Fig. 2. True plant height percentiles (•) and the intervals predicted (using a 10-fold cross validation) by a nondeterministic classifier. To ease the readability, the horizontal axis represents the indexes of barley samples ordered according to their predictions.

To reach a more useful prediction score, it is necessary to use more markers spread throughout several chromosomes (first column of Table 1). Thus the heritability of this trait is low although it may be predicted from a wide genomic description.

To illustrate the performance of nondeterministic classifiers, we included Table 3 and Figure 2. The size of predictions (first column) is the number of ranks (Equation 3) included in the intervals $h(\boldsymbol{x})$. Most of the times nondeterministic

Table 3. Performance of the nondeterministic classifier that predicts the rank of plant height. The rows show the proportion of successful and failed classifications of size 1 and 2 respectively.

$\|h(x)\|$	success	fail	total
1	0.42	0.14	0.56
2	0.35	0.09	0.44
Total	0.77	0.23	

classifiers predict only one rank (56%), and in that case the proportion of successful and failed classification is 0.42 and 0.14 respectively. On the other hand, when classifiers predict 2 ranks, almost always they are right.

Acknowledgements

The research reported here is supported in part under grant TIN2008-06247 from the MICINN (Ministerio de Ciencia e Innovación, of Spain). The authors also acknowledge the people who shared the barley data used in this paper, [18,19] and the software used in the experiments reported here.

References

1. Mauricio, R.: Mapping quantitative trait loci in plants: uses and caveats for evolutionary biology. Nature Reviews Genetics 2, 370–381 (2001)
2. Lee, S., van der Werf, J., Hayes, B., Goddard, M., Visscher, P.: Predicting Unobserved Phenotypes for Complex Traits from Whole-Genome SNP Data. PLoS Genetics 4(10) (2008)
3. Wray, N., Goddard, M., Visscher, P.: Prediction of individual genetic risk to disease from genome-wide association studies. Genome Research 17, 1520–1528 (2007)
4. Meuwissen, T., Hayes, B., Goddard, M.: Prediction of Total Genetic Value Using Genome-Wide Dense Marker Maps. Genetics 157, 1819–1829 (2001)
5. Dekkers, J.: Prediction of response to marker-assisted and genomic selection using selection index theory. Journal of Animal Breeding and Genetics 124(6), 331–341 (2007)
6. Kusabs, N., Bollen, F., Trigg, L., Holmes, G., Inglis, S.: Objective measurement of mushroom quality. In: Proc. New Zealand Institute of Agricultural Science and the New Zealand Society for Horticultural Science Annual Convention, Hawke's Bay, New Zealand, p. 51 (1998)
7. Goyache, F., Bahamonde, A., Alonso, J., López, S., del Coz, J.J., Quevedo, J., Ranilla, J., Luaces, O., Álvarez, I., Royo, L., Díez, J.: The usefulness of artificial intelligence techniques to assess subjective quality of products in the food industry. Trends in Food Science & Technology 12(10), 370–381 (2001)
8. Alonso, J., del Coz, J.J., Díez, J., Luaces, O., Bahamonde, A.: Learning to predict one or more ranks in ordinal regression tasks. In: Daelemans, W., Goethals, B., Morik, K. (eds.) ECML PKDD 2008, Part I. LNCS(LNAI), vol. 5211, pp. 39–54. Springer, Heidelberg (2008)

9. Bahamonde, A., Díez, J., Quevedo, J.R., Luaces, O., del Coz, J.J.: How to learn consumer preferences from the analysis of sensory data by means of support vector machines (SVM). Trends in Food Science & Technology 18(1), 20–28 (2007)
10. Joachims, T.: Optimizing search engines using clickthrough data. In: Proceedings of the ACM Conference on Knowledge Discovery and Data Mining (KDD) (2002)
11. Bahamonde, A., Bayón, G.F., Díez, J., Quevedo, J.R., Luaces, O., del Coz, J.J., Alonso, J., Goyache, F.: Feature subset selection for learning preferences: A case study. In: Proceedings of the International Conference on Machine Learning (ICML 2004), pp. 49–56 (2004)
12. Alonso, J., Bahamonde, A., Villa, A., Castañón, Á.R.: Morphological assessment of beef catle according to carcass value. Livestock Science 107, 265–273 (2007)
13. Luaces, O., Bayón, G.F., Quevedo, J.R., Díez, J., del Coz, J.J., Bahamonde, A.: Analyzing sensory data using non-linear preference learning with feature subset selection. In: Boulicaut, J.-F., Esposito, F., Giannotti, F., Pedreschi, D. (eds.) ECML 2004. LNCS (LNAI), vol. 3201, pp. 286–297. Springer, Heidelberg (2004)
14. del Coz, J.J., Bayón, G.F., Díez, J., Luaces, O., Bahamonde, A., Sañudo, C.: Trait selection for assessing beef meat quality using non-linear SVM. In: Saul, L.K., Weiss, Y., Bottou, L. (eds.) Advances in Neural Information Processing Systems (NIPS 2004), vol. 17, pp. 321–328. MIT Press, Cambridge (2005)
15. Díez, J., del Coz, J.J., Sañudo, C., Albertí, P., Bahamonde, A.: A kernel based method for discovering market segments in beef meat. In: Jorge, A.M., Torgo, L., Brazdil, P.B., Camacho, R., Gama, J. (eds.) PKDD 2005. LNCS (LNAI), vol. 3721, pp. 462–469. Springer, Heidelberg (2005)
16. Luaces, O., Quevedo, J.R., Pérez-Enciso, M., Díez, J., del Coz, J.J., Bahamonde, A.: Explaining the genetic basis of complex quantitative traits through reliable prediction models. Technical report, Centro de Inteligencia Artificial. Universidad de Oviedo at Gijón (2009)
17. Yu, L., Liu, H.: Efficient Feature Selection via Analysis of Relevance and Redundancy. Journal of Machine Learning Research 5, 1205–1224 (2004)
18. Wenzl, P., Li, H., Carling, J., Zhou, M., Raman, H., Paul, E., Hearnden, P., Maier, C., Xia, L., Caig, V., Ovesná, J., Cakir, M., Poulsen, D., Wang, J., Raman, R., Smith, K., Muehlbauer, G., Chalmers, K., Kleinhofs, A., Huttner, E., Kilian, A.: A high-density consensus map of barley linking DArT markers to SSR, RFLP and STS loci and agricultural traits. BMC Genomics 7(206) (2006)
19. Bedo, J., Wenzl, P., Kowalczyk, A., Kilian, A.: Precision-mapping and statistical validation of quantitative trait loci by machine learning. BMC Genetics 9(35) (2008)

Controlling Particle Trajectories in a Multi-swarm Approach for Dynamic Optimization Problems

Pavel Novoa[1], David A. Pelta[2], Carlos Cruz[2], and Ignacio García del Amo[2]

[1] University of Holguín, Holguín, Cuba
pnovoa@facinf.uho.edu.cu
[2] Models of Decision and Optimization Research Group,
University of Granada 18071 - Granada, Spain
{dpelta,carloscruz}@decsai.ugr.es

Abstract. In recent years, particle swarm optimization has emerged as a suitable optimization technique for dynamic environments, mainly its multi-swarm variant. However, in the search for good solutions some particles may produce transitions between non improving ones. Although this fact is usual in stochastic algorithms like PSO, when the problem at hand is dynamic in some sense one can consider that those particles are wasting resources (evaluations, time, etc). To overcome this problem, a novel operator for controlling particle trajectories is introduced into a multi-swarm PSO algorithm. Experimental studies over a benchmark problem shows the benefits of the proposal.

1 Introduction

Many real-life optimization problems are dynamic in some sense, which means that some elements of their models are time-varying (eg. objective function, restrictions, etc.). These kind of problems are known as dynamic optimization problems (DOPs) and performing optimization on DOP implies not only to find the best solutions, but also to track these global optimums in the search space as time goes by.

The standard particle swarm optimization (PSO) method has been adapted to deal with dynamic environments, mainly to overcome the problems of *outdated memory, diversity loss and linear collapse* [1]. A complementary line of adaptation focuses on the uses of several swarms simultaneously that are somehow forced to explore different regions of the search space.

However, there is also another issue in PSO related to the exploration process: it is well know that being a stochastic algorithm, PSO moves its solutions (particles) in the search space guided by (among others) random factors. The presence of these factors obviously enhances the diversity of the population but may lead some particles to follow a trajectory of worse solutions than the reference one. One may expect that the swarm self-organize and recover itself from that situation, but in the context of DOPs , this recovering time implies a potential loose of resources that could be profitable used to track for a better solution.

J. Mira et al. (Eds.): IWINAC 2009, Part I, LNCS 5601, pp. 285–294, 2009.

In this context, we propose a novel operator for controlling particle trajectories (CPT) in a multi-swarm approach. In one hand by virtue of the multiswarm, the diversity of solutions is maintained while in other hand, convergence is accelerated by CPT.

In order to show the benefits of the proposal, the contribution is structured as follows: in Section 2 we briefly review concepts about DOPs and PSO, emphasizing the role of multiswarm approached. Section 3 explains the proposed mechanism while the experimental studies and validations are shown in Section 4. Finally, Section 5 is devoted to conclusions.

2 Background and Related Works

Given a search space Ω and a set of objective functions $f^{(t)} : \Omega \rightarrow \mathrm{I\!R}(t \epsilon \mathrm{I\!N}_0)$, the goal in a dynamic optimization problem is to find the set of global optimums $\mathcal{X}^{(t)}$ at every time step t, where:

$$\mathcal{X}^{(t)} = \{x^* \in \Omega | f^{(t)}(x^*) \succeq f^{(t)}(x), \forall x \in \Omega\} \tag{1}$$

"$\succ$" is a comparison relation which means "is better than", hence "$\succeq$" $\in \{\leq, \geq\}$. In this work, we assume $\Omega \subseteq \mathrm{I\!R}^n$.

The reader should note that for a given time step t we can obtain an specific stationary optimization problem. However, we can not consider a DOP as a set of different stationary problems as one assumption is that the changes are correlated in some way. Otherwise, the best approach would be to randomly restart the search every time a change is detected. In general, most of the techniques available to deal with DOP belongs to the class of "population based" methods [10], but others like Stochastic Diffusion Search [7], agent-based cooperative strategies [9], and of course, particle swarm techniques have been also used.

Particle swarm optimization is a versatile and very popular optimization technique that was first introduced by Kennedy and Eberhart in 1995 [5]. In PSO each particle is a candidate solution that it is represented by a position ($\mathbf{x}$) and a velocity ($\mathbf{v}$). Both, $\mathbf{v}$ and $\mathbf{x}$ are updated according to the following expressions:

$$\mathbf{v}(t+1) = \omega\, \mathbf{v}(t) + \eta_1\, c_1\, (\mathbf{x}_{pbest} - \mathbf{x}(t)) + \eta_2\, c_2\, (\mathbf{x}_{gbest} - \mathbf{x}(t)) \tag{2}$$

$$\mathbf{x}(t+1) = \mathbf{x}(t) + \mathbf{v}(t+1) \tag{3}$$

where $\mathbf{v}(t)$ and $\mathbf{v}(t+1)$ are the previous and the current velocity. Similarly, $\mathbf{x}(t)$ and $\mathbf{x}(t+1)$ are the previous and the current position. The factor ω is an inertia weight that says how much of the previous velocity will be preserved in the current one. η_1 and η_2 are two acceleration constants, while c_1 and c_2 are random numbers in the interval $[0.0, 1.0]$. Finally, $\mathbf{x}_{pbest}$ and $\mathbf{x}_{gbest}$ are the best so far position discovered by the particle and the current best so far position of the whole swarm, respectively.

Theoretical analysis of the particle trajectory [4] show that, through the expressions (2) and (3), each particle converges to a weighted mean of $\mathbf{x}_{pbest}$ and $\mathbf{x}_{gbest}$. For that reason, these two positions are also known as attractors.

Employing several populations has been one of the most successful strategy to enhanced the diversity in dynamic environments. For example, the Self Organizing Scouts (SOS) presented in [2] is a multi-population EA that showed excellent results in some dynamic test problems. Parrot and Li [8] created an speciation based PSO (SPSO), which dynamically adjust the number and the sizes of swarms through an ordered list of particles. Also, an atomic approach for many swarms is used by Blackwell and Branke [1], where charged and quantum swarms ensure the intra-swarm diversity, while an exclusion principle permits that just one swarm can surround only one peak. Recently, a fast multi-swarm algorithm (FMSO) was proposed by Li and Yang [6] where a parent swarm keep exploring the entire search space using a global search operator, while many child swarms are dynamically created around the best solutions found by this parent swarm.

3 Controlling Particle Trajectories

As a stochastic search algorithm, PSO moves its particles employing random numbers (see expressions (2) and (3)), thus we expect that particles alternate between good (improving) and bad (not improving) solutions with respect to a reference value. We consider that a particle produces a failure when there is a transition from solution s_i to another s_j such that $f(s_j) > f(s_i)$ (considering a minimization problem). A problem may arise if a "chain" of failures occurs because in stationary environments, this behavior may help to keep the diversity into the swarm and to enhance the algorithm exploration; in the contrary for dynamic problems, this would imply a waste of resources (time, evaluations, etc).

The controlling mechanism we propose is aimed at controlling the length of such chain of failures: each particle has a failures counter ($num\_failures$) that it is incremented when a failure occurs, otherwise it is reset to zero. If this failures counter reaches some pre-specified threshold ($max\_failures$) then the particle position and velocity are updated through the following equations:

$$\mathbf{x}(t) = \mathbf{x}_{sbest} \quad (4)$$

$$\mathbf{v}(t) = c_3 \frac{\mathbf{v}(t)}{2} \quad (5)$$

where, $\mathbf{x}_{sbest}$ is the position of the best solution found in the swarm where the particle is located, and c_3 is a random number in the interval $[0.0, 1.0]$.

The implications of expressions (4) y (5) are very simple, those particles that are moving in trajectories with consecutive failures will be positioned around the best solution of its swarm ($sbest$). The half decreasing of $\mathbf{v}(t)$ guarantees a closer exploration around the $sbest$, while the random number empowers diversity in a promising region of the search space.

The controlling mechanism is included in a multiswarm approach whose main details are taken from [1], specifically, we considered the anti-convergence operator, the exclusion principle and the way of detecting environment changes. However, non atomic representations have been used.

Algorithm 1. mCPT-PSO

```
1: for each particle p in swarm s do
2:    Randomly initialize p.v, p.x = p.x_pbest
3:    p.num_failures = 0
4:    Update s.x_gbest
5: end for
6: while (!End) do
7:    Test for Exclusion
8:    Test for Convergence
9:    Test for Change
10:   for each particle p in swarm s do
11:      if p.num_failures = max_failures then
12:         Update p with expressions 4) y 5)
13:      else
14:         Update p with expressions 2) y 3)
15:         Evaluate p.x
16:         Update p.x_pbest
17:         p.num_failures = 0
18:      end if
19:      Update s.x_gbest
20:   end for
21: end while
```

The proposed algorithm (mCPT-PSO) is described in Alg. 1. For more details on statements 7), 8) and 9), please refer to [1]. If any of the previous test is positive, the current number of particle failures is reset to 0 as we can see in the statements 3) and 17).

4 Computational Experiments

With the aim of assessing the benefits of the CPT mechanism, we conducted a set of experiments over a benchmark problem. More specifically, we designed three experiments oriented to analyze:

1. the effect of varying the maximum number of failures $max\_failures$ (the length of the chain of failures).
2. the role of CPT when the frequency of problem changes is varied (Δe).
3. the performance of mCPT-PSO against other algorithms.

These test are done using the Moving Peaks Benchmark (MPB), which offers a multi-modal, time-varying landscape. In MPB, different problem families can be obtained when different parameters configurations are used. In this work, we focus on the so called Scenario 2 whose parameters are summarized in the Table 1. To evaluate the efficiency of the algorithm, we used the off-line error [3] which is the running average deviation of the best solution from the optimum before each problem change.

Table 1. Standard settings for the Scenario 2 of the Moving Peaks Benchmark

Parameter	Setting
Number of peaks (p)	10
Number of dimensions (d)	5
Peaks heights (H_i)	$\in [30, 70]$
Peaks widths (W_i)	$\in [1, 12]$
Change frequency (Δe)	5000
Change severity (s)	1.0
Correlation coefficient (λ)	0.0

In what follows, we perform 50 runs for every tested configuration. In every run, the position of the particles and peaks in the problem are randomly generated. Also, we used a multi-swarm configuration composed by 10 swarms with 10 particles each one. Note that for this configuration the relation between swarms and peaks is one to one. The parameters r_{excl} and r_{conv} were fixed in 30.0 and 40.0 as suggested in [1]. For the update expression 2) we use the following settings: $\omega = 0.729844$, $\eta_1 = \eta_2 = 1.496180$.

4.1 On the Maximum Number of Consecutive Failures Allowed

In this experiment, we want to analyze first the lengths of the chain of failures that may occur in a typical multiswarm approach. This value is named $number\_failures$. We have run mCPT-PSO 50 times and we collect the number of times the value of failures reach $1, 2, 3, \ldots$ and so on. For example, when $number\_failures = 1$ then, the particle has visited solutions s_1, s_2, s_3 such that $f(s_1) < f(s_2) > f(s_3)$.

The result is shown as a histogram in Figure 1. Note that here the mechanism proposed is not used. It is clear that having one failure is more frequent than having two. This frequency monotonically decreases as $num\_failures$ becomes higher. From this point of view, we clearly expect that better results may be obtained if we can "stop" those bad trajectories thus having available more resources to explore promising regions. However, one may also claim that such apparently bad trajectories may help to enhance diversity or to escape from local optima, so the question is where to put the limit?.

To answer these questions, we run our method 50 times for different values of parameter $max\_failures = \{1, 2, \ldots, 10, \infty\}$. The value ∞ means that the CPT operator is switched off and represents the baseline for comparison.

Table 2 shows the average Offline error, Standard Deviation and the Minimum and Maximum values for each value of $max\_failures$. The worst result is achieved when $max\_failures = 1$, which means that if the new visited solution is worse than the previous one, then, the particle "jumps" to a new position. When $max\_failures$ is in the interval [2,4] we can observe a remarkable improvement in the average error. Finally, the same results are obtained when $max\_failures = \infty$ and $max\_failures > 5$. This is somehow expected, (recall Figure 1), as it is almost impossible to obtain such long chain of failures.

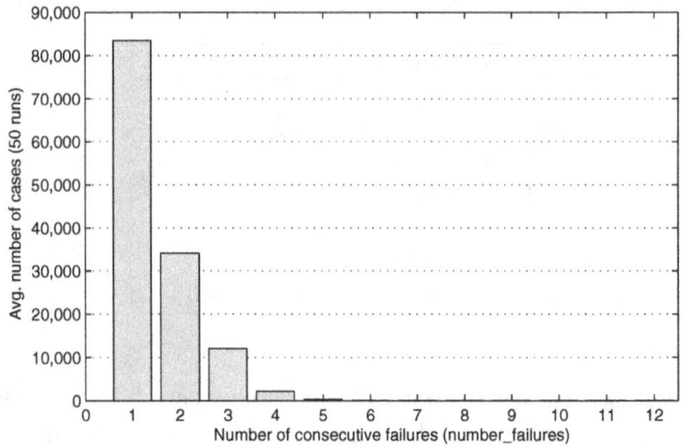

Fig. 1. Histogram of number of consecutive failures in particles

Table 2. Offline error summary for different values of $max\_failures$

$max\_failures$	Offline error(Std. Dev.)	Min.	Max.
∞	1.52(0.83)	0.51	5.00
1	3.06(1.44)	0.65	7.92
2	0.40(0.25)	0.13	1.11
3	0.52(0.39)	0.16	2.43
4	0.85(0.47)	0.29	2.07
5	1.18(0.54)	0.26	2.93
6	1.44(0.67)	0.51	3.09
7	1.45(0.69)	0.49	3.15
8	1.50(0.76)	0.48	4.19
9	1.48(0.71)	0.51	3.38
10	1.52(0.83)	0.51	5.00

To verify the above statements we performed an ANOVA analysis considering $max\_failures$ as a factor. The results indicated that there are a significative difference among the variants ($p < 0.05$). Then, a post-hoc Tahmane test confirmed that the variants : $(\infty), (5), (6), (7), (8), (9), (10)$ are not significantly different, while (1), (2), and (3) are different to the rest. Finally, (4) is not different to (5).

The offline error gives a global idea of the algorithm performance, but hides the algorithm behavior over time. In order to show this behavior, Figure 2 plots the average of the best value obtained before each problem change for the variants $max\_failures \in \{\infty, 1, 2\}$. Similarly, Figure 3 shows the offline error evolution for these variants. For illustration purposes, we only represent 50 problems changes.

4.2 The Role of CPT at Different Change Frequencies

The main goal of this experiment is to observe if the results obtained in the previous experiment are still valid when different change's frequencies (Δe) are

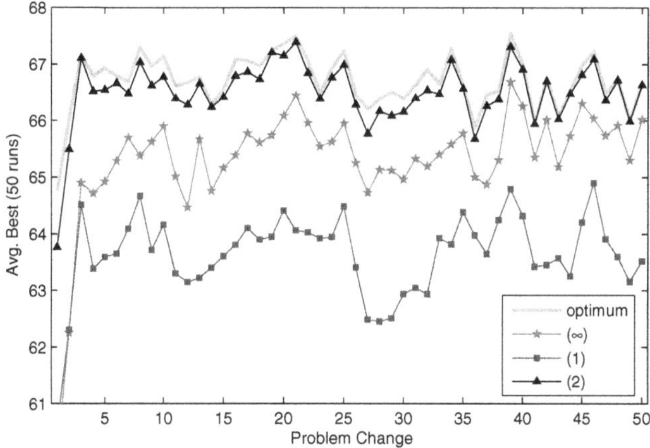

Fig. 2. mCFP-PSO accuracy over time for different values of $max\_failures$

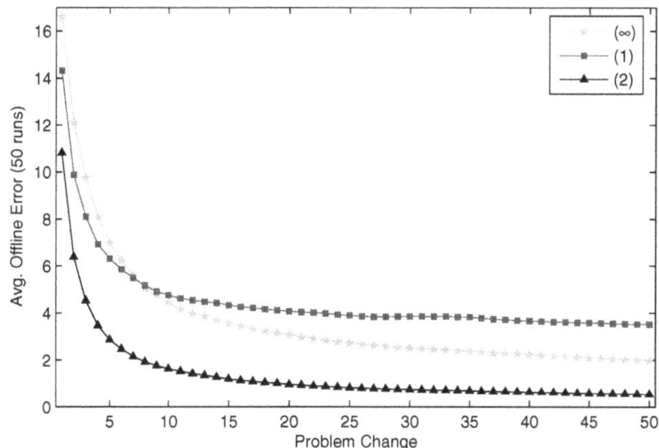

Fig. 3. Offline error evolution over time for different values of $max\_failures$

considered. Now, for each combination of $max\_failures \in \{\infty, 1, 2, \ldots, 10\}$ and $\Delta e \in \{500, 1000, 2000, 5000, 10000\}$ we run 50 times our method. Note that for a small value of Δe the algorithm has less evaluations to track the optimum, so we could confirm if including the CPT operator allows the algorithm to better use the evaluations available.

Table 3 and Figure 4 provide insights into the answers. If we do not consider the case $max\_failures = 1$ which is somehow senseless, one can conclude that any setting for the CPT lead to (on average) better or at least equal results to those obtained when the proposed mechanism is not present $max\_failures = \infty$. The best setting is $max\_failures = 2$. Using this value, the rate of improvement with respect to $max\_failures = \infty$ ranges from 75% when $\Delta e = 2000$ to 65%

Table 3. Offline Error (Std. Dev.) for $max\_failures$ vs. different values of Δe

$max\_failures$	$\Delta e = 500$	$\Delta e = 1000$	$\Delta e = 2000$	$\Delta e = 5000$	$\Delta e = 10000$
∞	15.68(4.16)	7.09(2.30)	3.56(1.49)	1.52(0.83)	0.95(0.55)
1	14.92(4.12)	9.73(3.44)	6.55(2.63)	3.06(1.44)	1.91(0.96)
2	5.48(1.57)	2.01(0.60)	0.89(0.32)	0.40(0.25)	0.28(0.25)
3	9.44(2.22)	3.06(0.87)	1.35(0.52)	0.52(0.39)	0.32(0.35)
4	15.62(4.42)	4.81(1.26)	2.13(0.75)	0.85(0.47)	0.42(0.29)

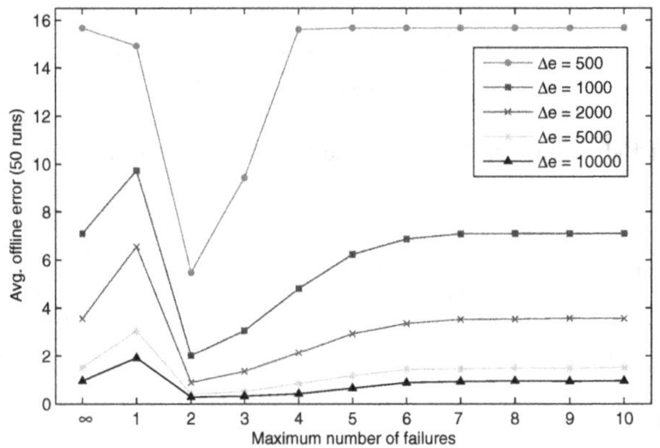

Fig. 4. Avg. Offline error of $max\_failures$ vs. Δe

when $\Delta e = 500$. We confirm these suppositions through statistic tests. A simple ANOVA analysis was performed for each Δe value. In all cases there were significant differences among the variants ($p < 0.05$), and very similar results to the previous experiment were obtained when the Tahmane test was applied. However, for $\Delta e = 10000$ there is not difference between (2) and (3).

Another interesting point is the following. It is reasonable to think that if a larger value of Δe is available, then a better performance can be achieved. The results plotted in Figure 4 shows that better results can be obtained using CPT and less evaluations than not using the mechanism and having more evaluations. Specifically, lets consider the cases when CPT is ON ($max\_failures = 2$) and OFF ($max\_failures = \infty$). Then one can clearly observe that it is better to have $\Delta e = k$ and CPT ON, than $\Delta e = 2 \times k$ and CPT OFF. In short, the control mechanism proposed allows to better use the evaluations available.

4.3 Comparison with Other Algorithms

In this section we will compare our proposal with two similar state-of-the-art algorithms introduced in [1]: mCPSO (multi-Charged Particle Swarm Optimization)

Table 4. Comparison among mCPSO, mQSO and mCPT-PSO

Configuration	mCPSO	mQSO	mCFP-PSO
5(10+10)	3.74(0.14)	3.71(0.15)	2.89(0.16)
10(5+5)	2.05(0.07)	1.75(0.06)	0.40(0.04)
14(4+3)	2.29(0.07)	1.93(0.06)	0.61(0.06)
20(3+2)	2.89(0.07)	2.35(0.07)	0.83(0.06)
25(2+2)	3.27(0.08)	2.69(0.07)	1.35(0.08)

and mQSO (multi-Quantum Swarm Optimization). For a fair comparison, we have used the same multi-swarm configurations employed by mCPSO and mQSO under the same test bed. In [1] these multi-swarm configurations had the following form $M(N_1+N_2)$, where M represents the total number of swarms, each one with N_1 neutral particles and N_2 charged particles for mCPSO or N_2 quantum particles for mQSO. In our case (mCPT-PSO) there is no distinction between particles, so this configuration will be interpreted as M swarms each one with $N_1 + N_2$ particles. Also, the $max\_failures$ was stated to value 2. Table 4 shows the average offline error (over 50 runs in our case) and the typical error for the three algorithms. The results clearly show the benefits of our proposal: the average error is much lower for every configuration tested. Another interesting point is that the best configuration for all the methods is the one using ten swarms with ten particles each. This can be understood as a confirmation of the idea that the number of swarms should match the number of peaks.

5 Conclusions

In this work we presented a simple mechanism to control the particle trajectories in the context of a multi-swarm approach for dynamic optimization problems. Essentially, it is aimed to control and react on the number of steps that a particle can do without improvements.

From the experiments we observed the positive influence of this CPT operator in the convergence of the algorithm, particulary when the maximum number of consecutive steps without improvements allowed to a particle is 2. When such "bad" trajectory is interrupted, then some cost function evaluations can be saved and not wasted in unpromising regions of the search space. Moreover, we showed that better results are obtained using this mechanism instead of giving the system more evaluations to track the optimum. We also show that our proposal is better than two similar state-of-the-art algorithms.

The results are very encouraging as not only the error is drastically improved, but they also indicate that a simple mechanism as the one proposed can lead to improvements not achieved by far more complex strategies like having different types of particles.

Acknowledgments

Research supported in part by Projects TIN2008-01948 (Spanish Ministry of Science and Innovation) and P07-TIC02970 (Andalusian Government).

References

1. Blackwell, T.M., Branke, J.: Multiswarms, exclusion, and anti-convergence in dynamic environments. IEEE Trans. on Evol. Computation 10(4), 459–472 (2006)
2. Branke, J., Kauler, T., Schmidt, C., Schmeck, H.: A multi-population approach to dynamic optimization problems. Adaptive Computing in Design and Manufacturing, 299–308 (2000)
3. Branke, J., Schmeck, H.: Designing evolutionary algorithms for dynamic optimization problems. In: Theory and Application of Evolutionary Computation: Recent Trends, pp. 239–262. Springer, German (2002)
4. Clerc, M., Kennedy, J.: The particle swarm: Explosion, stability and convergence in a multi-dimensional complex space. IEEE Trans. on Evol. Computation 6, 58–73 (2002)
5. Kennedy, J., Eberhart, R.: Particle swarm optimization. In: Proceedings of the IEEE International Conference on Neural Networks, pp. 1942–1948 (1995)
6. Li, C., Yang, S.: Fast multi-swarm optimization for dynamic optimization problems. In: Fourth International Conference on Natural Computation, IEEE Computer Society, Los Alamitos (2008)
7. Meyer, K., Nasut, S., Bishop, M.: Stochastic diffusion search: Partial function evaluation in swarm intelligence dynamic optimisation. Studies in Computational Intelligence 31 (2006)
8. Parrott, D., Li, X.: A particle swarm model for tracking multiple peaks in a dynamic environment using speciation. In: Proc. of the 2004 IEEE Congress on Evolutionary Computation, pp. 98–103 (2004)
9. Pelta, D., Cruz, C., Verdegay, J.: Simple control rules in a cooperative system for dynamic optimization problems. International Journal of General Systems (2008)
10. Yang, S., Ong, Y.-S., Jin, Y.: Evolutionary Computation in Dynamic and Uncertain Environments. Studies in Computational Intelligence, vol. 51. Springer, Heidelberg (2007)

Clustering Ensembles Using Ants Algorithm

Javad Azimi, Paul Cull, and Xiaoli Fern

EECS Department, Oregon State University,
Corvallis, Oregon, 97330, USA
{azimi,pc,xfern}@EECS.oregonstate.edu

Abstract. Cluster ensembles combine different clustering outputs to obtain a better partition of the data. There are two distinct steps in cluster ensembles, generating a set of initial partitions that are different from one another, and combining the partitions via a consensus functions to generate the final partition. Most of the previous consensus functions require the number of clusters to be specified a priori to obtain a good final partition. In this paper we introduce a new consensus function based on the Ant Colony Algorithms, which can automatically determine the number of clusters and produce highly competitive final clusters. In addition, the proposed method provides a natural way to determine outlier and marginal examples in the data. Experimental results on both synthetic and real-world benchmark data sets are presented to demonstrate the effectiveness of the proposed method in predicting the number of clusters and generating the final partition as well as detecting outlier and marginal examples from data.

Keywords: Cluster ensembles, ant colony, consensus function, outlier and marginal samples.

1 Introduction

There is no clustering algorithm that performs best for all data sets. Choosing a single clustering algorithm and appropriate parameters for a data set requires both clustering expertise and insights about the data set itself. Instead of running the risk of picking an inappropriate clustering algorithm, we can leverage the different options available by applying all of them to the data and then combining their clustering results. This is the basic idea behind cluster ensembles [1].

There are two major steps in cluster ensembles: first generating a set of different partitions and then combining them using a consensus function. A major difficulty in cluster ensembles is the design of the consensus function for combining the partitions into a final clustering solution. The combination of multiple partitions can be viewed as finding a median partition with respect to the given partitions in the ensemble, but determining this median partition has been shown to be NP-complete [2].

There are many types of consensus functions that solve this problem heuristically.Most of them require the number of clusters to be specified a priori, but in practice the number of clusters is usually unknown. In this paper, we propose

J. Mira et al. (Eds.): IWINAC 2009, Part I, LNCS 5601, pp. 295–304, 2009.

a new consensus function for cluster ensembles based on swarm intelligence [3] that addresses this problem. In particular, given a set of partitions, we apply ant clustering to the co-association matrix computed from the ensemble to produce the final partition, and automatically determine the number of clusters.

Ant clustering algorithms are inspired by how ants organize their food in their nests. Ant clustering typically involves two key operations: picking up an object from a cluster and dropping it off into another cluster. At each step, some ants perform pick-up and drop-off based on some notions of similarity between an objects and the clusters. In our method, the similarity measure is defined based on the co-association matrix. The clustering process is completely decentralized and self-organized, allowing the clustering structure to emerge automatically from the data. As a result, we can accurately determine the number of clusters in the data. The experimental results show that the proposed consensus function is very effective in predicting the number of clusters and also achieves reliable clustering performance. In addition, by introducing some simple heuristics, we can detect the marginal and outlier samples in the data to improve our final clustering.

The rest of the paper is organized as follows. Section 2 describes the basic concepts for cluster ensembles. The proposed consensus function and the heuristics for outlier and marginal sample detection are presented in Section 3. In Section 4, experimental results of the proposed method are compared with other methods. The paper is concluded in Section 5.

2 Cluster Ensemble Basics

Cluster ensembles have become increasing popular in the clustering community in recent years. A cluster ensemble framework usually consists of two stages. First, an initial set of partitions is generated to form the ensemble. It is typically required that the partitions are diverse, i.e., they need to be different from one another. Once a diverse set of partitions are generated, a consensus function is used to combine them and produce a final partition. Existing consensus functions can be grouped into the following categories: *Graph based methods* [1,9], *Voting approaches* [4,5], *Feature-based approaches* [6], *Co-association based methods* [7].

A common problem of the existing consensus functions is that, to apply techniques such as above, it is required that the number of clusters in the data be pre-specified. This can be problematic in practice because such information is rarely available. The method proposed in this paper addresses this problem by employing ant clustering to the co-association matrix, which requires no prior knowledge of the number of clusters. Let D be a data set of N data objects in a d-dimensional space. Let $P = \{P_1, \cdots, P_B\}$ be a set of partitions, where P_i is a partition of a bootstrap sample of D (or D itself) represented by X_i. In other words, each component partition in P is a set of clusters, i.e. $P_i = \left\{C_1^i, C_2^i, ..., C_{k(i)}^i\right\}$, where $C_1^i \cup C_2^i \cdots \cup C_{k(i)}^i = X_i \quad (X_i \subset D)$ and $k(i)$ is the number of clusters in the i_{th} partition.

The co-association matrix is computed as follows:

$$co - association(x, y) = \frac{1}{B}\sum_{i=1}^{B} \phi\left(P_i(x), P_i(y)\right)$$
$$\phi(a, b) = \begin{cases} 1 & \text{if} \quad a = b \\ 0 & \text{if} \quad a \neq b \end{cases} \tag{1}$$

Note that, in this paper we use the k-means algorithm with different initialization as our base clustering algorithm to generate the initial partitions and co-association matrix.

3 The Proposed Method

In ant based clustering approaches, each ant is an agent with simple probabilistic behaviors of picking up and dropping down objects (analogous to organizing their food in their nests) [10]. These probabilities depend on the similarity between the objects and other objects. Note that ant clustering has been typically applied using Euclidean distance to compute the similarity between objects. In our work, instead, we use the co-association matrix as the similarity measure among objects to perform ant clustering.

3.1 Ant Clustering Based on Co-association Matrix

Ant clustering begins by considering each object as a singleton cluster. The algorithm then proceeds by letting a fixed number of ants randomly move the objects around according to some basic probabilistic rules. There are two types of operations that the ants perform: picking up and dropping off objects. During clustering, each ant randomly selects a cluster from all clusters and picks up a sample from it and then drops it into an appropriate cluster. Below we explain the two types of operations.

Picking up. To perform a pick up operation, an ant first selects a cluster randomly from all existing clusters. If the selected cluster is empty, the ant will continue selecting until a non-empty cluster is found. Examining the selected cluster, there will be three possibilities: 1) the cluster contains only one object, 2) the cluster contains two objects and 3) the cluster contains more than two objects. In the first case, the ant picks up that object. In the second case, the ant selects one of the two objects randomly. In the final case, the ant identifies the object that is most dissimilar to the other objects in the cluster, which is defined as follow: For each sample x_i in cluster j we calculate:

$$S(x_i) = \frac{1}{\mid C_j \mid}\sum_{t \in C_j} co - association(x_i, t) \tag{2}$$

where $x_{Dissimilar}$ minimize S, i.e:

$$x_{Dissimilar} = argmin\ S(x_i) \qquad x_i \in C_j \tag{3}$$

If $S(x_{Dissimilar})$ is less than P_{remove}, the ant pick up $x_{Dissimilar}$. We set P_{remove} to be 0.5 which means that the ant will pick up $x_{Dissimilar}$ if it was clustered together with the other objects in the cluster for fewer than half of the original partitions of the ensemble. If the ant fails to find a qualifying object to pick up from the selected cluster, it will continue to select another cluster. Conceptually, the above probabilistic pick-up rule ensures that the ant always picks up objects that do not fit in the cluster well according to the co-association matrix, leading to clusters with the highest inner similarity.

Dropping off. After an ant picks up an object, it will look for an appropriate cluster to drop the object off. To do so, it first randomly selects a cluster from all available clusters. Depending on the number of objects contained in that cluster, there will be again three possible cases: 1) the cluster is empty, 2) the cluster contains one or two objects, and 3) the cluster contains more than two objects. In the first case, the ant does not drop the object and continues looking for another cluster. In the second case, the ant will only drop off the object if its average similarity with the objects(s) in the cluster in question is greater than P_{Create}. Here we set the P_{Create} to be 0.5, which requires the object to have been clustered together with the objects of the current cluster in more than half of the initial partitions of the ensemble. In the final case where there are more than three objects, the ant will only drop off the object if the object is more similar to the cluster than its current most dissimilar object. More specifically, its average similarity with the objects contained in the cluster in question needs to be greater than the average similarity between the most dissimilar object $x_{Dissimilar}$ and the other objects in the cluster. When this is true, it indicates that the carried object is more similar to the cluster than the most dissimilar object of the cluster is, providing evidence that the object should be part of that cluster. In all cases, if the ant fails to drop off the object in the selected cluster, the ant continues to look for another cluster. If the ant could not drop it in any cluster, it returns the object to its original cluster.

Sweeping. Due to the stochastic nature of the algorithm, it is possible to have some objects that have never been selected by any ant by the end of the clustering procedure. Further, some clusters may contain only a few objects, and should really be merged with other clusters to form bigger clusters. To address this issue, we include in our ant clustering algorithm a procedure called *sweeping* that examines all objects once and assigns them to an appropriate clusters. In particular, for all clusters that are singletons or contain too few examples (less than 5% of the examples), we assigns each of their objects into its most similar cluster based on the co-association values. Sweeping can effectively eliminate single-object clusters as well as clusters that consist of only a few objects. Note that in total, our method uses *6*n* ants, where n equals the number of objects in the data. We also interleave the regular ant clustering stage with the sweeping stage to address the problems caused by singleton and small clusters.

Algorithm Parameters. Traditional applications of ant clustering require many initial parameters which need insights about the data [10], and the clustering performance is highly sensitive to these parameters. In contrast, our method avoids this problem by working with the co-association matrix, whose element values are from a fixed range and have natural interpretations, making it easier to set the parameters in a meaningful manner such as described above. In addition, our method has only two critical parameters P_{Remove} and P_{Create}.

Recall that P_{Remove} is the threshold for deciding if an ant can picks up an object from a cluster of two or more objects. This parameter is intended to recognize the fact that sometimes all objects in a cluster are very close to one another and even the most dissimilar object is not far from the other objects. In such cases, it is intuitive to avoid picking up any object from such a tight cluster. Here P_{Remove} is set to 0.5, meaning that the most dissimilar object will not be picked up if it has been clustered together with other objects in more than half of the initial clusterings.

Similarly, P_{Create} is the threshold for deciding whether the ant can drop off an object to a cluster or not. We set P_{Create} to 0.5 which conceptually requires the object to have been clustered together with the other objects in more than half of the initial clusterings. The value 0.5 is completely understandable since we expect to cluster the sample with the objects that have been clustered together in more than half of the initial partitions.

It should be noted again that determining the initial parameters for traditional ant clustering algorithms can be challenging because the parameter values are based on the Euclidean distances and different data sets may have drastically different value ranges. Our method avoids this problem by working with the co-association matrix, whose element values are from a fixed range and have natural interpretations, making it easier to set the parameters in a meaningful manner such as described above.

3.2 Outlier and Marginal Objects Detection

We distinguish between outlier and marginal examples in data. In particular, we refer to the objects that are far from the center of their clusters and also other objects in their clusters as outlier examples. In contrast, the marginal objects are those objects that border two or more clusters and as a result change their cluster memberships frequently in the ensemble. The outlier and marginal objects can be detrimental to clustering because they may mislead the algorithm. For example, outlier examples may make the estimation of cluster center inaccurate, and marginal objects may cause two or more clusters to be merged together incorrectly. Therefore, appropriately detecting the marginal and outlier objects can be useful in obtaining a better final partition. There have been numerous studies on detecting outlier and marginal objects in data mining applications [8, 11]. In this paper we introduce some simple heuristics based on the ant clustering algorithm to effectively identify both outlier and marginal examples appropriately to help improve the final clustering.

Recall that the marginal objects are located at the intersecting areas of clusters and they frequently change their cluster associations in a cluster ensemble. Therefore, it is expected that their average similarity values with other objects in their own cluster to be low in comparison to other objects in their cluster. As a result, the marginal objects will likely be picked up frequently by ants during ant clustering. Furthermore, the ants will also likely fail to find an alternative cluster to place these marginal objects because they tend to be dissimilar to other clusters as well. We design our heuristic based on the above intuition. In particular, for each object we computes a marginality index that records how many times the object was picked up and then returned to the original cluster during ant clustering. The larger the marginality index, the more likely we consider the object to be marginal. It is clear that this procedure can be done along with the proposed algorithm without introducing any additional computational cost.

The outlier objects are those that are far from their cluster centers. They may lead to biased estimation of the cluster centers, resulting in some errors in clustering. The behavior of outlier objects is different from marginal examples in the sense that they usually are consistently assigned to specific clusters and rarely change their clusters. This means that if we investigate the co-association matrix, we expect an outlier object to have high similarities with most of the objects in its clusters, which is different from what we expect for the marginal examples. To detect the outlier objects, we add some virtual ants which deploy a different pick up strategy. In particular, these virtual ants pick up a sample from a cluster which is far from the center of the cluster based on the Euclidean distance instead of the co-association matrix values. These ants, however, will use the same drop off procedure as described before. If an object that is picked up by an virtual ant could not be dropped in any cluster and has to be returned to its original cluster it is likely an outlier. To capture this, we compute an outlier index for each object that records the number of times that it is picked up by a virtual ant and returned to its original cluster. The higher the value, the more likely we consider it an outlier.

In our experiments, the marginality indices are computed during the regular ant clustering procedure. The outlier indices are computed by applying the same ant clustering algorithm with the revised pick-up strategy with $6{*}n$ ants, where n is the number of objects in the data set. To designate an object as marginal or outlier, we require its corresponding index to be greater than a given threshold. Note that we observe a strong modal distribution of the index values, with most of the objects having index value zero and some with significantly higher value. This makes it easy to determine a good threshold. In our experiments, we set the threshold to be six for detecting both marginal and outlier examples. It should also be noted that the results of our algorithm are not sensitive to this threshold.

4 Experimental Results

We performed experiments on eight data sets. Four of them are from the UCI machine learning data sets [12] including the Iris, Wine, Soybean and Thyroid data sets and a real data set O8X from image processing data sets [13]. We also artificially generated three synthetic data sets. All data sets are preprocessed to normalize the feature values to a fixed range. This is done by dividing the value of each feature by the maximum value of that feature in all samples. Note that all data sets are labeled, but the labels are omitted during clustering and only used for evaluation purposes. In addition, to build the co-association matrix we use 100 independent runs of the k-means algorithm as initial clustering algorithm with different initializations.

Note that when evaluating the first two aspects, we do not apply the outlier/marginal sample detection heuristic and cluster all of the data. So the results presented in the following two subsections (4.1 and 4.2) are all based on the original data with no outlier/marginal examples removed. Furthermore, all reported results are averaged across 100 independent runs.

4.1 Identifying the Number of Clusters

In Table 1, we present the results of our method on predicting the number of clusters. In particular, the first column lists the name of each data set and the second column provides the real number of clusters (classes) of each data set. The third column shows the average number of clusters obtained by our algorithm and the fourth column shows the most frequent number of clusters from 100 independent runs. From these results, it can be seen that our method can usually obtain the number of clusters correctly.

We also compare our method with the ISODATA algorithm [13] which has been designed to obtain the number of clusters automatically. The ISODATA is a single clustering algorithm which can also detect the outlier samples based on its initial parameter values. The results of the ISODATA algorithm are presented in Table 2. The comparison indicates that our method is more effective at correctly determining the number of clusters than the ISODATA algorithm on the data sets used in our experiments. This is possibly due to the fact that ISODATA requires some initial parameters to be set appropriately to have a good clustering result while the proposed method has only two parameters and requires no fine tuning.

Table 1. The proposed method results

Name	# of classes (Real)	# of classes (Average)	# of classes (most Frequent)	Proposed method error %	CSPA error %	ALA error %
O8X	3	2.92	3	9.33	11.2	8.7
Thyroid	3	3.1	3	14.50	15.86	48.65
Iris	3	2.9	3	5.80	4	5.5
Wine	3	2.85	3	7.10	9.88	10.11
Soybean	4	3.8	4	6.30	13.51	4.51
Artificial 1	2	2.2	2	4.50	5.5	4.2
Artificial 2	3	3.5	4	4.20	6.5	5.1
Artificial 3	3	2.9	3	6.30	15.91	8.6

Table 2. The ISODATA algorithm results

Name	# of classes (Real)	# of classes (Average)	Miss classification error %
O8X	3	5.51	9.5
Thyroid	3	7.24	8.23
Iris	3	6.52	6.93
Wine	3	9.36	5.42
Soybean	4	8.34	2.72
Artificial 1	2	4.2	5.3
Artificial 2	3	5.5	7.3
Artificial 3	3	6.3	8.4

4.2 Clustering Accuracy

In Table 1, we show the misclassification error rates of the introduced method in column five. Also provided in the table are the results of two competitive consensus functions, namely the CSPA method (column six) and the average link agglomerative (ALA) method (column seven). Note that CSPA and ALA require the number of clusters to be set *a priori*. In our experiments, for these two methods, we set k to be the true number of classes in the data. The comparison indicates that the proposed method produced highly competitive clustering results. Note that we also present the error rates of the ISODATA algorithm in Table 2 (column four). While it appears that the ISODATA method outperforms all the other methods including the proposed method in a few data sets, this is due to the bias of the performance measure we used. In particular, the clustering results produced by ISODATA generally contain significantly larger number of clusters (as suggested by column 3 of Table 2) than the proposed method. Note that the classification error rate will always decrease as we increase the number of clusters in the clustering result. (Consider the extreme case where each point is in its own cluster, the error rate will be decreased to 0.) The readers are advised to interpret the results of ISODATA with caution.

4.3 Outlier and Marginal Detection

In this section, we evaluate the effectiveness of the introduced heuristics for identifying outlier and marginal examples from data. In Table 3 we show the scatter plot of the three artificial data sets and the Iris data, where the identified outlier and marginal examples are highlighted as dark points. For Iris, we set feature 2 of the original data as the X axis and feature 3 as the Y axis to produce a 2-dimensional scatter plot.

The first row of Table 3 (a-d) highlights the samples that are identified as outlier in each data set. We can see that samples that are far from the center of the clusters have been selected as outliers. In the Iris data set this is less evident possibly because we only used two of the four features to produce the scatter plot.

We can see the identified marginal samples highlighted in the second row of Table 3 (e-h). We can see that the examples that are identified by our algorithm

Table 3. The Outlier detected samples (a-d) and marginal detected samples (e-h) in related data sets

Artificial 1	Artificial 2	Artificial 3	Iris
a	b	c	d
e	f	g	h

as marginal are primarily located in the area that borders multiple clusters. In the Iris data set, which consists of three clusters, the marginal sample are mainly located in the area between two of the clusters since the third cluster is linearly separable from the other two clusters. In the Artificial 1 and 2 dataset the samples which are located between clusters have been selected as marginal samples and in Artificial 3 data set the samples which have been located in the conjunction of the three clusters have been selected as marginal samples. It can be seen that the selected marginal samples are different from the outlier samples.

5 Conclusion

In this paper, we presented an ant colony based algorithm which forms clusters from the results of other clustering methods. The input to our method is an ensemble of various clusterings. The proposed method applies ant clustering to the co-association matrix of the ensemble to produce the final partition. Our method has the unique ability to automatically determine the number of clusters. Furthermore, it produces reliable clustering performance. Compared to other applications of ant clustering, our method uses fewer parameters than other ant based clustering methods . The use of the co-association matrix instead of Euclidean distance allows us to avoid fine tuning of the parameters. Finally, a by-product of our ant clustering algorithm is that we can introduce very simple heuristics that can effectively identify both marginal and outlier examples, which in turn further improves the performance of the final partition. Our experimental results provide clear evidence that the proposed method can: 1) reliably predict the number of clusters in the data, 2) produce highly competitive clustering results in comparison to other ensemble methods, and 3) effectively identify marginal and outlier examples.

References

1. Strehl, A., Ghosh, J.: Cluster ensembles-a knowledge reuse framework for combining partitioning. In: IJCAI, Edmonton, Alberta, Canada, pp. 93–98 (2002)
2. Barthelemy, J.P., Leclerc, B.: The median procedure for partition. Partitioning Data Sets, AMS DIMACS Series in Discrete Mathematics, 3–34 (1995)
3. Kennedy, J., Russell, S.: Swarm Intelligence. Morgan Kaufmann, San Francisco (2001)
4. Fern, X.Z.: Brodley: Random projection for high dimensional data clustering: a cluster ensemble approach. In: ICML, Washington, DC, pp. 186–193 (2003)
5. Weingessel, A., Dimitriadou, E., Hornik, K.: An ensemble method for clustering. Working paper (2003), `http://www.ci.tuwien.ac.at/Conferences/DSC-2003`
6. Topchy, A., Jain, A.K., Punch, W.: A mixture model for clustering ensembles. In: Proceedings of SIAM Conference on Data Mining, pp. 379–390 (2004)
7. Duda, R.O., Hart, P.E., Stork, D.G.: Pattern Classification, 2nd edn. John Wiley and Sons Inc., New York (2001)
8. Topchy, A., Minaei-Bidgoli, B., Jain, A.K., Punch, W.: Adaptive Clustering ensembles. In: International Conference on Pattern Recognition, Cambridge, UK, pp. 272–275 (2004)
9. Fern, X.Z., Brodley, C.E.: Solving cluster ensemble problems by bipartite graph partitioning. In: 21th International Conference on Machine Learning, pp. 281–288 (2004)
10. Monmarch'e, N., Silmane, M., Venturini, G.: AntClass: discovery of clusters in numeric data by an hybridization of an ant colony with k-means algorithm. Internal Report no. 213, Laboratoire Informatique de l'Universite (1999)
11. Jain, A.K., Murty, M.N., Flynn, P.J.: Data clustering: A review. ACM Comp. Surveys 31(3), 264–323 (1999)
12. Blake, C., Merz, C.: UCI repository of machine learning databases (1998), `http://archive.ics.uci.edu/ml/`
13. Gose, E., Johnsbaugh, R., Jost, S.: Pattern Recognition and Image Analysis. Prentice Hall, Englewood Cliffs (1996)

The kNN-TD Reinforcement Learning Algorithm

José Antonio Martín H.[1], Javier de Lope[2], and Darío Maravall[2]

[1] Dep. Sistemas Informáticos y Computación, Universidad Complutense de Madrid
jamartinh@fdi.ucm.es
[2] Perception for Computers and Robots, Universidad Politécnica de Madrid
javier.delope@upm.es,dmaravall@fi.upm.es

Abstract. A reinforcement learning algorithm called kNN-TD is introduced. This algorithm has been developed using the classical formulation of temporal difference methods and a k-nearest neighbors scheme as its expectations memory. By means of this kind of memory the algorithm is able to generalize properly over continuous state spaces and also take benefits from collective action selection and learning processes. Furthermore, with the addition of probability traces, we obtain the kNN-TD(λ) algorithm which exhibits a state of the art performance. Finally the proposed algorithm has been tested on a series of well known reinforcement learning problems and also at the Second Annual RL Competition with excellent results.

1 Introduction

Reinforcement Learning (RL) [1] is a paradigm of Machine Learning (ML) in which rewards and punishments guide the learning process. In RL there is an Agent (learner) which acts autonomously and receives a scalar reward signal which is used to evaluate the consequences of its actions. The framework of RL is designed to guide the learner in maximizing the average reward that it gathers from its environment in the long run.

Let us start with a classical Temporal Difference (TD) learning rule [2]:

$$Q(s,a)_{t+1} = Q(s,a)_t + \alpha[x_t - Q(s,a)_t], \tag{1}$$

where $x_t = r + \gamma \max_a Q(s_{t+1}, a)_t$.

This basic update rule of a TD-learning method can be derived directly from the formula of the expected value of a discrete variable. We know that the expected value (μ) of a discrete random variable x with possible values $x_1, x_2 \ldots x_n$ and probabilities represented by the function $p(x_i)$ can be calculated as:

$$\mu = \langle x \rangle = \sum_{i=1}^{n} x_i \, p(x_i), \tag{2}$$

thus for a discrete variable with only two possible values the formula becomes:

$$\mu = (1-\alpha)a \; + \; b\alpha \;\; , \tag{3}$$

J. Mira et al. (Eds.): IWINAC 2009, Part I, LNCS 5601, pp. 305–314, 2009.

where α is the probability of the second value (b). Then taking a as a previously stored expected value μ and b as a new observation x_t, we have:

$$\mu = (1 - \alpha)\mu \ + \ x_t\alpha \ , \tag{4}$$

and doing some operations we get:

$$\begin{aligned} \mu &= (1 - \alpha)\mu \ + \ x_t\alpha \\ &= \mu - \alpha\mu + \alpha x_t \\ &= \mu + \alpha[-\mu + x_t] \\ &= \mu + \alpha[x_t - \mu] \end{aligned} \tag{5}$$

Then we can see that the parameter α, usually described as the learning rate, express the probability that the random variable x get the value of the observation x_t and that replacing μ by $Q(s,a)_t$ and then replacing x_t by the expected cumulative future reward (this value is usually settled as the the learning target):

$$r_t + \gamma \max_a Q(s_{t+1}, a)_t \tag{6}$$

we finally get again the Q-Learning update rule (1).

In this paper we present a reinforcement learning algorithm called kNN-TD. This algorithm has been developed using the classical formulation of TD-Learning and a k-nearest neighbors scheme for the action selection process, expectations memory as well as for the learning process.

The k-nearest neighbors (k-NN) technique [3,4,5] is a method for classifying objects based on closest training examples in the feature space. The training examples are mapped into a multidimensional feature space. The space is partitioned into regions by class labels of the training samples. A point in the space is assigned to a class if it is the most frequent class label among the k-nearest samples used in the training phase. The determinant parameters in k-NN techniques are the number k which determines how many units are to be considered as the neighbors and the distance function, generally the euclidian distance is used.

As it has been proved [4] there is a close relation between nearest neighbors methods and the bayesian theory of optimal decision making. One of the most relevant results on this line is the bayesian error bound for the nearest neighbors method which establishes that:

$$P(e_b|x) \leq P(e_{nn}|x) \leq 2P(e_b|x), \tag{7}$$

where $P(e_b|x)$ is the Bayes error rate and $P(e_{nn}|x)$ is the nearest neighbor error rate.

So we see that the upper bound on the probability of error for the NN-Rule is less than twice the Bayes error rate. Of this form we can note that the k-NN rule is an empirical approach to optimal decision making according to the Bayesian Theory and that a well established theoretical error bound is guaranteed.

There has been other uses of the k-NN rule, i.e. in Dynamic Programming (DP), as a function approximator with successful results [6]. Also, we must mention that the there is a relation between the kind of algorithms that we propose and the so called "Locally Weighted Learning" (LWL) methods surveyed by Atkeson and Moore [7]. Indeed LWL approximations can be viewed as a particular kind of artificial neural network (ANN) for approximating smooth functions [8]. While the survey by Atkeson and Moore [7] covers a wide range of characteristics and relevant parameters of LWL and Bosman [8] follows an approach to present LWL as a kind of neural network, we focus our approach in the relation between our methods and a probabilistic approach supported by the domain of non-parametric statistics and bayesian inference. Thus, there are several interpretations about these kinds of learning methods, each of them supported by some particular approach. Our approach has been developed independently starting with a preliminary work [9] proposing the use of k-NN as a perception scheme in RL.

Here we improve the initial proposed scheme by developing the kNN-TD(λ) which is a state of the art reinforcement learning algorithm. The algorithm is described in detail and a pseudo-code is presented. Also, public implementations of this algorithm will be published on the web site: `http://www.dia.fi.upm.es/~jamartin/download.htm` in order that these tools be used, evaluated or improved by the researchers interested in RL.

2 The kNN-TD Reinforcement Learning Algorithm

Figure 1 shows the agent-environment interaction cycle in RL. This interaction cycle could be described by four basic steps:

1. Perceive the state s of the environment.
2. Emit an action a to the environment based on s.
3. Observe the new state s' of the environment and the reward r received.
4. Learn from the experience based on s, a, s' and r.

The first task to address is the definition of the points (classifiers[1]) which will act as the neighbors of the observations (s). Although there are many ways of covering the space, a simple procedure is to generate points uniformly over the

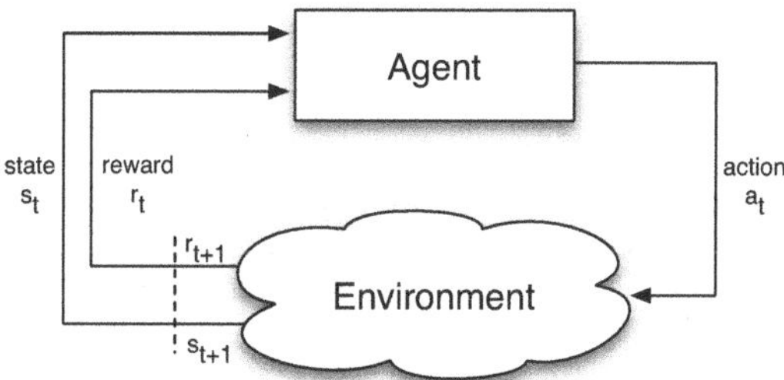

Fig. 1. Agent-Environment interaction cycle in RL

[1] Although the term experts is often used for the same purpose we prefer to use the term classifier which has more meaning in statistical machine learning.

complete state space. Despite this technique works for some problems it may happen that the absolute magnitudes of some dimensions be greater than others leading to a biased selection of the k-nearest neighbors over the dimension whose magnitude is smaller. For avoiding this potential bias it is necessary the correct standardization of all the dimensions in such a way that its absolute magnitudes don't biases the selection of the k-nearest neighbors. This standardization could be achieved by mapping all the points over a same range space, for instance, by applying the rule (8) which maps all points from the state space X to a space $\dot{x}$ in the range $[-1, 1]$:

$$\dot{x} = 2\left(\frac{x - \min(X)}{\max(X) - \min(X)}\right) - 1, \tag{8}$$

Also the observations s should then be mapped into the same space when determining the k-nearest neighbors set.

Finally for each defined classifier cl_i in the state space will be an associated action-value predictor $Q(i, a)$.

2.1 Perception

The first task is to determine the k-nearest neighbors set (knn). Figure 2 illustrates a time step of the perceptual process for a value of $k = 5$ and the 5-nearest neighbors of an observation s. Columns $Q(a1)$ and $Q(a2)$ of figure 2 represent the reward prediction for each action (1 or 2) for each classifier $c_i \in [1, ..., 5]$, column d is the euclidian distance from the current mapped observed point s to each classifier c_i and the values of column w represents the weights of each classifier based on the euclidian distance following the formula:

$$w_i = \frac{1}{1 + {d_i}^2} \quad \forall i \in [1, ..., k] \tag{9}$$

Thus less distance imply more weight (activation) and then for each time step will be k active classifiers whose weights are inverse to the euclidian distance from the current observation (s).

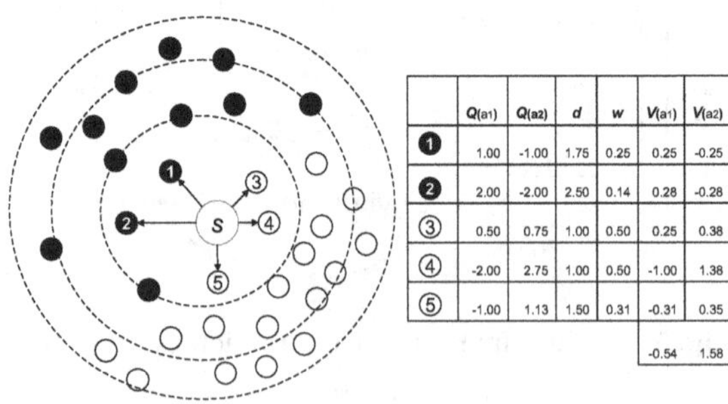

	Q(a1)	Q(a2)	d	w	V(a1)	V(a2)
1	1.00	-1.00	1.75	0.25	0.25	-0.25
2	2.00	-2.00	2.50	0.14	0.28	-0.28
3	0.50	0.75	1.00	0.50	0.25	0.38
4	-2.00	2.75	1.00	0.50	-1.00	1.38
5	-1.00	1.13	1.50	0.31	-0.31	0.35
					-0.54	1.58

Fig. 2. The k-nearest neighbors rule at some time step for a value of $k = 5$

Then we can state that a perceptual representation, or simply a perception, is uniquely and completely defined by two components (knn, w) obtained from a currently observed state (s):

1. A set (knn) which contains the k-nearest neighbors of the observation s.
2. A vector (w) which contains the weights (or activations) of each element in the knn set.

2.2 Action Selection

Once obtained a perceptual representation (knn, w) of an observed state (s) the Agent should emit an action (a) to the Environment. For doing so, the Agent has to apply some "rational" action selection mechanism based on its expectations about the consequences of such action (a). These consequences are represented by expectations of future rewards. The discussion about exploration vs. exploitation and the corresponding methods like greedy, ϵ-greedy or soft-max methods can be found in the literature [1]. We will center the attention on calculating the expectations for each action by means of a collective prediction process which can then be used directly by any classical action selection mechanism.

The objective of the prediction process consist of calculating the expected value of the predictions for each available action a_i in the observed state s. Since this process involves many different predictions, one for each element in the knn set, the procedure gets reduced to estimate an expected value, that is:

$$\mu = \sum_{i=1}^{n} x_i \, p(x_i) \; , \tag{10}$$

In this case the probabilities of each value are implied by the current weights vector (w). This vector should be normalized in order to express a probability distribution $p(.)$, thus the probabilities for each element in the knn set are calculated by the equation (11).

$$p(i) = \frac{w_i}{\sum w_i} \quad \forall i \in knn, \tag{11}$$

In this way, the expected value of the future reward for a given action a is estimated by (12).

$$\langle V(a) \rangle = \sum_{i=1}^{knn} Q(i,a) p(i) \; , \tag{12}$$

hence $p(i)$ is the probability $P(Q(i,a) = V(a)|s)$ that $V(a)$ takes the value $Q(i,a)$ given the state of the environment s. Then the action selection mechanism can directly derive a greedy policy from the expectations for each perceptual representation as shown in (13).

$$\max V(a) \tag{13}$$

This can be understood, in k-NN terminology, as that the Agent will select the most "frequent" recommended action in the neighborhood of the observed state s or, in bayesian terms, the action for which there is more evidence of a high future reward.

2.3 Learning

The learning process involves the use of the accumulated experience and thus it will rely on past and current experienced states, actions performed and received rewards (s, a, s' and r). Since we restrict the RL problem to a Markov Decision Process (MDP) we only need to deal with the immediate experienced states and performed actions.

Given that the Agent have experienced the state s and performed an action a and then observed a new state s' and received a reward r there are two perceptual representations, once for each state respectively. These perceptual representations are the tuples (knn, w) for state s and (knn', w') for state s'.

Following the TD-Learning scheme we need an estimation of the future reward. This estimation is generally obtained by the value of the current state s' represented by $\max_a Q(s', a)$ in off-policy learning methods and by $Q(s', a')$ in on-policy learning.

For whichever (off/on policy) method be followed, an expected (predicted) value is calculated:

$$\langle V'(a) \rangle = \sum_{i=1}^{knn'} Q(i, a) p'(i) \,, \tag{14}$$

where knn' is the set of the k-nearest neighbors of s', $p'(i)$ are the probabilities of each neighbor in the set knn' and the TD error (δ) can be calculated as shown in (15) or (16).

$$\delta = r + \gamma V'(a') - V(a) \tag{15}$$

$$\delta = r + \gamma \max_a V'(a) - V(a) \tag{16}$$

Thus we can simply update the prediction of each point in the knn set by applying the update rule (17).

$$Q(i, a)_{t+1} = Q(i, a)_t + \alpha \delta p(i) \;\; \forall i \in knn \tag{17}$$

2.4 Probability Traces $e(i) \leftarrow p(i)$: The k-NN$Q(\lambda)$ Algorithm

The use of eligibility traces [1, p.161] is a useful technique for improving the performance of RL algorithms. Eligibility traces are controlled by a parameter λ that takes values in the range $[0, 1]$. This parameter makes that the methods based on eligibility traces behaves between two extremes: one-step TD-Learning ($\lambda = 0$) and Monte Carlo methods ($\lambda = 1$) [1, p.163]. A pseudo-code of the kNN-TD(λ) RL algorithm is presented in algorithm 1.

In order to implement eligibility traces it is necessary to define a matrix e in which the traces for each classifier will be stored. A consequence of the use of eligibility traces is that the update rule should be applied to all the classifiers and not only to the knn set. Thus the update rule for the case of eligibility traces becomes:

$$Q_{t+1} = Q_t + \alpha \delta e \tag{18}$$

Algorithm 1. The kNN-TD(λ) reinforcement learning control algorithm

```
Initialize the space of classifiers cl
Initialize Q(cl, a) arbitrarily and e_i(cl, a) ← 0
repeat {for each episode}
  Initialize s
  knn ← k-nearest neighbors of s
  p(knn) ← probabilities of each cl ∈ knn
  V(a) ← Q(knn, a) · p(knn) for all a
  Chose a from s according to V(a)
  repeat {for each step of episode}
    Take action a, observe r, s'
    knn' ← k-nearest neighbors of s'
    p(knn') ← probabilities of each cl ∈ knn'
    V'(a) ← Q(knn', a) · p(knn') for all a
    Chose a' from s' according to V'(a)
    Update Q and e:
      e(knn, ···) ← 0 {optional}
      e(knn, a) ← p(knn)
      δ ← r + γV'(a') − V(a)
      Q ← Q + αδe
      e ← γλe
    a ← a', s ← s', knn ← knn', p ← p'
  until s is terminal
until learning ends
```

At least two kinds of traces can be defined: cumulating traces and replacing traces.

Cumulating traces are defined in a simple form:

$$e(knn, a)_{t+1} = e(knn, a)_t + p(knn), \tag{19}$$

where knn is the set of k-nearest neighbors of the observation s and $p(knn)$ are its corresponding probabilities.

In general, replacing traces are more stable and perform better than cumulating traces [10,1]. Replacing traces in the kNN-TD(λ) algorithm are defined in a simple way by establishing the value of the trace for each classifier as its probability $p(i)$ for the performed action (a) while for actions different from a the traces reset to 0. This form of trace resetting is a specialization of replacing traces that exhibits better performance than classical replacing traces schemes [1, p.188]. The expression (20) shows the way in which this kind of traces are defined for the kNN-TD(λ) algorithm.

$$e(knn, j) = \begin{cases} p(knn) & j = a \\ 0 & j \neq a \end{cases} \tag{20}$$

where knn is the set of k-nearest neighbors of the observation s, $p(knn)$ are its corresponding probabilities and a is the last performed action. The traces always decay following the expression (21).

$$e_{t+1} = \gamma\lambda e_t \quad (21)$$

As we can see, the eligibility traces scheme is replaced by a probability traces scheme given the fact that the stored values are just probabilities and not the value 1 as is used in classical eligibility traces theory.

3 Performance and Evaluation

The algorithm kNN-TD(λ) has been intensively tested on various kinds of problems with very robust results. These problems range from the classical Mountain-Car, Acrobot and Cart-Pole to some specific problems in robot control.

Figure 3 shows the graphical representation of two classical control problems: the Mountain-Car problem (taken from [1]) and the Cart-Pole balancing task (taken from the ANJI web site `http://anji.sourceforge.net/`).

First we present the result obtained for the Mountain-Car problem in order that it could be compared with the more standard results [1, p.214] in the literature.

In figure 4 the final solution obtained by the kNN-TD(λ) algorithm is shown. This result was obtained with a value of $\lambda = 0.95$, $\alpha = 0.9$ and for a value of $k = 4$. Each dimension is defined by a set of points in the usual ranges with a linear partitioned space with 20 points for the position dimension and 15 points for the velocity dimension.

Figure 3 shows the learning convergence graph for method kNN-TD(λ) in the Cart-Pole problem. As can be seen the algorithm solves the problem in only 29 episodes.

Even more, this algorithm was tested at the Second Annual RL Competition hosted at `http://www.rl-competition.org/` (RL2008) with excellent results. The kNN-TD(λ) algorithm won the PolyAthlon event (JAMH team) showing that is a method of general applicability as well as of high performance. Also it got the second best mark on the Mountain-Car domain without a significant statistical difference with the best mark. This result is very significative since this problems where generalized and altered in rare ways indeed with a highly noisy state variables. In that problems we tested the kNN-TD(λ) algorithm with a huge state space and more than 200 nearest-neighbors, that is with a value of $k \geq 200$, while achieving good performance in computing cost terms.

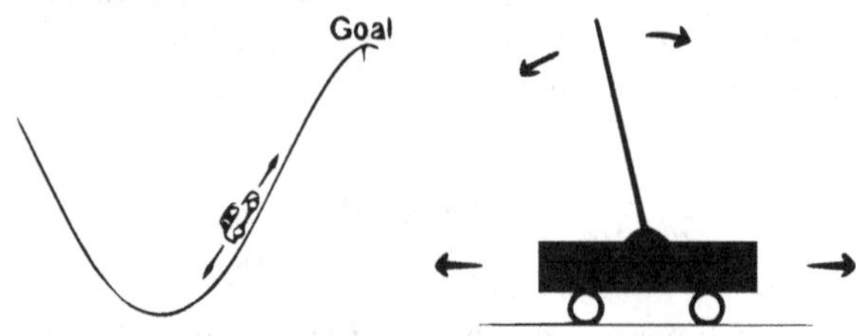

Fig. 3. The Mountain-Car problem and the Cart-Pole balancing task

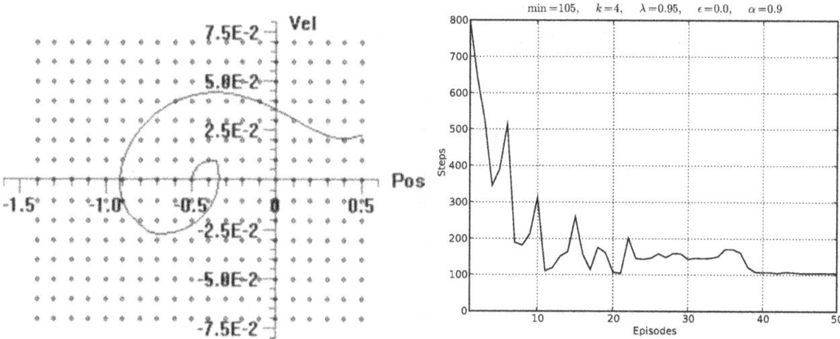

Fig. 4. State space and optimal solution for the Mountain-Car problem: the algorithm is stabilized from the episode 44 at 105 steps

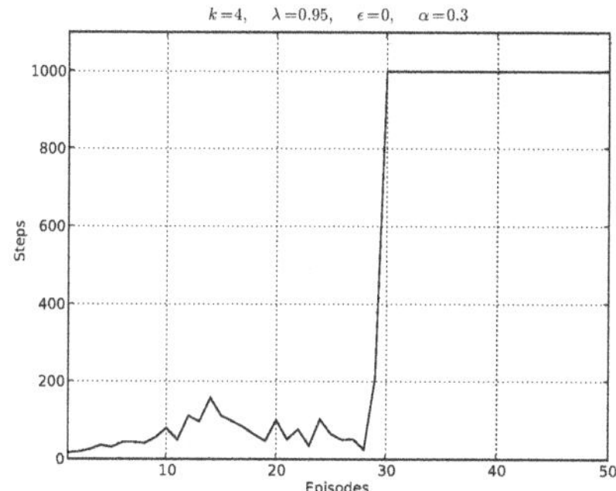

Fig. 5. Learning convergence graph for the method kNN-TD(λ) in the Cart-Pole problem

4 Conclusions and Further Work

We have introduced a reinforcement learning algorithm called kNN-TD. We have extended the basic algorithm with a technique called probability traces, which is a natural specialization of eligibility traces in this context, obtaining the kNN-TD(λ) algorithm. This new algorithm can be compared with the state of the art reinforcement learning algorithms both in performance and simplicity terms as shown by the experimental results.

The action selection mechanism of the proposed method can be understood, in k-NN terminology, as that the Agent will select the most "frequent" recommended action in the neighborhood of the observed state s or, in bayesian terms, the action for which there is more evidence of a high future reward. The the learning mechanism can be understood as a kind of temporal difference back-propagation which assigns the right TD error to each classifier according to its respective influence in the decision making and action selection procedures, which is represented by its probability of being the nearest neighbor of the observed states.

Finally, immediate lines of future work can be identified, for instance adaptive classifier space determination, adaptive action space determination and, as a mandatory line of research, to provide improvements on the state space representations in order to find the k-nearest neighbors efficiently for avoiding the so called "curse of dimensionality". Along the last line of future research, we can mention the promising techniques of approximate nearest neighbors search [11] which could lead to some compromise solutions between the accuracy of a k-NN expectations memory and its computational efficiency in complexity terms.

Acknowledgments. This work has been partially funded by the Spanish Ministry of Science and Technology, project DPI2006-15346-C03-02.

References

1. Sutton, R., Barto, A.: Reinforcement Learning, An Introduction. MIT Press, Cambridge (1998)
2. Watkins, C.J., Dayan, P.: Technical note Q-learning. Machine Learning 8, 279 (1992)
3. Cover, T.M., Hart, P.E.: Nearest neighbor pattern classification. IEEE Transactions on Information Theory IT-13(1), 21–27 (1967)
4. Duda, R.O., Hart, P.E.: Pattern Classification and Scene Analysis. Wiley, Chichester (1973)
5. Dudani, S.A.: The distance-weighted k-nearest-neighbor rule. IEEE Transactions on Systems, Man and Cybernetics SMC-6(4), 325–327 (1976)
6. Gordon, G.J.: Stable function approximation in dynamic programming. In: ICML, pp. 261–268 (1995)
7. Atkeson, C., Moore, A., Schaal, S.: Locally weighted learning. AI Review 11, 11–73 (1997)
8. Bosman, S.: Locally weighted approximations: yet another type of neural network. Master's thesis, Intelligent Autonomous Systems Group, Dep. of Computer Science, University of Amsterdam (July 1996)
9. Martin, H., Antonio, J., de Lope, J.: A k-NN based perception scheme for reinforcement learning. In: Moreno Díaz, R., Pichler, F., Quesada Arencibia, A. (eds.) EUROCAST 2007. LNCS, vol. 4739, pp. 138–145. Springer, Heidelberg (2007)
10. Singh, S.P., Sutton, R.S.: Reinforcement learning with replacing eligibility traces. Machine Learning 22(1-3), 123–158 (1996)
11. Indyk, P., Motwani, R.: Approximate nearest neighbors: Towards removing the curse of dimensionality. In: STOC, pp. 604–613 (1998)

Recombination Patterns for Natural Syntax

Vincenzo Manca[1] and M. Dolores Jiménez-López[2]

[1] Department of Computer Science, University of Verona
Ca Vignal 2- Strada Le Grazie 15, 37134 Verona, Italy
vincenzo.manca@univr.it

[2] Research Group on Mathematical Linguistics, Rovira i Vigili University
Av. Catalunya, 35, 43002 Tarragona, Spain
mariadolores.jimenez@urv.cat

Abstract. The main goal of this paper is to show that the mechanism of recombination, found in biology, also works in natural languages. We aim to demonstrate that it may be possible to generate a natural language, or at least an important part of it, by starting with a base composed of a finite number of simple sentences and words, and by applying a small number of recombination rules. *Recombination patterns* are introduced as a simple and efficient formalism –based on the behavior of DNA molecules– to describe the syntax of natural languages. In this paper we deal with the description of complex sentences and we have used English, Italian and Spanish to define our formalism.

1 Introduction

Methods and ideas from linguistics and formal language theory have been used to study genetics and other issues in biology (cf. [8], [16], [17], [18], [23]). Ideas from genetics and biology are used in the fields of linguistics and formal languages (cf. [18]). So, on the one hand, there are attempts to improve computation and the study of natural languages by using the structure and behavior of DNA. On the other hand, there are other approaches that try to improve the understanding of genetics and heredity by using computational and natural language models. Following these ideas, two different directions would seem to be possible: (1) from DNA to computation and natural languages (cf. [5], [6], [7], [19], [20], [25], [2], [3]), and (2) from computation and natural languages to DNA (cf. [23]). In this paper we would like to contribute to the most recent direction and thus to use DNA as a support for the study of languages – in our case natural languages.

Our goal in this paper is to show that the fundamental mechanism by which genetic material is merged, i.e. recombination, works and is also valid in natural languages. We believe that recombination may explain the way in which speakers combine linguistic elements (words, phrases, sentences) that they know or have already heard, in order to construct new linguistic structures. More specifically, we focus on the possibility of generating a natural language (or an important part of it) by using a finite number of words, phrases, sentences etc., and by only applying recombination rules. Note, however, that our main aim is not so

J. Mira et al. (Eds.): IWINAC 2009, Part I, LNCS 5601, pp. 315–324, 2009.

much to explain the human processing of language as to be able to construct a generation mechanism for natural language which is as simple and efficient as possible and capable of, for example, improving human-computer interfaces in natural language.

Some naive intuitions behind our goal are related to the way speakers seem to generate new structures in some specific situations. Let us consider students of whatever foreign language. In the early stages of the learning process, such students tend to use phrases, sentences or structures that they have already heard in order to construct their own new sentences. Their linguistic strategy may be said to consist just in (1) considering the sentences of the language that they have already learned, heard or used previously (i.e. a limited number of sentences); (2) *cutting* up the parts of each sentence that fit with the idea they want to communicate; and, finally, (3) *recombining* each of those fragments in order to generate a new sentence. Therefore, what such language students are doing here seems to be nothing more than applying a recombination rule, i.e. *cutting and pasting*, recombining phrases, sentences, words etc. that belong to their knowledge corpus of the new language in order to generate new sentences for new situations.

These ideas are to some extent a motivating factor in our research. However, it is the problems that one faces when trying to describe natural languages and their processing using current approaches that led us to search for new formalisms to account in a simpler and more natural way for the syntax of natural languages. Some of these problems have been discussed in [2]:

1. The rewriting methods used in a large number of natural language approaches are not very adept, from a cognitive perspective, for accounting for the processing of language.
2. Natural language sentences cannot be placed in any of the families of the Chomsky hierarchy in which current computational models tend to be based.

Recombination methods may be a valid alternative, since:

1. Genetics offers a *natural* model for the study of *natural* languages.
2. DNA computation provides simpler alternatives to the rewriting rules.
3. Languages that have been generated following a molecular computational model are placed in-between the CS and CF families.

In addition, the idea of using recombination rules in the description of natural language syntax is backed up by a long tradition of interchanging methods in biology and natural/formal language theory. The three main areas behind our research are:

1. Recent developments in formal language theory that provide new generative devices that allow close simulation of molecular recombination processes by generative processes acting on strings [5], [6], [7] .
2. The field of molecular computing, i.e. using biomolecules as a support for computations and for devising computers [19], [20].
3. The well-known analogies between genetic and natural languages (cf. [24], [18]).

The same basic idea we propose here can be found in [2], where molecular methods are used to describe natural languages and in [3], where a new framework in linguistics called *biosyntax* is proposed.

2 Recombination Patterns

As already mentioned, in this paper we propose to apply *recombination rules* to the syntax of natural language. So far in our research, we have focused on studying *complex sentences*. We have defined different patterns for the generation of complex structures accepted in English, Italian and Spanish. We initially considered coordinate sentences in the three languages studied. We then analyzed subordinate sentences in English, Italian and Spanish, describing independently the three traditional types considered in grammar i.e. noun and relative and adverbial clauses. Finally we defined patterns for the generation of complex sentences in which non-finite forms of verbs appear.

The method we used to define the patterns was as follows: (1) we considered independently each of the different types of complex sentences we wanted to formalize; (2) we studied the accepted constructions for such complex types of sentences in each of the three languages considered; (3) we analyzed a total of around 1000 examples of those constructions; (4) finally, we defined patterns that generate the complex sentences by using recombination rules reported in the literature of DNA computing. As will be shown below, the patterns we defined are very simple devices that can account in an efficient way for complex sentences in natural language and that can be easily implemented.

2.1 Basic Categories

In order to define patterns we assume some basic categories (i.e. verb, noun, substantive, proposition) and some basic axioms referring to basic syntactic operations (i.e. conjugation, determination, predication, modification and complementation). We will define these basic categories and operations in this section. Note however that although the terms proposition, noun, verb, substantive are taken from traditional grammatical, logical, and semantic analysis, the definition of our basic types is completely formal. It is based on the assumption of some initial operations, and is not aligned with the traditional senses of these terms.

First of all, we assume the following three basic operations as preliminary data for our analysis: Conjugation ($\backslash conjug$); Determination ($\backslash determ$); Predication ($\backslash pred$).

Intuitively, *conjugation* ($\backslash conjug$) adds temporal and dynamical parameters to syntactic elements that can play *verbal* roles. *Determination* ($\backslash determ$) adds spatial and contextual information to elements that can play *nominal* roles. The information that is added in both cases is supported by some elements that are called *conjugation parameters* and *determination parameters*, collected into two sets –*conjugative* and *determinative*, respectively. *Predication* ($\backslash pred$) is the basic sentence-building construction.

We assume that we are able to know when these operations can be applied to some arguments and, in this case, to get the results they yield.

We use four basic syntactic categories: Verb; Noun; Substantive (*Subst*); Proposition (*Prop*).

We define the class *verb* of elements x such that, for some y, $\backslash conjug(x, y)$ provides a result: $Verb = \{x \mid \exists\ y\ z\ \backslash conjug(x, y) = z\ ,\ y \in Conjugative\}$.

Likewise: $Noun = \{x \mid \exists\ y\ z\ \backslash determ(x, y) = z\ ,\ y \in Determinative\}$.

The result of a *conjugation* operation is a *(definite) verb* or *verbative.* We differentiate here between *verb* and *verbative*, but for the sake of simplicity hereafter we will use *verb* to refer to both cases. The result of a *determination* operation is called *substantive* (Subst)

$$Verbative = \{z \mid \exists\ x\ y\ \backslash conjug(x, y) = z\ ,\ y \in Conjugative\}.$$
$$Subst = \{z \mid \exists\ x\ y\ \backslash determ(x, y) = z\ ,\ y \in Determinative\}.$$

Predication takes as its arguments a substantive and a verb, and provides a *proposition* (Prop): $Prop = \{z \mid \exists\ x\ y\ \backslash pred(x, y) = z\ ,\ x \in Subst\ ,\ y \in Verb\}$.

Proposition is the category of the minimal units of a discourse that can be considered, in its abstract semantic role, as the localization of an event in a space. The localization component is referred to as an object, that is, a static entity determined by spatial-contextual attributes. This component hosts a dynamics specified by temporal-dynamical attributes.

We use *Cat* to indicate any of the previous categories (substantive, noun, verb, proposition). To these four basic categories we add four *ad categories* (*AdCat* indicates any of them): AdProp; AdSubst; AdNoun; AdVerb.

The eight categories above are the syntactic categories of *full* or *lexical* linguistic elements. We will add a further category of *empty* linguistic elements which we call *particles* (prepositions, articles, etc.). Some of the operations we will introduce need some additional parameters, which we call *grammemes.* Examples of grammemes are: Singular (`SING`), Plural (`PLU`), Present (`PRES`)... It is important to distinguish grammemes from particles. In fact, particles are strings that are inserted in a text as a consequence of the application of a syntactic operation with some grammemes. In other words, particles are strings that are provided in the surface syntactic realization of the operations, while grammemes are kinds of information that the grammar of L requires in order to provide the morphological realization corresponding to the syntactic representation.

2.2 Structure of Patterns

Every pattern of the framework that we introduce here is a very simple formalism divided into three different parts:

1. The first one, called *pattern recognition*, defines in terms of some basic categories (proposition, substantive, noun, etc.) the two propositions that we always have available in order to generate a complex sentence.

2. The second part, *pattern recombination*, is where the application of different recombination rules (insertion, deletion, transposition, transformation, etc.) takes place. Taking into account that in every pattern we start by having two propositions available, this phase starts by placing proposition 2 somewhere with reference to proposition 1. There are three types of insertion, according to the place –beginning, middle, end– where the second proposition is situated with respect to the first one. This is achieved by performing:
   (a) either an *initial insertion*. In this type of insertion, proposition 2 is situated before proposition 1. The typical instruction here says: `Insert Prop.2 before Prop.1`;
   (b) or a *median insertion*. Here, proposition 2 is placed after the subject of proposition 1, since the instruction is something like: `Insert Prop.2 after subject in Prop.1`;
   (c) or a *final insertion*. Final insertion places proposition 2 after proposition 1. So, the instruction here says: `Insert Prop.2 after Prop.1`

   After the positioning of proposition 2 with respect to proposition 1, some recombination rules are applied.
   If we analyze patterns by taking into account the type of insertion performed in the *pattern recombination* phase, we realize that the most frequent type of insertion in all the three languages considered is the so-called *final insertion*, followed a long way after by *median insertion*. A very small number of cases in which *initial insertion* is performed have been found.
3. The third and final part, called *grammatical adaptation*, carries out the changes (e.g. insertion of elements, transformation of verbal tenses and modes) needed in order to obtain grammatical sentences.

One of our goals in defining such formulae is to do without the traditional grammatical division of complex sentences and to think simply in terms of patterns that recombine some acceptable structures and give some other acceptable structures. This means that the patterns are not defined with the usual terminology (relative clauses, noun clauses and the like) and are organized taking into account the following criteria:

- *Variables used.* We define patterns by specifying the variables used. We specify the variables available when we start the application of the sequence of rules (in) and the result obtained after the application of this sequence (out).
- *Type of insertion.* Taking into account that in every pattern we start by having two propositions, by type of insertion we mean the place –beginning, middle, end– where the second proposition is situated with respect to the first one.
- *Sequence of rules.* The sequence of rules applied in order to obtain a complex sentence from the two simple sentences considered.
- *Element introduced.* The word we introduce in order to generate a complex sentence from the recombination of two simple ones.

2.3 Types of Rules

Most of the recombination rules applied during the *pattern recombination* and the *grammatical adaptation* phases of our formalism are the same as those proposed in DNA computing literature, i.e. rules that formalize the rearrangement, mutations or recombinations at a genomic level. We only refer to five rules, which can be formulated as follows:

1. *Insertion*, where some element or structure is inserted into some point of the string.
2. *Deletion*, where some elements of the structure are deleted.
3. *Transposition*, where some elements are shifted to another location in the string.
4. *Inversion*, where elements in the string can be reversed.
5. *Transformation*, where some elements change their form.

If we consider the frequency of application of these different types of rules in our patterns, we see that the commonest rule is *Insertion*, followed by *Deletion*. Then a long way after we have *Transformation, Transposition* and *Inversion*.

3 An Example of Pattern

The following example show the features of patterns outlined above. To show how patterns work we apply it to a complex sentences, one for each of the patterns presented.

The pattern contains the three parts mentioned in the above section: 1) Pattern recognition (roman typeface) which presents the elements (propositions, substantives, etc.) we have at hand in order to generate complex sentences; 2) pattern recombination (italics), where we apply different recombination rules to the elements provided in the first part; and finally, 3) grammatical adaptation (bold) where we perform the changes needed to obtain a grammatical sentence. Moreover, we consider that each of the simple sentences we present in the so-called *pattern recognition* belongs to our corpus.

In the example presented in this section, the criteria used in the organization of patterns can also be seen. The pattern is specified by referring to the following issues: 1) *variables used* (*in*: variables we have available when we start the application of rules; *out*: the result obtained at the end); 2) *sequence of rules*, that is, the sequence of rules applied in order to obtain a complex sentence from two simple ones; 3) *language* for which the pattern has been defined (English, Spanish, Italian); and 4) *element introduced*, that is, the word inserted in order to generate the complex sentence. Note that the number written has no specific meaning: it appears in the heading of every example because while defining the patterns we numbered them for practical/working reasons.

The following a pattern that presents *initial insertion* in the *pattern recombination* step. It is an example of the most recurrent type of sequence of rules, i.e. the one that consists four rules. It shows how the most frequent types of rules –*deletion, insertion* and *transformation*– are applied.

PATTERN 340: (in: x, y, w, z, r, t, s, u, s_2; out: w'S s_2 u r t)
`Rules Sequence: Ini. Ins., Del., Transf., Ins.`
English: 'S

xy = Prop. 1: Proposition
wz = Prop. 2: Proposition
x: Subject
w: Subject
y = r t: Predicate
z = s u: Predicate
r: Verb
t: Substantive
s: Verb, Indicative
u: Substantive
s_2: Verb, Present Participle (PP)

(w s u) x r t: Quasi Proposition `Insert Prop.2 before Prop.1`
(w s u) r t: Quasi Proposition `Delete` x `in Prop.1`

(w s_2 u) r t: Quasi Proposition `Transform` s `in Prop.2 into PP`
w'S s_2 u r t: Proposition `Insert 'S after` w `in Prop.2`

Using the above pattern, and having as simple sentences *'Something annoyed Sally'* and *'Martha gets a splinter in her toe'*, we can generate *'Martha's getting a splinter in her toe annoyed Sally'* in the following way:

- Something annoyed Sally:= x y; y = r t
- Martha gets a splinter in her toe:= w z; z = s u

something:= x	annoyed Sally:= y
annoyed:= r	Sally:= t
Martha:= w	gets a splinter in her toe:= z
gets:= s	a splinter in her toe:= u

(Martha gets a splinter in her toe) Something annoyed Sally:=(w s u) x r t
(Martha gets a splinter in her toe) annoyed Sally:= (w s u) r t

(Martha getting a splinter in her toe) annoyed Sally:= (w s_2 u) r t
Martha'S getting a splinter in her toe annoyed Sally:= w'S s_2 u r t

The example above highlight how simple our formalism is. By using recombination patterns we do not need to postulate complicated rules in order to generate (or parse) complex sentences, nor do we need to make a difference between types of complex sentences, and to postulate different types of rules in order to account for subordinate sentences or for coordinate ones. By using recombination patterns, every type of complexity in natural language is explained in the same way:

recombining accepted simple sentences in order to generate other accepted, this time complex, sentences. Classical problems such as determining when *that* is a conjunction or a relative pronoun do not arise in a formalism where complexity is obtained by only applying five simple rules based on the natural operation of recombination. So, in some sense, the formalism we have defined reconsiders complexity in language and offers a new classification of complexity where no important difference between coordination and subordination exists. In fact as shown by the formal framework presented in this paper, if we cut and paste pieces of simple sentences in order to obtain complex sentences, that is, if we use recombination techniques, the process for obtaining subordinate and coordinate sentences is very similar.

4 Final Remarks

Patterns defined in this new framework show how to combine and modify simple sentences in order to generate complex ones. What is important in our formalism is that we can generate complex structures by using simple sentences and by applying *only five types of very simple rules: insertion, inversion, deletion, transposition,* and *transformation.* This implies a very important simplification of the mechanism of generation of natural languages. From the patterns defined it may be not difficult to define an algorithm that specifies the steps needed to generate complex sentences from simple ones. In addition, we believe that such an algorithm may be language independent, since the patterns already defined for the three languages considered are very similar, they only have significant differences in the so-called grammatical adaptation. An interesting idea that follows from our research is thus the possibility of defining formal methods to generate natural language independently on specific languages. The definition of an algorithm of these characteristics may have important implications in the field of natural language generation as well as in cognitive science or machine learning theory since, as pointed out by [2], intuitively speaking recombination rules may be more feasible than rewriting systems to explain how humans process language.

By introducing recombination patterns what we want to do is to reconstruct syntax with molecular methods by formulating a system capable of generating most syntactic structures of language. Our goal is to define a formalization that can be implemented and may be able to describe and predict the behavior of syntactic structures. This means that we want to provide a method of linguistic manipulation that would be useful for natural language processing. Although this paper is just an initial and rough approach to the topic of using recombination mechanisms in linguistics that needs further research, we feel that this approach may be suitable for the syntax of natural languages since it presents interesting advantages with respect to current methods such as:

- It is simple. By using recombination we can model syntax with only five rules and, in general, we can affirm that recombination patterns are simpler than rewriting methods for the syntax of natural languages.

- Traditional syntactic classifications of complexity can be redefined thus providing a simpler model in which fewer concepts are needed in order to deal with classical syntactic problems.
- It represents an explicative and generative approach to linguistics. It is able to account for a number of problems, but it is also able to generate correct sentences of a language.

Following the difference presented in [8] about the *adequacy* and *relevance* of a model:

> *"...there is a substantial difference between the adequacy of a conceptual tool, i.e. the ability of the model to fit the features of the modeled object, and its relevance i.e. the possibility to obtain non-trivial insights into the object by studying the model, insights which cannot be obtained without using the model. Generally speaking, adequacy is ensured by a well-inspired definition, but relevance entails making a special effort in order to be proved, along with mathematical investigations which may last a significant period."*

we think that the formalism we are proposing is *adequate* for natural language syntax since it seems to fit the basic features of the object modeled, i.e. natural languages. But is it relevant? Our future work will focus on demonstrating the relevance of such a recombination mechanism in natural language generation. We need to relate our formalism with other grammar models based on rewriting in order to establish whether or not recombination patterns offer a generation device that is simpler and more natural than current grammatical approaches. Whatever the answer, we would like to stress that by introducing recombination patterns we are not trying to offer an approach to linguistics that replaces already existing models, but we simply want to offer an *alternative* way to approach natural languages that can easily be placed alongside current accepted theories.

References

1. Barwise, J. (ed.): Handbook of Mathematical Logic. North-Holland, Amsterdam (1977)
2. Bel Enguix, G.: Molecular Computing Methods for Natural Language Syntax, GRLMC Report, 18/01, Tarragona (2001)
3. Bel Enguix, G., Jiménez López, M.D.: Is a Biosyntax Possible? In: Angelova, G., Bontcheva, K., Mitkov, R., Nicolov, N., Nikolov, N. (eds.) Recent Advances in Natural Language Processing, Incoma, Bulgaria, pp. 46–50 (2003)
4. Bell, J.L., Machover, M.: A Course in Mathematical Logic. North-Holland, Amsterdam (1977)
5. Dassow, J., Păun, G.: Remarks on Operations Suggested by Mutations in Genomes. Fundamenta Informaticae 36, 183–200 (1998)
6. Head, T.: Formal Language Theory and DNA: An Analysis of the Generative Capacity of Specific Recombinant Behaviours. Bulletin of Mathematical Biology 49(6), 737–759 (1987)

7. Head, T., Păun, G., Pixton, D.: Language Theory and Molecular Genetics: Generative Mechanisms Suggested by DNA Recombination. In: Rozenberg, G., Salomaa, A. (eds.) Handbook of Formal Languages, vol. 2, pp. 295–360 (1997)
8. Kelemen, J., Kelemenova, A., Mitrana, V.: Towards Biolinguistics. Grammars 4, 187–203 (2001)
9. Manca, V.: A Metagrammatical Logical Formalism. In: Manca, V. (ed.) Metagrammatical Representations. VI Tarragona Seminar on Formal Syntax and Semantics, GRLMC Report 11/97, Tarragona, pp. 77–91 (1996)
10. Manca, V.: A Metagrammatical Logical Formalism. In: Martín-Vide, C. (ed.) Mathematical and Computational Analysis of Natural Language. John Benjamins, Amsterdam (1998)
11. Manca, V.: Logical Splicing in Natural Languages. In: Martín Vide, C. (ed.) Issues in Mathematical Linguistics, pp. 131–143. John Benjamins, Amsterdam (1999)
12. Manca, V.: Logical String Rewriting. Theoretical Computer Science (2001)
13. Manca, V.: Splicing Normalization and Regularity. In: Calude, C., Păun, G. (eds.) Finite versus Infinite. Contributions to an Eternal Dilemma, pp. 199–215. Springer, London (2000)
14. Manca, V.: String Rewriting and Metabolism: A Logical Perspective. In: Păun, G. (ed.) Computing with Bio-Molecules. Theory and Experiments, pp. 36–60. Springer, Singapore (1998)
15. Manca, V., Martino, D.: From String Rewriting to Logical Metabolic Systems. In: Salomaa, A., Păun, G. (eds.) Grammatical Models of Multi-Agent Systems, pp. 297–315. Gordon and Breach, London (1999)
16. Marcus, S.: Linguistics as a Pilot Science. In: Sebeok, T. (ed.) Current Trends in Linguistics, vol. 12. Mouton, The Hague (1974)
17. Marcus, S.: Linguistic Structures and Generative Devices in Molecular Genetics. Cahiers de Linguistique Theorique et Appliquee 11(1), 77–104 (1974)
18. Marcus, S.: Language at the Crossroad of Computation and Biology. In: Păun, G. (ed.) Computing with Bio-Molecules, pp. 1–35. Springer, Singapore (1998)
19. Păun, G. (ed.): Computing with Bio-Molecules. Springer, Singapore (1998)
20. Păun, G., Rozenberg, G., Salomaa, A.: DNA Computing. New Computing Paradigms. Springer, Berlin (1998)
21. Rozenberg, G., Salomaa, A. (eds.): Handbook of Formal Languages. Springer, Berlin (1997)
22. Salomaa, A.: Formal Languages. Academic Press, New York (1973)
23. Searls, D.B.: The Computational Linguistics of Biological Sequences. In: Hunter (ed.) Artificial Intelligence and Molecular Biology, pp. 47–120 (1993)
24. Shanon, B.: The Genetic Code and Human Language. Synthese 39, 401–415 (1978)
25. Yokomori, T., Kobayashi, S.: DNA Evolutionary Linguistics and RNA Structure Modelling: A Computational Approach. In: Proceedings of the 1st International Symposium on Intelligence in Neural and Biological Systems, pp. 38–45. IEEE, Herndon (1995)

Computing Natural Language with Biomolecules: Overview and Challenges*

Gemma Bel-Enguix

[1] GRLMC-Research Group on Mathematical Linguistics
Rovira i Virgili University
43002 Tarragona, Spain
gemma.bel@urv.cat

[2] Language, Evolution and Computation Research Unit
School of Philosophy, Psychology and Language Sciences,
University of Edinburgh
Edinburg, UK

Abstract. This paper aims to present the formal framework in which the interdisciplinary study of natural language is conducted by integrating linguistics, computer science and biology. It provides an overview of the field of research, conveying the main biological ideas that have influenced research in linguistics. Especially, this work highlights the main methods of molecular computing that have been applied to the processing and study of the structure of natural language. Among them, DNA computing, membrane computing and NEPs are the most relevant computational architectures that have been adapted to account for one of the most unknown cognitive capacities of human beings.

1 The Relevance of the Triangle Computer Science-Linguistics-Biology in the Science of the XXIth Century

In the science of XXIth century, as well as in the society and culture, two chief phenomena can be easily detected: specialization, and convergence. These two movements could seem to be paradoxical, but they are not. On one hand, the more human beings know, the more we have to restrict the field of our activities. But on the other hand, the more we learn, the more we comprehend the similarity among underlying mechanisms in many states of nature. Consequently, although disciplines are more and more specialized, research areas become more and more interdisciplinary; traditional boundaries fall down in order to achieve a new understanding of both, human capabilities and the behaviour of different types of life. Moreover, we know the reality can be approached from multiple perspectives, most of them complementary, that provide new views of the same phenomenon.

Within this interdisciplinary framework, this paper presents an example of research that has been developed in a shared space between the boundaries

* This work has been partially supported by the Spanish Ministry of Science and Technology, projects MTM-2007-63422 and JC2008-00040.

J. Mira et al. (Eds.): IWINAC 2009, Part I, LNCS 5601, pp. 325–335, 2009.

of three areas, that seem to have crucial impact in the future of the society: linguistics, biology and computer science.

The final goal of the research is linguistics; the main model comes from biology; computer science is the way to transfer biological concepts to formal linguistics and, at the same time, it provides a new methodology of research.

Molecular biology has been the best and more increasingly developed area in the last 50 years, and it seems that most of the future goals of the humanity have something to do with the deciphering of DNA and genetic engineering. Currently, biology –especially molecular biology– has become a pilot science, so that many disciplines have formulated their theories under models taken from biology. Computer science has become almost a bio-inspired field thanks to the great development of natural computing and DNA computing. As for linguistics, the relationship is even earlier, provided that the publication of Darwin's work The origin of the species, in 1859, was the first step towards an integrative view of language as a biological construct.

Computer science is not only transforming classical methods of science, but also our everyday life, by means of theoretical approaches and applications. Nowadays, in a process which is similar to the one undergone in biology, informatics provides to science revolutionary methods of research, models that can simulate unknown states of the nature and new perspectives in both, experimental and basic research.

However, linguistics has still the challenge to understand how natural language is acquired, produced and processed, in a multi-lingual society that needs some tools for linguistic interaction, automatic translation and human-computer interfaces. Up to now, linguistics has not been able to solve these challenges, partly, because of the fail in the models adopted. Indeed, it has been proofed that natural language does not fit in the classical Chomskyan hierarchy of languages, that constitutes the base of mathematical models for linguistics. Moreover, the metaphor of the mind as a computer seems to be almost exhausted. Finally, psychological models have a lack of formalization that prevents them to be conveniently implemented.

Surprisingly, and despite the fact that genetic code has been considered a code from its discovery [37], linguistics has not attempted to construct a new paradigm taking advantage of the great developments in molecular biology. Nevertheless, it is an especially suggestive idea, because of the interesting similarities between natural language and the genetic code in different levels.

Natural Language Processing (NLP) can take great advantage of the structural and "semantic" similarities between these codes. Specifically, taking the systemic code units and methods of combination of the genetic code, the methods of such entity can be translated to the study of natural language. Therefore, NLP could become another "bio-inspired" science, by means of theoretical computer science, that provides the theoretical tools and formalizations which are necessary for approaching such exchange of methodology.

Another key issue that biology can help to approach is the emergence and evolution of language. If morphological and cognitive features of humans have to be approached from inside the darwinian paradigm, it seems clear that the configuration of cognitive and physiological capacities that allow human beings to talk, as well as the process that leads to the arising of human language, have to be tackled from a perspective that integrates biology. However, the lack of data in paleontology and historical linguistics leaves the use of computer science and computational simulations as the only available tool to figure out how the human started to use the symbolic capacity to design a language.

In this way, a theoretical framework can be drawn, where biology, NLP and computer science exchange methods and interact. And, for linguistics, the bio-inspired computational paradigm can be a powerful tool for explaining language capacity from two different perspectives; synchrony – or the veiw of language as a complex system – and diachrony – or language understood as an evolutionary system.

2 Inside the Triangle

In this section, a short overview is provided of the main models that have been introduced on the interaction of disciplines inside the triangle linguistics-biology-computer science. We refer only those that have involved linguistics and biology, living aside the most developed areas, computational linguistics and bioinformatics, where either linguistics or biology are not represented.

2.1 Explaining the Structure of Natural Language by Bio-inspired Methods

After the chomskian revolution the relationship between natural and formal languages gave rise to a new way to understand language studies. Natural language was thought as an object of the natural world that could be approached from a rationalist science as any other object in the nature. The attempts to describe natural language with mathematics were important milestones in early post-chomskian linguistics.

Nowadays, most current linguistic schools prefer the search of new formalisms to account for languages in a simpler and more natural way. Two main facts lead to look for a more natural computational system to give a formal description of natural languages: a) natural language sentences cannot be placed in any of the families of the Chomsky hierarchy in which current computational models are basically based, and b) rewriting methods used in a large number of natural language approaches seem to be not very adequate, from a cognitive perspective, explain how natural language is generated and processed.

Now, if to these we add 1) that languages that have been generated following a molecular computational model are placed in-between CS and CF families; 2) that genetic model offers simpler alternatives to the rewriting rules; 3) and that genetics is a natural informational system as natural language is, we have the ideal scene to propose biological models in NLP.

Three main bio-inspiered methods are pointed out in this paper to account for the structure of natural language: DNA Computing [30], Membrane Computing

[29] and NEPs [7]. The contributions of bio inspired computing to the description and understanding of natural language will be developed more in depth in section 3.

2.2 Understanding the Genetic Code as a Linguistic Informational System

Since the discovery of the basic structural configuration of DNA [37] many researchers pointed out the communicative nature of the genetic code. This topic, that caused a philosophical discussion for years, has been tacked by authors like Jakobson [19], Jacob and Monod [18], Marcus [25], Ji [20], López García [22], Collado [9], Searls [33] and Victorri [36]. Among them, was Jakobson who introduced the most spread mapping between units in genetic code and their correspondence in natural language.

Therefore, some authors as Searls [33] and Brendel & Busse [5], have successfully used formal languages for the description of DNA. An specially important work for linguistics is the approach of Collado-Vides [9,10], who was one of the pioneers in the application of methods of analysis taken form generative grammar to explain the functioning of genetic processes of regulation. Perhaps this is the first example of the exportation of models which are usually used in linguistics to molecular biology.

Formal language theory has also devoted some effort to explain the structure of Genetic Code. Some examples can be (1) Pawlak [31] dependency grammars as an approach in the study of protein formation; (3) stochastic context-free grammars for modeling RNA [32]; (4) definite clause grammars and cut grammars to investigate gene structure and mutations and rearrangement in it [33]; (5) tree-adjoining grammars used for predicting RNA structure of biological data [35].

2.3 Explaining the Evolution of Natural as a Darwinian Process

August Schleicher [34], in 1953, introduced one of the most fruitful ideas in language evolution, the use of trees to illustrate language descent. The idea, that is identified as darwinian, was raised before the publication of the Origin of the Species, what means that both, Schleicher and Darwin [15], came up with the same idea in two distant disciplines: biology and linguistics. The research, since that, has been focused on insvestigating whether or not some relationship can be established between the trees of both, the descent of languages and species.

The work by Cavalli-Sforza [8] has been determinant in this sense. The geneticist has established the relationship between the trees of human and language diversity. Maynard-Smith [23] has also tackled the problem of the relationship between transitions in evolutionist biology and human language, combining the approach with game theory.

Maynard-Smith [24] considered the emergence of language as one of the major transitions in evolution. The problem of language emergence is crucial in linguistics, but it is also one of the major problems in evolutionary biology, psychology and biolinguistics. The major contributions to the topic of language emergence have been obtained by combining darwinism, linguistics and computer science. The main exponent of this line of research is Kirby [21], who introduced the computational model of iterated learning, that could demonstrate the emergence of compositionality despite the lack of physical data.

3 Bio-inspired Methods for Natural Language Processing

Here, we present an overview of different bio-inspired methods that during the last years have been successfully applied to several NLP issues, from syntax to pragmatics. Those methods are taken mainly from computer science and are basically the following: DNA computing, membrane computing and networks of evolutionary processors.

3.1 DNA Computing

One of the most developed lines of research in natural computing is the named molecular computing, a model based on molecular biology, which arose mainly after [1]. An active area in molecular computing is DNA computing [30] inspired in the way that the DNA perform operations to generate, replicate or change the configuration of the strings.

Application of molecular computing methods to natural language syntax gives rise to molecular syntax [2]. Molecular syntax takes as a model two types of mechanisms used in biology (especially in genetic engineering) in order to modify or generate DNA sequences: mutations and splicing. Mutations refer to changes performed in a linguistic string, being this a phrase, sentence or text. Splicing is a process carried out involving two or more linguistic sequences. It is a good framework for approaching syntax, both from the sentential or dialogical perspective.

Methods used by molecular syntax are based on basic genetic processes: cut, paste, delete and move. Combining these elementary rules most of the complex structures of natural language can be obtained, with a high degree of simplicity.

This approach is a test of the generative power of splicing for syntax. It seems, according to the results achieved, that splicing is quite powerful for generating, in a very simple way, most of the patterns of the traditional syntax. Moreover, the new perspectives and results it provides, could mean a transformation in the way syntactic mechanisms have been conceived so far.

From here, bio-NLP, applied in a methodological and clear way, is a powerful and simple model that can be very useful to a) formulate some systems capable of generating the larger part of structures of language, and b) define a formalization that can be implemented and may be able to describe and predict the behavior of natural language structures.

3.2 Membrane Computing

Membrane systems –introduced in [29]– are models of computation inspired by some basic features of biological membranes. They can be viewed as a new paradigm in the field of natural computing based on the functioning of membranes inside the cell. Membrane systems can be used as generative, computing or decidability devices. This new computing model has several intrinsically interesting features such as, for example, the use of multisets and the inherent parallelism in its evolution and the possibility of devising computations which can solve exponential problems in polynomial time.

This framework provides a powerful tool for formalizing any kind of interaction, both among agents and among agents and environment. One of key ideas of membrane systems is that generation is made by evolution. Therefore, most of evolving systems can be formalized by means of membrane systems.

Linguistic Membrane System (LMS) [3] aim to model linguistic processes, taking advantage of the flexibility of membrane systems and their suitability for dealing with some fields where contexts are a central part of the theory. LMS can be easily adapted to deal with different aspects of the description and processing of natural languages. The most developed applications of LMS are semantics and dialogue. Membrane systems are a good framework for developing a semantic theory because they are evolving systems by definition, in the same sense that we take meaning to be a dynamic entity. Moreover, membrane systems provide a model in which contexts, either isolated or interacting, are an important element which is already formalized and can give us the theoretical tools we need. LMS for semantics deal with the main idea that meaning is something dynamic. From that perspective, semantic membranes may be seen as an integrative approach to semantics coming from formal languages, biology and linguistics. Taking into account results obtained in the field of computer science as well as the naturalness and simplicity of the formalism, it seems the formalization of contexts by means of membranes is a promising area of research for the future.

A topic where context and interaction among agents is essential is the field of dialogue modeling and its applications to the design of effective and user-friendly computer dialogue systems. Taking into account a pragmatic perspective of dialogue and based on speech acts, multi-agent theory and dialogue games, Dialogue Membrane Systems have arisen, as an attempt to compute speech acts by means of membrane systems. Considering membranes as agents, and domains as a personal background and linguistic competence, the application to dialogue is almost natural, and simple from the formal point of view.

3.3 NEPS-Networks of Evolutionary Processors

Networks of evolutionary processors (NEP) are a new computing mechanism directly inspired in the behavior of cell populations. Every cell is described by a set of words (DNA) evolving by mutations, which are represented by operations on these words. At the end of the process, only the cells with correct strings will survive. In spite of the biological inspiration, the architecture of the system is directly related to the Connection Machine (Hillis, 1985) and the Logic Flow paradigm [16]. Moreover, the global framework for the development of NEPs has to be completed with the biological background of DNA computing [30], membrane computing [29] – that focalizes also in the behavior of cells – and specially with the theory of grammar systems [11], which share with NEPs the idea of several devices working together and exchanging results.

First precedents of NEPs as generating devices can be found in [14] and [13]. The topic was introduced in [7] and [27], and further developed in [6], [12]. A new approach to networks of evolutionary processors as accepting devices has started in [26].

With all this background and theoretical connections, it is easy to understand how NEPs can be described as agential bio-inspired context-sensitive systems. Many disciplines are needed of these types of models that are able to support a biological framework in a collaborative environment. The conjunction of these features allows applying the system to a number of areas, beyond generation and recognition in formal language theory. NLP is one of the fields with a lack of biological models and with a clear suitability for agential approaches.

NEPs have significant intrinsic multi-agent capabilities together with the environmental adaptability that is typical of bio-inspired models. Some of the characteristics of NEPs architecture are the following: Modularization. NEPs are a distributed system of contributing nodes, each one of them undergoing just one type of operation. Each one of these specialized processors may be modular as well. Contextualization and redefinition of agent capabilities during the computation. Synchronization. It is necessary to define protocols for the timing of the computation. Evolvability, agents/nodes/modules can change their definition during the computation. Learnability, the changes in the definition of the agent can be given by the context, by the elements inside or by the learning in their modules.

Inside of the construct, every agent is autonomous, specialized, context-interactive: and learning-capable. Social competences among processors are specified by a Graph-Supported structure and Filtering-Regulated Communication and Coordination.

In what refers to the functioning of NEPs, two main features deserve to be highlighted:

- *Emergence*: The behavior of the nodes is not necessarily pre-established. It depends on both, the evolution of the system and the conditions of the environment.
- *Parallelism*: Different tasks can be performed at the same time by different processors.

Because of those features, NEPs seems to be a quite suitable model for tackling natural languages [4]. One of the main problems of natural language is that it is generated in the brain, and there is an important lack of knowledge of the mental processes the mind undergoes to, finally, bring about a sentence. While expecting new and important advances in neuro-science and neuro-language, we are forced to use the models that seem to fit better to language generation and recognition. Modularity has shown to be a very important idea in a wide range of fields: cognitive science, NLP, computer science and, of course, NLP. From this point of view, we think that NEPs provide a quite suitable theoretical framework for the formalization of modularity in NLP.

Another chief problem for the formalization and processing of natural language is its changing nature. Not only words change through the time. Also rules, meaning and even phonemes can take different shapes during the process of computation. Formal models based only in mathematical (an extremely stable) language have a lack of flexibility to describe natural language. Biological models seem to be better to this task, since biological entities share with languages the concept of "evolution" as one of the main features of the system. From this perspective, NEPs offer enough flexibility to model any change at any moment in any part of the system. Besides, as a bio-inspired method of computation, they have the capability of simulating natural evolution in a highly pertinent and specialized way.

Some linguistic disciplines, as pragmatics or semantics, are context-driven areas, where the same utterance has different meanings in different contexts. To model such variation, a system with a good definition of environment is needed. NEPs offer some kind of solution to approach formal semantics and formal pragmatics from a natural computing perspective.

Finally, the multimodal approach to communication, where not just production, but also gestures, vision and supra-segmental features of sounds have to be tackled, refers to a parallel way of processing. NEPs allow to work in the same goal, but in different tasks, to every one of the modules. The autonomy of every one of the processors and the possible miscoordination between them can also give account of several problems of speech.

4 Closing Remarks and Challenges

The progressive change in the models of computation implies also a change in the problems they can tackle, as well as in the models of cognition they suggest. The first Chomskian languages were based in rewriting, and caused the generalization of the rewriting and tree-like models for describing and explaining natural language. Later on, with the arising of contextual grammars and new typologies of languages not exactly corresponding to the Chomskian hierarchy, new perspectives were started in the formalization of natural language.

A further step in the development of formal languages was the idea of several devices collaborating for achieving a common goal. This new model, besides giving new horizons to formal linguistics, allowed, for the first time, the possibility of approaching with formal methods, not only syntax, but the whole process of language generation. With the interaction of several independent, but communicated modules, grammar systems favored the emergence both of cognitive modeling and language generation simulations in a consistent formal framework.

In the nineties, the emergence of biology as a pilot science had a direct impact in theoretical computer science by means of Molecular Computing, DNA Computing and Cell Computing. The computability models taken directly from nature, more precisely, from the behavior of DNA replication and communities of cells, provided a natural formalized model for explaining a natural communication system.

The natural axis in computation has developed in the last years in two different lines: a) the introduction in symbolic systems of the concept of net, and b) the improving of the notion of evolution that, being in the model from the beginning is gaining more importance.

The idea of net was implicit in grammar systems, but it failed in arising due to the lack of development of the initial network theory in symbolic systems. The concept of net allows a different type of interaction between the agents of systems, since them can interact not only with the environment, but among them. On the other hand, the relationship between the processors is more flexible, in a way that many components can be added to the system, and many situations can be designed for every one of the cooperating devices.

Evolution in formal language theory is synonym for "change by the time". The main aspects of evolution: random change by replication and natural selection, still await to be developed.

For the future, formal languages need to go deeper in the ideas of net and context, which are currently not so well developed as in other branches of computing. Swarm intelligence, for example, exploits the possibilities of interaction among agents and environment in a more optimal way. Therefore, the full development of networking theories and context theories in bioinspired formal languages, can help to tackle some parts of linguistics, like pragmatics, semantics or dialogue, which are crucial for artificial intelligence, and that do not have a suitable model for development so far.

It would be also interesting to be able to generate and accept languages including the idea of natural selection. The goal would be to formalize the rules of nature in a given environment for generating and erasing symbols in a system. This could help linguistics to approach the topics of language change, language death and language interaction.

Finally, it is a current challenge for biology, linguistics and computer science, to be able to formalize the theory of evolution in order to model language evolution. In this way, it would be possible to go from the nature to mathematics and computers and, from here, again to the nature, closing a circle that can help to understand and simulate our amazing linguistic capacity.

References

1. Adleman, L.M.: Molecular Computation of Solutions to Combinatorial Problems. Science 226, 1021–1024 (1994)
2. Bel-Enguix, G., Jiménez-López, M.D.: Byosyntax. An Overview. Fundamenta Informaticae 64, 1–12 (2005a)
3. Bel-Enguix, G., Jiménez-López, M.D.: Linguistic Membrane Systems and Applications. In: Ciobanu, G., Păun, G., Pérez Jiménez, M.J. (eds.) Applications of Membrane Computing, pp. 347–388. Springer, Berlin (2005b)
4. Bel-Enguix, G., Jimnez-López, M.D.: Analysing Sentences with Networks of Evolutionary Processors. In: Mira, J., Álvarez, J.R. (eds.) IWINAC 2005. LNCS, vol. 3562, pp. 102–111. Springer, Heidelberg (2005c)
5. Brendel, V., Busse, H.: Genome structure described by formal languages. Nucleic Acids Research 12(5), 2561–2568 (1984)

6. Castellanos, J., Leupold, P., Mitrana, V.: Descriptional and Computational Complexity Aspects of Hybrid Networks of Evolutionary Processors. Theoretical Computer Science 330(2), 205–220 (2005)
7. Castellanos, J., Martín-Vide, C., Mitrana, V., Sempere, J.M.: Networks of Evolutionary processors. Acta Informatica 39, 517–529 (2003)
8. Cavalli-Sforza: Cultural Transmission and Evolution. Princeton University Press, Princeton (1981)
9. Collado-Vides, J.: A transformation-grammar approach to the study of regulation of gene expression. Journal of Theoretical Biology 136, 403–425 (1989)
10. Collado-Vides, J., Gutirrez-Rios, R.-M., Bel-Enguix, G.: Networks on transcriptional regulation encoded in a grammatical model. BioSystems 47, 103–118 (1998)
11. Csuhaj-Varjú, E., Dassow, J., Kelemen, J., Păun, G.: Grammar Systems. Gordon and Breach, London (1994)
12. Csuhaj-Varjú, E., Martín-Vide, C., Mitrana, V.: Hybrid Networks of Evolutionary Processors are Computational Complete. Acta Informatica 41(4-5), 257–272 (2005)
13. Csuhaj-Varjú, E., Mitrana, V.: Evolutionary Systems: A Language Generating Device Inspired by Evolving Communities of Cells. Acta Informatica 36, 913–926 (2000)
14. Csuhaj-Varjú, E., Salomaa, A.: Networks of Parallel Language Processors. In: Păun, G., Salomaa, A. (eds.) New Trends in Formal Languages. LNCS, vol. 1218, pp. 299–318. Springer, Heidelberg (1997)
15. Darwin, C.: The origin of the species (1859)
16. Errico, L., Jesshope, C.: Towards a New Architecture for Symbolic Processing. In: Plander, I. (ed.) Artificial Intelligence and Information-Control Systems of Robots, vol. 94 (31-40). World Scientific Publisher, Singapore (1994)
17. Hillis, W.D.: The Connection Machine. MIT Press, Cambridge (1985)
18. Jacob, F., Monod, J.: Genetic repression, allosteric inhibition and cellular differentiation. In: Locke (ed.) Cytodifferentiation and Macromolecular Synthesis, pp. 30–64. Academic Press, Inc., New York (1963)
19. Jakobson, R.: Essais de Linguistique Générale. 2. Rapports Internes et Externes du Language. Les Éditions de Minuit, Paris (1973)
20. Ji, S.: Microsemiotics of DNA. Semiotica 138(1/4), 15–42 (2002)
21. Kirby, S.: Spontaneous evolution of linguistic structure: an iterated learning model of the emergence of regularity and irregularity. IEEE Transactions on Evolutionary Computation 5(2), 102–110 (2001)
22. López García, A.: Fundamentos genéticos del lenguaje, Madrid, Cétedra (2002)
23. Maynard Smith, J., Szathmáry, E.: The Major Transitions in Evolution. Oxford University Press, New York (1997)
24. Maynard Smith, J., Szathmáry, E.: The Origins of Life: From the Birth of Life to the Origin of Language. Oxford University Press, Oxford (1999)
25. Marcus, S.: Language at the Crossroad of Computation and Biology. In: Păun, G. (ed.) Computing with Bio-Molecules, pp. 1–35. Springer, Singapore (1998)
26. Margenstern, M., Mitrana, V., Pérez-Jiménez, M.: Accepting Hybrid Networks of Evolutionary Processors. In: Ferreti, C., Mauri, G., Zandron, C. (eds.) DNA 10. Preliminary Proceedings, pp. 107–117. University of Milano-Biccoca, Milan (2004)
27. Martín-Vide, C., Mitrana, V., Pérez-Jiménez, M., Sancho-Caparrini, F.: Hybrid Networks of Evolutionary Processors. In: Cantó-Paz, E., Foster, J.A., Deb, K., Davis, L., Roy, R., O'Reilly, U.-M., Beyer, H.-G., Kendall, G., Wilson, S.W.,

Harman, M., Wegener, J., Dasgupta, D., Potter, M.A., Schultz, A., Dowsland, K.A., Jonoska, N., Miller, J., Standish, R.K. (eds.) GECCO 2003. Part I, LNCS, vol. 2723, pp. 401–412. Springer, Heidelberg (2003)
28. Monod, J.: Le hasard et la ncessit. ditions du Seuil, Paris (1970)
29. Păun, G.: Computing with Membranes. Journal of Computer and System Sciences 61, 108–143 (2000)
30. Păun, G., Rozenberg, G., Salomaa, A.: DNA Computing. New Computing Paradigms. Springer, Berlin (1998)
31. Pawlak, Z.: Gramatyka i Matematika. Panstwowe Zakady Wydawnietw Szkolnych, Warzsawa (1965)
32. Sakakibara, Y., Brown, M., Underwood, R., Saira Mian, I., Haussler, D.: Stochastic Context-Free Grammars for Modeling RNA. In: Proceedings of the 27th Hawaii International Conference on System Sciences, pp. 283–284. IEEE Computer Society Press, Honolulu (1994)
33. Searls, D.: The Linguistics of DNA. American Scientist 80, 579–591 (1993)
34. Schleicher Die ersten Spaltungen des indogermanischen Urvolkes. Allgemeine Zeitung fuer Wissenschaft und Literatur (August 1853)
35. Uemura, Y., Hasegawa, A., Kobayashi, S., Yokomori, T.: Tree Adjoining Grammars for RNA Structure Prediction. Theoretical Computer Science 210(2), 277–303 (1999)
36. Victorri, B.: Analogy between language and biology: a functional approach. Cogn Process 8, 11–19 (2007)
37. Watson, J., Crick, F.: A structure for deoxyribose nucleic acid. Nature 171, 137 (1953)

A Christiansen Grammar for Universal Splicing Systems*

Marina de la Cruz Echeandía and Alfonso Ortega de la Puente

Universidad Autónoma de Madrid, Campus de Cantoblanco
Departamento de Ingeniería Informática, Escuela Politécnica Superior, España
{marina.cruz,alfonso.ortega}@uam.es

Abstract. The main goal of this work is to formally describe splicing systems. This is a necessary step to subsequently apply Christiansen Grammar Evolution (an evolutionary tool developed by the authors) for automatic designing of splicing systems. Their large number of variants suggests us a decisions: to select a family as simple as possible of splicing systems equivalent to Turing machines. This property ensures that the kind of systems our grammar can generate is able to solve any arbitrary problem. Some components of these universal splicing systems depend on other components. So, a formal representation able to handle context dependent constructions is needed. Our work uses Christiansen grammars to describe splicing systems.

1 Motivation

One of our main topics of interest is the formal specification of complex systems that makes it possible to apply formal tools to their design, or study some of their properties. Our research group has successfully applied this approach to other formal computational devices (L systems, cellular automata [8,9,3,4]) and proposed a new evolutionary automatic programming algorithm (Christiansen Grammar Evolution or CGE [10]) as a powerful tool to design complex systems to solve specific tasks.

CGE wholly describes the candidate solutions, with Christiansen grammars, both syntactically and semantically, and hence, improves the performance of other approaches, because CGE reduces the search space by excluding non-promising individuals with syntactic or semantic errors.

Splicing systems are abstract devices with a complex structure, because some of their components depends on others. This dependence makes it difficult to use genetic techniques to search splicing systems because, in this circumstance, genetic operators usually produce a great number of incorrect individuals (both syntactically and semantically).

This paper describes a first step to formally tackle the design of splicing systems. The next step will be to apply Christiansen grammar evolution to the design of splicing systems.

* This work was partially supported by DGUI CAM/UAM, project CCG08-UAM/TIC-4425.

J. Mira et al. (Eds.): IWINAC 2009, Part I, LNCS 5601, pp. 336–345, 2009.

There are a lot of variants in the way in which researchers define and use splicing systems. We are interested in the family of splicing systems equivalent to Turing machines. We are focused only on the formal description of these sytems. Our final goal is to test CGE for automatically designing of splicing systems to solve given tasks. We hope that our results could result in the proposal of a methodology to automatically design splicing systems.

2 Syntax and Semantics of High Level Programming Languages

Chomsky grammars [2] provide a powerful tool that allows the development of systematic algorithmic techniques for designing programming languages, compilers and interpreters, that is one of the main topics in Theoretical Computer Science.

The Chomsky hierarchy defines the expressive power of each grammar type: type 2 (or context free) and type 0 grammars, for example, are respectively related to pushdown automata and Turing machines. Type 0 languages can be found for any computable problem, while there is a proper subset of this set of problems that may be represented by type 2 languages, because pushdown automata cannot be considered universal computers. Chomsky type 0 grammars are very difficult to handle (design and parse). Therefore, in spite of their greater power, they have been little used in computer science to formally specify high level programming languages. Context free grammars are used instead.

This approach is the reason for the rather artificial distinction between the syntax and the semantics of high level programming languages: since programming languages must be able to describe algorithms for any problem that a digital computer can solve (that is, any computable problem) it seems clear that something must be added to context free grammars to keep the expressiveness of Chomsky type 0 grammars.

Informally *syntax* is associated with the context free constructos while *semantics* is related to those aspects that depends on the context. The mandatory declarations of variables before their use, or the proper correspondence of type and number in the arguments of a function, are examples of constructions that depend on the context.

Several approaches to efficiently express every computable problem by means of extended context-free grammars have been published since the sixties. They can be grouped according to their possession of the *adaptability property* [12]: (a grammar is said to be *adaptable* if it can be modified while it is being used). *Attribute grammars* [7] are the best-known non-adaptable grammars more complex than context free grammars. This paper uses *Christiansen grammars*, which are adaptable.

All these formalisms have the same expressive power, thus, which one to use is just a question of comfort.

3 Christiansen Grammars

Christiansen grammars [12,1] are an extension of attribute grammars where the first attribute associated to every symbol is a Christiansen grammar.

One of the most used notation is described in [12] which uses respectively the symbols $\downarrow$ and $\uparrow$ before the name of inherited and synthesized attributes. It is slightly more declarative than that used in attribute grammars. It could be summarized as follows:

1. Nonterminals are written in angled brackets and are followed by the parenthesized list of their attributes.
2. In the production rules, the names of the attributes are implicit. Their values are used instead. This syntax is similar to that of variables in logic programming (Prolog, for example), where the names actually used stand for constraints between the variables. Semantic actions follow their corresponding production rule in brackets, where { } stands for no semantic action.

We will use an example borrowed from [12] that describes a toy programming language in which is written the following code:

{int i;int j; i=j;}

These programs need two sections enclosed between brackets:

- The declaration of all the variables used in the second section. Only integer variables are allowed.
- The sequence of statements. Each statement ought to be an assignment between previously declared identifiers.

The following Christiansen grammar generates correct programs, that satisfy also the context-dependent feature about the mandatory declaration of variables before their use, without any auxiliary symbol table.

The set of non terminals is

$$\{program(\downarrow g), dec\_list(\downarrow g_i, \uparrow g_o), dcl(\downarrow g_i, \uparrow g_o), alpha\_list(\downarrow g_i, \uparrow word), alpha(\downarrow g_i, \uparrow word), stm\_list(\downarrow g), stm(\downarrow g), id(\downarrow g)\}$$

The set of terminals is

$$\{a, ..., z, int, "\{", "\}"\}$$

The production rules are listed bellow:

$$program(\downarrow g_0) \rightarrow "\{"\ decl\_list(\downarrow g_0, \uparrow g_1)\ stm\_list(\downarrow g_1)\ "\}"\ \{\ \}$$
$$decl\_list(\downarrow g, \uparrow g) \rightarrow \lambda\ \{\ \}$$
$$decl\_list(\downarrow g_0, \uparrow g_2) \rightarrow decl(\downarrow g_0, \uparrow g_1)\ decl\_list(\downarrow g_1, \uparrow g_2)\ "\}"\ \{\ \}$$
$$decl(\downarrow g, \uparrow g \cup \{id \downarrow g_{id} \rightarrow w\}) \rightarrow int\ alpha\_list(\downarrow g, \uparrow w)\ \{\ \}$$
$$alpha\_list(\downarrow g, \uparrow w) \rightarrow alpha(\downarrow g, \uparrow w)\ \{\ \}$$
$$alpha\_list(\downarrow g, \uparrow w_1 w_2) \rightarrow alpha(\downarrow g, \uparrow w_1)\ alpha\_list(\downarrow g, \uparrow w_2)\ \{\ \}$$
$$alpha(\downarrow g, \uparrow "a") \rightarrow "a"\ \{\ \}$$

$$\ldots$$
$$alpha(\downarrow g, \uparrow "z") \rightarrow "z" \;\{\;\}$$
$$stm\_list(\downarrow g) \rightarrow \lambda \;\{\;\}$$
$$stm\_list(\downarrow g) \rightarrow stm(\downarrow g)\; stm\_list(\downarrow g) \;\{\;\}$$
$$stm(\downarrow g) \rightarrow id(\downarrow g) \;\{\;\}$$

Notice that

- g_0 (the initial grammar for the list of declarations) has no rule for the nonterminal *id*.
- g_1 holds the changes made by the list of declarations and becomes the initial grammar for the list of statements.
- In this way, after executing the action $\uparrow g \cup \{id \downarrow g_{id} \rightarrow w\}$, g_1 contains one rule for nonterminal *id* with the name of each declared variable as its right hand side.
- The name of the identifier is syntethized as the value of the second attribute of *alpha_list*.

Figure 1 shows the semantically annotated parse tree for the program. It only shows the value of the grammar attributes. Continuous arrows are used for synthesized attributes. Changes in the grammars are highlighted by means of italic

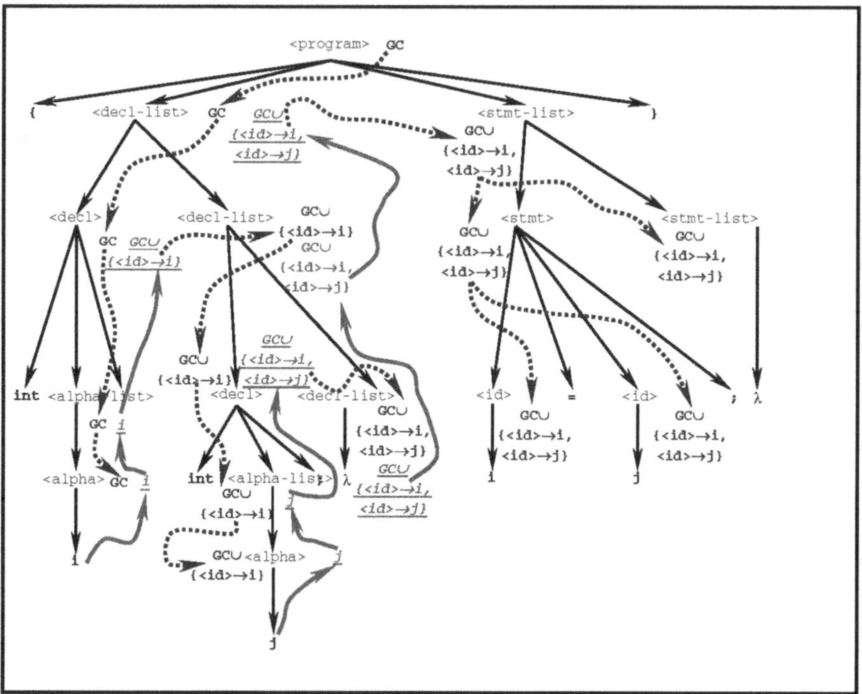

Fig. 1. Semantic tree for the toy program

and underlined fonts. The initial value of the grammar inherited by the axiom (GC) is the whole grammar itself.

For the shake of clarity, we will follow from this point on, the most popular attribute grammar notation, that is, we will explicitly write all the sentences needed to compute the value of all the attributes within the semantic actions of each rule. We will include in the name of the attributes the proper symbols $\downarrow$ and $\uparrow$. This corresponds with the first convention of the list previously described and taken from [12].

4 Splicing Systems

Splicing systems were introduced by Head in [6] as a DNA inspired computing device. Splicing systems formalize the DNA recombination operation.

A splicing rule on strings is formally defined by means of four patterns u_1, u_2, u_3 and u_4 and, when applied to two strings $x = x_1u_1u_2x_2$ and $y = y_1u_3u_4y_2$, it can produce two resulting strings $z = x_1u_1u_4y_2$ and $z' = y_1u_3u_2x_2$; although the second one (z') usually is discarded because the simmetric rule has the same result.

Formally they are represented as $u_1\#u_2\$u_3\#u_4$

A splicing system (H) consists of a set of splicing rules (R) that is applied to a set of strings (L), both sets share the same alphabet (V).

Extended splicing systems are one of the best studied variants of the basic model. They use a distinguished subset of the alphabet (T) named *set of terminals* with the same meaning than in Chomsky grammars.

Formally $H = (V, T, L, R)$

The *language generated by* a splicing system ($L(H)$) contains all the possible results of the splicing rules to the initial language (L) The languages generated by extended splicing systems contain strings of terminal symbols.

For example. It is easy to demonstrate that the following system generates strings with the structure $c_1(ab)^n(ab)^m...c_2$

$$H_0 = \{V_0, L_0, R_0\}$$

- $V_0 = \{a, b, c_1, c_2, c_3, c_4\}$
- $L_0 = \{c_1abc_2, c_3ac_4, c_4bc_3, c_1c_4, c_4c_2\}$
- $R_0 = \{c_3a\#c_4\$c_1\#a^+b^+c_2, c_3a^+b^+\#c_2\$c_4\#bc_3, c_1\#c_4\$c_3\#a^+b^+c_3,$ $c_1a^+b^+\#c_3\$c_4\#c_2, c_1a^+b^+\#\$c_1\#a^+b^+c_2\}$

The complexity of a splicing system can be studied in terms of the complexity of the initial set and the complexity of the rules in the Chomsky Hierarchy. It is known [11] that the extended splicing system with a finite initial language and with rules whose patterns are specified using regular expressions is equivalent to Turing machines.

5 Christiansen Grammar for Universal Splicing Systems

5.1 Informal Description

In the following paragraphs we will design a Christiansenn grammar able to generate the family of universal splicing systems described before.

Symbols in the alphabet (V) are the only permitted symbols in the set of initial strings (L) and in the patterns of the rules (R). This is the main dependence on the context in the formal definition of splicing systems. This semantic constrain will be solved in the same way in which the variables of the first example (the toy programming language) are declared before their use. We will use different non terminal symbols for elements in V (*string*) and in the rest of sets (*symbol*). Our initial grammar has only derivation rules for *string*. Once the alphabet is fully derived, the grammar contains the proper right hand sides to derive from *symbol* only the elements of V.

We will formally describe this grammar in the next subsection.

5.2 Formal Description

The set of non terminals is

$$\{H(\downarrow g, \uparrow g), V(\downarrow g, \uparrow g), T(\downarrow g, \uparrow g), L(\downarrow g, \uparrow g), R(\downarrow g, \uparrow g), string(\downarrow g, \uparrow s), E(\downarrow g, \uparrow g), symbol(\downarrow g)\}$$

Where g stands for a grammar and s for the string represented by the symbol *string*.

The set of terminals is

$$\{a, ..., z, A, ..., Z, 0, ..., 9, +, *, \lambda, \cup, (,), \#, \$\}$$

$H \rightarrow VTLR$
{
$V. \downarrow g = H. \downarrow g;$
$T. \downarrow g = V. \uparrow g;$
$L. \downarrow g = V. \uparrow g;$
$R. \downarrow g = V. \uparrow g;$
}
$V_1 \rightarrow string\ V_2$
{
$string. \downarrow g = V_1. \downarrow g;$
$V_2. \downarrow g = V_1. \downarrow g;$
$V_1. \uparrow g = V_2. \uparrow g \cup \{symbol \rightarrow string. \uparrow s\};$
}
$V \rightarrow string$
{
$string. \downarrow g = V_1. \downarrow g;$
$V. \uparrow g = V. \downarrow g \cup \{symbol \rightarrow string. \uparrow s\};$
}

$\forall x \in \{a, ..., z, A, ..., Z, 0, ..., 9\}$
$\quad string_1 \rightarrow x\ string_2\ \{\ string_1.\uparrow s = "x".string_2.\uparrow s;\ \}$
$\quad string \rightarrow x\ \{\ string.\uparrow s = "x";\ \}$
$T_1 \rightarrow symbol\ T_2 \mid symbol\ \{\ \}$
$L \rightarrow EL \mid L\ \{\ \}$
$R \rightarrow rule\ R \mid rule\ \{\ \}$
$rule \rightarrow E\#E\$E\#E\ \{\ \}$
$E \rightarrow \lambda \mid symbol \mid symbol\ E \mid (E) \mid E \cup E \mid E\ E \mid E^{+} \mid E^{*}\ \{\ \}$

We want to highlight the following comments:

– In the first rule the non terminals T, L and R inherit their first attribute (Christiansen grammar) from the grammar syntethized by the alphabet V. V has to add the proper right hand sides for *symbol*

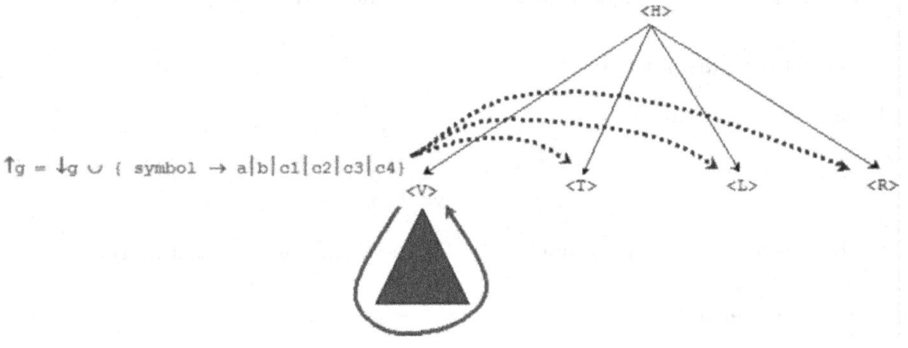

Fig. 2. First step to derive H_0

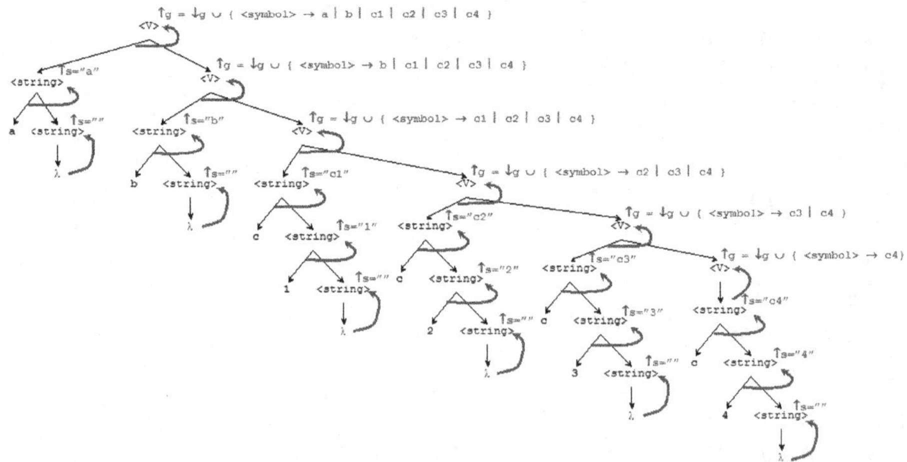

Fig. 3. Derivation of V_0

- Each production rule for V adds as a new derivation rule for the nonterminal *symbol*. The right hand side of this rule is the string derived by the subtree of the nonterminal *string*.
- Two sets represent the production rules for *string*. They are particularized with an universally quantified variable ($\forall\ x$)
- The semantic actions for the derivation rules of the non terminals T, L, R, E and *rule* are empty for space reasons. They just copy (inherit) the Christiansen grammar (first attribute) from the left hand side of the rule

5.3 Example

We will show how this grammar generates the system H_0 described in the section in which splicing systems were introduced.

Figure 2 shows the first level of the derivation tree. The arrows show how T, L and R inherits their first attributes from the second one of V.

Figure 3 describes in detail how V_0 is derived. The alphabet contains five elements. There is a subtree for each. Without loss of generality we can assume

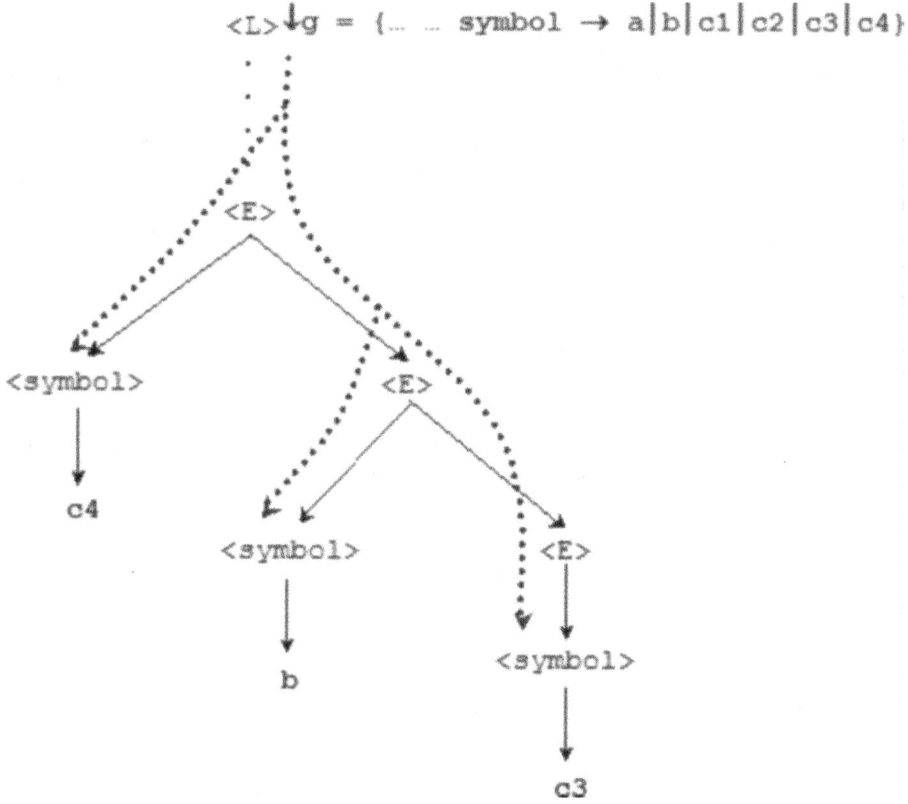

Fig. 4. Derivation of c_4bc_3

that deeper trees are processed first so the following rules are added to the grammar (in this same order)

1. $symbol \rightarrow c_4$
2. $symbol \rightarrow c_3$
3. $symbol \rightarrow c_2$
4. $symbol \rightarrow c_1$
5. $symbol \rightarrow b$
6. $symbol \rightarrow a$

At this point the grammar is complete because it also contains the proper derivation rules for *symbol*, thus, no more changes are needed. From this point on the non terminals T, L and R do not modify their initial grammar. They find the right choices for the non terminal *symbol* when they have to derive it. Figura 4 how the third string of the initial language (c_4bc_3) is derived.

The reader could easily see that the rest of components is generated in a similar way.

6 Conclusions and Future Research

This paper uses Christiansen grammars to formally describe an interesting family of splicing systems, those that are equivalent to Turing mahines. We have shown, in this way, that Christiansen grammar are a natural, elegant and easy to understand way of formally describe and represent universal splicing systems. In the future we plan to study the possibility of proposing a methodology to automatically design splicing systems to solve given problems by means of Christiansen Grammar Evolution (a evolutionary automatic programming algorithm proposed by the authors of this paper) In order to reach this goal we will need to take the following steps:

- To select a particular problem to be solved with splicing systems.
- To particularize the grammar described in this paper according to the particular family that could solve this problem.
- To develop several modules needed to incorporate splicing systems to our algorithm: the Christiansen grammar, a splicing system simulator to be included in the fitness function of the genetic engine.
- To design the set of experiments to find (design) the proper splicing system.

References

1. Christiansen, H.: A Survey of Adaptable Grammars. In: ACM SIGPLAN Notices, November 1990, vol. 25(11), pp. 35–44 (1990)
2. Chomsky, A.N.: Formal properties of grammars. In: Handbook of Math. Psych., vol. 2, pp. 323–418. John Wiley and Sons, New York (1963)
3. Abu, D.A.L., Ortega, A., Alfonseca, M.: Cellular Automata equivalent to PD0L Systems. In: International Arab Conference on Information Technology (ACIT 2003), Alexandria, Egipto. Pub: Proceedings, December 20-23, pp. 819–825 (2003)

4. Abu, D.A.L., Ortega, A., Alfonseca, M.: Cellular Automata equivalent to D0L Systems. In: 3rd WSEAS International Conference on Systems Theory and Scientific Computation, Special Session on Cellular Automata and Applications, Rodas, Grecia. Pub: Proceedings en CDROM, November 15-17 (2003)
5. Head, T.: Formal language theory and DNA: an analysis of the generative capacity of specific recombinant behaviors. Bull. Math. Biology 49, 737–759 (1987)
6. Head, T.: Splicing schemes and DNA. In: Rozenberg, G., Salomaa, A. (eds.) Lindenmayer Systems; Zmpacts on Theoretical Computer Science and Developmental Biology, pp. 371–383. Springer, Berlin (1992)
7. Knuth, D.E.: Semantics of Context-Free Languages. Mathematical Systems Theory 2(2), 127–145 (1968)
8. Ortega, A., Dalhoum, A.A., Alfonseca, M.: Grammatical evolution to design fractal curves with a given dimension. IBM Jr. of Res. and Dev. (SCI JCR 2.560) 47(4), 483–493 (2003)
9. Ortega, A., Abu, D.A.L., Alfonseca, M.: Cellular Automata equivalent to DIL Systems. In: 5th Middle East Symposium on Simulation and Modelling (MESM 2003), Eurosim, 5-7 Ene. 2004, Sharjah, Emiratos Arabes Unidos. Pub: Proceedings, pp. 120–124 (2003) ISBN: 90-77381-06-6
10. Ortega, A., Cruz Echeandía, M., Alfonseca, M.: Christiansen Grammar Evolution: grammatical evolution with semantic (in press)
11. Păun, G.: Regular extended H systems are computationally universal. Journal of Automata, Languages, Combinatorics 1(1), 27–36 (1996)
12. Shutt, J.N.: Recursive Adaptable Grammars. A thesis submitted to the Faculty of the Worcester Polytechnic Institute in partial fulfillment of the requirements for the degree of Master of Science in Computer Science, August 10 (1993) (emended December 16, 2003)

DNA Replication as a Model for Computational Linguistics

Veronica Dahl[1,2] and Erez Maharshak[1,2]

[1] Departament of Computing Science
Simon Fraser University
Burnaby, B.C., Canada
[2] Departament de Filologies Romaniques
Universidad Rovira i Virgili
43002 Tarragona, Spain

Abstract. We examine some common threads between biological sequence analysis and AI methods, and propose a model of human language processing inspired in biological sequence replication and nucleotide bindings. It can express and implement both analysis and synthesis in the same stroke, much as biological mechanisms can analyse a string plus synthesize it elsewhere, e.g. for repairing damaged DNA substrings.

Keywords: computational linguistics, molecular biology, constraint handling rules, analysis, synthesis, DNA replication, long distance dependencies.

1 Introduction

Biological sequence analysis is resorting more and more to AI methods, given the astounding rate at which such information grew over the last decade. Old methodologies for processing it can no longer keep up with this rate of growth.

In particular, applying to molecular biology AI methods such as logic programming and constraint reasoning constitutes a fascinating interdisciplinary field which, despite being relatively new, has already proved quite fertile. Some examples of applying logic programming techniques to molecular biology problems are the description and analysis of protein structure [9], protein secondary structure prediction [8], drug design [7] and predicting gene functions [6].

As well, methodologies that pertain to the natural language processing field of AI are now being exploited to analyze biological sequences, which is uncovering similarities between the languages of molecular biology and human languages.

On the other hand, the influence of computational molecular biology over natural language processing, has been much less studied.

In this article we identify some of the forms that tend to repeat in both human and molecular biology languages, and emphasize their uniform treatment through constraint reasoning, regardless of the area of application. We then propose a model of human language processing inspired in biological sequence replication and nucleotide bindings. We exemplify its uses around the language

J. Mira et al. (Eds.): IWINAC 2009, Part I, LNCS 5601, pp. 346–355, 2009.

processing phenomenon of long distance dependencies, which also presents in molecular biology since it involves relating two substrings (of either human or biological language text) which might be arbitrarily far apart from each other. Our proposed model is suitable to those language processing frameworks known as *compositional*, where the representations obtained for the whole are composed out of partial representations obtained for the parts.

2 Background

2.1 Some Forms That Are of Interest Both in Molecular Biology and in Natural Languages

DNA sequences are made up of four different compounds called nucleotides, or bases, each noted with one of the letters A (Adenine), C (Cytosine), G (Guanine) and T (thymine). From a language processing point of view, therefore, the lexicon of DNA has four "words": A, C, T and G. Within strands of these bases, biologists have found it useful to identify certain forms, some of which are also found in natural languages. Some of these forms are relatively simple yet not necessarily easy to parse: e.g. *palindromes* (sequences that read the same from left-to-right or from right-to-left, as the Spanish sentence, modulo blank spaces: "Dabale arroz a la zorra el abad", or as the sequence A C C T G G T C C A): their length can vary, and their position within in a string is unpredictable. *Tandem repeats* (where a substring repeats again right away) also appear in both types of languages, as "tut" does in "Tut, tut, it looks like rain", or as C G A within the sequence C C A T C G A C G A U A). David Searls proved that the grammar of nuclear acid language is in fact non-deterministic and ambiguous and moreover not context-free [10].

In addition to these basic structures, which can be found in linear sequences, pairings of nucleotides which attract each other form more complex structures, where the sequences fold into three dimensions: the nucleotide A tends to pair with T, and C with G. These are called *Watson-Crick* or *canonical base pairs.* These base pairings result in structures or motifs of a variety of forms, such as helix, hairpin loop, bulge loop and internal loop, etc (Figure 1).

One of the widely occurring complex structures in molecular biology is the *pseudoknot* which has been proved to play an important role for the functions of

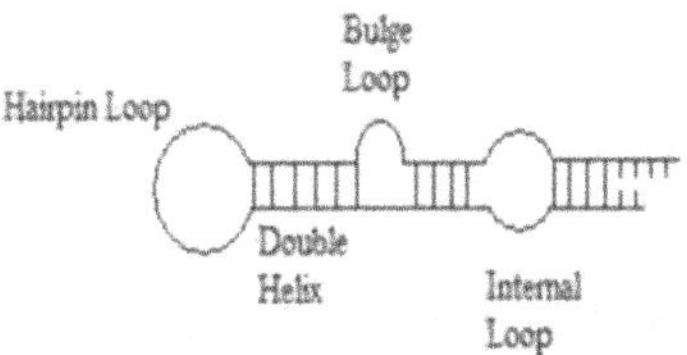

Fig. 1. Common motifs in RNA secondary structure are: hairpin loop, bulge loop, internal loop, etc

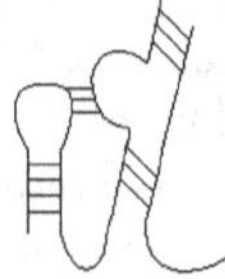

Fig. 2. A simple pseuodoknot

RNA. A simple pseudoknot is formed by pairing some of the bases in a hairpin loop that are supposed to stay unpaired, with bases outside the loop (Figure 2). If we draw for natural language sentences some of the links between for instance a clause's antecedent and the clause itself, we obtain similarly shaped figures.

Another phenomenon of interdisciplinary interest is that of string and information replication: to pass their genetic information on to new cells, cells must be able to replicate the DNA to be passed on to offspring, and to repair damaged DNA, they must be able to replicate the affected subsequence. It is also required that fragments of DNA (genes) be copied to code for particular bodily function. We will present a natural language processing mechanism that has been inspired by DNA replication and can be used to simulate it as well.

2.2 Constraint Handling Rules (CHR), Constraint Handling Rule Grammars (CHRG)

Constraint handling rules (CHR) provide a simple bottom-up framework which has been proved to be useful for algorithms dealing with constraints [5]. Because logic terms are used, grammars can be described in human-like terms and are powerfully extended through (hidden) logical inference. CHR rules have the form:

```
Head ==> Guard|Body
```

`Head` and `Body` are conjunctions of atoms and `Guard` is a test constructed from (Prolog) built-in or system-defined predicates. The variables in `Guard` and `Body` occur also in `Head`. If the `Guard` is the constant "true", then it is omitted together with the vertical bar. Its logical meaning is the formula $(Guard \rightarrow (Head \rightarrow Body))$ and the meaning of a program is given by conjunction.

CHR works on constraint stores with its rules interpreted as rewrite rules over such stores. A string to be analyzed such as *"leucine tryptophan phenylalanine"* is entered as a sequence of constraints

```
{token(0,1,leucine),
 token(1,2,tryptophan),
 token(2,3,phenylalanine)}
```

that comprise an initial constraint store. The integer arguments represent word boundaries. A grammar for this language can be expressed in CHR as follows.

```
token(X0,X1,tryptophan)==> codon(X0,X1,[u,u,g]).
token(X0,X1,leucine)==> codon(X0,X1,[u,u,a]).
```

```
token(X0,X1,leucine)==> codon(X0,X1,[u,u,c]).
token(X0,X1,phenylalanine)==> codon(X0,X1,[u,u,u]).
```

Ambiguity is inherently treated because all possibilities resulting from ambiguous input are expressed in the constraint store. In the above example, both a codon [u,u,a] and a codon [u,u,c] will be found between points 0 and 1.

CHR Grammars, or CHRGs for short, are based on Constraint Handling Rules [5] and were introduced in [4] as a bottom-up counterpart to definite clause grammars (DCGs) defined on top of CHR in exactly the same ways as DCGs take their semantics from and are implemented by a direct translation into Prolog. CHRGs are executed as CHR programs that provide robust parsing with an inherent treatment of ambiguity.

The input and output arguments of the above translation example can be spared (i.e., left implicit and invisibly managed) if using CHRG, which uses `::>` for the rewrite symbol. Hence, the first CHR rule above is equivalent to the CHRG rule:

```
token(tryptophan)::> codon([u,u,g])
```

We use the version of CHR embedded in SICStus Prolog, and notation with capital letters for variables, etc., is as in Prolog. Here we restrict ourselves to the subset of CHR consisting of propagation rules only. Other types of CHR rules are simplification rules which remove from the store the constraints used to match their left hand side; and simpagation rules, which remove the constraints past a symbol in their right hand side.

3 Parsing Biological or Human Language Sequences through Constraint Based Reasoning

3.1 Tandem Repeats

The following CHRG grammar determines whether a sequence is a tandem repeat or not. It can also identify a tandem repeat inside another sequence.

```
[X], string(Y) ::> string([X|Y]).
[X] ::>string([X]).

string(X),string(X)::>
tandem_repeat(X).
```

Sample parses and their results follow. Note that inner tandem repeats are also found, e.g. the second result for the first string shown. Note also that, although the nucleotide positions within the string has not explicitly been given, CHRG manages them invisibly and even outputs them, so that for instance it notes that the tandem repeat G T T A occurs between positions 1 and 9 of the second input string below.

```
?-  parse([a,c,c,g,t,a,c,c,g,t]).

    tandem_repeat(0,10, [a,c,c,g,t]);
    tandem_repeat(1,3,[c])

?- parse([c,g,t,t,a,g,t,t,a]).

    tandem_repeat(1,9,[g,t,t,a])

?- parse([a,g,c,t,c]).      % No tandem repeats are found

    no
```

Other than their appearances in molecular biology, tandem repeats show up as well in human language structures, both in the same form (as literal repetition of surface strings, as in "Tut, tut, it looks like rain"), or in more involved phenomena such as full conjunctive clauses, where the surface forms are not the same, but the structure repeats around some coordinating word like "and", "or", "but" (as in "Slowly but surely, ...", where the tandem repeat is between two adverbs, and a mediating conjunction intervenes).

3.2 Palindromic Structures – Heuristics

Simple palindromic structures can be treated similarly. These include hairpin loops, which occur when two regions of RNA, usually palindromic in nucleotide sequence, base-pair to form a double helix that ends in an unpaired loop. For instance, the sequence C C A A _ _ _ _ _ U U G G, where the underscores represent an arbitrary sequence (the loop), forms a hairpin loop when the subsequences u=C C and v=A A base-pair with their complements v'= UU and u'= G G. Thus a hairpin loop can be symbolized by u v loop v' u', with the primes denoting complementarity.

More complex structures, such as pseudoknots, cannot be treated as directly, because the base pairing is not well nested: base pairs occur that overlap one another in sequence position, e.g. the sequence C C A A G G U U is a pseudoknot, which, using the same names as above for the substrings, can be symbolized by u v u' v' - close to a palindrome, but with bindings that cross. These overlaps make pseudoknots impossible to predict either by the standard method of dynamic programming (which uses a recursive scoring system to identify paired nucleotides and in consequence cannot detect non-nested pairs) or by the newer method of stochastic context-free grammars.

In fact it has been shown that allowing generalized pseudoknots in problems such as RNA secondary structure prediction (i.e., determining which secondary structure will be adopted by a given sequence of nucleotides) makes it NP-hard. Using CHR we have developed a method for the inverse problem – that of RNA secondary structure design, consisting of finding a sequence which folds onto a given secondary structure – which can be revisited around the analysis plus synthesis metaphor we develop here (although at the time we were not conscious of the analogy), and which solves this problem and obtains approximate but still useful solution in O(n) time. It does however complement the CHR rules with heuristically obtained probability values, to reflect the fact that the number of GC pairs has an important role in stabilizing a certain structure, so we must inform CHR rules with a heuristic function representing the probabilities that are believed to govern the proportion of base pairs within RNA sequences, or we might end up with a sequence that may not actually fold into the input structure. The interested reader is referred to [2] for further details.

Similarly, some complex natural language structures present palindromic structures that often need heuristics. E.g., for completing implicit elements in coordinated structures we may adopt the heuristics that closer scoped coordinations will be attempted before larger scoped ones. Thus in Woods' example [11], "John drove his car through and demolished a window", the heuristic rule implies trying to conjoin two verb phrases before trying to conjoin two sentences.

4 A Biologically Inspired Model of Human Language Processing: Analysis Plus Synthesis

4.1 The Main Idea

Natural language processing schemes usually focus on a single processing mode: either analysis (i.e., producing meaning representation from syntactic form such as a sentence), or synthesis (generating syntactic form from meaning representation). There are very few exceptions to this view of language processing, unless the application targeted itself is one of translation. However even this task is usually not performed as simultaneous analysis and synthesis, but rather, it is tackled by analyzing a sentence in one language into an interlingua meaning representation of it, and then synthesizing, from this interlingua meaning obtained, the same sentence expressed in another language. The analyzing and synthesizing modules of a translator do not in general communicate or intermingle.

Biological machinery, in contrast, very often resorts simultaneously to the analysis of a string and the replication, or synthesis, of some substring of interest, as in the case of repairing damaged DNA. These copy-and-paste mechanisms also involve the binding of some nucleotides into pairs that attract each other.

We can view some natural language processing problems under the same metaphor, saying for instance that pronouns attract proper nouns that can serve to complete their meaning, and that this meaning can be obtained basically through replication of the noun into the position the pronoun occupies. Thus, the sentence "Leonardo portrayed Gioconda while she smiled", the pronoun "she" can be viewed in a way as attracting the proper name "Gioconda" into the position the pronoun occupies, where it will be replicated to make the sentence's meaning more explicit ("Leonardo portrayed Gioconda while she/Gioconda smiled").

Just as in molecular biology there will be conditions as to which pairs do get formed, since for instance the number of GC pairs has an important role in stabilizing a certain structure, in human language pairings we can place appropriate conditions to string pairings and replications, to ensure for instance that "she" will bind with "Gioconda" and not with "Leonardo". While pronoun reference is a complex problem for which no one-size-fits-all solution has yet been found, but an array of complementary yet partial solutions have been proposed, for our purposes here let's just exploit the fact that associating gender and number features to each word allows us in many cases to find a pronoun's referent on the basis of feature unification. For instance, by allowing only feminine and singular proper names to bind with feminine and singular pronouns we can discard Leonardo as a potential referent for "she" in the sentence given.

Replication of a proper noun into the position a pronoun occupies is easily achieved by having both atoms occupy the same substring, i.e., by noting both as starting and ending at the same word boundaries. The relevant grammar fragment is shown below. Note that we take advantage of the fact that grammar symbols actually compile into constraints with start and end points invisibly added by CHRG, which means that we can write CHR rules to use these constraints whenever we want to manipulate the word boundaries explicitly [1]. Thus rule marked as (1) below (a CHR rule, as indicated by its rewrite symbol) says that if there is a pronoun between positions Start and End, and a proper name Name of same gender (G) and number (N) has also be found, we can superimpose the noun phrase N at the same position occupied by the pronoun, i.e., between the points Start and End.

```
    token(gioconda) ::> name(gioconda,fem,sing).
    token(leonardo) ::> name(leonardo,mas,sing).
    token(she) ::> pronoun(she,fem,sing).
    ...

(1)       pronoun(Start,End,Pro,G,N), name(S1,E1,Name,G,N) ==>
      name(Start,End,Name,G,N).
```

This superimposing one more string (the proper name Name) at the position already occupied by the pronoun involves two cooperating processes: the *analysis* of the pronoun and of the noun phrase it refers to, followed by the *synthesis*, from both of them (which involves matching of the gender and number in both), of another copy of the name, at the position that the pronoun occupies.

We are of course simplifying for explanatory purposes: further conditions need in fact to be tested to ensure the proposed noun phrase is the one the pronoun refers to. But regardless of which tests one could add, our point is that the parsing process is now treated as a dual process, as in molecular biology: the left hand side of our CHR rule *analyses* two substrings of interest, as its right-hand side *synthesizes* a copy of one of them by superimposing it where its meaning is needed.

[1] There is a CHRG facility to make the word boundaries explicit within a CHRG rule, but we prefer here to show the last rule as a CHR rule for clarity.

4.2 Syntactic Form vs. Meaning Representation

In our previous example, we can consider the meaning of the proper noun for which the pronoun stands as represented by the constant "gioconda". Thus replicating the grammar symbol that contains it (namely, name(Start,End,Name, G,N)) into the space occupied in the string by the pronoun will have the desired effect of producing at this point the meaning of the pronoun, which in this case coincides with the internal representation Name of the proper name.

However, more complex examples might need distinguishing the surface form of a string that appears overtly at one site and implicitly at another, from the string's meaning. Most often, it is the meaning that we are really after, so when introducing meaning representations that are more complex we will have the choice between synthesizing both the surface string and its meaning representation, or just the latter. We next discuss these options through concrete, more complex examples which address an interesting phenomenon both for molecular biology and for human languages.

4.3 Long Distance Dependencies

Pronoun resolution is one of the simpler instances (while being, in its full generality, by no means simple) of a general problem affecting both computational molecular biology and computational linguistics: that of long distance dependencies, or in other words, the need to relate given substrings of (molecular or human) text with other substrings which might be arbitrarily far apart.

A more involved example for natural language than the simple pronoun reference one we just saw is where we must relate a relative pronoun with the relative clause's antecedent, but reconstruct its meaning for some constituent that is missing *at some other point*, rather than simply superimpose it with the pronoun's. For instance, in "This is the house that Jack built", "the house" is implicit at the position after "built", and its meaning representation must be identified as the meaning of the missing direct object of "built".

Here again we can apply our DNA inspired analysis-plus-synthesis model to advantage. The following rough grammar fragment exemplifies.

```
token(the) ::> det(the).
token(house) ::> noun(house).
token(that) ::> rel_pronoun(that).
token(jack) ::> name(jack).
token(built) ::> verb(built).

name(R) ::> noun_phrase(R).
```

```
(2)    noun_phrase(P0,P1,N), verb(P1,P2,V) ==> missing_noun_phrase(P2,P2).

(3)    det(P0,P1,D), noun(P1,P2,N),  rel_pronoun(P2,P3,_),
        missing_noun_phrase(P4,P4,[D|N]) ==>  P5=P4+1, P6=P5=1|
                                          det(P4,P5,D), noun(P5,P6,N).
```

The last two rules relate two long distance constituents through analysis plus synthesis: rule (2) synthesizes a new constituent, "missing noun phrase", to indicate a missing, or non-overt, noun phrase right after the verb (since it both starts and ends at P2, which is the end point of the verb); rule (3) synthesizes the implicit string at the point where it belongs (i.e., from point P4 onwards), after analysing it from its overt position right before the relative pronoun.

This first approximation does show our methodology in action, but is only concerned with syntactic form replication. More interesting is the case where we replicate the correct *meaning representation* at the point where the overt string that would give rise to it is missing. The following grammar fragment exemplifies this case. Note that we no longer need to synthesize the surface form of the missing string: we merely synthesize the meaning representation resulting from our analysis of a string, rather than the literal string itself, and fit it into the appropriate place in the overall meaning representation.

```
token(the) ::> det(the).
token(house) ::> noun(house).
token(that) ::> rel_pronoun(that).
token(jack) ::> name(jack).
token(built) ::> bi_transitive-verb(X,Y,built(X,Y)).

name(Meaning) ::> noun_phrase(Meaning).


(2')    noun_phrase(P0,P1,X), verb(P1,P2,X,Y,M) ==> missing_noun_phrase(P2,P2,Y).

(3')    det(P0,P1,_), noun(P1,P2,N),  rel_pronoun(P2,P3,_),
                missing_noun_phrase(P4,P4,N) ==> true.
```

Rule (2') now synthesizes a place-holder Y for the meaning representation of the missing noun phrase, which is replicated by the verb rule into its appropriate place inside the skeleton representation induced by the verb (in this case, built(X,Y)). Rule (3') disregards the determiner's meaning (a simplistic choice) and identifies the missing noun phrase's meaning with the noun itself (another simplistic choice, just for explanation purposes), so that the final representation obtained by parsing the relative clause will be "built(jack,house)".

Of course, we have left out for presentation and clarity purposes many small technical details, but this implies no loss of generality since they are easy, albeit tedious, to incorportate [2].

---

[2] For instance, rigorously speaking, these rules are an abuse of notation, since CHR does not allow the unification of variables inside constraints already in the constraint store, whereas our two CHR rules unify them freely. Other technicalities would need to be addressed as well in a full solution, e.g. rules (2) and (2') must be made to apply only when there is no overt noun phrase following the verb.

5 Concluding Remarks

We have identified some forms that are common to both molecular biology and human language sequences. We have shown that for the simpler of these forms a uniform treatment through constraint reasoning is adequate in both disciplines, whereas more complex forms might require the complement of heuristic rules.

We have also taken inspiration from the replicating mechanisms of nuclear acids- in particular, the simultaneous use of analysis and synthesis to propose a similarly simultaneous constraint based method of analysis and synthesis for natural language parsing, and we have shown its adequacy for modeling the important natural language processing feature of long distance dependencies.

References

1. Barranco-Mendoza, A.: Stochastic and Heuristic Modelling for Analysis of the Growth of Pre-Invasive Lesions and for a Multidisciplinary Approach to Early Cancer Diagnosis. Ph.D. Thesis, Simon Fraser University, Burnaby, BC (2005)
2. Bavarian, M., Dahl, V.: Constraint-Based Methods for Biological Sequence Analysis. Journal of Universal Computing Science 12(11), 1500–1520 (2006)
3. Barranco-Mendoza, A., Persaoud, D.R., Dahl, V.: A property-based model for lung cancer diagnosis. RECOMB poster, 27–31 (2004)
4. Christiansen, H.: CHR as Grammar Formalism, a First Report. In: Apt, K.R., Bartak, R., Monfroy, E., Rossi, F. (eds.) Sixth Annual Workshop of the ERCIM Working Group on Constraints, Prague (2001)
5. Fruhwirth, T.: Theory and Practice of Constraint Handling Rules. In: Stuckey, P., Marriot, K. (eds.) Journal of Logic Pro., Special Issue on Constraint Logic Programming 37(1-3), 95–138 (1998)
6. King, R.D.: Applying inductive logic programming to predicting gene function. AI Mag. 25(1), 57–68 (2004)
7. King, R.D., Muggleton, S., Lewis, R.A., Sternberg, M.J.E.: Drug design by machine learning. Proc. Natl. Acad. Sci. 89, 11322–11326 (1992)
8. Muggleton, S., King, R.D., Sternberg, M.J.E.: Protein secondary structure prediction using logic-based machine learning. Protein Eng. 5, 647–657 (1992)
9. Rawling, C.J., Taylor, W.R., Nyakairo, J., Fox, J., Sternberg, M.J.E.: Reasoning about protein topology using the logic programming language PROLOG. J. Mol. Bio. 3(4), 151–157 (1985)
10. Searls, D.B.: The computational linguistics of biological sequences. In: Artificial intelligence and molecular biology, pp. 47–120. American Association for Artificial Intelligence (1993)
11. Woods, W.A.: An experimental parsing system for transition network grammars. In: Rustin, R. (ed.) Natural Language Processing, pp. 145–149. Algorithmics Press, New York (1973)

jNEPView: A Graphical Trace Viewer for the Simulations of NEPs*

Emilio del Rosal[1] and Miguel Cuéllar[2]

[1] Escuela Politécnica Superior Universidad San Pablo C.E.U, Spain
emilio.rosalgarcia@ceu.es
[2] Universidad Autónoma de Madrid, Spain
miguel.cuellar@estudiante.uam.es

Abstract. jNEP, a Network of Evolutionary Processors (NEP) simulator, has been improved with several visualization facilities. jNEPView display the network topology in an friendly manner and shows the complete description of the simulation state in each step. Using this tool, it is easier to program and study NEPs, whose dynamic is quite complex, facilitating theoretical and practical advances on the NEP model.

1 Introduction

NEP stands for *Network of Evolutionary Processors*. NEPs are an abstract model of distributed/parallel symbolic processing presented in [3,4]. The model is inspired by biological cells. These are represented by nodes in a graph, each one containing words which describe the cell's DNA sequences. The nodes are simple string processors that are able to change their words in a predefined way. They also filter the words and communicate them to the other processors of the graph.

Despite the simplicity of each processor, the entire net can carry out very complex tasks efficiently. Many different works demonstrate the computational completeness of NEPs [10] and their ability to solve NP problems with linear or polynomial resources [2][4]. The emergence of such a computational power from very simple units acting in parallel is one of the main interests of NEPs.

Programing such devices for solving non-trivial problems is a quite difficult task. In the same manner, it is difficult to understand the NEPs dynamic and find possible bugs and other kind of mistakes. In order to overcome these difficulties, jNEP[1] [5] was developed. jNEP is a simulator of NEPs that permits to put in practice what was only theoretical discussion before it appeared. Besides, it provides a framework to exploit the parallel nature of NEPs to solve hard problems efficiently [1].

However, many improvements in terms of visualization and monitoring of the simulation can still be made on jNEP. jNEPView[2] fills this gap by displaying the topology of the NEP and its contents and offering the possibility of moving throughout the simulation at user request.

* This work was partially supported by CAM/UAM, project CCG08-UAM/TIC-4425.

[1] http://jnep.e-delrosal.net

[2] http://jnepview.e-delrosal.net

J. Mira et al. (Eds.): IWINAC 2009, Part I, LNCS 5601, pp. 356–365, 2009.

2 jNEP

A lot of research effort has been devoted to the definition of different families of NEPs and to the study of their formal properties, such as their computational completeness and their ability to solve NP problems with polynomial performance. However, no relevant effort, apart from [6], has tried to develop a NEP simulator or any kind of implementation. Unfortunately, the software described in this reference gives the possibility of using only one kind of rules and filters and, what is more important, violates two of the main principles of the model: 1) NEP's computation should not be deterministic and 2) evolutionary and communication steps should alternate strictly. Indeed, the software is focused in solving decision problems in a parallel way, rather than simulating the NEP model with all its details.

jNEP [5] tries to fill this gap in the literature. It is a program written in Java which is capable of simulating almost any NEP in the literature. In order to be a valuable tool for the scientific community, it has been developed under the following principles:

a) It rigorously complies with the formal definitions found in the literature.
b) It serves as a general tool, by allowing the use of the different NEP variants and is ready to adapt to future possible variants, as the research in the area advances.
c) It exploits as much as possible the inherent parallel/distributed nature of NEPs.

The jNEP code is freely available in http://jnep.e-delrosal.net.

2.1 jNEP Design

As shown in figure 1, the design of the NEP class mimics the NEP model definition. In jNEP, a NEP is composed of evolutionary processors and an underlying graph (attribute *edges*) to define the net topology and the allowed inter processor interactions. The *NEP* class coordinates the main dynamic of the computation and rules the processors (instances of the *EvolutionaryProcessor* class), forcing them to perform alternate evolutionary and communication steps. It also stops the computation when needed. The core of the model includes these two classes, together with the *Word* class, which handles the manipulation of words and their symbols.

We keep *jNEP* as general as possible by means of the following mechanisms: Java interfaces and the develop of different versions to widely exploit the parallelism available in the hardware platform.

jNEP offers three interfaces:

a) *StoppingCondition*, which provides the method *stop* to determine whether a *NEP* object should stop according to its state.
b) *Filter*, whose method *applyFilter* determines which objects of class *Word* can pass it.

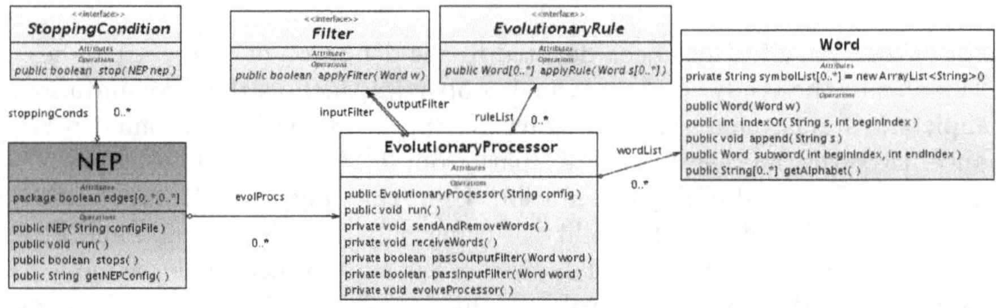

Fig. 1. Simplified class diagram of jNEP

c) *EvolutionaryRule*, which applies a *Rule* to a set of *Word*s to get a new set.

jNEP tries to implement a wide set of NEPs' features. The *jNEP user guide* (http://jnep.e-delrosal.net) contains the updated list of filters, evolutionary rules and stopping conditions implemented.

Currently *jNEP* has two list of choices to select the parallel/distributed mode on which it runs. For the first list, concurrency is implemented by means of two different Java approaches: *Thread*s and *Process*es. Concerning the second list, the supported platforms are standard JVM and clusters of computers (by means of JavaParty [7]).

Depending on the operating system, the Java Virtual Machine used and the concurrency option chosen, jNEP will work in a slightly different manner. The user should select the best combination for his needs.

2.2 jNEP Config Files

In this section, we want to focus on the configuration file which has to be written before running the program, since it has some complex aspects important to be aware of the potentials and possibilities of jNEP.

The configuration file is an XML file specifying all the features of the NEP. Its syntax is described below in BNF format, together with a few explanations. Since BNF grammars are not capable of expressing context-dependent aspects, context-dependent features are not described here. Most of them have been explained informally in the previous sections. Note that the traditional characters <> used to identify non-terminals in BNF have been replaced by [] to avoid confusion with the use of the <> characters in the XML format.

- [configFile] ::= <?xml version="1.0"?> <NEP nodes="[integer]"> [alphabetTag] [graphTag] [processorsTag] [stoppingConditionsTag] </NEP>
- [alphabetTag] ::= <ALPHABET symbols="[symbolList]"/>
- [graphTag] ::= <GRAPH> [edge] </GRAPH>
- [edge] ::= <EDGE vertex1="[integer]" vertex2="[integer]"/> [edge]
- [edge] ::= λ

- [processorsTag] ::= <EVOLUTIONARY_PROCESSORS> [nodeTag] </EVOLUTIONARY_PROCESSORS>

The above rules show the main structure of the NEP: the alphabet, the graph (specified through its vertices) and the processors. It is worth remembering that each processor is identified implicitly by its position in the processors tag (first one is number 0, second is number 1, and so on).

- [stoppingConditionsTag] ::= <STOPPING_CONDITION> [conditionTag] </STOPPING_CONDITION>
- [conditionTag] ::= <CONDITION type="MaximumStepsStoppingCondition" maximum="[integer]"/> [conditionTag]
- [conditionTag] ::= <CONDITION type="WordsDisappearStoppingCondition" words="[wordList]"/> [conditionTag]
- [conditionTag] ::= <CONDITION type="ConsecutiveConfigStoppingCondition"/> [conditionTag]
- [conditionTag] ::= <CONDITION type="NonEmptyNodeStoppingCondition" nodeID="[integer]"/> [conditionTag]
- [conditionTag] ::= λ

The syntax of the stopping conditions shows that a NEP can have several stopping conditions. The first one which is met causes the NEP to stop. The different types try to cover most of the stopping conditions used in the literature. If needed, more of them can be added to the system easily.

At this moment jNEP supports 4 stopping conditions, the *jNEP user guide* explains their semantics in detail:

1. **ConsecutiveConfigStoppingCondition:** It produces the NEP to stop if two consecutive configurations are found as communication and evolutionary steps are performed.
2. **MaximumStepsStoppingCondition:** It produces the NEP to stop after a maximum number of steps.
3. **WordsDisappearStoppingCondition:** It produces the NEP to stop if none of the words specified are in the NEP. It is useful for generative NEPs where the lack of non-terminals means that the computation have reached its goal.
4. **NonEmptyNodeStoppingCondition:** It produces the NEP to stop if one of the nodes is non-empty. Useful for NEPs with an output node.

- [nodeTag] ::= <NODE initCond="[wordList]" [auxWordList]> [evolutionaryRulesTag] [nodeFiltersTag] </NODE> [nodeTag]
- [nodeTag] ::= λ
- [auxWordList] ::= λ | auxiliaryWords="[wordList]"
- [evolutionaryRulesTag] ::= <EVOLUTIONARY_RULES> [ruleTag] </EVOLUTIONARY_RULES>
- [ruleTag] ::= <RULE ruleType="[ruleType]" actionType="[actionType]" symbol="[symbol]" newSymbol="[symbol]"/> [ruleTag]
- [ruleTag] ::= <RULE ruleType="splicing" wordX="[symbolList]" wordY="[symbolList]" wordU="[symbolList]" wordV="[symbolList]"/> [ruleTag]

- [ruleTag] ::= <RULE ruleType="splicingChoudhary" wordX="[symbolList]" wordY="[symbolList]" wordU="[symbolList]" wordV="[symbolList]"/> [ruleTag]
- [ruleTag] ::= λ
- [ruleType] ::= insertion | deletion | substitution
- [actionType] ::= LEFT | RIGHT | ANY
- [nodeFiltersTag] ::= [inputFilterTag] [outputFilterTag] | [inputFilterTag] | [output-FilterTag] | λ
- [inputFilterTag] ::= <INPUT [filterSpec]/>
- [outputFilterTag] ::= <OUTPUT [filterSpec]/>
- [filterSpec] ::= type=[filterType] permittingContext="[symbolList]" forbiddingCon-text="[symbolList]"
- [filterSpec] ::= type="SetMembershipFilter" wordSet="[wordList]"
- [filterSpec] ::= type="RegularLangMembershipFilter" regularExpression="[regExpression]"
- [filterType] ::= 1 | 2 | 3 | 4

Above, we describe the elements of the processors: their initial conditions, rules, and filters. jNEP treats rules with the same philosophy as in the case of stopping conditions, which means that our systems supports almost all kinds found in the literature at the moment and, more important, future types can also be added.

jNEP can work with any of the rules found in the original model [3,11,4]. Moreover, we support splicing rules, which are needed to simulate a derivation of the original model presented in [13] and [10]. The two splicing rule types are slightly different. It is important to note that if you use Manea's splicing rules, you may need to create an auxiliary word set for those processor with splicing rules.

With respect to filters, jNEP is prepared to simulate nodes with filters based on random context conditions. To be more specific, any of the four filter types traditionally used in the literature since [12]. Besides, jNEP is capable of creating filters based in membership conditions. A few works use them, for instance [3]. They are in some way non-standard and could be defined as follows:

1. **SetMembershipFilter:** It permits to pass only words that are included in a specific set.
2. **RegularLangMembershipFilter:** This filter contains a regular language to which words need to belong. The language have to be defined as a Java regular expression.

We will finish the explanation of the grammar for our xml files with the rules needed to describe some of the pending non-terminals. They are typical constructs for lists of words, list of symbols, boolean and integer data and regular expressions.

- [wordList] ::= [symbolList] [wordList]
- [wordList] ::= λ
- [symbolList] ::= `a string of symbols separated by the character '_'`
- [boolean] ::= true | false
- [integer] ::= `an integer number`
- [regExpression] ::= `a Java regular expression`

The reader may refer to the *jNEP user guide* for further detailed information.

2.3 jNEP Logging System

jNEP has been modified to produce a sequence of log files, one for each simulation step. This sequence of files will be read by jNEPView to show the successive configurations of the NEP. These logs are in a very simple format that contains a line for each processor in the same implicit order in which they appear in the configuration file. Each line contains the strings of the corresponding processor. This little extension of jNEP makes it simple to follow the trace of the simulation and manage it.

3 jNEPView Design

To handle and visualize graphs, we have used JGraphT [9] and JGraph [8] which are free Java libraries under the terms of the GNU Lesser General Public License.

JGraphT provides mathematical graph-theory objects and algorithms. It is used by jNEPView to represent formally the NEP underlying graph. Fortunately, JGraphT also allows to display its graphs using the JGraph library, which is graph visualization library with a lot of utilities.

We use those libraries to show the NEP topology. Once jNEPView is started, a window shows the NEP layout as clear as possible. We have decided to set the nodes in a circle, but the user can freely move each component. In this way, it is easier to interpret the NEP and study its dynamics.

Moreover, several action buttons have been placed to study the NEP state and progress. If the user clicks on a node, a window is open where the words of the node appear. In order to control the simulation development, the user can move throughout the simulation and the contents of the selected nodes are updated in their corresponding windows in a synchronize way.

Before running jNEPView, jNEP should have actually finished the simulation. In this way, jNEPView just reads the jNEP state logs and the user can jump from one simulation step to another fast, without worrying about the simulation execution times.

4 jNEPView Example

This section describes how jNEPView shows the execution of a NEP solving a particular case of the Hamiltonian path in an undirected graph. This NEP was already presented in [1] and its configuration file is delivered with the jNEP package.

Firstly, the user has to select the configuration file for jNEP which defines the NEP to simulate. After that, the layout of the NEP is shown like in figure 2.

At this point, the buttons placed in the main window to handle the simulation are activated and the user can select the nodes whose content wants to inspect during the simulation. Besides, the program allows the user to move throughout the simulation timeline by stepping forward and backward. Figures from 3 to 6 display the contents of all the nodes in the NEP in three different moments: the three first step and the final one. The user can also jump to a given simulation step clicking on the appropriate button.

5 Conclusions and Further Work

jNEPView provides a visual and friendly way to follow and study the dynamic of NEPs. During our experience with the NEP model, we have realized that the parallel nature of the model and the simplicity of its nodes cause that non-trivial problems are quite difficult to *translate* and interpret in terms of the model components. Our tool could facilitate the tasks of those working on NEPs.

As the community uses jNEPView, we pretend to improve the tool with the user advices. Besides, given the fast development of new NEPs variants, jNEPView should be prepared to change and adapt as the community advance.

References

1. del Rosal, E., Rojas, J.M., Núñez, R., Castañeda, C., Ortega, A.: On the solution of NP-complete problems by means of JNEP run on computers. In: Proceedings of International Conference on Agents and Artificial Intelligence (ICAART 2009), Porto, Portugal, January 19-21, 1999, pp. 605–612. INSTICC Press (2009)
2. Manea, F., Martin-Vide, C., Mitrana, V.: All np-problems can be solved in polynomial time by accepting networks of splicing processors of constant size. DNA Computing, 47–57 (2006)
3. Castellanos, J., Martin-Vide, C., Mitrana, V., Sempere, J.M.: Networks of evolutionary processors. Acta Informatica 39(6-7), 517–529 (2003)
4. Castellanos, J., Martin-Vide, C., Mitrana, V., Sempere, J.M.: Solving NP-Complete Problems With Networks of Evolutionary Processors. In: Mira, J., Prieto, A.G. (eds.) IWANN 2001. LNCS, vol. 2084, p. 621. Springer, Heidelberg (2001)
5. del Rosal, E., Nuñez, R., Castañeda, C., Ortega, A.: Simulating NEPs in a cluster with jNEP. International Journal of Computers, Communications & Control; Supplementary Issue: Proceedings of ICCCC 2008, vol. III, pp. 480–485 (2008)
6. Diaz, M.A., Gomez Blas, N., Santos Menendez, E., Gonzalo, R., Gisbert, F.: Networks of evolutionary processors (nep) as decision support systems. In: Fith International Conference on Information Research and Applications, vol. 1, pp. 192–203. ETHIA (2007)
7. `http://wwwipd.ira.uka.de/JavaParty/`
8. `http://www.jgraph.com/jgraph.html`
9. `http://jgrapht.sourceforge.net/`
10. Manea, F., Martin-Vide, C., Mitrana, V.: Accepting networks of splicing processors: Complexity results. Theoretical Computer Science 371(1-2), 72–82 (2007)
11. Martin-Vide, C., Mitrana, V., Perez-Jimenez, M.J., Sancho-Caparrini, F.: Hybrid networks of evolutionary processors. In: Cantú-Paz, E., Foster, J.A., Deb, K., Davis, L., Roy, R., O'Reilly, U.-M., Beyer, H.-G., Kendall, G., Wilson, S.W., Harman, M., Wegener, J., Dasgupta, D., Potter, M.A., Schultz, A., Dowsland, K.A., Jonoska, N., Miller, J., Standish, R.K. (eds.) GECCO 2003. LNCS, vol. 2723, pp. 401–412. Springer, Heidelberg (2003)
12. Martin-Vide, C., Mitrana, V.: Solving 3CNF-SAT and HPP in linear time using WWW. In: Margenstern, M. (ed.) MCU 2004. LNCS, vol. 3354, pp. 269–280. Springer, Heidelberg (2005)

13. Choudhary, A., Krithivasan, K.: Network of evolutionary processors with splicing rules. In: Mira, J., Álvarez, J.R. (eds.) IWINAC 2005. LNCS, vol. 3561, pp. 290–299. Springer, Heidelberg (2005)
14. Castellanos, J., Martin-Vide, C., Mitrana, V., Sempere, J.M.: Networks of evolutionary processors. Acta Informatica 39(6-7), 517–529 (2003)

A jNEPView Figures

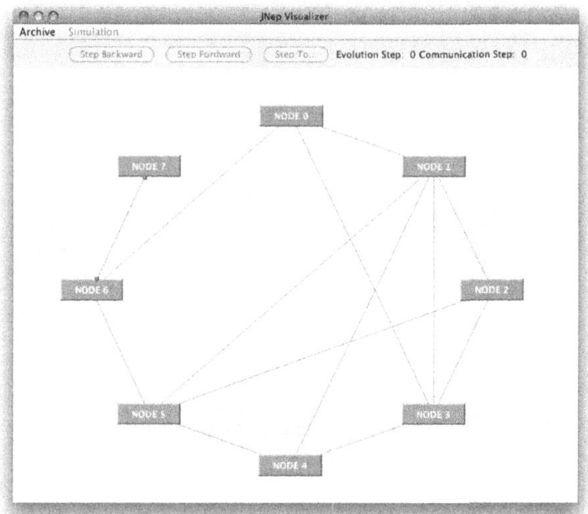

Fig. 2. Window that shows the layout of the simulated NEP

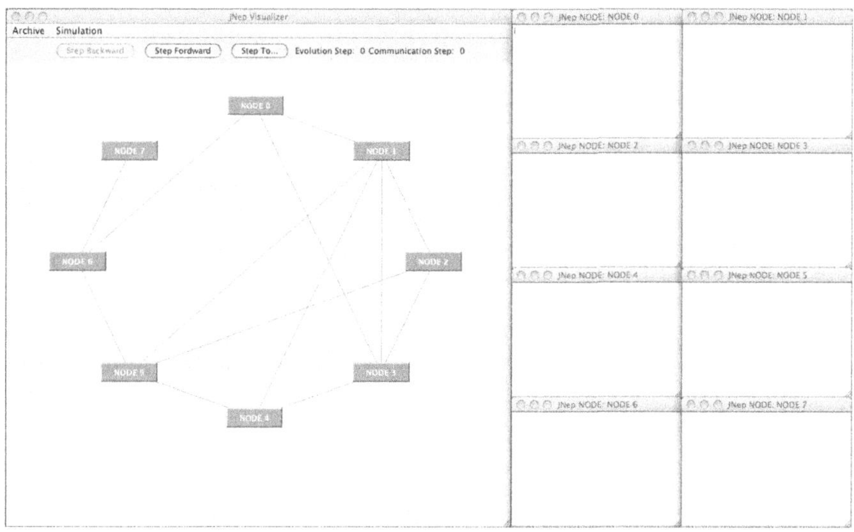

Fig. 3. Initial simulation step

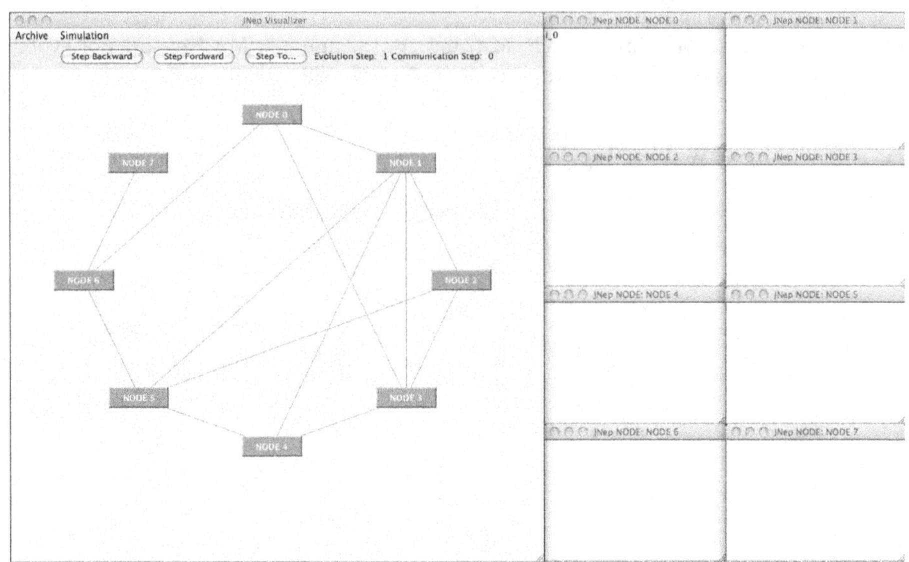

Fig. 4. Next simulation step

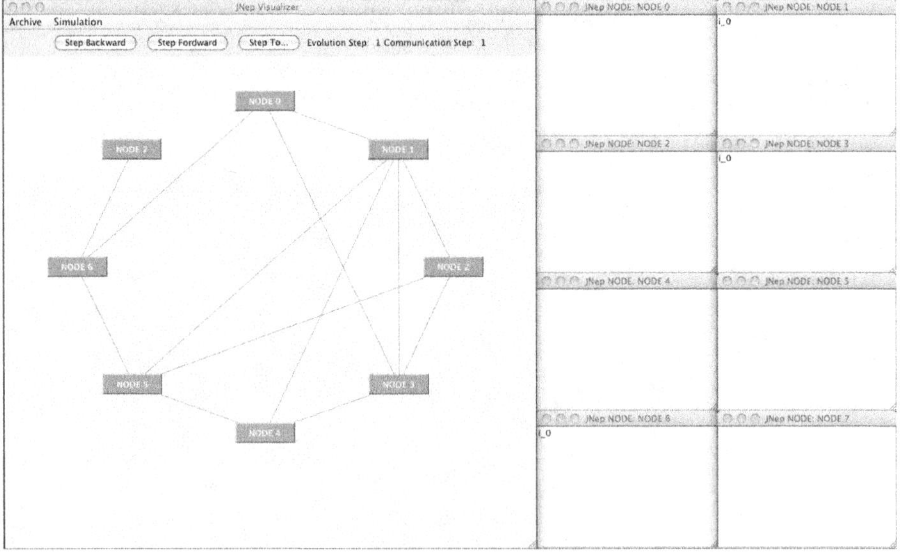

Fig. 5. Second simulation step

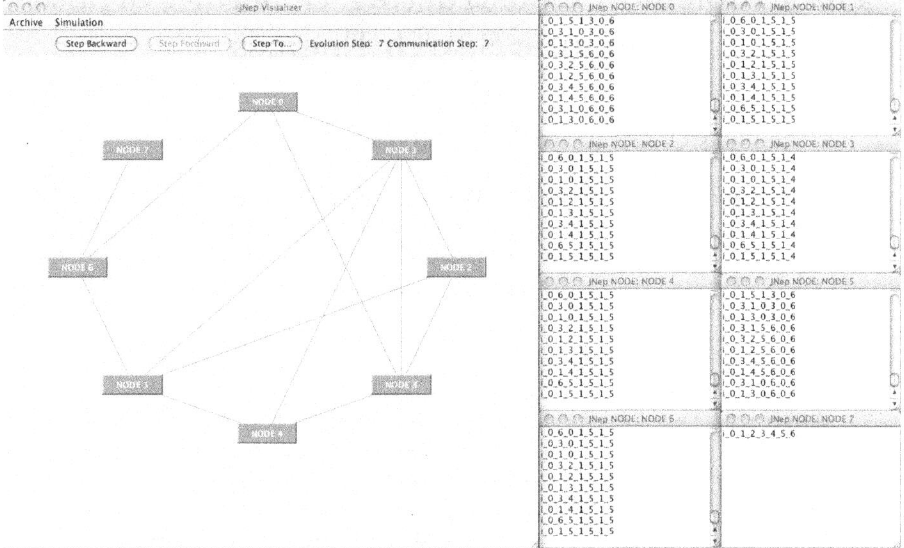

Fig. 6. End of simulation

The Problem of Constructing General-Purpose Semantic Search Engines

Luis Criado Fernández and Rafael Martínez-Tomás

Dpto. Inteligencia Artificial. Escuela Técnica Superior de Ingeniería Informática, Universidad Nacional de Educación a Distancia, Juan del Rosal 16, 28040 Madrid, Spain
rmtomas@dia.uned.es

Abstract. This work proposes the basic ideas to achieve a really semantic search. For this, first the number of semantic web sites must be increased which, on the one hand, maintain compatibility with the current Web and, on the other, offer different interpretations (now with semantics) of the same information according to different ontologies. Thus, the design of tools is proposed that facilitate this translation of HTML contents into OWL contents, as we say, possibly, according to different ontologies. The article continues by analysing the possible functionalities that we consider a semantic search engine based on the Semantic Web paradigm must support and by presenting a general-purpose search engine prototype on the two structures, current and semantic.

1 Introduction

There are currently specialist semantic search engines that operate effectively, like Saragossa City Council search engine (Naveganza) developed by iSOCO (2006). However, this search engine has been designed for particular cases. So, this solution [15] uses just one ontology, designed for processing specific information that Saragossa City Council uses and generates, and thus the search is better. Therefore, it is not a solution for general cases that use any kind of content. Naveganza incorporates Semantic Web techniques. In other words, semantic annotation for an ontology to represent Saragossa City Council content and then a user interface where natural language processing (NLP) is incorporated to improve querying to the search engine database (DB), where the representation is ontology based. Thus, solutions for intranets are feasible, since the information depends on the same organisation and can be semantically annotated according to specially designed ontologies. Another issue is to have general-purpose semantic search engines that are able to support any kind of ontology-based information. It is here where the real problem lies and it is the reason why the Semantic Web has still not been implemented. In 2007, several semantic search engines appeared and today, Hakia [7], Powerset [8] and True Knowledge [9] are

J. Mira et al. (Eds.): IWINAC 2009, Part I, LNCS 5601, pp. 366–374, 2009.

very well known. These semantic search engines try to obtain better results than Google and Yahoo. However, they are not significantly better than the traditional search engines and proof of this is that they have not yet been imposed on them. Probably, it is because these semantic search engines only incorporate, as a novelty, user interface NLP to improve DB querying, but in the DB the information is still represented like in any traditional search engine and the criterion is still to search for key words. Ten years ago Tim Berner-Lee (September 1998) published "Semantic Web Road Map" and "What the Semantic Web can represent" [1]. Two famous articles that presented the Semantic Web idea as an extension of the current Web with meaning. The Semantic Web is a space where information would have a well-defined meaning, so that it could be interpreted both by human agents and computerised agents. The first version of OWL appeared in 2004 [6] and SPARQL in 2006 [4]. As well as format specifications, we also have tools to help construct ontologies and development APIs. Thus the idea of constructing the Semantic Web dates back to 2006. Yet today it still does not exist. Why? What's the problem? We have to admit that the set objective is ambitious. We are saying that computational agents understand the Web content. Secondly, two fundamental problems have to be solved. The first is the Web volume. Currently, the Web is routine for many of the millions of Internet users that there are in the world. Its uses are diverse; its impact on the world economy is considerable. There are not only documents with text: there are images, videos, music, you can buy things, book travel tickets, a virtual world has been created that offers access to information, services and leisure. The volume is enormous and its growth is exponential. According to Netcraft [10], in February 2009 the number of web sites was 216.000.000. It is very difficult to estimate the size of the Web, since a web site has several web pages, but in 2001, Bergman did a study, where it was calculated that the volume of static web pages was 4*109 (equivalent to between 14 and 28 million books) [2]. Bearing in mind the number of web sites (according to Netcraft) and Bergman's work, it is deduced that approximately $35*10^6$ web sites are $4*10^9$ static web pages, so considering the same ratio, in February 2009 the estimated static web pages could be $24.6*10^9$. Yet this is just the surface web. It has been calculated that the deep web, the web that is generated dynamically from accessing the DB content, may have information several hundredfold the static web and it is growing at an even greater rate [5]. This massive development has made the process of accessing information an increasingly critical problem. Search engines worked very well during those years; however, the exceptional growth of the Web is beginning to show its first frustrating symptoms. Every day users have to devote more time to filter the web pages from a search, i.e., the quality of information is increasingly worse. As a strategy to solve the future situation, the World Wide Web Consortium (W3C) has proposed using ontologies and, in 2004, it developed the OWL-DL specifications, which will help advance towards the Semantic Web. In other words, implementing the Semantic Web is not an alternative, in fact, if it is not implemented, the moment will arrive when current technology will not be able to support its correct functioning. So, implementing the Semantic Web is a necessity to solve the problem that is already occurring. In this article we propose first (section 2), a way of extending the Semantic Web from editing the OWL contents, then (section 3), the functionalities that we consider a semantic search engine must support are described and we also show with a prototype how a true semantic search engine must operate. Finally, section 4 sets forth the conclusions of all this work and the lines of future work where we believe effort is necessary to achieve a true general-purpose Semantic Web.

2 Towards a True Semantic Web

Since the Semantic Web still does not exist, web sites do not incorporate OWL annotations. General-purpose semantic search engines cannot incorporate the functionalities inherent to the new Web paradigm, so they have to resort to user interface NLP as we indicated at the beginning of the article. In other words, the reason why the semantic search engines that have appeared do not obtain better results than Google and Yahoo, is fundamentally because they do not operate with semantic annotations. There are simply no semantic web sites.

We are going to specify the precise meaning of semantic web site. We understand by "semantic" that it has a meaning for computing systems. In theory, it could be said that a site that incorporates labelling to classify information could be considered a semantic web site. Under this definition those sites based on microformats could be considered [11], labels that attempt to enrich current HTML with semantics; in particular, it is interesting to mention the Dublin Core Metadata. This microformat tries to introduce semantic content, taking advantage of the characteristics of the attributes "cid" or "class" used by some HTML labels. However, HTML was devised to present information and not to structure it, so it is unlikely that HTML with semantics can be obtained with minor changes. We do not think it is a good solution to construct semantic search engines based solely on Dublin Core Metadata.

Another two examples in pursuit of the Semantic Web are RSS and FOAF. RSS [12] is an RDF XML-based vocabulary that classifies news and events so that it is possible to find the precise information adjusted to user preferences. FOAF [13] is a project that describes people, links between them and things that they do and believe. It is also an RDF vocabulary. However, both incorporate elementary semantics that current search engines already use and which is still a long way off the Semantic Concept proposed by Tim Berner-Lee in 1998. Thus, for the Web to be a Semantic one, more features are necessary. For this reason, the previous definition has to be completed with the concept of logic. We could say that a semantic web site is a site that as well as incorporating labelling to classify information, also represents this labelling in a language supported by descriptive logic, like OWL. By incorporating OWL DL into the labelling, it is possible to use DL reasoners, which differentiate TBOX from ABOX. TBox includes all the terminology, i.e., the vocabulary of an application domain according to:

- Concepts: they denote classes or a set of individuals.
- Roles: they denote binary relations between individuals.
- A set of complex descriptions on this vocabulary (restricted, of course, by the description language)

Whereas ABox contains affirmations on individuals named in vocabulary terms, a knowledge base based on TBox and ABox permits inferences from which "implicit" knowledge is derived from "explicit" knowledge in the knowledge base. If we accept the previous definition of the concept of semantic web site, we must now address the problem of how to transform a current web site (HTML) into a semantic web site (HTML alongside OWL DL).

First, the compatibility task has to be tackled; we cannot construct a semantic web site that cannot be indexed by current search engines. The aim is to improve on what there is, so it is essential to complement what already exists. Consequently, approaches that produce incompatibilities have to be avoided. Therefore, it seems sensible to attack the problem with external annotation, so that it does not affect the already existing annotation that has a well-defined function. External annotation has the advantage that it will only be used by applications designed to interpret it, thereby avoiding unwanted consequences of any kind for current applications like search engines.

We consider semantic annotation as the task of identifying and representing a concept in accordance with the structure defined in an ontology, for example, let us assume that on a web page's content we find the following sentence "The Boxer is an intelligent dog and easy to train", thus, the fact that it is possible to identify and represent "Boxer" as belonging to the class "dogs" means that it has been annotated semantically. Secondly, it has to be borne in mind that before doing semantic annotations it is necessary to check that the content of a

```
<rdf:RDF
      xmlns:j.0="http://www.criado.info/owl/vertebrados_es.owl#"
      xmlns:protege="http://protege.stanford.edu/plugins/owl/protege#"
      xmlns:rdf="http://www.w3.org/1999/02/22-rdf-syntax-ns#"
      xmlns:xsd="http://www.w3.org/2001/XMLSchema#"
      xmlns:rdfs="http://www.w3.org/2000/01/rdf-schema#"
      xmlns:owl="http://www.w3.org/2002/07/owl#"

xmlns="http://www.criado.info/owl/stand_boxer_3F972AA91EEE3451B64A67FCDA136ACB.owl#"

xml:base="http://www.criado.info/owl/stand_boxer_3F972AA91EEE3451B64A67FCDA136ACB.owl">
   <owl:Ontology rdf:about="">
     <owl:imports rdf:resource="http://www.criado.info/owl/vertebrados_es.owl#"/>
   </owl:Ontology>
    <j.0:perros rdf:ID="Boxer"/>
    <owl:AllDifferent>
      <owl:distinctMembers rdf:parseType="Collection">
        <perros rdf:about="#Boxer"/>
      </owl:distinctMembers>
    </owl:AllDifferent>
 </rdf:RDF>

<!-- Creado por [sw2sws]  http://www.luis.criado.org -->
```

Fig. 1. Example of semantic annotation

specific web page can be represented according to a specific ontology. Thus, if we try to transform a complete web site, using automatic or semi-automatic tools, an initial stage of "identification" must be defined, whose purpose is to ascertain the content topics for each web page, i.e., associating it with an ontology. So, for automatic processes, the identification process will consist of determining what ontologies can be related to a web page's content. Of course, everything that has been explained for a web page has to be generalised for all the web pages on the web site. The identification process can be relatively quick when we work with a small number of ontologies, because, in fact, it does not require NLP, it compares the ontology identifiers with the web page's content. However, these identifiers could in theory be in a different language to the web site content that is to be transformed, so a prior stage is also necessary for extracting all the ontology identifiers and translating them into the web site's content language. A tool that aims to be commercial must try to identify a web site's content with the greatest number of ontologies possible and this is the main difficulty, since contrasting the web site content with a greater number of ontologies implies raising the computational cost and, consequently, the processing time. Therefore, it is interesting to use a repository of ontologies in this stage, which provide access to the direct use of identifiers in different languages. The result of this stage is to provide information for each web page indicating, whenever possible, whether the content is associated with one or several ontologies. Thirdly, assuming that the previous stage is successful and the content of each of the web pages has been associated with an ontology or a set of ontologies, then the moment has probably come to "extract and analyse" the natural written language and classify the terms and represent it formally. To do this NLP techniques must be used, since a morphosyntactic analysis is necessary. At this point, we can do the semantic annotation. This last process (interpretation) is based, on the one hand, on the formal representation of the language in the previous stage and, on the other, on one or several ontologies associated with the content, and the end user can participate and help construct the Semantic Web generating annotations in a suitable language. The annotation that we propose is based on user cooperation and is done automatically or semi-automatically in OWL DL, because it is the standard language for describing semantics on the Web [6] and it allows inferences inherent to the descriptive logic SROID(D) supporting it.

3 Towards a True Semantic Search

As we mentioned at the beginning of this article Hakia, Powerset and True Knowledge try to offer better links to the user compared with traditional search engines, since they use NLP techniques, which in theory improve searches in their databases. However, these databases do not incorporate semantic annotations for web sites, we know this because there are no semantic web sites based on OWL-DL annotations. We consider that a true semantic search engine should incorporate the following functionalities:

1. Lexicographic analysis: with this the search engine can extract all the elements from the user's question so that when the question is analysed, the search engine will try to optimise its "comprehension". Ideally, the lexicographic analyser will send all the extracted information to an NLP layer. This is the technique that search engines like Hakia, Powerset and True Knowledge use.
2. Context selection: this layer allows the semantic search engine to decide with what ontology or sets of ontologies to work.
3. OWL-DL Inferential Motor: it searches for a response, in accordance with an appropriate ontology. The response can be the result of an inference based on descriptive logic SROID(D).
4. The capacity to respond: a semantic search engine must be able to respond, offer a direct response. In other words, to the question, like for example, "Which is more dangerous, a poodle or a Doberman?" A semantic search engine should respond that the Doberman is more dangerous and then offer links in case the user wishes to contrast the conclusion. The advantage compared with just offering links is that the user does not waste time discovering the response; the semantic search engine can offer any information directly.

This capacity to respond can be constructed when semantic web sites exist. That is why we said at the beginning of this article that it is difficult for the semantic search engines that have appeared to obtain better results than Google and Yahoo, since they have nowhere to search. There are no OWL-DL annotations linked to web sites and lexicographic analysis is not enough to construct a true semantic search engine. Nevertheless, in order to be able to construct search engines with these functionalities semantic annotations have to be generated automatically. The way to do this [3] could be to follow the stages presented in this article consecutively and then group them into three stages; identification, extraction and interpretation. Identification determines the ontology or ontologies that are associated with the same web page. This selection of ontologies is fundamental so that in the next stage, which is responsible for extraction and analysis, the web page information is obtained to represent it formally; this stage processes the text morphologically and syntactically. Finally, the last stage that we have called interpretation is responsible for semantic annotation based on OWL DL. To perform this transformation or migration, a prototype tool was implemented (sw2sws) that automated the three stages that we have presented according to a few example ontologies. The annotation obtained depends on several factors; such as the very onotology compared with what is to be annotated, the content quality and the capacity to extract and analyse, conditioned by NLP. After executing "sw2sws" semantic web sites were obtained, this tool is devised for use by any actor participating in contents, like webmasters. Although the prototype only operates with static information and a few ontologies, the aim is that these kinds of tools support a large number of ontologies and use both static and dynamic information. Thus, user collaboration could lead to web sites with OWL-DL annotations appearing. Once we had obtained semantic web sites

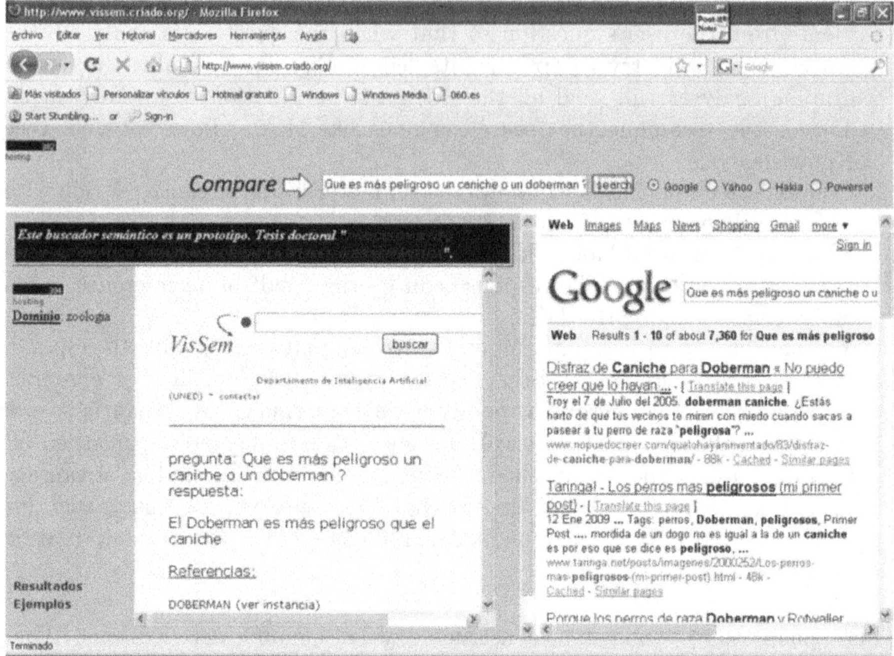

Fig. 2. Vissem and Google compared

generated with sw2sws, we were able to implement our search engine Vissem [14]. Vissem is a real semantic search engine although limited basically by NLP simplicity and the low number of ontologies with which it works, but it allows questions or searches with promising results that are better than those of other search engines. For example, it can answer questions correctly like "Which is more dangerous, a poodle or a Doberman?"

The comparison (see Figure 2) was drawn between our prototype and Google, since current semantic search engines do not support Spanish. Note how Vissem responds and also offers links so that the user can contrast the conclusion. The prototype provides two links; one for people (the web page, the other, the link "(see example)" which is what Vissem understands on the web page. Note too how Google proposes links not very related to the question like "Poodle Fancy Dress for Doberman" and "Taringa! – The most dangerous dogs". Therefore, users have to access and read to obtain the response that they are searching for.

4 Conclusions

In this work we have tried to highlight the basic ideas to advance towards a really semantic search engine. First, for this, semantic web sites have to be developed which, on the one hand, have compatibility with the current Web and, on the other, have flexibility to obtain different interpretations according to different ontologies. The procedure is general purpose, although in our prototype and examples, for simplification, we have tested it in static content according to a few ontologies and the web sites that we have transformed have been on vertebrates in order to obtain sufficient instances for Vissem. It should also be explained that the web sites used were downloaded from the Internet and were not altered, although they were chosen for their simple language, since the tool is clearly restricted by the NLP module and access to existing ontologies.

As regards future works, there is still much to be done, but perhaps the most pressing problem is the automatic selection of contexts for semantic applications and investigating strategies for the massive processing of ontologies in the identification process which enable semantic annotation tools to be developed, such as sw2sws, which support a large number of ontologies. Of course, the semantic Web breakthrough will occur when we can annotate the dynamic content semantically.

References

1. Berners-Leel, T.: Semantic Web Road map (1998), Disponible en Web: http://www.w3.org/DesignIssues/Semantic.html
2. Bergman, M.K.: Obra original: The Deep Web: Surfacing Hidden Value. Journal of Electronic Publishing 7(1) (2001), http://quod.lib.umich.edu/cgi/t/text/text-idx?c=jep;view=text;rgn=main;idno=3336451.0007.104, http://hdl.handle.net/2027/spo.3336451.0007.104
3. Criado Fernández, L.: Tesis doctoral, Procedimiento semi-automático para transformar la Web en Web Semántica; Rafael Martinez Tomás (director). Universidad Nacional de Educación a Distancia. Escuela Técnica Superior de Ingeniería Informática, Madrid (2009)
4. Prud'hommeaux, E., Seaborne, A.: SPARQL Query Language for RDF (2006), Disponible en Web: http://www.w3.org/TR/rdf-sparql-query/
5. O'Neill, E.T., Lavoie, B.F., Bennett, R.: Obra original: Trends in the Evolution of the Public Web. Web: D-Lib Magazine 9(4) (2003) ISSN 1082-9873, Disponible en Web: http://www.dlib.org/dlib/april03/lavoie/04lavoie.html
6. Smith, M.K., Welty, C., McGuinness, D.L.: OWL Web Ontology Language Guide, online: W3C (2004), Disponible en Web: http://www.w3.org/TR/owl-guide/
7. Hakia Semantic Search Engine, http://www.hakia.com/
8. Powerset Semantic Search Engine, http://www.powerset.com/
9. True Knowledge Semantic Search Engine, http://www.trueknowledge.com/
10. Netcraft, http://news.netcraft.com/
11. Webposible: Microformatos Dublin Core, http://www.webposible.com/microformatos-dublincore/
12. RSS 2.0 Specification, http://blogs.law.harvard.edu/tech/rss, http://www.w3c.es/Divulgacion/Guiasbreves/WebSemantica

13. FOAF, http://www.foaf-project.org/,
http://f14web.com.ar/inkel/2003/01/27/foaf.html,
http://www.w3c.es/Divulgacion/Guiasbreves/WebSemantica
14. Vissem search engine prototype, http://www.vissem.criado.org/
15. Information on ISOCO search engine,
http://www.isoco.com/buscador_tramites.htm

Computational Agents to Model Knowledge - Theory, and Practice in Visual Surveillance

José Mira[1,⋆], Ana E. Delgado[1], Antonio Fernández-Caballero[2], José M. Gascueña[2], and María T. López[2]

[1] Departamento de Inteligencia Artificial, E.T.S.I. Informática, Universidad Nacional de Educación a Distancia, 28040-Madrid, Spain
adelgado@dia.uned.es

[2] Universidad de Castilla-La Mancha, Departamento de Sistemas Informáticos & Instituto de Investigación en Informática de Albacete, 02071-Albacete, Spain
{caballer,jmanuel,mlopez}@dsi.uclm.es

Abstract. In this work the concept of computational agent is located within the methodological framework of levels and domains of description of a calculus in the context of different usual paradigms in Artificial Intelligence (symbolic, situated, connectionist, and hybrid). Emphasis in the computable aspects of agent theory is put, leaving open the possibility to the incorporation of other aspects that are still pure cognitive nomenclature without any computational counterpart of equivalent semantic richness. These ideas are currently being implemented on semi-automatic video-surveillance.

Keywords: Computational Agents, Knowledge Engineering, Artificial Intelligence, Visual surveillance.

1 Introduction

The concept of agent comes from the persistent attempt in Science and Engineering to modularize the knowledge necessary to specify a calculus, and of the later attempt to progressively increase the level of complexity and autonomy, making them re-usable. Agent theory takes from Artificial Intelligence (AI) the general intention to approach the functionality of biological systems. Thus, there are adaptive, intelligent, intentional agents, with learning capacity and equipped with a certain level of social organization that allows cooperation in the accomplishment of "group tasks". In this sense, the objectives of multi-agent systems (MAS) agree with distributed AI (DAI) [27]. "Nature-inspired computation" [18] is also practically isomorphic to agent and MAS theory. This is true at the level of individual agents ("organisms") as well as at the level of social organizations in MAS (ants, bees, human societies, collective games, etc.). In this last case, to the functionalities demanded for the individual behavior, it is necessary to

⋆ This article is dedicated to the memory of Professor José Mira, a great researcher, a wise man, a loving husband, and a close friend; but who sadly passed away.

J. Mira et al. (Eds.): IWINAC 2009, Part I, LNCS 5601, pp. 375–385, 2009.

add the use of an interaction language among agents that allows to share goals and to coordinate the collective plans used to reach those goals. But, usually the specification and the modeling of the environment is forgotten. However, its richness, diversity and other characteristics are fundamental to understand the dynamics of the agent-system interaction.

In this work we approach the general concept of agent from a computational perspective. We consider that an agent starts being a conceptual model, later it is reduced to a formal model and finally to a physical machine with sensors, effectors and a control program. The agent concept is located within the methodological framework of description levels and domains of a calculus. Independently of the cognitive characteristics being assigned to an agent (intentions, purposes, beliefs, desires, emotions, or conscience), to our opinion an agent is a "computational agent", which performs a calculation in a physical machine having an interface (human or electromechanical) with an external environment.

2 Computational Agents in Perspective

Historically, Cybernetics introduced the concept of system, clearly analogous to the agent one. Physics and Engineering continue preferring to use the system concept, to which less cognitive characteristics are assigned. On the other hand, it is much more precise and operational, since its dynamics can be described in terms of differential equations - usual in Physics and Control Theory - and of combinatory and sequential logic - proper of computation and integrated in finite state automata (FSA) theory. In the end, when reformulating any definition in natural language of an agent in a computable form, it is also ended up using algorithms and automata. The initial cybernetical proposal consisting in studying the representation, calculation, communication and control systems, can now be rewritten in terms of agents to obtain a first general agent classification: representation, calculation, communication and control agents. Each one of these functions is characterized to have a clear objective (a goal), and to implicitly take the distinction between information (message) and energy (signal). Wiener [28] established the distinction between signal and information, and associated "purposes" to goal states, or "consigns" in feedback loops. This way, the concepts described in cognitive language are anchored in Physics, Engineering and Mathematics. McCulloch [13] and von Neumann [26] constructed the modular automata theory and applied it to the formal description of neuronal networks and the synthesis of computers. Additionally, von Neumann addressed self-reproduction, self-reparation and tolerance to failure problems, which are nowadays also associated to agents (e.g. [2],[6],[7]).

With the arrival of AI, the initial formulations of connectionism (artificial neuronal networks and modular theory of deterministic and probabilistic automata) are partially obscured by new symbolic formulations based in rules. Here the inference is understood as a search process in a states space. Thus, the idea of "actors" appears, like a concurrent computation model in distributed systems [1]. Along with object-oriented programming [15] these are the antecedents of the agents. Conceptually, it is Minsky [16] who raises the social idea of agency, essentially basing on "personification" of the verbs used in natural language to describe the necessary processes for the execution of a certain activity ("build", "see", "grasp", "move", "release"). The strategy to describe in natural language an agent's "beliefs", "desires" and "intentions" and later to develop the formal counterpart of the linguistic terms, is still used at the present time [23]. In fact, a complete agent language is dominant in Software Engineering (SE) [19], and its importance also grows in Knowledge Engineering (KE) [3]. It is advisable to indicate that, as with the term of AI, in the agents field there is usually an abuse of excessively loaded cognitive nomenclature of anthropomorphous semantics. Finally, during implementation, there is no remedy than to reduce to algorithms and automata, and to entities and relations of the existing programming languages.

3 Levels and Domains in Agent Models

Since the introduction of the knowledge level by Newell [17] and Marr [10] - called "theory of calculus" -, it is usual to describe the knowledge necessary to understand any calculation in three levels: physical (PL), symbols (SL) and knowledge (KL). Or, in a simpler way: machine hardware, programs and models, and algorithms. Starting from the idea of reference systems in Physics and the proposals by Maturana [12] and Varela [25] in Biology, in 1978 [14] the figure of the external observer in computation is introduced. This gives rise to two description domains of the organizations and relations at each level: (1) the own domain of the level (OD), where the causality is intrinsic and the semantics comes imposed by the structure and dynamics of the level, and, (2) the external observer domain (EOD) to the computation carried out in that level, where the semantics is arbitrary and the interpretation of the calculation depends of the observer and, in general, of the application domain.

When superposing both domains (OD, EOD), at the three levels ((KL, SL, PL), we obtain a building of three plants and six apartments (two by plant), in which the agents reside (see Fig. 1). Thus, we have agents (OD-PL), (EOD-PL), (EOD-SL), etc. That is to say, the knowledge necessary to specify an agent may be decomposed into six elements: models (EOD, OD), programs (EOD, OD) and machines EOD, OD). When one or more of these components is the relevant one (assuming the existence of the others), this one gives the generic name to the agent. Thus, for example, robots are essentially physical agents because it is assumed that the greater complexity is associated to the design of its sensors and effectors, independently of their navigation programs. In an analogous manner, a great part of agents usual in IS is symbolic because it is assumed that most of its complexity is associated to the construction of a program whose interface is human (through screen and keyboard) and we do not need to worry about the design of the body of this type of agents.

Finally, the majority of the current most ambitious proposals are conceptual agents, at level of knowledge, and, more exactly, knowledge agents in the external

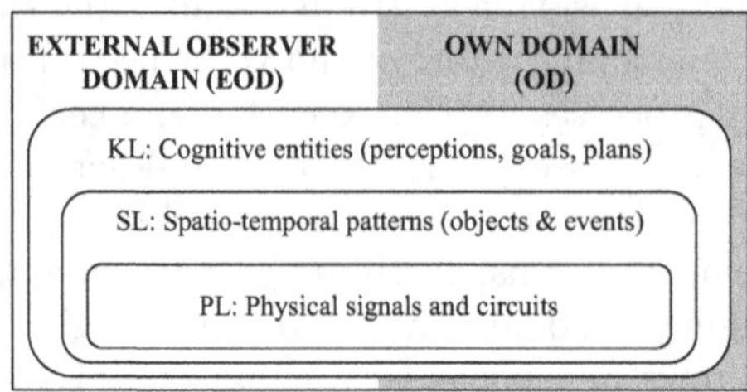

Fig. 1. The three nested levels (*PL*, *SL*, *KL*) and the two domains (*EOD*, *OD*) of description of the calculus performed by an agent

observer domain (*KL-EOD*) because their implementation only exists in natural language. Those components of a *KL-EOD* agent that may be formalized in terms of algorithms, ANN or FSA (that is to say, of its underlying formal model), already are *KL-OD* agents, leaving all the own semantics of the *EOD*. For example, the intentions and desires are now sets (classes) and the transition diagrams among states (elements of a class) are now matrixes. Analogously, the beliefs (the knowledge base), when passing from *EOD* to *OD*, are reduced to inferential rules, frames, logical entities, graphs or causal networks. And, most of the current architectures of "social agents" and "intentional agents" (BDI [22], for instance), are essentially developments at knowledge level and in the external observer domain.

4 Computational Agent Conceptual Model (*KL-EOD*)

Like in AI and Robotics, in the agency three basic architectures are distinguished - symbolic, situated and connectionist -, to approach different solutions to a problem, by modeling data and knowledge and later operating the inferences of the model. An agent is symbolic when it uses declaratory and explicit knowledge in natural language to describe the constituent organizations ("concepts") and the inference rules. In Robotics this is associated to "deliberative architectures", which spend time and a high number of computational resources in the decision process. The associated tasks usually are diagnosis, planning, and inductive and case-based reasoning (CBR) learning. Most knowledge-based systems (KBS) follow this paradigm. An agent is reactive or situated when knowledge representation is within two configuration tables of precalculated input and output, usually called "perceptions" and "actions". Here, the inference procedure is of "reflex" type (stimulus-response), very fast and adapted for real-time applications. The perception-action link is also given by a table or a automata with few states. It is proper for monitoring tasks and for the execution phase of motor planning, where a command is decomposed into a set of precalculated elementary actions that execute in "efficient" time, without having to deliberate. An agent is connectionist when knowledge representation is given in terms of labeled numerical lines, as much for the inputs as for the outputs, and the inference functions are adjustable numerical associators.

It is difficult to have all the necessary knowledge for an application. For that reason, the most frequent situations demand hybrid solutions, with reactive and deliberative, and with symbolic and connectionist parts. For that reason, in the agency paradigm, there are also hybrid architectures that combine agents of reactive and deliberative type. The reactive part reacts to the events of the environment without investing reasoning, whereas the deliberative support plans (it distributes the simplest goals) and performs tasks of superior abstraction level. Hybrid architectures are organized horizontally, so that the layers have access to sensors and actuators, and vertically, where a layer acts as an interface to sensors and actuators.

Finally, it is our conjecture that only two basic types of agent exist: (1) the ones based on descriptions in EOD that use declarative knowledge in natural language, and, (2) those based on mechanisms of the OD, causal in the implementation of the three mentioned levels. Whichever the final version of the agent at knowledge level (KL-EOD) should be, the following phase is to operationalize its entities and relations, its data and inferences. The formal model most used for the description of abstract agent architectures is automata theory.

5 A Case Study in Visual Surveillance

The previous concepts are being applied in semi-automatic visual surveillance tasks [5],[8],[9],[11],[21],[24] composed of a set of collaborative cameras installed in a building and a camera-mounted mobile robot to offer pre-alarms and/or alarms detected indoor and outdoor. The video images captured by each of the cameras enable segmenting and tracking objects of interest (obtained as image blobs) with the objective of providing meaningful events and suspicious activities. The cameras collaborate in the sense of obtaining richer surveillance observations that are only available through the fusion of information captured on various places. The mobile robot may be used if necessary to navigate to zones not monitored by fixed cameras or to dangerous places. People roaming or abandoning an object are some typical suspicious surveillance situations. After explaining this initial specification, an Analysis Overview Diagram [20] as shown in Fig. 2 is gotten.

The original diagram has been adapted to clearly show the three levels (PL, SL and KL) exposed previously. In this scheme there are physical agents *Camera*, *Robot* and *Alarm Center* at PL. On the next level, SL, software agents (*Camera Agent, Situation Agent, Social Agent, Coordinator Agent* and *Historic Agent*), message events (*Inform Event, Inform Situation, Command Camera Agent, Command Situation Agent* and *Alert Alarm Central*), as well as believes data *Image Database, Situation Database* are shown. Lastly, the KL includes goals, perceptions and actions, as described earlier. At top level, the main goal is *Detect anomalous situation* from which all subgoals are derived. Perceptions are also simplified into one general percept called *Behavior annotated video.*

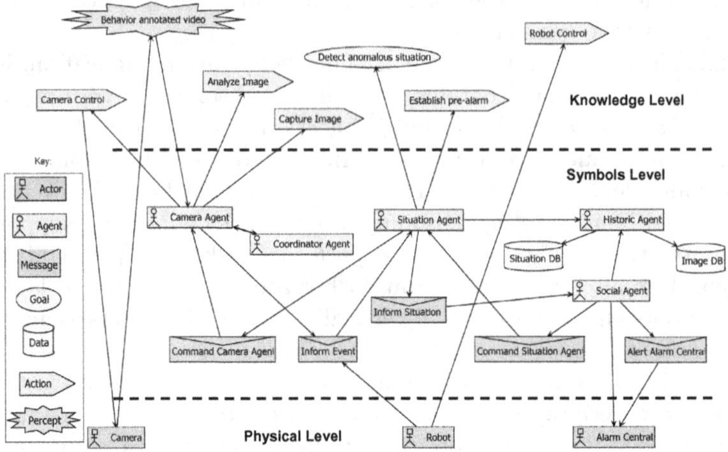

Fig. 2. Analysis Overview Diagram for a semi-automatic video-surveillance system

The actions shown in the figure are proper of any surveillance task faced from computer vision and robotics.

A more precise system specification is detailed next. A physical *Camera* provides the system (concretely to *Camera Agent* and *Mobile Robot Agent*) with the environment information captured. The *Camera Agent* has the ability to control the camera to move it in horizontal and vertical direction, and even change the zoom. This control is represented by means of action *Camera Control.* Similarly, *Mobile Robot Agent* has the ability to control the movements of the physical *Robot* (for example to advance, to turn down, to turn right or left, to accelerate, to stop, etc.). This ability is determined by means of action *Robot Control.* The *Camera Agent* analyzes input videos (it is represented as an action) to later send it to *Historic Agent*, which will store it in the *Image Database.* For example, an annotated image could be the face of a suspicious person. The *Camera Agent* is also in charge of studying an image or a sequence of images to understand what is happening in the field of vision. By means of this study, behaviors, actions, trajectories, etc., are obtained. The *Mobile Robot Agent* also incorporates these abilities (represented by actions *Analyze Image* and *Capture Image*) because the robot is also equipped with a built-in camera. The *Camera Agent* communicates the *Situation Agent* the events that it has detected. Now, *Situation Agent* has the objective to detect anomalous situations. In addition, it sends commands to the *Mobile Robot Agent* or *Camera Agent* (through message *Command Camera Agent*). It sends a command to the *Camera Agent* to communicate a need of coordination (when an event happens through several cameras), to request that it captures an image, or that it performs a zoom, etc. It sends a command to the mobile robot to navigate to the place where a possible suspicious event, which has fallen out of the visual field of the cameras, is taking place. The *Situation Agent* communicates with *Historic Agent* to ask for the events stored in the *Situation Database.* In this data base the situations and the behaviors considered

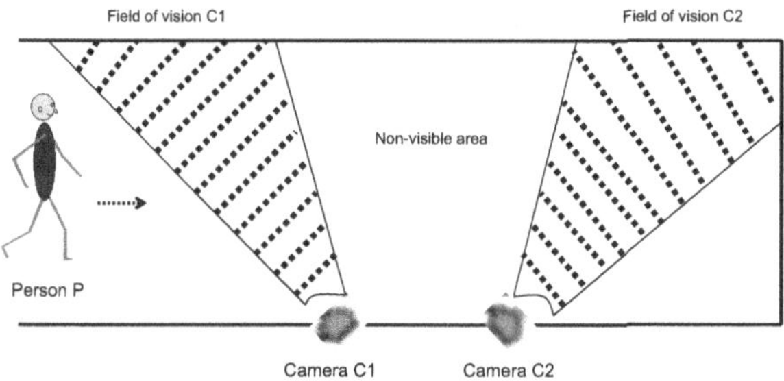

Fig. 3. A sidewalk with two non-overlapping cameras

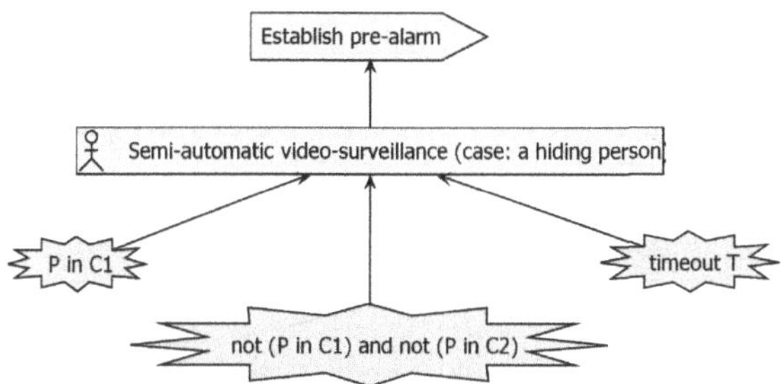

Fig. 4. Goal "Detect hiding person"

suspicious are stored. The *Situation Agent* communicates to *Social Agent* that a suspicious situation has happened (message *Inform Situation*). In the *Situation Database* the detected suspicious situations are registered.

Let us now approach one possible suspicious situation detection at KL and represented in Fig 3. Two cameras are focusing on a sidewalk and a person is walking along it. One sub-goal of "Detect anomalous situation" is "Detect hiding person". As shown in Fig. 4, action *Establish pre-alarm* comes from consecutively analyzing perceptions "P in C1", "not (P in C1) and not (P in C2)" and "timeout T" at knowledge level and in the external observer domain. The meanings for these perceptions are "the person P is present in the field of vision of camera 1", "the person P is neither present in the field of vision of camera 1, nor in the field

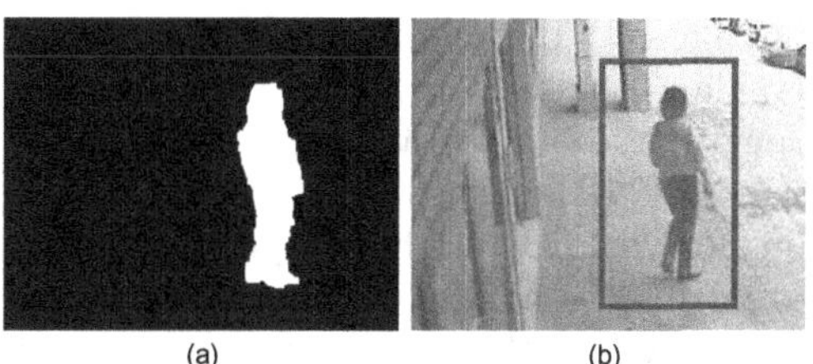

Fig. 5. Relation between *OD* and *EOD* in image segmentation. (a) Blob in the *OD*. (b) Person in the *EOD*.

Table 1. *OD-EOD* relation for goal "Detect hiding person" (simplified to a single camera)

OD	EOD
No blob at $t-1$ No blob at t	No person in camera 1 field of vision [ ]
No blob at $t-1$ Blob at the left part of image at t	Person P arrives to camera C1 field of vision from the left [Arrives (P,L,C1)]
No blob at $t-1$ Blob at the right part of image at t	Person P arrives to camera C1 field of vision from the right [Arrives (P,R,C1)]
Blob at the left part of image at $t-1$ Blob at any central part of image at t	Person P walks in camera C1 field of vision [Walks (P,C1)]
Blob at any central part of image at $t-1$ Blob at any central part of image at t	Person P walks in camera C1 field of vision [Walks (P,C1)]
Blob at the right part of image at $t-1$ Blob at any central part of image at t	Person P walks in camera C1 field of vision [Walks (P,C1)]
Blob at any central part of image at $t-1$ Blob at the right part of image at t	Person P leaves camera C1 field of vision to the right [Leaves (P,R,C1)]
Blob at any central part of image at $t-1$ Blob at the left part of image at t	Person P leaves camera C1 field of vision to the left [Leaves (P,L,C1)]

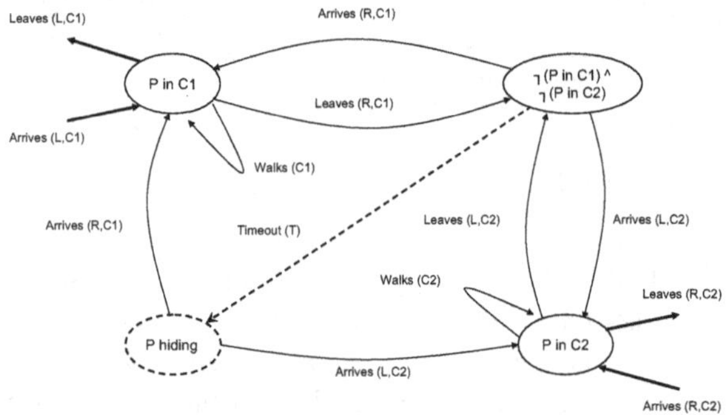

Fig. 6. Automata including state suspicious activity "Person hiding"

of vision of camera 2", and "a period superior to T has elapsed since the person P left the field of vision of camera 1 in the direction of camera 2" respectively. An "intelligent" system should consider this as a suspicious situation. Evidently, this representation may only be described at *EOD*, where the information has passed from the *OD* through a domain dependent knowledge injection.

Fig. 5 shows a simple example of the relation between *OD* and *EOD* in a typical visual surveillance task. In the *OD*, there is a segmented blob with parameters height, length, ..., obtained at instant t. (b) In the external observer domain, there is a person P "walking", tracked by camera C. In this sense, Table 1 shows both domains for goal "Detect hiding person".

Also, as explained in this paper, the operationalization of the computational agent conceptual model (*KL-EOD*) may be formally described as finite state automata. Fig. 6 shows the four states necessary to model all possibilities for a person walking through the sidewalk described in Fig. 3. The automaton covers

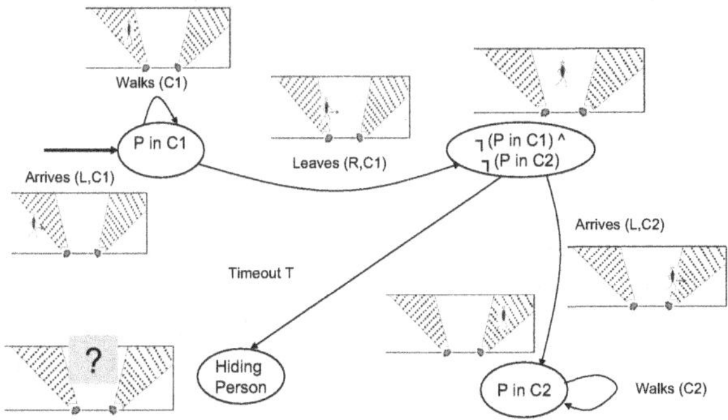

Fig. 7. Sub-automata illustrating the detection of activity "Person hiding"

all the *EOD* semantic richness previously explained. Also Fig. 7 is offered illustrating one possible path towards detecting a hiding person in the scenario described.

6 Conclusions

In this paper the agent concept has been faced from a computational perspective. Therefore, it has been shown that a computational agent must specify its conceptual model, its formal model and its implementation, starting from the set of functional specifications available on its goals, activities and tasks. We are currently engaged in modeling the visual surveillance task. Surveillance systems consist of a great diversity of entities that have to cooperate in highly dynamic and distributed environments. The use of agents for their control allows a greater degree of autonomy and response because of their capabilities to adapt and to cooperate.

Acknowledgements

This work was partially supported by Spanish Ministerio de Ciencia e Innovación TIN2007-67586-C02-02 grant, and by Junta de Comunidades de Castilla-La Mancha PII2I09-0069-0994, PII2I09-0071-3947 and PEII09-0054-9581 grants.

References

1. Agha, G.: Actors: A Model of Concurrent Computing in Distributed Systems. The MIT Press, Cambridge (1986)
2. Bertier, M., Marin, O., Sens, P.: Performance analysis of a hierarchical failure detector. In: Proceedings of the International Conference on Dependable Systems and Networks, DSN 2003, pp. 635–644 (2003)

3. Cuena, J., Demazeau, Y., Garcia-Serrano, A., Treur, J.: Knowledge Engineering and Agent Technology. Frontiers in Artificial Intelligence and Applications, vol. 52. IOS Press, Amsterdam (2003)
4. Franklin, S., Graesser, A.: Is it an agent, or just a program?: A taxonomy for autonomous agents. In: Jennings, N.R., Wooldridge, M.J., Müller, J.P. (eds.) ECAI-WS 1996 and ATAL 1996. LNCS, vol. 1193, pp. 121–135. Springer, Heidelberg (1997)
5. Gascueña, J.M., Fernández-Caballero, A.: The INGENIAS methodology for advanced surveillance systems modelling. In: Mira, J., Álvarez, J.R. (eds.) IWINAC 2007. LNCS, vol. 4528, pp. 541–550. Springer, Heidelberg (2007)
6. Guessoum, Z., Briot, J.P., Marin, O., Hamel, A., Sens, P.: Dynamic and adaptive replication for large-scale reliable multi-agent systems. In: Proceedings of the 1st International Workshop on Software Engineering for Large-Scale Multi-Agent Systems, SELMAS 2002, pp. 26–30 (2002)
7. Jiang, Y., Xia, Z., Zhong, Y., Zhang, S.: The construction and analysis of agent fault-tolerance model based on pi-calculus. In: Bubak, M., van Albada, G.D., Sloot, P.M.A., Dongarra, J. (eds.) ICCS 2004. LNCS, vol. 3038, pp. 591–598. Springer, Heidelberg (2004)
8. López, M.T., Fernández-Caballero, A., Fernández, M.A., Mira, J., Delgado, A.E.: Dynamic visual attention model in image sequences. Image and Vision Computing 25(5), 597–613 (2007)
9. López, M.T., Fernández-Caballero, A., Fernández, M.A., Mira, J., Delgado, A.E.: Visual surveillance by dynamic visual attention method. Pattern Recognition 39(11), 2194–2211 (2006)
10. Marr, D.: Vision. Freeman, San Francisco (1982)
11. Martínez, R., Rincón, M., Bachiller, M., Mira, J.: On the correspondence between objects and events for the diagnosis of situations in visual surveillance tasks. Pattern Recognition Letters 29(8), 1117–1135 (2008)
12. Maturana, H.R.: The organization of the living: A theory of the living organization. International Journal of Man-Machine Studies 7, 313–332 (1975)
13. McCulloch, W.S.: Embodiments of Mind. The MIT Press, Cambridge (1965)
14. Mira, J., Delgado, A.E., Moreno-Diaz, R.: Cooperative processes in cerebral dynamic. Applications of Information and Control Systems 3, 273–280 (1979)
15. Meyer, B.: Object-Oriented Software Construction. Prentice Hall, Englewood Cliffs (1997)
16. Minsky, M.L.: Steps towards Artificial Intelligence. Proceedings of the Institute of Radio Engineers 49, 8–30 (1961)
17. Newell, A., Simon, H.A.: Human Problem Solving. Prentice Hall, Englewood Cliffs (1972)
18. Nunes de Castro, L.: Fundamentals of Natural Computing: Basic Concepts, Algorithms, and Applications. Chapman & Hall/CRC, Boca Raton (2006)
19. Padgham, L., Zambonelli, F.: AOSE VII / AOSE 2006. LNCS, vol. 4405. Springer, Heidelberg (2007)
20. Padgham, L., Winikoff, M.: Developing Intelligent Agents Systems: A Practical Guide. John Wiley and Sons, Chichester (2004)
21. Pavón, J., Gómez-Sanz, J., Fernández-Caballero, A., Valencia-Jiménez, J.J.: Development of intelligent multi-sensor surveillance systems with agents. Robotics and Autonomous Systems 55(12), 892–903 (2007)
22. Rao, A.S., Georgeff, M.P.: Modeling rational agents within a BDI-architecture. In: Proceedings of the 2nd International Conference on Principles of Knowledge Representation and Reasoning, KR 1991, pp. 473–484 (1991)

23. Russell, S., Norvig, P.: Artificial Intelligence: A Modern Approach, 2nd edn. Prentice Hall, Englewood Cliffs (2002)
24. Valencia-Jiménez, J.J., Fernández-Caballero, A.: Holonic multi-agent systems to integrate independent multi-sensor platforms in complex surveillance. In: Proceedings of the IEEE International Conference on Advanced Video and Signal based Surveillance, AVSS 2006, p. 49 (2006)
25. Varela, F.J.: Principles of Biological Autonomy. North-Holland, Amsterdam (1979)
26. von Neumann, J.: The Computer and the Brain. Yale University Press, New Haven (1958)
27. Weiss, G.: Multiagent Systems: A Modern Approach to Distributed Artificial Intelligence. The MIT Press, Cambridge (1999)
28. Wiener, N.: Cybernetics: Or the Control and Communication in the Animal and the Machine. Wiley, Chichester (1961)

Knowledge and Event-Based System for Video-Surveillance Tasks

Rafael Martínez Tomás and Angel Rivas Casado

Dpto. Inteligencia Artificial. Escuela Técnica Superior de Ingeniería Informática, Universidad Nacional de Educación a Distancia, Juan del Rosal 16, 28040 Madrid, Spain
rmtomas@dia.uned.es, rivas.angel@gmail.com

Abstract. This work describes an event–based system supported by knowledge to compose high-level abstraction events from intermediate agent events. The agents are in a level that interprets multi-sensory signals according to the scenario ontology, particularly, from video-sequence identification and monitoring. The target task is surveillance understood in its entirety from the identification of pre-alarm signals to planned action. The work describes the system architecture for surveillance based on this composition knowledge, how the knowledge base is organised, tools for its management and examples of event inference/composition characterising a scene situation.

1 Introduction

Surveillance associated tasks are increasingly prevalent in different scenarios and services. The spectrum of possible target situations is enormous and of varying complexity, from simply detecting movement in a controlled space to more global surveillance where scenes are monitored with different cameras and sensors, the suspicious situation is studied and diagnosed, and the dynamic planning is done coherently, according to how the situation, activities and resources evolve. The surveillance task as a whole, therefore, has a similar structure to that of a control task,

1. Monitoring critical variables, whose deviations from normality is a sign of malfunction or a warning of possible subsequent malfunctions,
2. Diagnosing a problem, consisting either of the search for the cause of the malfunction or a prediction of failures, and
3. Planning coordinated action of the different agents that collaborate in solving the problem.

In any case, the fundamental problem is to understand or interpret appropriately what is happening in a surveillance target scenario and somehow solve the great semantic leap that occurs when passing from the physical level of the sensors, particularly in video-surveillance, from the pixels captured by the camera to the identification of subtle or complex movements and scenographies of different actors.

J. Mira et al. (Eds.): IWINAC 2009, Part I, LNCS 5601, pp. 386–394, 2009.

For this, conceptually decomposition in several intermediate description levels is used with an incremental degree of abstraction [1,2,3] which, moreover, enables knowledge to be injected into the appropriate level, particularly, information on the environment (physical, behavioural or social, knowledge of the task, etc.). This also implies the possibility of considering different feedback loops from the highest semantic levels to the lower levels to improve the specific tasks in these levels. There are specific references on the use of top-down feedback in the high-level vision in the works of ours groups [4]. The feedback includes a reflection on what is inferred, a search for inconsistencies, greater precision, etc. Particularly, following the proposal of [1] in our works we differentiate between pixel level, blob level, object level and activity level [3].

In this article we are particularly interested in this last level, which starts from the description, O of the identification of objects or classified scene elements and their monitoring in time and space in the object level. In particular, in computer vision High Level Vision [4] is precisely the interpretation of scenes beyond the mere recognition of objects: recognising situations, activities and interactions between different agents participating in a video sequence.

In this line, back in 1983 and 1984 Neumann and Novak [5,6] worked on a system to generate a natural language description of the activities observed in a traffic video sequence, using frames of cases based on locomotion verbs organised hierarchically for the representation. Bobick (1997) [7] characterised movement in terms of the consistency of entities and relations detected in a time sequence. Conversely, the concept of activity is understood as a composition of stereotyped movements, whose time sequence is characterised by statistical properties (e.g. hand gesture). Finally, he defined action as "semantic primitives relating to the context of the motion". In [8] a hierarchical ontology is structured (events, verbs, episodes, stories, etc.,). By contrast, Chleq and Thonnat (1996) [9]only differentiate between primitive and composed events.

Thus, generally, the most abstract activities or events are considered as a composition from other more primitive events inherent to lower semantic levels. This composition is done from spatio-temporal relations ([10] for example) or from common sense knowledge on hierarchies and concept relations ([6], for example).

When human observers interpret the meaning of a scene obviously they use their knowledge of the world, the behaviour of the things that they know, the laws of physics and the set of intentions that govern agent activity. All this additional knowledge that does not appear explicitly in the signals generated by the sensors enables observers to model the scene and use this model to interpret or predict, at least partially, what is happening or may happen in the scene. It is knowledge that must be made explicit and represented for its operationalisation.

In this article, which is a continuation of other works by the group [11,12,13], we focus on explicit knowledge to identify activities as event composition in the activity level and use the events from the identification and monitoring processes in a video surveillance system. The objective is also to develop some tools that facilitate the generation of new high-level interpretation systems and reuse standardised and recurrent events from different surveillance scenarios and situations. In the following section, an example of event composition is shown to identify an alarming situation and another example to solve a monitoring problem. The prototype will be shown below. It operationalises this knowledge, applies the composition mechanism on an event base, and has tools that facilitate the configuration and incorporation of new knowledge on activities, scenes and scenarios [3].

2 High-Level Event Composition and Knowledge-Base Structuring

Events from the identification and monitoring agents (agents-sensors) which meet specific spatio-temporal restrictions make it possible to infer, thereby, trigger events with a greater semantic level. Figure 1 schematically illustrates an example of an alarming situation. The first row includes simplified images in the scene instants. The second row contains the simple events that are generated from the segmentation, monitoring and identification of each frame of the sequence. The third row shows the pattern for the composition of events occurring at each instant. It is a knowledge unit for a composition: a set of events that must meet specific spatio-temporal relations and the consequent events with greater semantics.

Thus, following the example we can interpret the sequence as follows:

1. At instant t human1 is detected on the scene. The identification and monitoring agent does not recognise that the human is carrying an object, so the spatio-temporal location of the human is only represented with the event "At".

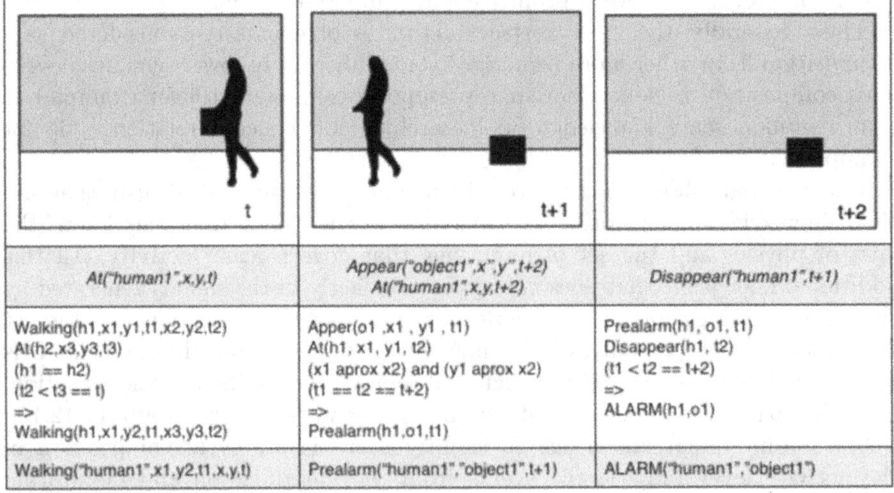

t	t+1	t+2
At("human1",x,y,t)	*Appear("object1",x",y",t+2)* *At("human1",x,y,t+2)*	*Disappear("human1",t+1)*
Walking(h1,x1,y1,t1,x2,y2,t2) At(h2,x3,y3,t3) (h1 == h2) (t2 < t3 == t) => Walking(h1,x1,y2,t1,x3,y3,t2)	Apper(o1 ,x1 , y1 , t1) At(h1, x1, y1, t2) (x1 aprox x2) and (y1 aprox x2) (t1 == t2 == t+2) => Prealarm(h1,o1,t1)	Prealarm(h1, o1, t1) Disappear(h1, t2) (t1 < t2 == t+2) => ALARM(h1,o1)
Walking("human1",x1,y2,t1,x,y,t)	Prealarm("human1","object1",t+1)	ALARM("human1","object1")

Fig. 1. Table showing input events, knowledge used and events inferred schematically between successive frames

2. At instant t+1 "object1" is detected near to the position of human1. This situation activates a pre-alarm of possible abandonment of an object with the event "Pre-alarm". Since there are no other humans nearby, it is inferred that "human1" has left the object. An association is created between the object and human and the pre-alarm is activated.
3. At instant t+2 "human1" is detected leaving the scene. This event and the active pre-alarm identify a situation of abandoning an object. The event "Alarm" goes off.

We group *knowledge units* into *packages* that identify a specific situation. In turn, the *packages* are organised into *composition levels.*

Each *package* is assigned a composition level. Each composition level sends the composed events that it has generated to their higher composition level. *Packages* in different *composition levels* are interdependent. Thus, if a *package* in a specific level is added to the knowledge base, all those *packages* in the lower *composition levels* that are necessary for its functioning will be added.

We pursue the objective of creating a library of *packages* rich enough to be able to configure a system with ease. Each library of *packages* has its own corresponding ontology of events.

In Figure 2 we can observe all the composition hierarchy between the different system elements. The knowledge base consists of *composition levels.* Each composition level has *packages.* Each *package* has *knowledge units*.

As shown in the second example, the knowledge base not only includes the precise knowledge for identifying alarming situations, but also knowledge that complements the information received from previous levels. The more expert knowledge there is the fewer the number of false alarms.

This other example shows that the activity level may recognise actions to enrich or complete the identification information. We have a human, who is walking behind a column. The system is aware of the situation of the column. We define the event Column as any scene element causing occlusion.

Figure 3 has a similar structure to the previous example. In the first row we find the schematised images between instants t and t+2 of the scene where a human passes behind a column. In the second row the simple events are represented that reach the activity level and trigger the *packages* that infer events (third row) of the next composition level.

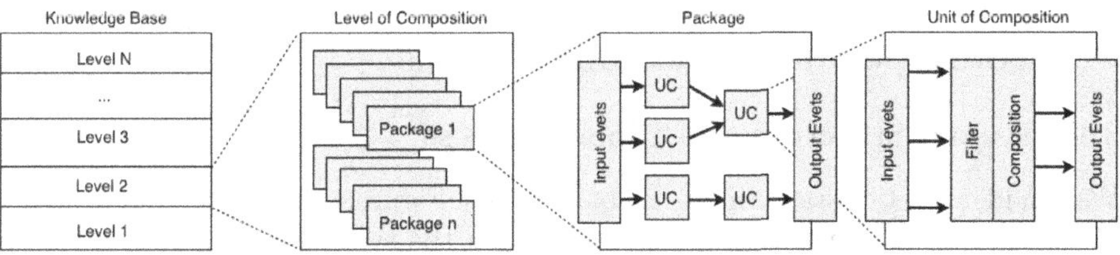

Fig. 2. Cascade of the knowledge base

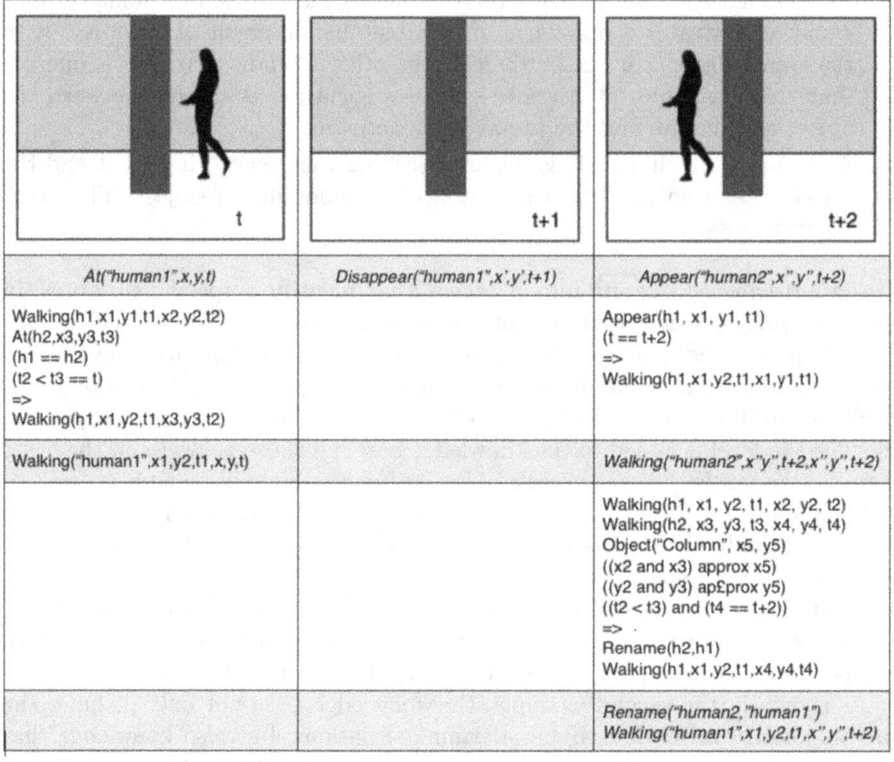

t
t+1
t+2
At("human1",x,y,t)
Disappear("human1",x',y',t+1)
Appear("human2",x",y",t+2)
Walking(h1,x1,y1,t1,x2,y2,t2)
At(h2,x3,y3,t3)
(h1 == h2)
(t2 < t3 == t)
=>
Walking(h1,x1,y2,t1,x3,y3,t2)
Appear(h1, x1, y1, t1)
(t == t+2)
=>
Walking(h1,x1,y2,t1,x1,y1,t1)
Walking("human1",x1,y2,t1,x,y,t)
Walking("human2",x"y",t+2,x",y",t+2)
Walking(h1, x1, y2, t1, x2, y2, t2)
Walking(h2, x3, y3, t3, x4, y4, t4)
Object("Column", x5, y5)
((x2 and x3) approx x5)
((y2 and y3) ap£prox y5)
((t2 < t3) and (t4 == t+2))
=>
Rename(h2,h1)
Walking(h1,x1,y2,t1,x4,y4,t4)
Rename("human2,"human1")
Walking("human1",x1,y2,t1,x",y",t+2)

Fig. 3. Example of composition levels

The fourth row shows these events inferred in level one, the fifth row, the behaviour of a level-two *package* and, finally, the sixth row, the composed events generated from this level.

In t+2 a sequence was identified that makes it possible to infer (it is assumed), since no individuals participate in the scene other than the person that has entered and left the column space. This implies correcting the information that comes from the object level and renaming human2 as human1. Therefore, two *packages* in different levels were used to identify the occlusion situation, thanks to knowledge of the scenario.

3 Prototype and Development Tools

3.1 System Structure and Global Process

We can identify two stages in the global process. Figure 4 shows the system structure and this differentiation schematically. All the system components are connected via a local area network (LAN). The knowledge base and the identification and monitoring agent interfaces are designed in the first stage.

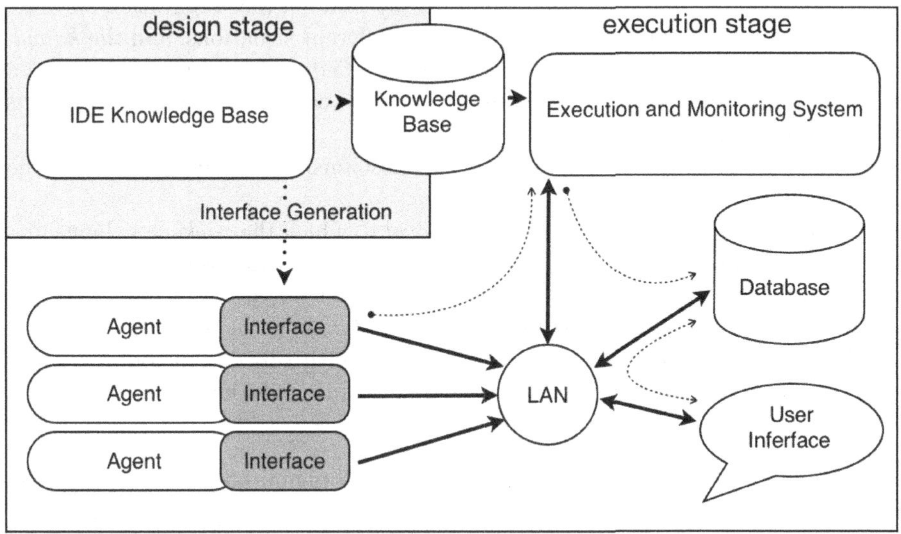

Fig. 4. System structure

An environment for the development (KB IDE) facilitates this process from the repository of reusable *packages*. The agents, in the execution stage, send the simple events that they have identified to the Execution and Monitoring System via the corresponding interface. This Execution and Monitoring system, using the knowledge base, infers composed events that it immediately stores in the Database. The user interface selects, organises and compiles the most relevant information.

The Execution System sends the inferred events that characterise alarm situations directly to the user interface. The system is multi-platform since it uses the LAN network as a connection medium between the components. Accordingly, for example, the Agents may be running under LINUX, the Database on Windows and the user interface with another operating system. This makes the system versatile unlike the previous version [12]. We have passed from an AllInOne system to a distributed system that minimises execution problems and requirements, since several interconnected machines are used to distribute the global calculation load. Logically, the execution time must be taken into account when the time requirements of an environment are highly demanding. Execution times can be adjusted to the work environment needs. The size of the knowledge base should also be considered. If this has a large number of *packages* the execution time will have to be increased to attain a time slice large enough to be able to perform all the necessary operations.

3.2 Knowledge Base IDE

This tool is used to create the knowledge base that is subsequently executed in the Execution and Monitoring System. The following tasks can be performed with this application:

- *Configuration Parameter Definition*: the system has a configuration parameter table that reconfigures the system for different situations, and the *knowledge units* or *packages* do not have to be modified. If, for example, the system is working with information from cameras, we can define their resolution as configuration parameters.
- *Event Ontology Definition*: we have a mechanism that helps define all the event ontology.
- *Package Definition*: the composition Level to which the *package* belongs and the dependences with other *packages* in the lower composition level can be defined with this tool. Thus the environment of this *package* is automatically filtered. Only certain events in the previous level can be accessed.
- *Knowledge Unit Definition in the packages*: with this design tool a composition can be constructed. Each *knowledge unit* only has access to events contained in its *package*.

Figure 5 shows the main window of this tool describing initially the simple events and the composed events that the user has. It has tabs that open windows to edit the different system components, such as editing new *packages* and *knowledge units*.

3.3 Execution and Monitoring System

The system activates the knowledge base and stores all the composed events generated in a historical record. We distinguish between the following components:

- *Net server*: it activates the socket for the Agents to connect to the system. The connection is bi-directional in order to request certain information from the Agent.
- *List of connected Agents*: there is a list of all the Agents that are connected to the system along with their description and related data.
- *Event synchronisation system*: it is in charge of generating time slices to receive and label events. Once the event has been received, it is labelled with the time associated with the time slice. Thus an asynchronous system is transformed into a synchronous one.
- *Composition Motor*: inference motor that evaluates the requirements of the *knowledge units* and adds the new inferred events to the event base.
- *Statistic and monitoring system*: it analyses the number of events that arrive from each Agent, the execution time of the Composition motor, the synchronisation system buffers and database connection response times. This information is really useful when calibrating and configuring all the execution system parameters.
- *Database connection to store the information*: it sends all the events inferred at each instant to the selected database. This means that the information can later be processed. When analysing these data, possible errors in the knowledge base can be debugged, which is really useful for refining the system.

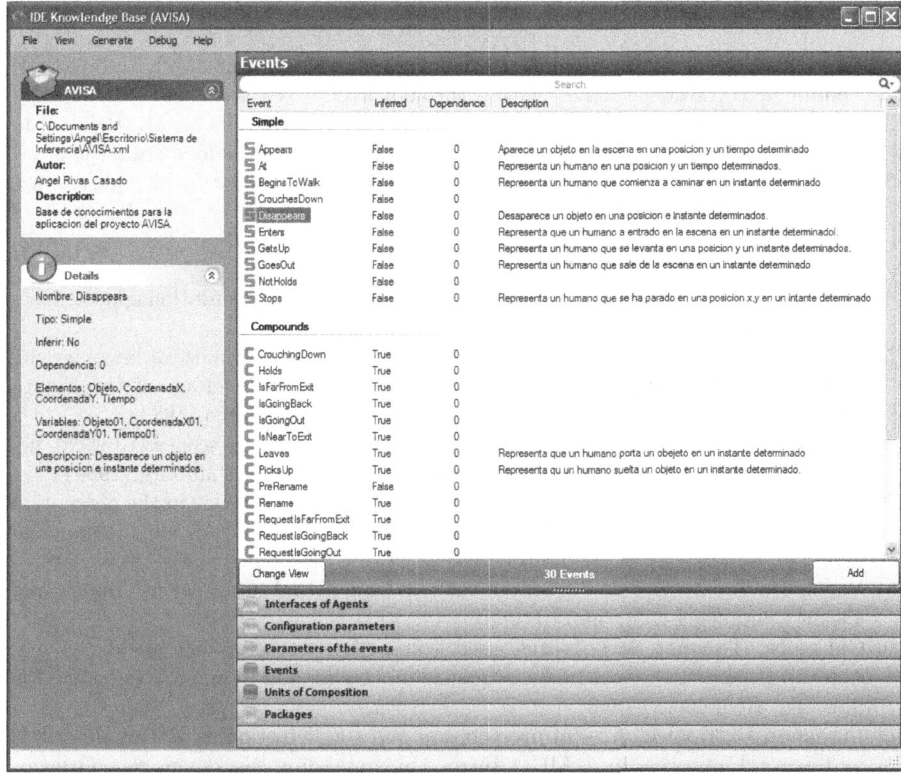

Fig. 5. Main window of Knowlendge Base IDE

4 Conclusions

This work highlights the ideas on which the our video-surveillance prototype is based. It is organised into description levels from the physical level to the activity level. This article focuses in particular on the activity level and its implementation as an event composition knowledge-based system, from the simplest events, which arrive at the activity level from identification and monitoring, to other events with a greater semantic load. The prototype is fully functional. It has two tools: a Development Interface for the Knowledge Base and an execution and monitoring system. The Development Interface pursues quite an ambitious objective to facilitate the configuration of the pattern of new alarming situations, based on standardised and habitual activities, and to build, repository of *knowledge packages* in video-surveillance identification and monitoring.

Acknowledgements

The authors are grateful to the CiCYT for financial aid on project TIN-2007-67586- C02-01.

References

1. Nagel, H.: Steps towards a cognitive vision system. AI Magazine 25(2), 31–50 (2004)
2. Neumann, B., Weiss, T.: Navigating through logic-based scene models for high-level scene interpretations. In: Crowley, J.L., Piater, J.H., Vincze, M., Paletta, L. (eds.) ICVS 2003. LNCS, vol. 2626, pp. 212–222. Springer, Heidelberg (2003)
3. Martínez-Tomás, R., Rincón-Zamorano, M., Bachiller-Mayoral, M., Mira-Mira, J.: On the correspondence between objects and events for the diagnosis of situations in visual surveillance tasks. Pattern Recognition Letters (2007), doi:10.1016/j.patrec.2007.10.020
4. Carmona, E.J.: On the effect of feedback in multilevel representation spaces. Neurocomputing 72, 916–927 (2009)
5. Neumann, B., Novak, H.: Events models for recognition and natural language descripcion of events in real-world image sequences. In: Proceedings of the Eighth IJCAI, Karlsruhe, pp. 724–726. Morgan Kaufmann, San Mateo (1983)
6. Neumann, B.: Natural language description of time-varying scenes. Brericht no. 105, FBI-HH-B-105/84, Fachberic Informatik, University of Hamburg (1984)
7. Bobick, A.: Movement, activity, and action: The role of knowledge in the perception of motion. In: Royal Society Workshop on Knowledge-based Vision in Man and Machine, London, pp. 1257–1265 (1997)
8. Nagel, H.: From imagine sequences towards conceptual descriptions. Image and Vision Computing 6(2), 59–74 (1988)
9. Chleq, N., Thonnat, M.: A rule-based system for characterizing blood cell motion. In: Huang, T.S. (ed.) Image sequence processing and dynamic scene analysis. Springer, Heidelberg (1996)
10. Pinhanez, C., Bobick, A.: Pnf propagation and the detection of actions described by temporal intervals. In: DARPA Image Understanding Workshop, New Orleans, Lousiana, pp. 227–234 (1997)
11. Mira, J., Tomás, R.M., Rincón, M., Bachiller, M., Caballero, A.F.: Towards a semi-automatic situation diagnosis system in surveillance tasks based. In: Mira, J., Álvarez, J.R. (eds.) IWINAC 2007. LNCS, vol. 4528, pp. 90–98. Springer, Heidelberg (2007)
12. Carmona, E., Cantos, J.M., Mira, J.: A new video segmentation method of moving objects based on blob-level knowledge. Pattern Recognition Letters 29, 272–285 (2008)
13. Rincón, M., Carmona, E.J., Bachiller, M., Folgado, E.: Segmentation of moving objects with information feedback between description levels. In: Mira, J., Álvarez, J.R. (eds.) IWINAC 2007, Part II. LNCS, vol. 4528, pp. 171–181. Springer, Heidelberg (2007)

ARDIS: Knowledge-Based Dynamic Architecture for Real-Time Surface Visual Inspection

D. Martín[1], M. Rincón[2], M.C. García-Alegre[1], and D. Guinea[1]

[1] Industrial Automation Institute, CSIC, 28500 Arganda del Rey, Madrid, Spain
{dmartin,maria,domingo}@iai.csic.es

[2] Dpto. de Inteligencia Artificial, ETSI Informtica, UNED, C/Juan del Rosal 16, 28040 Madrid, Spain
mrincon@dia.uned.es

Abstract. This work presents an approach to surface dynamic inspection in laminated materials based on the configuration of a visual system to obtain a good quality control of the manufacturing surface. The configuration task for surface inspection is solved as a Configuration-Design task following the CommonKADS methodology, which supports the proposed knowledge-based dynamic architecture (ARDIS).

The task is analysed at the knowledge level and is decomposed into simple subtasks to reach the inference level. All the generic knowledge involved in the surface inspection process is differentiated among environment, real-time, image quality and computer vision techniques to be integrated it in ARDIS.

An application has been developed to integrate four operation modes relating to visual system configuration and specific surface inspection. An example is shown for configuring a stainless steel inspection system and another one for wood inspection.

1 Introduction

The use of visual systems for inspecting surface defects has always been present in the laminated materials industry. Nevertheless these systems still present several drawbacks, such as reusing, as they are designed for a particular surface inspection application and do not offer the possibility of changing either the objectives or the inspection necessities.

Nowadays, some authors propose specific surface inspection systems focused on a particular inspection task but lacking the formulation of a general purpose framework for surface inspection systems. The computer vision wizards assist to select manually the inspection type without using configuration parameters and integration of domain knowledge [4].

The injection of surface inspection knowledge of the human experts is an essential issue in the configuration process of any visual inspection system. For example, the change of the surface thickness or lighting in a production line

J. Mira et al. (Eds.): IWINAC 2009, Part I, LNCS 5601, pp. 395–404, 2009.

implies a variation which is not considered in current inspection systems and need knowledge to adjust the surface inspection process. In case of surface thickness change, the distance between the camera and the surface changes and the camera gets out of focus generating blurred images, thus knowledge is necessary to solve it like human experts do.

Therefore, current visual systems are designed for a specific application and are not ready to hold unexpected variations. This fact entails a high cost as each new visual inspection task has to be redesigned from the surface defect analysis to the overall inspection system by a human expert. The solution here proposed points to reusing the generic knowledge on surface visual inspection. To this aim generic knowledge is previously differentiated in types. This would allow changing or reconfiguring the components related to the application that vary, in the production line, due to changes on either environment camera, surface or defect type.

The increase of the inspection systems and its complexity requires analysing complex tasks related to surface visual inspection to support surface quality control. The approach has to use the knowledge of the human experts to infer solutions (system configurations) to specific surface inspection problems (stainless steel or wood defects among other defects).

Cognitive architectures are defined as knowledge based models that assist to organize data, information and inferences to solve complex tasks [5]. Firstly, current work deals with the complex task of visual inspection of laminated surfaces in industrial environments with a high variability in lighting, reflectance and real-time defect detection conditions. Secondly, the generic expert knowledge on surface visual inspection and the deep knowledge on the inspection preferences, restrictions and defect characteristics that exhibit the "line inspectors", has to be used in the design of a visual inspection system.

The remainder of this paper is organized as follows: In Section 2, we provide an overview of the ARDIS architecture. In section 2.1, we explain system requirements that compose the dynamic knowledge. Section 2.2 gives some guidelines of the generic domain knowledge on surface inspection and its types and influences among them. Section 2.3 points out the dynamic configuration using ARDIS architecture that has been proposed in the former section. Section 3 performs a number of experiments in a variety of different situations to obtain empirically some parameters from images for evaluating the configured system. Section 4 shows the SIVA II application tool that integrates four operation modes to achieve better overall performance for system configuration task. Finally, conclusions are discussed in Section 5.

2 ARDIS a Cognitive Architecture

A dynamic cognitive architecture is proposed, namely ARDIS [2], to integrate expert knowledge for real-time defects detection which offers an adaptive behaviour for detecting different defects or inspecting different type of surfaces. To this aim the ARDIS architecture is designed so as to be able to configure on real-time a visual inspection system, adapted for each specific type of laminated surface and defect.

Accounting for the characteristics of the industrial environment — under rigid quality control regulations, different and random types of defect to be inspected, specific real-time conditions and suitable computer vision techniques — the configuration of a visual inspection system requires a detailed analysis. Consequently, the ARDIS architecture has been designed following the CommonKADS methodology [6], considering real-time surface dynamic inspection as a complex task of the paradigm of design-configuration.

The three terms relating to the ARDIS architecture are defined as follows:

- The architecture term, for referring to a model which is decomposed in different functional modules.
- The dynamic term, meaning that the inspection system depends on the environment, real-time, image quality and computer vision techniques requirements of the specific inspection application.
- The cognitive term, as it analyses at the knowledge level the inspection problem of surface defects in laminated materials.

The method used to solve the configuration task is Propose-&-Revise and is displayed in figure 1, showing the diagram of subtasks and inferences. This method shows how the ARDIS architecture operates in three steps:

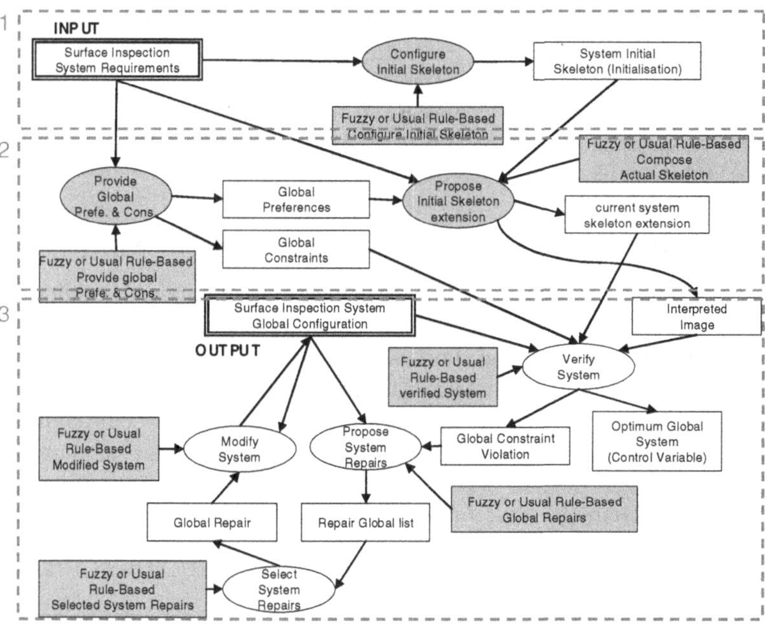

Fig. 1. Subtask and inference diagram of the method Propose-&-Revise to solve the system configuration task

The first is the "initialization process", which generates an initial skeleton of the visual inspection system that behaves as a seed for its subsequent extension into a more complex skeleton. The initial configuration is fast and allows the system to inspect with this basic configuration until the overall system is defined.

The second is the "extension of the initial inspection skeleton" which configures an inspection system that considers all the specific requirements of the inspection application. The configuration is slower but the system is set more accurately as all the inspection information on the application and system components are available.

The third is the "revision process" of the overall inspection system, configured in the former step. The revision will be explained in section 3.

The next three subsections give a brief description of the requirements (dynamic roles — white rectangles in figure 1 —) and the generic domain knowledge on surface inspection (static roles — grey rectangles in figure 1 —) and how they are utilized in the ARDIS architecture to dynamic system configuration.

2.1 Dynamic Knowledge: Requirements

The dynamic knowledge that exhibits the "line inspectors" is used in the configuration of a visual inspection system. The configuration process starts with the selection of the requirements by the "line inspector" where the requirements have been previously differentiated into global, initial, environment, real-time, image quality and computer vision techniques to be used in the ARDIS architecture. Next, the differentiated requirements are divided into preferences and restrictions.

For instance, environment requirements selected by "line inspector" and used in the inferences of the subtask "Propose Environment Skeleton Extension (PESE)"} — PESE is a subtask of "Propose Initial Skeleton extension", see figure 1 — to inspect stainless steel or wood are shown in table 1.

2.2 Static Knowledge: Domain Knowledge on Surface Inspection

The generic domain knowledge on surface inspection is composed of different types of generic knowledge. This differentiation of the type of knowledge makes possible to distinguish the knowledge used on each inference, making easier the specification of components and its interactions. Thus, the designed architecture has been provided with all the necessary functionalities to configure a specific complete system of visual inspection where knowledge is partitioned to be ease reused.

Table 1. Example of environment requirements selected by line inspector

ENVIRONMENT REQUIREMENTS USED IN THE INFERENCES OF THE SUB-TASK Propose Environment Skeleton Extension	
Stainless Steel	**Wood**
ProductionLineSpeed=1 $\frac{m}{s}$	ProductionLineSpeed=0.1 $\frac{m}{s}$
DefectOrientation={longitudinal}	DefectOrientation={random}
IlluminationType={GreenLaserSource}	IlluminationType={ConventionalLight}

Table 2. Generic knowledge related to environment

GENERIC KNOWLEDGE RELATED TO ENVIRONMENT USED IN THE INFERENCES OF THE SUBTASK Propose Environment Skeleton Extension
DiffuseIllumination = {true, false}
ExpositionTime = {long, medium, short}
CameraSensorGain = {automatic, minimum, medium, maximum}
PixelsPerDefect = {4pixels(2x2), 9pixels(3x3), ManualSelection}
Camera-IlluminationRelativePosition = {0deg , 45deg, 90deg, ManualSelection}
CameraPosition-InspectionPlane = {OverLaminate, UnderLaminate}

The generic domain knowledge on surface inspection is differentiated on global, initial, environment, real-time, image quality and computer vision techniques. Accordingly, the knowledge is distinguished to each inference and the knowledge can be reused. The generic knowledge, for instance related to the environment that will be used and reused in the configuration process, is presented in table 2.

The system configuration process requires knowledge about the relationships between knowledge types that point out the influence among them. The influence graph is described in figure 2. As an example, the environment knowledge is related to real-time knowledge, such as image acquisition rate or lighting components can influence real-time components in one or another way. So, during the configuration step of an inspection system if an environment component is configured this has to be taken into account in the configuration of the real-time components.

2.3 Dynamic Configuration Using ARDIS Architecture

The dynamic configuration process is based on (i) ARDIS architecture, (ii) generic domain knowledge on surface inspection and (iii) "line inspector" knowledge who selects the requirements of the inspection system. This dynamic configuration process allows the inspection of each specific surface in a production line. Consequently,

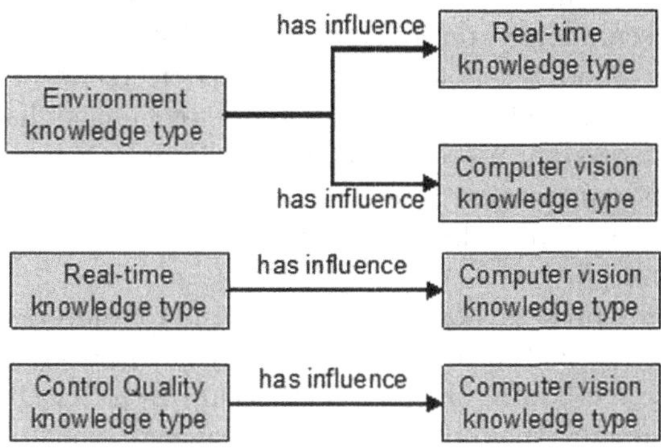

Fig. 2. Relations between domain knowledge types of surface inspection

Fig. 3. Stainless steel images with micro oxide defects

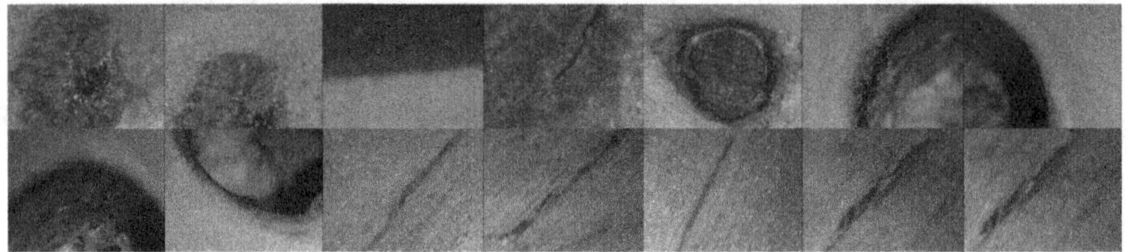

Fig. 4. Wood images with knots and others defects

Table 3. Example of the knowledge base which allows configuring the system

KNOWLEDGE BASE USED IN THE INFERENCES OF THE SUBTASK Propose Environment Skeleton Extension
IF ProductionLineSpeed is high THEN ExpositionTime is short
IF IlluminationSystem is ExtLaserIllumination THEN CameraSensorGain is auto
IF IlluminationSystem is ExtLaserIllumination THEN Camera-IllumiRelPos is 45deg
IF IlluminationType is GreenLaserSource THEN ImageChannel is Green

it is possible to inspect stainless steel, wood, paper and plastic, among others in the same production line. Next, we choose stainless steel and wood to show how it is possible to configure two inspection systems based on the former premises. The figure 3 shows a set of images of stainless steel where it is necessary to inspect micro residual oxide scale defects on cold stainless steel strip. Figure 4 displays a new laminate material, wood, where knots and others defects have to be inspected in the same production line. The wood and stainless steel images have been acquired by an experimental system based on green laser illumination — diffuse lighting technique for surface inspection which consists of a green laser diode — that allows inspecting from micro to five millimetres defects, as the acquisition system uses magnification 1 [2].

The configuration process, accounts for the former requirements (table 1) and domain knowledge (table 2), which are codified by means of crisp and fuzzy rules. Using these rules and following the control structure of the ARDIS architecture [2] it is possible to configure the components of the surface inspection system. Thus, the solution is a configured inspection system which depends on the type of surface and defect to be inspected. The table 3 shows a set of crisp rules which have been used to configure the environment components of the inspection system.

3 Parameters from Images for the Architecture Revision Process

The aim of the "revision process" is to validate the configuration of the system by means of the acquired images.

The analysis of the images and the numeric parameters for the revision process of the configured inspection system has been accomplished with the Image Processing Toolbox of MATLAB [3].

Parameters of environment (E), image quality (IQ), real-time (RT) and computer vision techniques (CVT) are differentiated and extracted from the original images.

Once the revision parameters are obtained from images it is necessary to perform its evaluation. The evaluation of these parameters will generate the additional dynamic knowledge that is required to establish the revision process of the configured system through the acquired image. The knowledge, extracted in such a way, will be injected in the revision process by means of a set of crisp and fuzzy rules.

In our architecture, we have selected the Mean-Shift algorithm [1] as a generic segmentation technique, due to its adaptive capacity to the different kind of images that can be presented for inspection. Mainly based on this segmentation method, Top-Hat filtering and grey-level co-occurrence matrix, the parameters chosen to evaluate the E, IQ, RT and CVT properties are the following (Table 4 summarizes the results obtained over two inspection configuration scenarios: stainless steel and wood):

1. Revision parameter of the environment (E): the lighting non-uniformity of the image is used to measure the influence of the environment in the image. The lighting non-uniformity is calculated by "Top-hat filtering" which is the equivalent of subtracting the result of performing a morphological opening operation — with a flat structuring element — on the input image from the input image itself. The numerical value of the environment revision parameter for each image is set as the difference between the standard deviation of the input image and the standard deviation of the filtered image.
2. Revision parameter of the Real-Time (RT): the computation time — in seconds — of the Mean-Shift algorithm is used to estimate a Real-Time parameter in the image, using the following Mean-Shift parameters: SpatialBandWidth(8), RangeBandWidth(4) and MinimumRegionArea(50).
3. Revision parameter of Image Quality (IQ): the image contrast property is used as Image Quality revision parameter and is calculated through the grey-level co-occurrence matrix of the image. This measure quantifies the intensity contrast between a pixel and its neighbour over the whole image.
4. Revision parameter of the segmentation technique (CVT): the Mean-Shift technique is used as the segmentation technique and the evaluation is based on the number of Mean-Shift regions located by the algorithm.

Table 4. Mean and standard deviation of the evolution of the variable lighting, Mean-Shift computation time, contrast and number of Mean-Shift regions

	Stainless Steel		**Wood**	
	Mean	Std	Mean	Std
Lighting non-uniformity	2.7623	3.4442	24.3636	12.1547
Computation time of the Mean-Shift	10.8926	0.8618	11.9844	1.0373
Contrast	16.8680	7.5442	4.7306	2.7405
Mean-Shift regions	2249.6	801.7189	100.5714	82.6975

4 The SIVA II Interface Tool

The SIVA II tool is developed as a friendly and complete interface. SIVA II has been developed and implemented for configuring and operating each surface inspection system and designed according to the premises of ARDIS architecture. The main window of the application is displayed in figure 5. Four operating modes can be selected:

- Mode 1. "Operation in the production line". This mode is selected when the configuration of the inspection system is finished to show the inspection results. This mode is used by the inspection operator to validate the inspection results of the configured system and to decide between cleaning or refusing the laminated material if defects are present.
- Mode 2. "Automatic configuration of the system". This mode automatically configures the surface inspection system under the requirements selected by the "line inspector", aided by the acquired image. This mode is used by the line inspector who knows the defect type and the specific requirements to carry out the defect inspection.

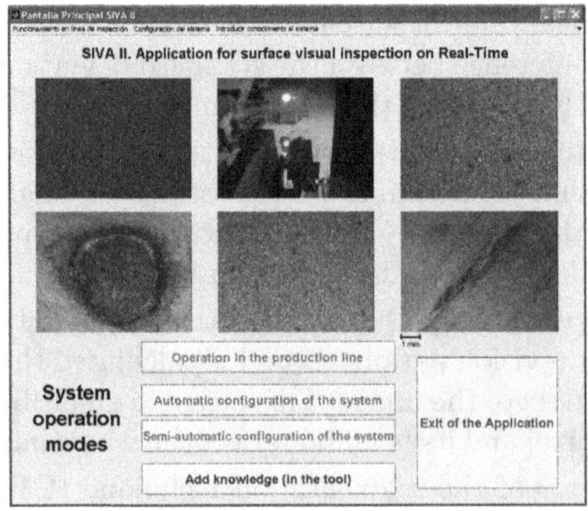

Fig. 5. SIVA II application main window

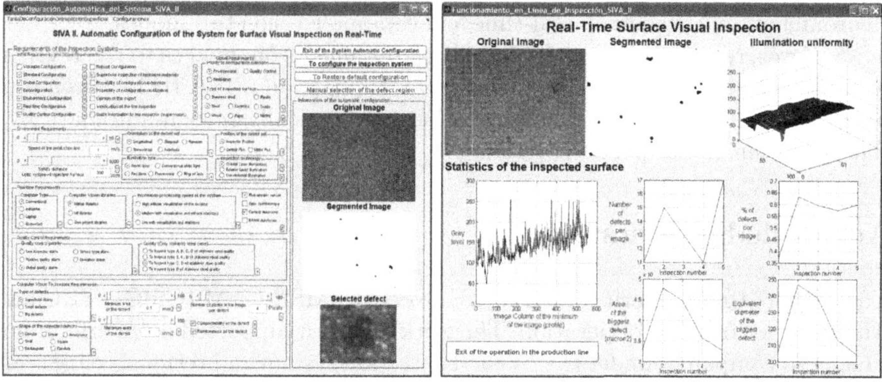

Fig. 6. (a) Window (left) for automatic configuration of the system. (b) Window (right) with inspection results obtained by the configured system.

- Mode 3. "Semi-automatic configuration of the system". This mode configures automatically the surface inspection system, but allows changing manually the components that have been automatically configured in the former mode. This mode is utilized by the line inspector and allows adjusting the configured components that require a change.
- Mode 4. "Add knowledge (in SIVA II)". This mode is used only by an expert to inject knowledge that will be used later within the frame of the ARDIS architecture. The role of the expert in the process is previous to the system configuration; however he can introduce or change knowledge at any time for reconfiguring the surface inspection system.

Finally, figure 6a displays the window of the application for automatic configuration of the system (Mode 2) and figure 6b displays the window with inspection results obtained by the configured system (Mode 1).

5 Conclusions

A knowledge-based architecture has been proposed to configure a specific surface inspection visual system to operate in dynamic environments. The architecture is instantiated to solve a specific surface inspection problem in a production line of laminated materials based on the requirements of the specific system and the identified generic knowledge on surface inspection.

The ARDIS architecture allows real-time changes due to the capability to reconfigure a new inspection system in the production line whenever the inspection goal changes. The structure stages of the proposed architecture offers a high flexibility for adaptation to different specific inspections.

The domain knowledge is organized so that it is possible to access to the specific knowledge used in each inference. The inferences use reusable generic knowledge on surface inspection which has been differentiated in several types related to the environment, image control, real-time and computer vision techniques.

Moreover, revision parameters have been obtained from images to the revision process of the configured system. Finally, an application (SIVA II) has been proposed and demonstrated to configure, either automatically or semi-automatically, each visual surface inspection system.

Acknowledgements

The authors thank Acerinox, Malaga University, IMSE-CNM (CSIC), TCC and Anafocus, for fruitful cooperation. The work has been supported by Acerinox and TCC S.A. companies under the "Visual detection system and residual oxide scale classification in stainless steel laminates" project.

References

1. Comaniciu, D., Meer, P.: Mean shift: a robust approach toward feature space analysis. IEEE Transactions on Pattern Analysis and Machine Intelligence 24, 603–619 (2002)
2. Martín, D.: Arquitectura dinámica para inspección visual de superficies en tiempo real. Ph.D thesis, UNED, Madrid, Spain (2008)
3. Matlab, The MathWorks Inc. Natick, MA, United States, http://www.mathworks.com/
4. NeuroCheck. Industrial Vision Systems. Version 5,1,1065 [SP9], NeuroCheck GmbH (2006)
5. Rincón, M., Bachiller, M., Mira, J.: Knowledge modeling for the image understanding task as a design task. Expert Systems with Applications 29, 207–217 (2005)
6. Schreiber, G., Akkermans, H., Anjewierden, A., de Hoog, R., Shadbolt, N., de Velde, W.V., Wielinga, B.: Knowledge Engineering and Management: The CommonKADS Methodology. The MIT Press, Cambridge (2000)

SONAR: A Semantically Empowered Financial Search Engine

Juan Miguel Gómez, Francisco García-Sánchez, Rafael Valencia-García, Ioan Toma, and Carlos García Moreno

Departamento de Informática, Escuela Politécnica Superior, Universidad Carlos III de Madrid, Leganés, Madrid, Spain
juanmiguel.gomez@uc3m.es
Departamento de Informática y Sistemas, Univeridad de Murcia, Spain
{frgarcia,valencia}@um.es
STI Innsbruck, University of Innsbruck, A6020 Innsbruck, Austria
ioan.toma@sti2.at
Indra Software Labs, Alcanto 11, 28045 Madrid, Spain
cgarciamo@indra.es

Abstract. The increasingly huge volume of financial information found in a number of heterogeneous business sources is characterized by unstructured content, disparate data models and implicit knowledge. As Semantic Technologies mature, they provide a consistent and reliable basis to summon financial knowledge properly to the end user. In this paper, we present SONAR, a semantically enhanced financial search engine empowered by semi-structured crawling, inference-driven and ontology population strategies bypassing the present state-of-the-art technology caveats and shortcomings.

1 Introduction

The enormous success of Google and similar search engines has demonstrated the value of crawling and indexing web resources. Nevertheless, knowledge intensive domains such as the financial information domain, where a huge number of business and companies hinge on, with a tremendous economic impact in our society, need more accurate and powerful strategies. Heedless of the complexity of the domain, financial companies and end users deem as absolutely necessary a fully-fledged integrated approach to cope with the ever-increasing volume of information outperforming current approaches such as Google Finance or Lexis-Nexis services.

Semantic Technologies [5] are currently achieving a certain degree of maturity. They provide a consistent and reliable basis to face the aforementioned challenges aiming at a fine-grained approach for organization, manipulation and visualization of the data. Consequently, the possibility of harvesting semi-structured data from financial information resources and using knowledge-oriented query answering to exploit the benefits of semantics has become top-class research challenge.

J. Mira et al. (Eds.): IWINAC 2009, Part I, LNCS 5601, pp. 405–414, 2009.

In this paper, we present a semantically enhanced financial search engine empowered by semi-structured crawling, inference-driven and ontology population strategies bypassing the present state-of-the-art technology caveats and shortcomings. The remainder of this paper is organized as follows. Section 2 describes how SONAR deals with financial data crawling and storage. Section 3 presents the financial inference and reasoning engine mechanism that SONAR implements. Section 4 depicts the Ontology Population Tool and techniques. The SONAR Web Platform for Accessing Integrated Financial Information is described and two case studies are discussed in section 6. Finally, related work and conclusions conclude the paper in sections 7 and 8.

2 Financial Data Crawling and Storage

SONAR needs to import data from a wide range of channels, being it structured or unstructured. Financial Data Crawling and Storage is based on a software component for the crawling, analysis and storage of financial data represented in several formats, extracting RDF metadata records to be used for search and retrieval, according to a Financial Ontology that serves as a unifying data model. This Financial Ontology is found with the requirements of the case studies and a number of alternatives are envisaged to make it consistent and self-contained. The scenario is shown in Fig. 1:

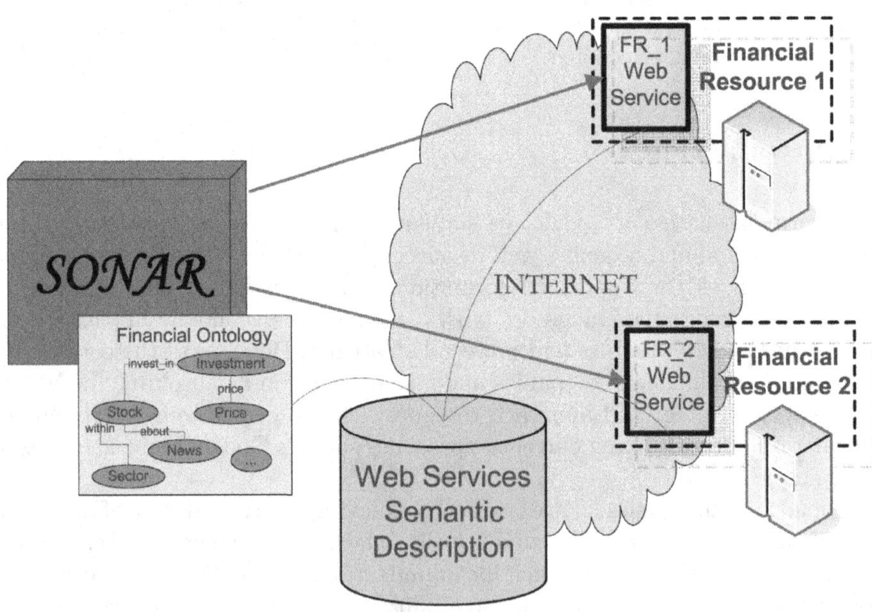

Fig. 1. Financial Data Crawling and Storage

In this component, we will build on the requirements derived from the analysis for automated or supervised knowledge extraction from web pages and other internet-based source, producing RDF metadata according to the defined ontology.

3 Financial Inference and Reasoning Engine

Reasoning is a research area intensively investigated in the recent years. Most of the techniques and inference engines developed for Semantic Web data are focusing either on reasoning over instances of an ontology with rules support (e.g. Rule based approaches [9]) or on reasoning over ontology schemas (DL reasoning [2]). Reasoning over instances of a ontology, for example, can derive a certain value for an attribute applied to an object. These inference services are the equivalent of SQL query engines for databases, but they provide more intelligent support such as handling of recursive rules, etc. Reasoning over concepts of an ontology, for example, can automatically derive the correct hierarchical location of a new concept in a given concept hierarchy. Nowadays also the integration of rule and DL based reasoning approaches gathered a lot of attention. However most of the approaches are considering rather static information, which does not apply to a dynamic domain such as the financial sector.

The Sonar financial inference and reasoning engine addresses the reasoning requirements specific for the financial domain. These includes reasoning with data which changes with a high frequency (e.g. stock exchange data), considerable amount of data produced by financial processes, processing financial data including rules of data manipulation, scalability, etc. The financial inference and reasoning engine provides a two level reasoning functionality. First, a lookup mechanism is provided which is mainly used in conjunction with RDF data. The RDF data is stored in RDBMS with a predefined schema for storing RDF triples. A SPARQL module uses the storage and exposes the query functionality as a SPARQL endpoint to other SONAR components or to external applications. On top of RDF lookup mechanism simple RDFs reasoning was implemented. The second module of SONAR financial inference and reasoning engine uses the semantic data available in richer semantic languages such as WSML [17]. The focus here was on integrating existing reasoning support for WSML-Rule, such as IRIS or KAON2 and to adapt it for reasoning with financial rules using as well the RDF data financial data collected by SONAR crawler.

4 Ontology Population Tool

Ontology population is the process through which a given ontology is populated with instances. Several works have been involved in the development of ontology population systems based on Natural Language Processing (NLP) tools, but these applications suffer from two main drawbacks: they are too language-dependent and they usually need human-expert supervision (i.e. semi-automatic) [11,14]. The here proposed Ontology Population Tool tries to contribute to the ontological engineering community by providing a new, language-independent,

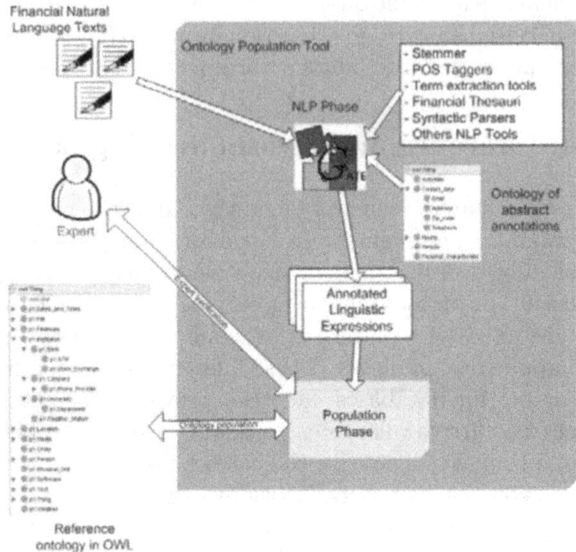

Fig. 2. Ontology Population Tool Architecture

automatic ontology population method from free text. Furthermore, this tool has been specifically trained and tailor-made for the financial domain so that the expected outcome is an extremely accurate application.

This tool gets the reference ontology and natural language free texts as inputs and, using a set of NLP tools, it obtains a populated ontology (see Fig. 2). It is based in the methodology for ontology learning presented in [18].

The ontology population process is comprised of two main phases: NLP phase and Population Phase. Next, each of these phases is described in detail.

4.1 NLP Phase

The aim of this phase is to annotate the most important linguistic expressions (i.e. set of words) that the financial natural language text contains with a set of general categories represented in the Ontology of abstract annotations. This annotation set is formed by both abstract semantic categories, which include person, region, location, city, telephone, price, and age among others, and syntactical categories such as common nouns, proper nouns or adjectives. Thus, in this phase, terms, attributes, values of attributes and other semantically-annotated information are obtained.

The NLP phase is composed of two main components: a set of NLP tools and the Ontology of abstract annotations.

NLP Tools. A set of NLP tools including a sentence detection component, tokenizer, POS taggers, stemmers and Syntactic Parsers has been developed and reused using the GATE framework . GATE is an infrastructure for developing and deploying software components that process human language. GATE helps scientists and developers in three ways: (i) by specifying an architecture, or organizational structure, for language processing software; (ii) by providing a framework, or class library, that implements the architecture and can be used to embed language processing capabilities in diverse applications; (iii) by providing a development environment built on top of the framework made up of convenient graphical tools for developing components.

Ontology of Abstract Annotations. GATE allows to annotate the most important linguistic expressions contained in a corpus. We have developed an ontology which describes a possible set of abstract semantic annotations for identifying possible individuals or these individuals' properties. For example, the linguistic expressions annotated as 'person' might be instances of the class 'Person' in the reference ontology and others linguistic expressions such as 'age' or 'telephone' could be the values of some attributes of the individuals that have been previously identified. The aforementioned NLP tools use the elements that are included in this ontology for carrying out the annotations.

4.2 Population Phase

In this phase, instances are created from the linguistic expressions annotated in the previous stage. The process that takes place during this phase is as follows. First, the most important linguistic expressions are identified using statistical approaches based on term extraction methods. Then, for each linguistic expression, the system tries to determine whether the expression under question is an individual of any of the classes that pertains to the reference ontology. Next, the system retrieves all the annotated knowledge that is situated next to the current linguistic expression in the text, and tries to create a fully-filled individual with this knowledge. For this, the classes, data properties and object properties of the reference ontology have been manually annotated. The system maps the semantically-annotated expressions into possible instances of determined classes or values of datatype properties. For example, if "'IBM"' has been identified as an Enterprise, and "'http://www.ibm.com"' has been identified as an URL and it appears next to the instance, the system would create a new individual of enterprise called IBM and would insert the value of the URL to the home_page datatype property of the new instance.

Besides, a lightweight user interface has been conceived based on the ontology population tool that processes natural language phrases and maps them into the financial reference ontology. Then, this interface is used for accessing financial data from natural language queries.

5 SONAR: A Web Platform for Accessing Integrated Financial Data

SONAR brings together all the components previously described in a combined framework (see Fig. 3). The main aim of the platform is to provide users quick

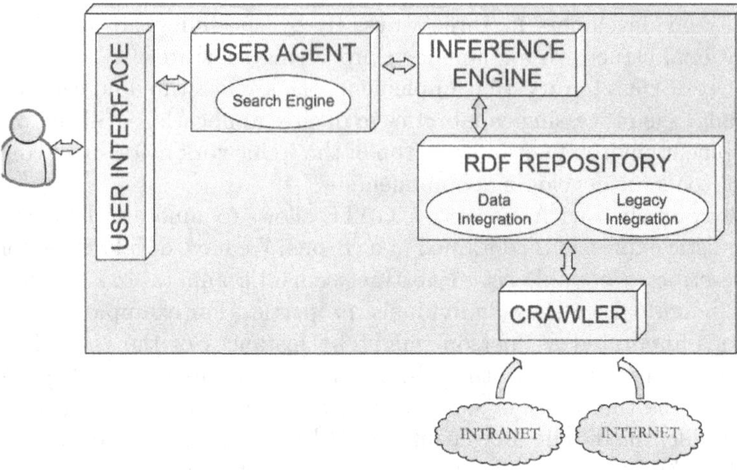

Fig. 3. SONAR Architecture

access to the (integrated) information they require. Accuracy, relevance and trustability are some of the parameters SONAR assesses during this search process.

The operation of the system is divided into two separate stages: (i) data gathering, and (ii) question answering. The components involved in the first stage are the crawler and the RDF repository. The crawler browses a predefined set of sites in the Web and gathers the relevant financial information. For this, the crawler makes use of the ontology population tool described before, which stores the collected instances in the RDF repository. For the system to reach its full potential, corporations' internal information must be made available to the crawler so that it is also analyzed and stored in the repository. The crawling process should be performed periodically so as to keep the repository up-to-date.

Once the semantically-enriched data are available, users can query the system. The user agent is responsible for analyzing the user query and starting the query-answering process. The user query is translated into OWL statements, which are employed during the reasoning stage. Once the system has inferred the expected results, they are sent back to the user.

6 Case Studies

Two software modules have been devised that lie on top of the SONAR Web Platform and serve as use case scenarios to validate the functionality and effectiveness of the platform. The "Analyst Information Assistant Module" (ANNE) is a component that, by processing the information gathered from public sources such as financial information web sites, economic opinion sources, financial information aggregators, blogs, etc., is able to decide what companies to invest in.

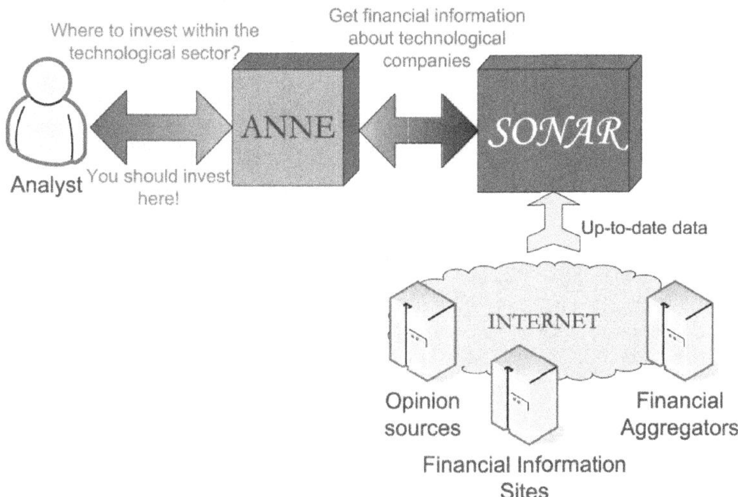

Fig. 4. ANNE mode of operation

On the other hand, the "Trader Information Decision Module" (TIME) deals with data concerning a specific company, and determines when to invest in that particular company.

6.1 ANNE: Analyst Information Assistant Module

ANNE is a software component built on top of SONAR. It is focused on the analysis of historical data, which can be accessed through the SONAR services. The main purpose of ANNE is to assist users (financial analysts for this case study) in identifying stocks which are more likely to increase their value. The result is a decision support system (DSS) for investment planning. The final goal is to build a system capable of determining stocks worth investing in the short, medium and long term taking into account expected returns, acceptable risk, market sectors, investment strategies and time constraints (see Fig. 4). The task is performed by applying new DSS algorithms, specifically engineered for this application, to be used with a dedicated DSS framework.

6.2 TIME: Trader Information Decision Module

The TIME software component module provides the means to decide when is the best time to invest in a company assuming that the goal is to maximize profits. In a similar fashion to ANNE, TIME is a DSS that feeds from the information SONAR possesses regarding companies' historical data, forecasts by experts, and other information sources (see Fig. 5). Thus, TIME can be seen as a DSS for investment scheduling that assists users (traders for this case study) in deciding when the best instant to buy a company's stocks, shares and bonds is. TIME is specifically designed to suit typical trader requirements, modelling the issue using mathematical optimization techniques.

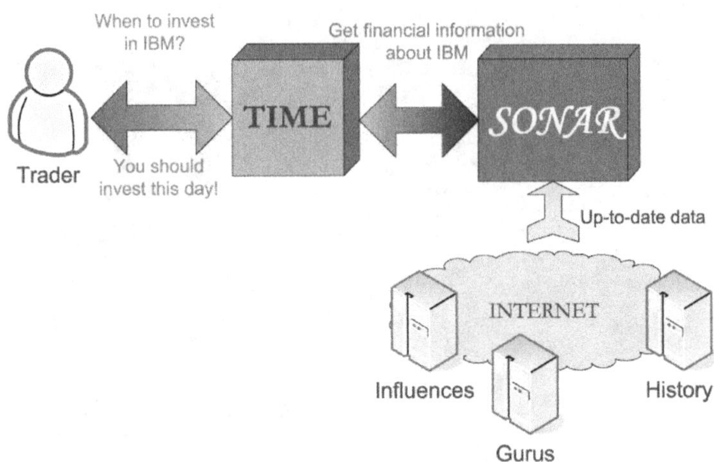

Fig. 5. TIME mode of operation

7 Related Work

Web 2.0 [12] is unleashing a number of possibilities that, combined with the Semantic Technologies, could result in significant success [1]. Since the work on improving search results spans over and binds together a number of research initiatives, in this section we briefly describe related work.

Searching has been subject of intensive research but a more concrete survey on filtering search results and optimizing results yields also a remarkable amount of efforts. Following research successfully implemented in the Google search engine [4], a number of search variants related to the work presented have been explored such as using faceted search [16], including its application to multimedia faceted metadata for image search and browsing [19] or navigating RDF data [13].

Collaborative filtering was coined by Goldberg in [7] and it has been extensively used for data-intensive recommendation systems for personalized recommendations for music albums and artists as can be found in Ringo [15]. Active Collaborative Filtering solutions such as the one discussed in [3] focus on one-to-one recommendations and a social collaborative filtering system where users have direct impact in the final process is described in [8]. A similar work has been intended in SITIO, a Social Semantic Recommendation System [6], which combines the use of semantics with socially-oriented collaborative recommendation systems for the discovery and location of Web resources. Also, Semantic Social Collaborative Filtering has been used with FOAF [10].

8 Conclusions and Future Work

Conventional wisdom holds that new Semantic Technologies promise a powerful paradigm for solving integration problems with current state-of-the-art IT infrastructure such as financial search engines and applications. However, there has not been significant progress in terms of real developments, particularly because of the so-called "chicken-egg" Semantic Technologies problem. In a nutshell, this problem relates to the fact that Web information or data source providers would always request for a good excuse or reason, a good application or benefit, from providing metadata. However, if the metadata is not generated, no application or value-added functionality can be achieved.

However, the explosive growth of a number of structured metadata formats in terms of blogs, wikis, social networking sites and online interoperability-aware communities has transformed the Web in recent years. Mainstream media has taken notice of the so-called Web 2.0 revolution and business success stories have gathered stream.

These technologies are blooming overnight and providing "metadata farms" whose potential can be unleashed by Semantic applications such as SONAR, which will gain momentum towards a Web generation in which Semantic and Web 2.0 technologies will end up meeting.

Our future work focuses on tuning and optimizing that crossway and enabling the transition from research and academic prototypes to practice and from standards to deployment. Tool vendors and manufacturers are reluctant to implement products until they see a market forming, but we envisage the market needs as a powerful driving force to make these solutions mature and fully-fledged business oriented.

Acknowledgements

This work is supported by the Spanish Ministry of Industry, Tourism, and Commerce under projects SONAR (FIT–340000-2007-212), SONAR2 (TSI-020100-2008-665) and GODO II (TSI-020100-2008-564), the Spanish Committee of Education & Science under projects PIBES (TEC2006-12365-C02-01) and MID-CBR (TIN2006-15140-C03-02).

References

1. Ankolekar, A., Krötzsch, M., Tran, T., Vrandecic, D.: The two cultures: Mashing up web 2.0 and the semantic web. J. Web Sem. 6(1), 70–75 (2008)
2. Baader, F., Calvanese, D., McGuinness, D.L., Nardi, D., Patel-Schneider, P.F. (eds.): The Description Logic Handbook. Cambridge University Press, Cambridge (2003)
3. Boutilier, C., Zemel, R.S., Marlin, B.: Active collaborative filtering. In: Meek, C., Kjærulff, U. (eds.) UAI, pp. 98–106. Morgan Kaufmann, San Francisco (2003)
4. Brin, S., Page, L.: The anatomy of a large-scale hypertextual web search engine. Computer Networks and ISDN Systems 30(1-7), 107–117 (1998)
5. Fensel, D.: Ontologies: A Silver Bullet for Knowledge Management and Electronic Commerce. Springer, Heidelberg (2002)
6. Gómez, J.M., Alor-Hernandez, G., Posada-Gómez, R., Abud-Figueroa, M.A., García-Crespo, A.: Sitio: A social semantic recommendation platform. In: Procs. of the 17th International Conference on Electronics, Communications and Computers, p. 29 (2007)

7. Goldberg, D., Nichols, D.A., Oki, B.M., Terry, D.B.: Using collaborative filtering to weave an information tapestry. Commun. ACM 35(12), 61–70 (1992)
8. Kautz, K., Selman, B., Shah, M.: Referral Web: Combining social networks and collaborative filtering. Communications of the ACM 40(3), 63–65 (1997)
9. Kifer, M., Lausen, G., Wu, J.: Logical foundations of object-oriented and frame-based languages. Journal of the ACM (May 1995)
10. Kruk, S.R., Decker, S.: Semantic social collaborative filtering with foafrealm. In: Decker, S., Park, J., Quan, D., Sauermann, L. (eds.) Proc. of Semantic Desktop Workshop at the ISWC, Galway, Ireland, November 6, vol. 175 (2005)
11. Maedche, A., Staab, S.: Ontology learning for the semantic web. IEEE Intelligent Systems 16(2), 72–79 (2001)
12. O'Reilly, T.: What is web 2.0 design patterns and business models for the next generation of software. Technical report(2005), `http://www.oreillynet.com/pub/a/oreilly/tim/news/2005/09/30/what-is-web-20.html`
13. Oren, E., Delbru, R., Decker, S.: Extending faceted navigation for rdf data. In: Cruz, I.F., Decker, S., Allemang, D., Preist, C., Schwabe, D., Mika, P., Uschold, M., Aroyo, L. (eds.) ISWC 2006. LNCS, vol. 4273, pp. 559–572. Springer, Heidelberg (2006)
14. Shamsfard, M., Barforoush, A.A.: Learning ontologies from natural language texts. Int. J. Hum.-Comput. Stud. 60(1), 17–63 (2004)
15. Shardanand, U., Maes, P.: Social information filtering: Algorithms for automating "word of mouth". In: CHI, pp. 210–217 (1995)
16. Suominen, O., Viljanen, K., Hyvönen, E.: User-centric faceted search for semantic portals. In: Franconi, E., Kifer, M., May, W. (eds.) ESWC 2007. LNCS, vol. 4519, pp. 356–370. Springer, Heidelberg (2007)
17. Toma, I., Steinmetz, N.: The web service modeling language wsml. Technical report, wsml, working draft d16.1v0.3 (2007), `http://www.wsmo.org/TR/d16/d16.1/v0.3/`
18. Valencia-García, R., Ruiz-Sánchez, J.M., Vicente, P.J.V., Fernández-Breis, J.T., Martínez-Béjar, R.: An incremental approach for discovering medical knowledge from texts. Expert Syst. Appl. 26(3), 291–299 (2004)
19. Yee, K.-P., Swearingen, K., Li, K., Hearst, M.A.: Faceted metadata for image search and browsing. In: Cockton, G., Korhonen, P. (eds.) CHI, pp. 401–408. ACM, New York (2003)

KBS in Context Aware Applications: Commercial Tools

Nayat Sánchez-Pi, Javier Carbó, and José Manuel Molina

Carlos III University of Madrid, Computer Science Department
Av. de la Universidad Carlos III, 22, 28270 Colmenarejo, Madrid
{nayat.sanchez,javier.carbo,jose.molina}uc3m.es

Abstract. Knowledge based systems are advanced systems of complex problems representation. Its architecture and representation formalisms are the base of nowadays systems. The nature of the knowledge is usually derived from the experience in specific areas and its validation requires a different methodology of the one used in the conventional systems because the symbolic characteristic of the knowledge. On the other hand, context-aware applications are designed to react to constant changes in the environment and to adapt their behavior to its users' situation, needs and objectives. In this contribution, we describe the design, definition and evaluation process of a knowledge-based system using CommonKADS methodology in order to represent the contextual information in a formal way for Appear platform. We also validate the prototype of the context aware system in different realistic environments: an airport, an intelligent home and elderly care which is a significant step into the formally-built applications of KBS.

1 Introduction

The conceptualization of the term "context-aware computing" has been an intention of several authors since there is a significant transformation in the way people look at their interaction with new technologies which are every time smaller and smarter. Context aware computing was firstly defined by [1] some years ago and he claimed that the main components of context were nearby. He was referring to: who you are, what are you doing, where you are, when and why. The resulting information to those questions was divided into three categories which contain concepts like: location, time, space, type of device, meteorological conditions, user activity, nearby people or devices, etc. There is also a more widely accepted and used definition of what context is, and it was given by [2] where he defines context as: "any information that characterizes a situation related to the interaction between humans, applications, and the surrounding environment."

Accordingly, there are several approaches developing systems such as platforms, frameworks and applications for offering context-aware services. The Context Toolkit proposed by [2] assist for instance developers by providing them with abstractions enabling them to build context-aware applications. The Context Fusion Networks [3] was introduced by Chen and Kotz in 2004. It allows

J. Mira et al. (Eds.): IWINAC 2009, Part I, LNCS 5601, pp. 415–425, 2009.

context–aware applications to select distributed data sources and compose them with customized data fusion operators into an information fusion graph. The graph represents how an application computes high level understandings of its execution context from low-level sensory data. The Context Fabric [4] is another toolkit which facilitates the development of privacy–sensitive, ubiquitous computing applications. There are previous approaches like Entree [5] which uses a knowledge base and case-based reasoning to recommend restaurant or for instance Cyberguide [6] project which provides user with context–aware information about the projects performed at GVU center in Atlanta. With TV remote controllers throughout the building to detect users locations and provide them with a map that highlights the project demos available in the neighboring area of the user. A recent one is Appear which is a context–aware platform designed to provide contextual information to users in particular and well defined domains. It has a modular architecture and we have already used it in a previous work [7].In other words, context–aware systems are expected to utilize contextual information in order to adapt their behavior, based on a predefined set of rules. Such rules are, most of the times, monitored by a system which dynamically adapts its operations based on this information. A good solution for the context management is given by Appear and that is why we selected it to model and test our prototype.

Consequently to the increasing development of these kinds of systems and applications, the development of knowledge-based systems are also being increased for being used in areas where context aware systems failures can be costly because of the losses in services, property, etc... There are several methodologies which tackle the development of KBS. CommonKADS [8], in advance CKADS, is the methodology we will use to represent in a formal way a case of use. A model set is the main product resulting from the application of this methodology. CKADS offers six models to represent context: organization, tasks, agents, communication, knowledge and design. The knowledge model is a very good approach to the knowledge representation [9]. It describes in three categories (domain knowledge, tasks knowledge and inference knowledge) what a specific agent has and what is relevant for the development of a particular task describing the structure based on its use. MIKE (Model-based and Incremental Knowledge Engineering) [10] is one of them that provides a methodology for the development of KBS following all the aspects of the process, from the knowledge acquisition. It provides a structured life-cycle facilitating the maintenance, the depuration and the reuse of the resulting product. Proteg [11,12,13] is another one that generates knowledge acquisition tools using reusable libraries. VITAL [14,15] is a less known methodology but it has similar points to CKADS and MIKE. It for instance, produces four products by result of its application. The Conceptual Model it's similar to CKADS and the Design Model it's similar to MIKE. We have chosen CKADS to model our problem because it covers the main points of the development of a KBS, from the very start analysis (in order to identify the problem and to establish the suitability of the solution based on a KBS) to the implementation of the project.

Our contribution in this paper comes from the design, definition and validation of a knowledge-based system using CKADS in order to represent the contextual information in a formal way for Appear platform. It is a centralized solution with a system core where all the received information is managed, allowing the correct interaction between system components. In the following section we briefly describe the fundamental features of the Appear platform and its architectural design. Section III gives a definition of the Appear's KBS using CKADS methodology. Later an overview of two scenarios is presented in order to validate the system and finally we give some conclusions.

2 Appear Platform

Appear is an application provisioning solution for a wireless network. Its solution enables the distribution of location-based applications to users with a certain proximity to predefined interest points. Appear just needs any IP based wireless network and any Java enabled wireless device. In order to locate devices and calculate its position, Appear uses an external positioning engine which is independent of the platform.

Appear platform consists of two parts: Appear Context Engine which is the core of the system and the Appear Client which is installed in the device. Applications distributed by the Context Engine are installed and executed locally in these wireless devices. The architecture of the Appear Context Engine is modular and separates the system responsibilities into: server, one or more proxies, and a client. Appear Context Server is part of the network management. It manages the applications distributed by the platform and the connections to one or more or proxies or positioning engines.

When a wireless device enters the network, it immediately establishes the connection with a local proxy which evaluates the position of the client device and initiates a remote connection with the server. Once the client is in contact with the server they negotiate the set of applications the user can access depending on his physical position. Appear's solution consists then of the Appear Context Engine and its modules: Device Management Module, Push Module and the Synchronization Module. The three modules collaborate to implement a dynamic management system that allows the administrator to control the capability of each device once they are connected to the wireless network. The Push or Provisioning Module manages the automatic distribution of applications and content to handheld devices. It pushes services on these devices using client–side intelligence when it's necessary to install, configure and delete user services. The Device Management Module provides management tools to deploy control and maintain the set of mobile devices. The Synchronization Module manages file–based information between corporate systems and the mobile handheld devices. The Device Management is continuously updated with up-to-date versions of the configuration files. All of these modules are made context aware using the Appear Context Engine.

In Appear, is the Appear Context Engine which gathers context of user data and builds a model based on the needs of the end user. It implements a rules engine, which determines which service is available to whom, and when and where it should be available. Services are filtered against a profile and when it is determined some data are relevant, the information is pushed to the device in a proactive way. As told Appear Context Engine gathers all the context information about the device and produces a context profile for that device. The main components of this model are Context Domain, Context Engine, Context Profile and Semantic Model.

The Context Domain is a set of context values the system can monitor. In the context domain all values are given without any internal relationship. It is fed with context parameters that measure real-world attributes that are transformed into context values. Context parameters include physical location, device type, user role, date/time, temperature, available battery.

The Semantic model is the Administrator model of the relationship between different context parameters and how these should be organized, using context predicates. The Context engine is the one that matches the context domain onto the semantic model and the result of it is the Context profile. To get into a more abstract level Appear creates more complex predicates combining and constraining the values of these context parameters and other context predicates. Context information in the system is used throughout the entire life-cycle of the service. The rules engine filters and determines the appropriate services to be pushed to the user, in the right time and at the right place. The provisioning of the services occurs automatically in the Appear Context Engine as the right context is found to each user: role, zone, location, time period, etc.

3 Appear's KBS Definition Using CommonKADS Methodology

Knowledge based systems are advanced systems of complex problems representation. Its architecture and representation formalisms are the base of nowadays systems. The nature of the knowledge is usually derived from the experience in specific areas and its validation requires a different methodology of the one used in the conventional systems because the symbolic characteristic of the knowledge.

The representation of knowledge in knowledge based systems is varied. CKADS is a structured approach to the development of knowledge based systems and as such is to be seen in contrast to unstructured approaches such as rapid prototyping. CKADS does not require a commitment to any specific implementation paradigm. According to KADS, the development of a knowledge-based system is to be seen as a modeling process, during which models of acquired knowledge at different levels of abstraction are developed. KADS identifies three levels of models:

- The process level identifies the tasks involved in the domain, the nature of data flows and stores, and the assignment of ownership of tasks and data stores to agents.

- The system level, the co-operation model describes in detail the interactions between the system and external agents, and how the internal agents interact. The co-operation model is used to separate the user task model from the system task model, and allows the knowledge wholly internal to the system to be readily identified.
- The expertise level corresponds to an expertise model. This divides the task of describing domain and expert knowledge and its use within the system into a number of supportive tasks. The layer-based framework for expertise consists of:
  - The domain layer is comprised of static or slowly changing knowledge describing concepts, relations, and structures in the domain.
  - The inference layer reformulates the domain layer in terms of the different types of inferences that can be made.
  - The task layer defines knowledge about how to apply the knowledge in these two layers to problem-solving activities in the domain.
  - The strategy layer concerns selecting, sequencing, planning and repairing tasks.

Within this framework knowledge engineering becomes a structured search for appropriate strategy, task, inference and domain models. The modeling of the layers can be made in three steps: a) to determine the static features or domain ontology, b) to obtain the inferences that describe the dynamical side, and c) to group the inferences in a sequential way to form tasks.

3.1 Domain Layer

In this section the representative knowledge of the domain is represented. It is here where the information and the static knowledge are described. Concepts in CKADS are used to define objects collections with similar characteristics. In the case of Appear, the concepts are identified as:

- Domain: It is the high level concept representing the domain on discourse.
- Location: Includes the domain location, their zones and the user location.
- (x;y): There are the coordinates of the airport location, their zones and the user position.
- Zone: It is a segment of the airport area distinguished from adjacent zones.
- User: Every person who has a role in the domain.
- Role: Role of each user in the domain.
- Offering: A class containing a set of services grouped by categories.
- Category: A class containing a set of grouped services.
- Services: Applications or notifications.

These concepts can be represented in different ways. In Figure 1 we can see the representation of the User concept using CML language where the characteristics are the represented by means of "attributes" unlike other approaches where as well as the characteristics, the functional information is included.

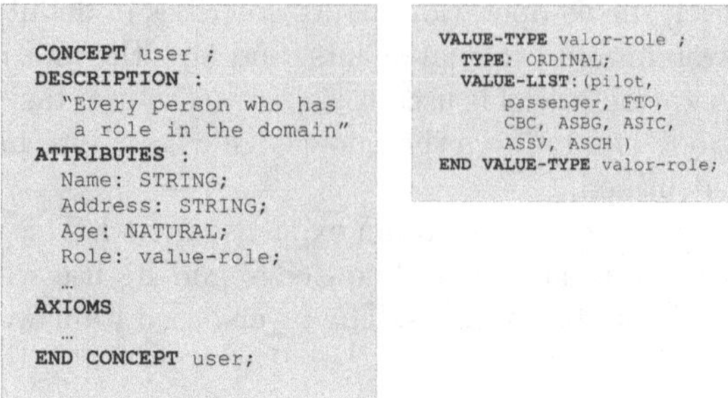

CONCEPT user ;
DESCRIPTION :
"Every person who has
a role in the domain"
ATTRIBUTES :
Name: STRING;
Address: STRING;
Age: NATURAL;
Role: value-role;
...
AXIOMS
...
END CONCEPT user;
VALUE-TYPE valor-role ;
TYPE: ORDINAL;
VALUE-LIST: (pilot,
passenger, FTO,
CBC, ASBG, ASIC,
ASSV, ASCH)
END VALUE-TYPE valor-role;

Fig. 1. CML definition of the User Concept

Once we have the concepts we can establish relationships between them. In order to define the different type of rules in our domain schema, we need to represent a previous/preamble and a consequent. It also necessary to represent the connection symbols used to connect the preamble to the consequent.

The input/output curve represents a global relation or a group of partial I/O relations. The Domain Knowledge dynamics is gathered from these relations between input ranges and output ranges.

There are concepts in Appear like predicates, context predicates and conditions that are combined to create rules which are part of the rules engine in the Appear Context Engine. It is here where it is determined which service is available to whom and where and when should it be pushed to the user's device. An example of the assessed to provide a service can be:

- **Appear Rule Inference:** If a user with a role is into a zone at a time it implies some offering will be offered to him.
- **Appear Rule Inference:** Given an offering, and category of the user, it implies that some services are provided to the user and others don't.

3.2 Inference and Task Layer

After getting the Domain Ontology, the domain dynamical component must be obtained. This dynamic aspect copes with the system input/output behavior that is stated as a set of production rules. In this layer we will describe how the static structures defined above will be used to develop the reasoning process.

Inferences are completely described through a declarative specification of its entries and exits (dynamic roles). Getting back to the representation using CML language, Figure 2 shows the CML specification of an inference where the input is a role "case" representing the knowledge elements of the domain and the output is the role "abstracted-case" representing the qualified description of the input.

When both Domain Ontology and Dynamics have been obtained, we have to group inference into tasks, where tasks are similar to traditional functions but the data manipulated by the task are described in a domain-independent way. They describe the input/output and how a task is realized through an

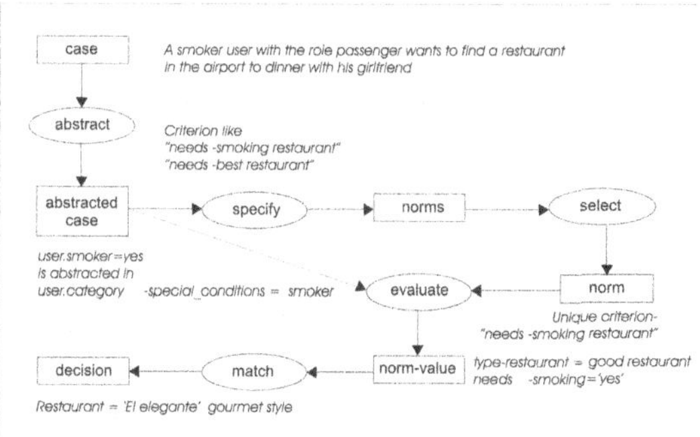

Fig. 2. Inference diagram for the assessment in the decision of choosing a restaurant in the Airport

ordered decomposition into sub-functions as: another task, inference and transfer function. They form a small program or algorithm that captures the reasoning strategy of the expert.

The particularity of CKADS is the partial reuse of knowledge models in new applications, proposing a catalog of task templates comprised by: a tentative inference structure, a typical control structure and a typical domain schema from task point-of-view which are specific for a task type.

Among the task types that CKADS suggests, we consider our problem as a particularization of an assessment task template; where an assessment problem consists of finding a decision category for a case based on domain specific norms.

4 Overview of Two Scenarios

In this section we will present two prototypes for two different scenarios: an airport and an intelligent home, where we made the validation of the KBS described in previous sections. It will also illustrate how concepts, relationships and inferences, described in the dynamic domain knowledge, are represented in the way the provisioning of services to each user occurs and it is represented in the user interface.

It is important to begin explaining we test our system in an HTC Touch terminal with a TI's OMAP 850, 200 MHz processor and Windows Mobile 6. We also deploy a WiFi positioning system in order to locate the user's terminals. Particularly, we used an Aruba WLAN deployment with a 2400 Mobility Controller with capacity of 48 access points and 512 users connected at the same time and several AP 60/ AP61 which are single/dual radio wireless access points. Aruba Networks [17] has a location tracking solution which uses an enterprise wide WLAN deployment to provide precise location tracking of any Wi-Fi device in the research facility. The RF Locate application can track and locate any Wi-Fi device within range of the Aruba mobility infrastructure.

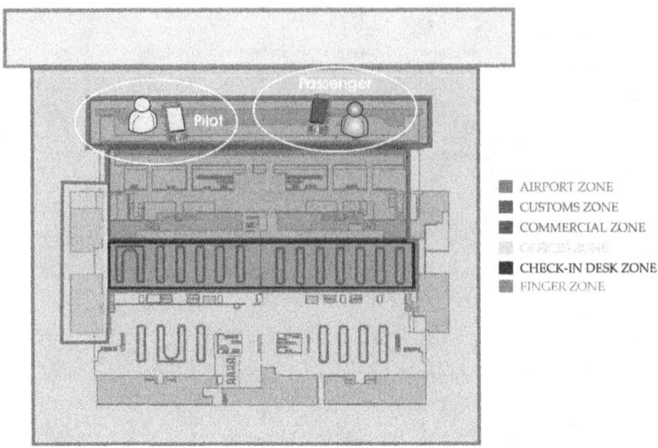

Fig. 3. Airport Environment. Pilot and Passenger users into customs zone.

4.1 Airport Scenario

Our first scenario in an airport WiFi covered, thought to assist the passengers and the workers of an airline. Airport environment is devided into several zones (see Figure 3).

- Scenario I: Airport + Pilot user + Customs Zone
- Event part: When John, carrying a PDA, is detected into the Customs Zone,
- Condition part: (and) he is a pilot of IBERIA Airlines
- Action part: (then) the PDA turns on, and receives FLIGHT RESOURCES SERVICES, NEWS SERVICES, MAPS SERVIÇES, LOAD CONTROL SERVICE, CONTROL REPORT SERVICE AND SUPPORT CONTROL SERVICE

The following rule is evaluated in order to offer the appropriate services to the pilot who is in the customs zone of the airport. Once this rule is evaluated, services are pushed to the pilot's device, Figure 4.

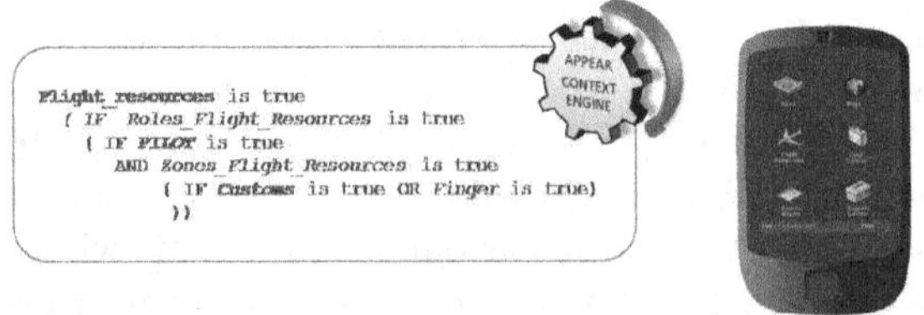

Fig. 4. Services offered to pilots in the customs zone

4.2 Intelligent Home Scenario

The second scenario is an intelligent home WiFi covered, thought to assist the members of a family. Our home environment is devided into several zones. In Figure 5, we can see the `TVroom_zone` and `Kitchen_zone`.

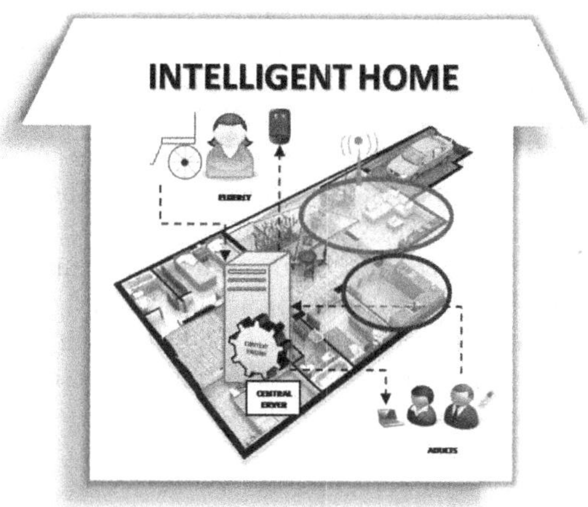

Fig. 5. Home environment. Elderly and adult users in different zones of the house: TV room and kitchen.

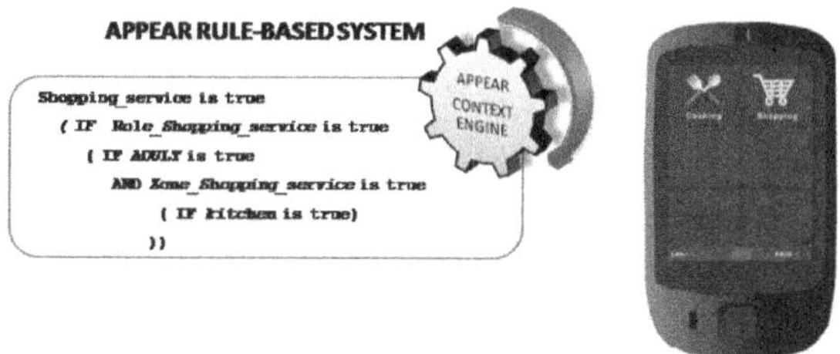

Fig. 6. Services offered to adult users in the kitchen

The provisioning of the services occurs automatically in the Context Engine as the right context is found to each user: Role, Zone, Location, etc. Our second scenario is an adult, Louise, who is entering the kitchen of the house and who usually do some online shopping every month. She's also carrying a WiFi PDA and the system:

- Scenario II: Intelligent Home + Adult + Food Shopping
- Event part: When Louise with a PDA is detected in the kitchen,
- Condition part: (and) it is the first time in the month,
- Action part: (then) turn on the PDA, and displays the SHOPPING'S ALERT on the PDA screen, as well as, the Shopping and the Cooking service.

Rule shown in Figure 6, is evaluated in order to offer the appropriate services to the adult the possibility of shopping through this system and also take some cooking advices and recipes.

5 Conclusions and Future Work

We have used Appear as an off-the-shelf platform that exploits the modular and distributed architecture to develop context-aware applications, in order to design the contextual information for an intelligent home. Appear platform was designed to alleviate the work of application developers. At some point it succeeds to do so, but the applications inherit the weaknesses that the system possesses, for instance, the Context domain is limited to a set of concepts and it is impossible to represent the real environment. So, this is not a realistic approach if this system is to be deployed for a testbed with real users. Among the issues that could be additionally improved, the platform could be extended in a manner that enables the consumer application to get information about the quality of the context data acquired.

Acknowledgments

This work was supported in part by projects CICYT TIN2008–06742–C02–02/TSI, CICYT TEC2008–06732–C02–02/TEC, SINPROB, CAM MADRINET S–0505/TIC/0255 and DPS2008–07029–C02–02.

References

1. Schilit, B.: A System Architecture for Context-Aware Mobile Computing. Ph.D thesis, Columbia University, Department of Computer Science (May 1995)
2. Dey, A., Abowd, G., Salber, D.: A conceptual framework and a toolkit for supporting the rapid prototyping of context-aware applications. Human-Computer Interaction 16, 97–166 (2001)
3. Chen, G., Li, M., Kotz, D.: Design and implementation of a large-scale context fusion network. MobiQuitous, 246–255 (2004)
4. Hong, J.: The context fabric: An infrastructure for context-aware computing. In: Minneapolis, A.P. (ed.) Extended Abstracts of ACM Conference on Human Factors in Computing Systems (CHI 2002), pp. 554–555. ACM Press, Minneapolis (2002)
5. Burke, R., Hammond, K., Young, B.: Knowledge-based navigation of complex information spaces. In: Proceedings of The National Conference On Artificial Intelligence, pp. 462–468 (1996)
6. Abowd, G., Atkeson, C., Hong, J., Long, S., Kooper, R., Pinkerton, M.: Cyberguide: A mobile context-aware tour guide. Wireless Networks 3(5), 421–433 (1997)
7. Sánchez-Pi, N., Fuentes, V., Carbó, J., Molina, J.: Knowledge-based system to define context in commercial applications. In: Proceedings of 8th International Conference on Software Engineering, Artificial Intelligence, Networking, and Parallel/Distributed Computing (SNPD), Qingdao, China (2007)

8. Schreiber, G., Akkermans, H., Anjewierden, A., Hoog, R., Shadbolt, N., Van de Velde, W., Wielinga, W.: Knowledge engineering and management, the commonkads methodology. The MIT Press, Cambridge (1999)
9. Wielinga, B., Akkermans, J., Hassan, H., Olsson, O., Orsvärn, K., Schreiber, A., Terpstra, P., Van de Velde, V., Wells, S.: Expertise model definition document. technical report espirit project p5248/ kads-ii/m2/uva/026/5.0. Technical report, University of Amsterdam and Free University of Brussels ans Netherlans Energy Research Centre ECN (June 1994)
10. Angele, J., Fensel, D., Studer, R.: Domain and task modeling in MIKE. Univ. (1996)
11. Puerta, A., University, S.: A multiple-method knowledge-acquisition shell for the automatic generation of knowledge-acquisition tools. In: Knowledge Systems Laboratory, Medical Computer Science. Stanford University (1991)
12. Eriksson, H., Shahar, Y., Tu, S.W., Puerta, A.R., Musen, M.A.: Task modeling with reusable problem-solving methods, vol. 79, pp. 293–326. Elsevier, Amsterdam (1995)
13. Tu, S., Eriksson, H., Gennari, J., Shahar, Y., Musen, M.: Ontology-based configuration of problem-solving methods and generation of knowledge-acquisition tools: Application of PROTÉGÉ-II to protocol-based decision support., vol. 7. Elsevier, Amsterdam (1995)
14. Shadbolt, N., Motta, E., Rouge, A.: Constructing knowledge-based systems. IEEE Software 10(6), 34–38 (1993)
15. Domingue, J., Motta, E., Watt, S.: The emerging VITAL workbench, p. 320. Springer, Heidelberg (1993)

An Architecture Proposal for Adaptive Neuropsychological Assessment

María M. Antequera[1], M. Teresa Daza[2], Francisco Guil[3,*], Jose M. Juárez[4], and Ginesa López-Crespo[5]

[1] Hospital Universitario Virgen de La Arrixaca - Murcia
mariam.antequera@carm.es

[2] Dept. Neurociencia y Ciencias de la Salud - Univeridad de Almería
tdaza@ual.es

[3] Dept. de Lenguajes y Computación - Universidad de Almería
francisco.guil@ual.es

[4] Dept. Ing. de la Información y las Comunicaciones - Universidad de Murcia
jmjuarez@um.es

[5] Dept. Psicología Básica y Metodología - Universidad de Murcia
ginesa.lopez.crespo@um.es

Abstract. In this work we present the architecture of an special sort of software capable of changing its presentation depending on the behavior of a particular user. Although the proposed solution is applicable as the base of a general adaptive application, we want to expose the peculiarities of the model designed specifically to work in a special domain, the cognitive neuropsychology. In this domain, one of the most important topic is the design of task for the assessment of patients with cerebral damage. Depending of the patient, and its particular characteristics, the therapists must adjust the exposition time of several stimulus in order to obtain the best parameter values, and therefore to obtain the profile in a precise way. This is a very important topic because correct assessment implies successful rehabilitation but, in practice, this process is time-consuming and difficult to be done. So, with the aim to help the therapists in the task tuning process, we propose the use of artificial intelligence-based techniques in order to detect the patient's profile automatically during the execution time, adapting the values of the parameters dynamically. In particular, we propose the foundation of temporal similarities techniques as the basis of design adaptive software.

1 Introduction

Cognitive science is an interdisciplinary field with contributors from several areas, including psychology, neuroscience, computer science, linguistics, philosophy of mind, anthropology, biology and physics. In other words, the field of cognitive science encompasses all of the different approaches to studying intelligent systems [11]. Its intellectual origins are in the mid-1950s when researchers in

* Corresponding author.

J. Mira et al. (Eds.): IWINAC 2009, Part I, LNCS 5601, pp. 426–436, 2009.

several fields began to develop theories of mind based on complex representations and computational procedures. However, its organizational origins are in the mid-1970s when the Cognitive Science Society was formed.

Cognitive psychology and cognitive neuropsychology are two branches of cognitive science which study the same intelligent system, the human brain. Cognitive psychologists seek to understand how people are able to carry out cognitive tasks as producing and understanding language, storing and retrieving information in memory, recognizing objects, etc. Typically, they do experiments with subjects fully competent at such tasks, testing predictions from their theories of cognition. Cognitive neuropsychologist study brain-damaged individuals to learn more about how the normal brain processes information. The cognitive neuropsychology uses data from people with impairments of cognitions (after brain damage), to develop or test theories about how the cognitive functions are normally carry out. In other words, the cognitive psychologists study normal brain function and cognitive neuropsychologists study individuals with brain dysfunction.

In clinical practice, the neuropsychologist is specialized in the assessment and treatment of patients with neurological damages, including those suffering from stroke, traumatic brain injury, dementia, epilepsy, and brain tumors, among others. The neuropsychologist needs to know the cognitive abilities and deficits in order to build a map of what is and not cognitively possible for the patient and to individualize the treatment or rehabilitation program [15]. In this context, a core part of neuropsychological assessment is the administration of specific tasks for the formal assessment of cognitive functioning. The major cognitive domains typically assessed in a neuropsychological examination include: perception, attention, memory, learning, language, thinking and executive functions.

The traditional paper-and-pencil neuropsychological tasks used to explore cognitive functioning in neurological patients have demonstrated some weaknesses, for example, lack timing precision and depend, in large part, on the motor and sensory skills of the examiner. Computerized neuropsychological tasks improve precision and accuracy, require less administration time and measure reaction time more precisely. But even with these standard computerized neuropsychological tasks, it is difficult to measure accurately cognitive functioning in patients with cerebral damage, specially when these patients show a general slowing of information-processing (or cognitive slowing). This "cognitive slowing" in each patient must be taken into careful consideration both during the administration of tasks and also during the process of analyzing the obtained results. The patients might show a deficient performance in a particular cognitive task because of a general slowing of information-processing even if the underlaying cognitive function remains intact. In this situation, and with the aim of controlling the effect of processing speed (the rate of overall information processing), it is necessary to adapt the timed-parameters associated with the tasks, establishing the right values according with the patient's profile. The adaptation process is iterative in nature and time-consuming: for each patient, the therapist must change the values of the parameters, studying the obtained results and setting the correct values once the tuning phase are concluded.

In order to solve this problem, we propose to incorporate intelligent-based techniques into the task design that allow to tune the parameters dynamically depending of the patient's behavior. In particular, the basic idea consist of the monitorization of the patient's action, determining its profile using temporal similarity techniques, and assigning the corresponding values to the parameters of the task. Similarity techniques are the core of case-based reasoning (CBR), a methodological approach based on the idea that a new problem can be solved using similar problems already solved [14]. In dynamic domains, which are time-varying in nature, these problems are modeled as temporal cases, represented by means of event sequences when the temporal features are heterogeneous [7]. A temporal case represents a previous experience describing the nature of the problem, when it occurred, and its temporal evolution. Usually, the case base (formed by a enumeration of temporal cases) is obtained directly from expert knowledge using knowledge acquisition tools. In our case, we propose to use temporal data mining techniques in order to obtain the set of temporal cases (patient profiles) from a historic database storing the results obtained from previous evaluated patients. In particular, we propose to use the sequential version of the $TSET - Miner$ algorithm [5], designed for mining frequent event-based sequences from relational and transactional databases. From the discovered set of sequences, we can obtain a set of temporal constraint-based possibilistic patterns [6] with the aim to combine discovered knowledge with expert knowledge.

The detection of the patient's profile will be carry out in a dynamic way, searching in the case base previous evaluated patients with similar temporal evolution and adopting the same parameter configuration for the current task. In order to provide an effective solution, it is necessary to use good temporal similarity measures, as the proposed in [8], which is based on the Minkowsky distance. An alternative similarity measure is defined in [7], which use a non-classical distance metric, based on a temporal model of possibilistic temporal constraint networks defined in [2].

In this work we present the architecture of an special sort of software capable of changing and adapting its presentation (its temporal parameters) depending on the particular characteristics of a given user (patient) in terms of cognitive slowing. In particular, an adaptive version of a particular task, the *Sternberg* task [13], is proposed in Section 2. In Section 3, we introduce the proposal architecture which is centered in a Web service that computes similarity event sequences measures in order to obtain the patient's profile. Section 4 formalizes the problem and presents the basic principles of event-based sequences similarity. Conclusions and future works are finally drawn in Section 5.

2 A Case Study: The Sternberg Task

The *Sternberg* task [13] is aimed to assess the capacity of an individual to codify and maintain information in working memory. Human memory is not the unitary system that the first psychologist thought; rather, it is composed of multiple systems each one subserving its own function [4]. Working memory is defined as the specific memory system that involves the temporary storage and manipulation of information [1]. Working memory function is necessary for a wide range of complex cognitive activities such as reasoning, language, comprehension, planning and spatial processing [3].

Briefly, a *Sternberg* task is a working memory computerized task in which the patient has to actively maintain a previously seen set of 3 to 9 consonants in his/her working memory because later he/she is required to inform if a given consonant is between this set or not. Figure 1 presents an item of the task. Each square represents a different computer screen. Each trial begin with the word "READY!" in the center of the first screen during 1 sec. Then, a string of 3 to 9 consonants appears on the center of the second screen. The duration of this screen is proportional to the number of consonants appearing (one sec per each consonant). In the example provided in Figure 1, the second screen last 4 sec. since there are 4 letters on it (*VCDR*). After 2 sec. in which a third screen remains blank, a single consonant appears in a fourth screen. The task of the participant is to inform if the single consonant was previously seen on the second screen or not. If his/her response is "YES" they are instructed to press the letter "M" on the keyboard. If his/her response is "NO", they have to press the letter "C". The single letter remains on the screen until the patient make a response on the keyboard. After that, the subject is required to press the space bar to continue with the next trial, i.e. to start with the next item. In total, the sequence is repeated 56 times.

As stated before, the patients with widespread cerebral damage show cognitive slowing. Let's imagine that we have a patient with cognitive slowing performing a *Sternberg* task. It is possible that the 3 to 9 sec that the second screen lasts is not enough for the patient to process the information contained on this screen, and

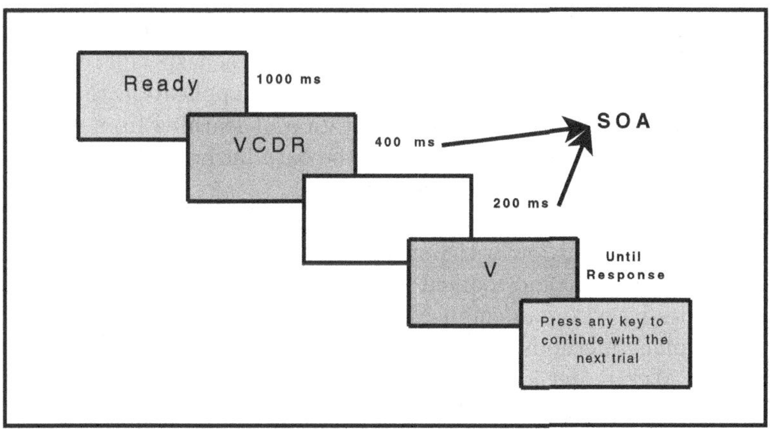

Fig. 1. An item in a typical Sternberg task

therefore a poor performance in the task is predictable. With only these data, we can establish a wrong diagnosis, concluding that the patient has a working memory dysfunction. But if the same patient is given an appropriate amount of processing time, it is possible to observe a performance similar to a pair (a control or intact cognitively subject). To solve this difficulty, we propose an adaptive version of the *Sternberg* task in which the second screen duration is adapted to the particular characteristics of each patient. It is of outstanding importance that the duration match exactly the speed of processing of this patients, since a shorter duration will preclude an appropriate processing, but a longer [11] duration will allow the patient to engage other memory systems (e.g. a long-term memory system) in such way that they can solve the task correctly but through a system different to the working memory, and therefore, with the risk of emitting a wrong diagnosis.

It is important to note that the same architecture can be adapted to further neuropsychological tasks sharing the same architecture than the *Sternberg* one. For example, it can be adapted to short-term memory neuropsychological tasks that follows a similar event sequence. In a typical short-term memory procedure, the subject is exposed briefly to a stimulus (for example, a photograph), and after a delay longer enough to exceed the temporal capacity of working memory, a set of several photographs are presented. The task of the subject is to inform if the first photograph is contained between this set of photographs or not. As we can see, both task shared a basic common architecture, allowing our proposed software to be adapted from one task to another. In general, our proposal is designed to deal with (neuro)psychological protocols, where a protocol is formed by a enumeration of interrelated tasks.

3 Adaptive Application Architecture

Learning and evolution are two fundamental forms of adaptation [16]. The adaptive Web emerges as a solution to satisfy, depending on his/her interaction with the Web application, the particular necessities of each specific user [10]. An adaptive Web application (or site) is defined as an application that automatically improve their organization and presentation by learning from visitor access patterns. Adaptive Web applications mine the data buried in Web server logs to produce more easily navigable Web applications.

We can find in the literature several approaches for adapting the Web structure depending of the particularities of the user. These approaches can be divided into two groups: explicit and implicit methods. The explicit methods determine the user profiles based on an enumeration of predetermined rules. On the other hand, the implicit methods extract the user profiles from a set of discovered patterns from Web server logs [9].

Following the same logic as the implicit adaptive methods, we propose an architecture of an adaptive application for neuropsychological assessment. However, the solution differs from the previous proposal in two ways. First, instead of mining the Web server log, the user profiles is extracted from a complex

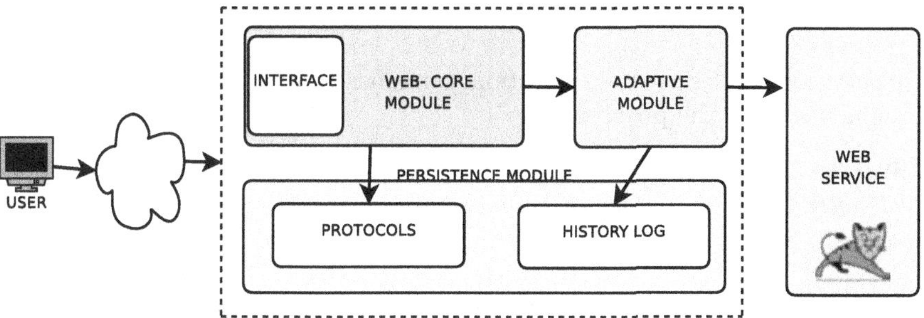

Fig. 2. Basic architecture of the adaptive Web

database storing the data of patient's evaluation. Second, in the data analysis we take into account the temporal dimension. Neuropsychological assessment is a dynamic domain, and therefore, if we want to obtain useful patterns, we must incorporate the temporal dimension in the data mining techniques. So, we propose to use the $TSET - Miner$ [5] algorithm, a temporal data mining algorithm that uses a unique tree-based structure for mining a set frequent sequences from databases. The adaptive nature of the Web application is characterized, precisely, both by the temporal data mining and the similarity used technique.

The proposal architecture (see Figure 2) is composed of four modules: consistency module, core-application module, adaptive module and the similarity Web service.

- The module of persistence includes the protocol (static) and history (dynamic) databases. The protocol database holds the static data of the psychological protocols, that is, the set of tasks (also called tests) and their multimedia sources (called stimuli). The history log registers the answers given by the users and its response times, that is, the temporal sequence of answers.
- The core application module is a Web-based application to show the tasks (loaded from the protocol database) driven by a set of temporal parameters (e.g. exposing time of a stimuli or blank screen) and to gather the responses given by the user (registered in the history log).
- The adaptive module tunes the timed-parameters of the core-application module to personalize the tasks depending of the particular characteristics of each patient. To this end, the adaptive system monitors the answers and time spent by the user (temporal behavior) and tries to adapt the becoming event sequences based on past experiences.
- To this end, the similarity Web service is used to find those records in the history log that are analogue to the temporal behavior of the user. The most similar record is selected and their timed-parameters used in that case are now used to tune, on-line, the current configuration of the task.

4 Adaptation Based on Similarity

In this section we describe our notation, some basic definition, and we introduce the foundations of the proposal.

Definition 1 (Neuropsychological protocol). *A neuropsychological protocol, denoted by Prot, is formed by an enumeration of interrelated task, that is,* $Prot = \{T_1, T_2, \ldots, T_n\}$*, where n is the number of tasks.*

Definition 2 (Assessment task). *An assessment task T is composed of a set of m items, that is,* $T_i = \{I_1, I_2, \ldots, I_m\}$*. Usually, the first items (selected by the therapist) corresponding to the tuning phase.*

As we can see in Figure 1, an item I has associated a set of four events corresponding with different computer screen. Each screen has associated different content and timed-parameters. In particular, each item is compound of a ready screen, an stimulus screen, a blank screen, and a response screen, each one associated with the t_{ready}, $t_{stimulus}$, t_{blank}, and $t_{response}$ timed-parameters, respectively. In the neuropsychology context, it is common to use the parameter SOA (*Stimulus Onset Asynchrony*) to measure the time passed from the exposition of the stimulus until the user response is registered. In short, $t_{SOA} = t_{stimulus} + t_{blank}$.

Definition 3 (Item). *For simplicity in the notation, we will consider an item represented by a unique event* $e(I) = (te, t_{response}) = te_{t_{response}}$*, where te is the type event associated with the item, and* $t_{response}$ *is the elapsed time between the end of the blank screen and the beginning of a response.*

Definition 4 (Type event). *A type event is the* $4-tuple$

$$te = (t_{ready}, stimuli, t_{SOA}, response),$$

where stimuli is formed by two multimedia files, the sample and the comparative stimulus, and response indicates if the answer was correct of wrong.

Example 1. Let's consider the task configuration showed in Table 1.

Table 1. An example task configuration

$Parameters$	$Item_1$	$Item_2$	$Item_3$
$ready$	1000	1000	1000
$sample$	VSR	$VEDS$	$VCDR$
$t_{stimulus}$	3000	4000	4000
t_{blank}	2000	2000	2000
$comparative$	V	D	X
$response$	S	N	N
$t_{response}$	1000	2000	2000

For this configuration, we can extract the three next type events:

$$a = te(item_1) = (1000, (VSR, V), (3000, 2000), ?)$$

$$b = te(item_2) = (1000, (VEDS, D), (4000, 2000), ?)$$

$$c = te(item_3) = (1000, (VCDR, X), (4000, 2000), ?)$$

The ? symbol denotes that any response has been realized by the user. Let's suppose that the user runs the task, obtained the results that showed in Table 1. As we can see, there is only one wrong answer, corresponding with second trial. The task execution generates 3 events:

$$e(item_1) = (1000, (VSR, V), (3000, 2000), Correct, 1000) = (a, 1000) = a_{1000}$$

$$e(item_2) = (1000, (VEDS, D), (4000, 2000), Wrong, 2000) = (b, 2000) = b_{2000}$$

$$e(item_3) = (1000, (VCDR, X), (4000, 2000), Correct, 2000) = (c, 2000) = c_{2000}$$

An finally, if we express the temporal unit of the events in seconds, the generated events are:

$$e(item_1) = (a, 1) = a_1$$

$$e(item_2) = (b, 2) = b_2$$

$$e(item_3) = (c, 2) = c_2$$

Definition 5 (Orden relation). *Given two events $e_i = e(item_i)$, $e_j = e(item_j)$, we define the $<$ relation as follows:*

$$e_i < e_j \textit{ iff } e_i.t_{response} < e_j.t_{response}$$

Definition 6 (Event-based sequence). *A event-based sequence S is an ordered set of events, $S = \{e_1, e_2, \ldots, e_k\}$, where $e_i < e_{i+1}$, for all $i = 1, \ldots, k-1$.*

Definition 7 (Derived sequence). *Let T be a task formed by m items. Its execution generates a set of m events, $\{e(item_1), e(item_2), \ldots, e(item_m)\}$. Let be S^T the sequence derived from the executed task. We define S^T as the sequence of $m+1$ events $S^T = \{e_0, e_1, \ldots, e_m\}$, where:*

1. *The e_0 event is an special event which represents the time origin. Its type event associated is empty and the response time is equal to 0 temporal unit.*
2. *For all $i > 0$:*
   (a) *$e_i.te = e(item_i).te$, and*
   (b) *$e.t_{response} = e(item_i).t_{response} + e(item_{i-1}).t_{response}$.*

Example 2. From the events generated in the Example 1, we obtain the derived sequence $S^T = \{e_0, e_1, e_2, e_3\}$, where $e_0 = (\emptyset, 0)$, $e_1 = a_1$, $e_2 = b_3$, and $e_3 = c_5$. That is, $S^T = \{\emptyset_0, a_1, b_3, c_5\}$.

After the tuning phase, which is usually associated with the first items of the task, the system asks the similarity Web service for the most similar patient in order to determine the patient's profile, and therefore, the most suitable parameters setting (specially the t_{SOA} values). The Web service, which accepts as input the derived sequence, searches for the most similar in the mined profiles database. The used similarity technique is the proposed in [8], where a temporal extension of the edit distance [12] is used. The intuitive idea of the technique is that the similarity (or distance) measure should somehow reflect the amount of work needed to transform a sequence to another. The definition of similarity is formalized as an edit distance $d(S, T)$, and for counting it, the authors define a set of transformation operations:

1. $S.ins(te, t)$ that adds the pair (te, t) to the sequence S;
2. $S.del(te, t)$ that removes the pair (te, t) from the sequence S, and
3. $S.move(te, t, t')$ that moves an existing event (te, t) from time t to time t'.

Each operation o_i has an associated cost, denoted by $cost(o_i)$. If we obtain the total cost as the sum of the costs of the whole set of necessary operations to transform a sequence S to a sequence T, the distance $d(S, T)$ is defined as the sum of costs of the cheapest sequence of operations for computing the transformation. The authors demonstrate that d is a metric and propose an efficient dynamic programming algorithm.

Definition 8 (Profiles database). *Let BD be a database formed by n mined patient profiles, that is, $BD = \{P_1, P_2, \ldots, P_n\}$. Each profile represents a temporal case describing the temporal evolution of a representative patient. Also, each one has associated the most suitable values for the timed-parameters of the task, that is, the best values for the t_{SOA} parameter. So, a patient's profile can be denoted as the pair $P_i = (S_i, t^*_{SOA})$, where t^*_{SOA} is the recommended for the profile P_i.*

Let S^T be the derived sequence from the task performed by the patient P. The similarity searching operation consist of obtaining the profile with the minimum edit distance with S^T. That is,

$$profile(S^T) = P_i | d(S^T, P_i) = \min_i \{d(S^T, P_i)\}$$

Once the profile is determined, the values of the t_{SOA} parameter is the associated with P_i, that is, t^*_{SOA}.

5 Conclusions and Future Works

In this paper we have presented a general architecture for an adaptive Web application. The adaptability is based on a similarity measure that determines the most suitable user's profile depending of his/her behavior. The aim is to build dynamic applications that personalize its content to the particular characteristics of the subject. The proposal is centered in the neuropsychological assessment domain which is time-varying in nature. In order to introduce this domain, a particular task is described, the *Sternberg* task, serving as the basis of the formal problem description. In this particular case, to adapt the content implies to set the best values of the timed-parameters associated with the task. As future work we propose to extend the framework to deal not only with *Sternberg*-like assessment tasks, but else with others specialized cognitive tasks that can be used to develop and improve "cognitive training programs", and also to develop specific adaptive learning methods for users with cognitive disabilities and learning difficulties (users with special needs education).

Acknowledgments

This work is supported by the Grant P07-SEJ-03214 from Consejería de Innovación, Ciencia y Empresa de la Junta de Andalucía (Spain).

References

1. Baddeley, A.: Working Memory and Language: and Overview. Journal of Communication Disorders 36, 189–208 (2003)
2. Dubois, D., HadjAli, A., Prade, H.: A Possibility Theory-Based Approach to the Handling of Uncertain Relations between Temporal Points. Intelligent Systems 22, 157–179 (2007)
3. D'Esposito, M.: From Cognitive to Neural Models of Working Memory. Phil. Trans. R. Soc. B. 362, 761–772 (2007)
4. Gabrielli, J.D.E.: Cognitive Neuroscience of Human Memory. Annu. Rev. Psychol. 49, 87–115 (1998)
5. Guil, F., Bosch, A., Marín, R.: *TSET*: An Algorithm for Mining Frequent Temporal Patterns. In: Proc. of the ECML/PKDD Workshop on Knowledge Discovery in Data Streams (KDDS 2004), pp. 65–74 (2004)
6. Guil, F., Marín, R.: Extracting Uncertain Temporal Relations from Mined Frequent Sequences. In: Proc. of the 13th Int. Symposium on Temporal Representation and Reasoning (TIME 2006), pp. 152–159 (2006)
7. Juárez, J.M., Guil, F., Palma, J., Marín, R.: Temporal Similarity by Measuring Possibilistic Uncertainty in CBR. Fuzzy Sets and Systems 160(2), 214–230 (2009)
8. Mannila, H., Moen, P.: Similarity between Event Types in Sequences. In: Mohania, M., Tjoa, A.M. (eds.) DaWaK 1999. LNCS, vol. 1676, pp. 271–280. Springer, Heidelberg (1999)
9. Mobasher, B., Cooley, R., Srivastava, J.: Automatic Personalization Based on Web Usage Mining. Communications of the ACM 43(8) (2000)
10. Perkowitz, M., Etzioni, O.: Towards Adaptive Web Sites: Conceptual Framework and Case Study. Artificial Intelligence 118, 245–275 (2000)
11. Simon, H.A., Kaplan, C.A.: Foundations in Cognitive Science. In: Posner, M.I. (ed.) Foundation of Cognitive Science, pp. 1–48. Bradford/MIT Press, Cambridge (1989)
12. Stephen, G.A.: String Searching Algorithms. World Scientific Publishing, Singapore (1994)
13. Sternberg, S.: Memory-Scanning: Mental Processes Revealed by Reactiontime Experiments. American Scientific 57, 421–457 (1969)

14. Watson, I.: Case-based Reasoning is a Methodology not a Technology. Knowledge-Based Systems 12, 303–308 (1999)
15. Wilson, B.A.: Cases Studies in Neuropsychological Rehabilitation. Oxford University Press, New York (1996)
16. Yao, X.: Evolving Artificial Neural Networks. Proc. of the IEEE 87(9), 1423–1447 (1999)

A Study of Applying Knowledge Modelling to Evidence-Based Guidelines

M. Taboada[1], M. Meizoso[1], D. Martínez[2], and S. Tellado[2]

[1] Dpto. de Electrónica e Computación, Universidad de Santiago de Compostela, 15782 Santiago de Compostela, Spain
chus@dec.usc.es
http://aiff.usc.es/ elchus/

[2] Dpto. de Física Aplicada, Universidad de Santiago de Compostela, 27002 Lugo, Spain
fadiego@usc.es
http://www.usc.es

Abstract. This paper reports on a case-study of applying the general purpose and widely accepted methodology CommonKADS to a clinical practice guideline. CommonKADS is focussed on obtaining a compact knowledge model. However, guidelines usually contain incomplete and ambiguous knowledge. So, the resulting knowledge model will be incomplete and we will need to detect what parts of the guideline knowledge are missing. A complementary alternative, which we propose in this work, is to reconstruct the process of knowledge model construction, proposed by CommonKADS, in order to force the knowledge engineer to keep the transformation paths during knowledge modeling. That is to say, we propose to establish explicit mappings between original medical texts and the knowledge model, storing these correspondences in a structured way. This alternative will reduce the existing gap between natural language representation and the corresponding knowledge model.

Keywords: CommonKADs methodology, clinical practice guideline, knowledge modeling, ontologies.

1 Introduction

Over the years, many Knowledge Engineering (KE) methodologies and representation languages have been proposed and successfully applied to the medical domain. An example of a methodology oriented to develop, in a structured way, knowledge-intensive systems is CommonKADS [1]. One of the central construction blocks of CommonKADS is the knowledge model, which provides a means of specifying in detail the types and structures of knowledge existing in a system. It also supplies an implementation-independent description. At the same time, several languages have been proposed to formally represent evidence-based guidelines designed in the health care domain [2]. These languages are very expressive, providing a set of predefined primitives, such as actions and decisions,

J. Mira et al. (Eds.): IWINAC 2009, Part I, LNCS 5601, pp. 437–446, 2009.

which make it possible to precisely and unambiguously describe the content of a clinical guideline.

The codification of a clinical guideline directly in these representation languages is an arduous and complex task, and the final resulting model is illegible for medical specialists, making difficult its validation. To overcome these drawbacks, at least two types of solutions have emerged. The first one involves to describe guidelines in a higher level of abstraction. For example, Clercq et al. [3] have proposed to model guidelines by combining typical components of a knowledge model (such as, ontologies and problem-solving methods) with guideline representation primitives. Another approach was presented by Vollebregt et al. [4]. They carried out a study on re-engineering a realistic system comparing the general purpose AI methodology CommonKADS [1] and the special purpose methodology PROforma [5]. From the results of this study, they suggest to use CommonKADS during the first stages concerning to the analysis phase and later, to use a methodology or language more specific to the guideline representation.

The second solution can be found in more recent works, focused on reducing the existing gap between natural language and formal guideline representations. In [6,7], they focus on keeping traces of the transformation process from natural language to formal representation. This alternative stores the modeling effort with the aim of facilitating validation (inconsistences in the formal representation can be used to detect inconsistences in the original guideline) and maintenance (changes in the original guideline will quickly evolve into changes in the formal representation). A new approach in this line is the semiformal language named MHB [8], which provides a means of designing a guideline in a intermediate representation between natural and a formal language.

The purpose of this work is combine the two types of mentioned solutions to the problem of acquiring knowledge from a clinical guideline. Such a combination is important for a number of reasons. First, using a general methodology, like CommonKADS, widely accepted in practical applications, will provide a model with all types of knowledge existing in the clinical guideline. Secondly, as guidelines usually contain incomplete and ambiguous knowledge, the resulting knowledge model will be incomplete, but we could make a note of all parts of the guideline where knowledge is missing. The identification of these omissions are required to produce a successfully implementation. Thirdly, in most of the cases, knowledge models are formulated by the knowledge engineer supported by interviews to the domain experts. Then, CommonKADS is focussed on obtaining a compact knowledge model. A complementary alternative, which we propose in this work, is to reconstruct the process of knowledge model construction, proposed by CommonKADS, in order to force the knowledge engineer to keep the transformation paths. That is to say, we propose to establish explicit correspondences (*mappings*) between original medical texts and the knowledge model, storing these correspondences in a structured way. We will provide methodological strategies that systematize, as far as possible, the knowledge modeling and acquisition process on medical diagnosis.

The structure of the article is the following one. We will begin presenting our case-study and the three main stages in the construction of a knowledge model. Next, we will detail the set of required activities for each stage, including examples of the actual case-study we performed. Finally, we will present the conclusions.

2 Stages in the Construction of a Knowledge Model

This section presents a set of strategies for the construction of a knowledge model in the domain of medical diagnosis. Its an extension of the construction process of CommonKADS, which allows the knowledge engineer to keep the transformation paths during knowledge modeling. All examples we will use in this section and the next one came from our case-study: the knowledge modeling on medical diagnosis in Conjunctivitis from the textual documentation provided by a clinical practice guideline[1].

Our approach includes three main stages and an ordered set of activities to carry out in each stage as shown in Fig. 1.

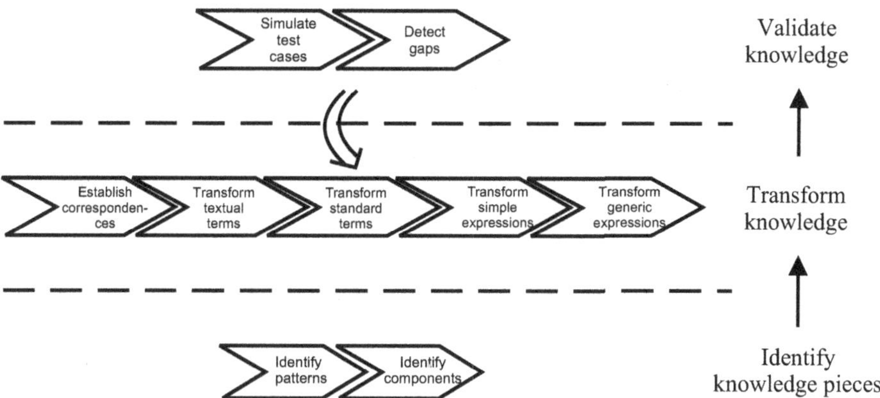

Fig. 1. Main stages and activities in the knowledge model construction preserving transformation traces

3 Identify Knowledge Pieces

In this stage, all the pieces that will be handled during the modeling process are specified. This stage includes two activities.

1. *Identify knowledge patterns.* This activity consists in analyzing sources of the available knowledge and identifying the medical knowledge patterns to be modeled. In our case-study, we have identified the following knowledge patterns:

[1] http://www.aao.org/aao/education/library/ppp

- *Isolated nouns* and *nouns with adjectives.* Examples are *diagnosis*, *cause*, *visual function* or *discharge.*
- *Vague* expressions that reflect the state of clinical variables that can be measured or quantified. For example, an expression like *decrease of the visual acuity* or *Rapid development of severe hyperpurulent conjunctivitis.*
- *Generic linguistics expressions*, including
  - Verbs that describe structures or parts. For example, in the expression *The patient population includes individuals of all ages who present with symptoms suggestive of conjunctivitis, such as red eye or discharge*, the verb *include* refers to the set of patients, which the clinical practice guideline aims at.
  - Verbs that describe medical actions. For example, the set of goals in the guideline include clinical actions, such as *Establish the diagnosis of conjunctivitis* or *Establish appropriate therapy.*
  - Decisions or causalities. An example is the expression *Questions about the following elements of the patient history may elicit helpful information: ...*, that expresses an indication on what information to compile during the interrogation of the patient.

2. *Identify knowledge components to be reused.* This activity is involved with reviewing reusable modeling components, such as models of stereotypical tasks, ontologies, etc. These components can exist as conceptual components or software components. This way, the construction of a knowledge model can be viewed as an activity consisting in assembling these components [9]. In our application, we have used:
   - Benjamins' library [10], which provides a compilation of problem solving methods (PSMs) in the domain of diagnosis. They are reasoning components that can be assembled with domain knowledge to create application systems. Benjamins' library has identified and collected them for the specific high-level task of diagnosis.
   - The UMLS Metathesaurus [11], a source of controlled vocabulary in the medical domain.
   - The UMLS Semantic Network, an ontology in the medical domain.

4 Transform Textual Knowledge into Knowledge Components

In this stage, it is where properly the explicit correspondences between textual knowledge and the different types of knowledge pieces are set, at the same time the knowledge model is constructed (Figure 1). It includes the following activities:

1. *Establish correspondences between textual knowledge.* As the knowledge engineers do not understand the semantics of medical documents, the first activity that we propose in this stage is the revision of documents on the part

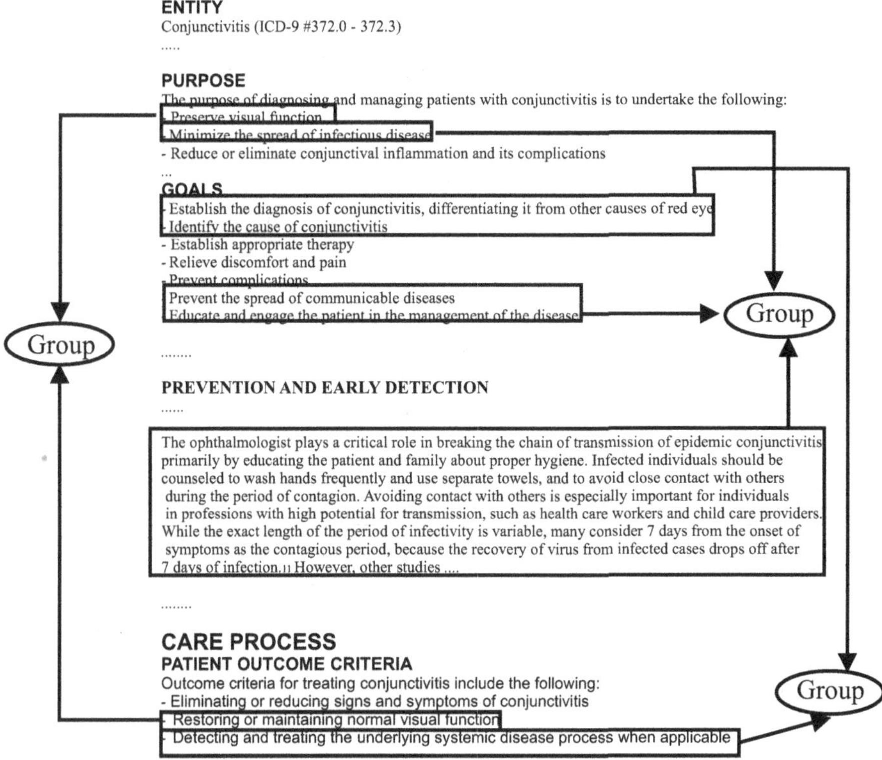

ENTITY
Conjunctivitis (ICD-9 #372.0 - 372.3)
.....
PURPOSE
The purpose of diagnosing and managing patients with conjunctivitis is to undertake the following:
- Preserve visual function
- Minimize the spread of infectious disease
- Reduce or eliminate conjunctival inflammation and its complications
...
GOALS
- Establish the diagnosis of conjunctivitis, differentiating it from other causes of red eye
- Identify the cause of conjunctivitis
- Establish appropriate therapy
- Relieve discomfort and pain
- Prevent complications
- Prevent the spread of communicable diseases
- Educate and engage the patient in the management of the disease
Group
Group
....
PREVENTION AND EARLY DETECTION
......
The ophthalmologist plays a critical role in breaking the chain of transmission of epidemic conjunctivitis primarily by educating the patient and family about proper hygiene. Infected individuals should be counseled to wash hands frequently and use separate towels, and to avoid close contact with others during the period of contagion. Avoiding contact with others is especially important for individuals in professions with high potential for transmission, such as health care workers and child care providers. While the exact length of the period of infectivity is variable, many consider 7 days from the onset of symptoms as the contagious period, because the recovery of virus from infected cases drops off after 7 days of infection.11 However, other studies ....
....
CARE PROCESS
PATIENT OUTCOME CRITERIA
Outcome criteria for treating conjunctivitis include the following:
- Eliminating or reducing signs and symptoms of conjunctivitis
- Restoring or maintaining normal visual function
- Detecting and treating the underlying systemic disease process when applicable
Group

Fig. 2. Correspondences marked between parts of the clinical practice guideline on conjunctivitis published by the American Academy of Ophthalmology

of clinical experts with the purpose of grouping and relating noncontiguous textual portions (chunks) that make reference to the same knowledge. For example, in figure 2 different parts from the text have been grouped, as they refer to the same item or they detail, more in depth, other paragraphs. We have used the DELTA tool[2], as it provides an easy way to establish links between chunks and create an XML file with these links.

2. *Transform textual terms into standard terms.* Nouns included in texts must be extracted and replaced by standard terms. This activity is commonly known as *Acquisition of standard medical terminology.* Using the resources provided by some controlled vocabulary can help us with the semi-automation of this process. For example, in our case of study, we have used the utilities that the server of the UMLS[3] provides. Thus, we have been able to establish direct correspondences between textual terms as *diagnosis* or *therapy* and its corresponding standard terms (Table 1). In addition, it is necessary to consider

[2] http://www.ifs.tuwien.ac.at/ votruba/DELTA/
[3] http://www.nlm.nih.gov

Table 1. Examples of transforming textual terms into standard terms

Textual Term	Type of Correspondence	Standard Term
diagnosis	Equivalent	Diagnosis (CUI: C0011900)
therapy	Equivalent	Therapeutic procedure (CUI: C0087111)
discharge	Refinement	Discharge from eye (CUI: C0423006)
mild mucous discharge	Generalization	EYE DISCHARGE, MUCOID (CUI: C0239425)
duration	Equivalent	Duration (CUI: C0449238)

Table 2. Examples of transforming standard terms into knowledge pieces

Standard Term	Type of Correspondence	Knowledge piece
Diagnosis (CUI: C0011900)	Mapping	Diagnosis Task
Therapeutic procedure (CUI: C0087111)	Mapping	Assessment Task
Discharge from eye (CUI: C0423006)	Mapping	Sign or Symptom (Domain Concept)
Duration (CUI: C0449238)	Mapping	Temporal Concept

that nouns must be extracted outside their context without losing knowledge. This transformation can entail a process of *Refinement of terms*. For example, the term *discharge* was refined on *Discharge from eye*. On the other hand, too much specialized terms (missing in the controlled vocabulary) must be replaced by more general terms (*Generalization of terms*) and represent the special characteristics as attributes of more general terms (in the following activity). For example, *mild mucous discharge* is a very specialized term, so it is not in the UMLS Metathesaurus. Nevertheless, we can find *EYE DISCHARGE, MUCOID*, which is a more general term.

3. *Transform standard terms into knowledge pieces.* In order to carry out this activity, we have used the resources provided by the UMLS Semantic Network. This network provides a medical ontology that classifies each Metathesaurus concept. Therefore, the extracted concepts from the Metathesaurus in the previous activity have one or more assigned semantic types, which provides a track about the knowledge piece that the concept is referred to. For example, *Diagnosis (CUI: C0011900)* is a *Diagnostic Procedure* and *Therapeutic procedure (CUI: C0087111)* is a *Therapeutic and preventive procedure.* So, they can be modeled as stereotypical tasks (Table 2). On the other hand, *Discharge from eye (CUI: C0423006)* is a *Sign or Symptom* and *Duration (CUI: C0449238)* is a *Temporal Concept.* So, they correspond to domain concepts.

4. *Transform simple expressions* on the state of clinical variables. This activity involves to decide what attributes should be added to the concepts, based on the analyzed expressions. It also assigns them ranges of values. For example, the expression *Rapid development of severe hyperpurulent conjunctivitis* shows the need of adding the attribute *Severity (CUI:C0449294)* to graduate symptoms and the attribute *Onset* to describe the way how they are established. Possible values for the attribute *Severity* are *Severe (CUI: C0205082)* and the siblings in the Metathesaurus: *Mild (CUI: C0205080)* and *Moderate (CUI: C0205081)*. Possible values for the attribute *Onset* are *Fast (CUI: C0456962)* and the siblings in the Metathesaurus: *Gradual (CUI: C0439833)* and *Sudden (CUI: C0439832)*, etc. As the term *hyperpurulent* does not exist in the Metathesaurus, we refined it on *purulent*. In addition, purulent is an attribute that describes discharge, not the conjunctivitis disease, so we refined *severe hyperpurulent conjunctivitis* on *severe hyperpurulent discharge*. Finally, the expression *Rapid development* was refined on the concrete time that it means: 1-3 days.
5. *Transform generic linguistic expressions* into knowledge components. Generally, this entails knowledge refinement. Examples of such expressions are:

   - Verbs describing structures or parts can be transformed into domain standard concepts or relations. For example, in the expression *The initial eye examination includes measurement of visual acuity, external examination, and slit-lamp biomicroscopy*, the verb *include* describes the parts that an ocular exploration must consist of. It is, therefore, a relation among components of the domain ontology.
   - Verbs describing actions can be transformed into PSMs, inferences or transfer functions. For example, in the expression describing the goal *Establish the diagnosis of conjunctivitis, differentiating it from other causes of red eye*, the noun *diagnosis* indicates a stereotypical task (Table 2). In addition, this noun is clarified by means of the verb *differentiate*, which matches the Metathesaurus concept *Differential Diagnosis (CUI: C0011906)*. As well, this concept can match the PSM called *prime diagnostic method* [10]. Following this PSM, the diagnostic task is decomposed into three sub-tasks: 1) symptom detection, 2) hypothesis generation, and 3) hypothesis discrimination.
   - Decisions or causalities (*if ..., but ..., it would have to be ..*, etc.) can be transformed into tasks or PSMs. For example, in the description *The external examination should include the following elements: ...*, the verb *include* describes the parts that, in ideal conditions of work, an external ocular exploration must consist of. It is, therefore, a relation among components of the domain ontology. But, the clarification *should include* also indicates that the provided list is too exhaustive and the doctor must decide what elements are the most important for each patient. So, the selection of the parts will be dynamic and depend on the particular data of the patient. Therefore, this expression matches an assessment task,

consisting of collecting additional data, based on the current diagnosis hypotheses as well as the costs implied in the external examination.

5 Validate Knowledge

This stage includes two activities:

1. *Validate*, as much as possible, the carried out *knowledge transformations*. An important technique is to make computer-based simulations of real scenes of the medical diagnosis task.
2. *Detect knowledge gaps.* From the validation stage, we will be able to detect whether the knowledge model is complete. In our example, we have found that the clinical practice guideline includes knowledge on some dimensions characterizing the conjunctivitis diagnosis, such as
   - *Differential diagnosis* with respect to other causes of *Red Eye.*
   - *Etiologic diagnosis*, which is used to identify the conjunctivitis cause.
   - *Multiple fault diagnosis*, which indicates what type of PSMs should be selected to generate and discriminate hypotheses. For example, classification-based PSMs are not suitable.

   It also includes domain knowledge on disease course, and symptoms and signs associated to each type of pathology. Nevertheless, it does not contemplate specific knowledge on the PSMs to be used in:

   - Generation of diagnostic hypotheses from the patient symptoms.
   - Decision of how to carry out the ocular exploration.
   - Evaluation of diagnostic hypotheses, in order to confirm or reject them.

6 Conclusions

In this paper, we have applied CommonKADS in the analysis phase of a clinical practice guideline, with the aim of systematizing knowledge acquisition. Our proposal extends the knowledge model construction of CommonKADS, as it provides a methodological support that helps to detect and document all the transformations from natural language to the structured representation of a knowledge model. When forcing to the knowledge engineer to keep these transformations, the knowledge modeling becomes more gradual. In addition, we have provided a limited number of transformations in each stage, so from the validation stage, we are able to detect what parts of the knowledge model are complete and what are missing in the guideline.

In the article, we have presented an evaluation study on a specific clinical practice guideline. The knowledge elicitation has been carried out in terms of re-usable knowledge components: controlled vocabulary, stereotypical tasks, domain ontologies and PSMs. Nevertheless, this way of acquiring clinical practice guidelines has not been the usual one in the last decades. Initially, guidelines were codified directly in some representation language specifically designed for them. In [2], a revision of the most important proposals can be found. In general, these languages are very expressive, allowing to precisely and unambiguously describe the content of a guideline, in terms of a set of primitives, such as actions and decisions. Nevertheless, the codification of a guideline directly in these languages is an arduous and complex task, and the final resulting model is illegible for medical specialists, making difficult its validation. In the last years, proposals to describe guidelines in a higher abstraction level have emerged. For example, Clercq et al. [3] have proposed to combine ontologies, PSMs and guideline primitives. This approach is similar to ours, but we have clearly separated the analysis from the representation phase, as Vollebregt et al. [4] have suggested.

Other interesting proposals are Stepper [6] and Delta [7] tools. These are oriented to mark free text (in chunks), and fragment it in interconnected component models (in the way of DTDs), using XML technology. For example, the components in the Stepper tool are four predefined ones: procedural, causality, objective and definition of concepts; while the Delta tool allow us to define our own components. We have used the DELTA tool in our case-study, as it provides an easy way to establish links between chunks and create a XML file with these links. In this way, we were able to fragment the text in interconnected reusable knowledge components: standard vocabulary, stereotypical tasks, domain ontologies and PSMs.

Acknowledgements. This work was supported by the Ministerio de Educacin y Ciencia, under Grant TIN2006-15453-C04-02.

References

1. Schreiber, G., Akkermans, H., Anjewierden, A., de Hoog, R., Shadbolt, N., Van de Velde, W., Wielinga, W.: Knowledge Engineering and Management. In: The CommonKADS Methodology. The MIT Press, Cambridge (1999)
2. de Clercq, P., Blom, J., Korsten, H., Hasman, A.: Approaches for creating computer-interpretable guidelines that facilitates decision support. Artificial Intelligence in Medicine 31, 1–27 (2004)
3. de Clercq, P., Hasman, A., Blom, J., Korsten, H.: The application of ontologies for the development of shareable guidelines. Artificial Intelligence in Medicine 22, 1–22 (2001)
4. Vollebregt, A., ten Teije, A., van Harmelen, F., van der Lei, J., Mosseveld, M.: A study of PROforma, a development methodology for clinical procedures. Artificial Intelligence in Medicine 17, 195–221 (1999)
5. Sutton, D.R., Fox, J.: The Syntax and Semantics of the PROforma guideline modelling language. J. Am. Med. Inform. Assoc. 10(5), 433–443 (2003)
6. Svatek, V., Ruzicka, M.: Step-by-step formalisation of medical guideline content. International Journal of Medical Informatics 70(2-3), 329–335 (2003)
7. Votruba, P., Miksch, S., Kosara, R.: Facilitating Knowledge Maintenance of Clinica Guidelines and Protocols. In: Fieschi, et al. (eds.) Proc. of MEDINFO. IOS Press, Amsterdam (2004)

8. Seyfang, A., Miksch, S., Polo, C., Wittenberg, J., Marcos, M., Rosenbrand, K.: Facilitating Knowledge MHB - A Many-Headed Bridge between Informal and Formal Guideline Representations. In: Proc. of 10th European Conference on AIME, Aberdeen, Scotland (2005)
9. Taboada, M., Des, J., Mira, J., Marín, R.: Development of diagnosis systems in medicine with reusable knowledge components. IEEE Intelligent Systems 16, 68–73 (2001)
10. Benjamins, V.R.: Problem Solving Methods for Diagnosis, Ph.D thesis, University of Amsterdam (1993)
11. Lindberg, D., Humphreys, B., Mc Cray, A.: The Unified Medical Language System. Methods of Information in Medicine 32, 281–291 (1993)

Fuzzy Classification of Mortality by Infection of Severe Burnt Patients Using Multiobjective Evolutionary Algorithms

F. Jiménez, G. Sánchez, J.M. Juárez, J.M. Alcaraz, and J.F. Sánchez

Dept. Ingeniería de la Información y las Comunicaciones
Universidad de Murcia

Abstract. The classification of survival in severe burnt patients is an on-going problem. In this paper we propose a multiobjective optimisation model with constraints to obtain fuzzy classification models based on the criteria of accuracy and interpretability. We also describe a multiobjective evolutionary approach for fuzzy classification based on data with real and discrete attributes. This approach is evaluated using three different evolutive schemas: pre-selection with niches, NSGA-II and ENORA. The results are compared as regards efficacy by statistical techniques.

1 Introduction

The Intensive Care Unit (ICU) in hospitals is responsible for providing medical attention to patients in a critical state. Death from general infection in Spanish ICUs involves 15000.patients per year, and is a common cause of death in patients suffering severe burns. One of the main problems in this respect is to be able to categorise the gravity of patients given the difficulties involved in their diagnosis. Recent SEMYCYUC[1] studies suggest that early classification and the application of suitable therapeutic measures could greatly help survival [16].

On of the ways to approach this problem is to use fuzzy modelling, a technique in which the main problem involves identifying a fuzzy model using input-output type data. Given a set of data for which we assume some functional dependence, the problem consists of deriving (fuzzy) rules from the data that characterise the unknown function as accurately as possible. Several approaches have been proposed to generate such if-then rules automatically from numerical data. Due to the complexity of the problem, evolutive computation is one of the most widely used techniques.

Evolutive computation [5] has successfully been applied to learn the fuzzy model [6,9], leading to many complex algorithms and, as described in [15] and [18], insufficient importance has been given to the interpretability of the resulting rules. In such cases, the fuzzy model is converted into a black box and the application of fuzzy modelling rather than other techniques maybe questioned. On the other hand, evolutionary algorithms are recognised as suitable for multiobjective optimisation since they search for multiple solutions in parallel [2,3]. Modern evolutive approaches for multiobjective

[1] SEMYCYUC: Spanish Society of Intensive Care Medicine and Coronary Units.

J. Mira et al. (Eds.): IWINAC 2009, Part I, LNCS 5601, pp. 447–456, 2009.

optimisation consist of multiobjective evolutionary algorithms based on Pareto optimality, in which all the objectives are optimised simultaneously in the search for non-dominated solutions with one run of the algorithm. The decision-maker can then choose the most suitable solution according to the setting of the current decision after executing the algorithm. Moreover, if the setting of the decision changes, it is not always necessary to run the algorithm again, and another solution can be taken from the set of non-dominated solutions already obtained.

Fuzzy modelling can be considered from a multiobjective evolutive optimisation point of view [8]. Most lines of research into fuzzy modelling attempt to improve the accuracy of descriptive models and to improve the interpretability in approximative models [1]. This article is set in the second approach, tackling the problem by means of multiobjective optimisation, considering the criteria of accuracy and interpretability simultaneously.

We propose an evolutive multiobjective optimisation approach to generate fuzzy classification models taking into account criteria of accuracy and interpretability. Section 2 describes the fuzzy classification model and the criteria taken into account in the optimisation process. A multiobjective optimisation model with constraints is proposed. Section 3 shows the principal components of the three multiobjective evolutionary algorithms used in the article, while Section 4 shows the experiments carried out and the results obtained for the problem of classifying infection-related mortality in patients suffering from severe burns. Conclusions are described in Section 5.

2 Fuzzy Classification Model

2.1 Model Identification

Let us consider a set of N learning data in the following form:

- A vector of real input attributes $x = (x_1, \ldots, x_p)$, $x_i \in [l_i, u_i] \subset \Re$, $i = 1, \ldots, p$, $p \geq 0$.
- A vector of discrete input attributes $w = (w_1, \ldots, w_q)$, $w_i \in \{1, \ldots, v_i\}$, $i = 1, \ldots, q$, $q \geq 0$, where v_i is the number of classes for the discrete input attribute i,
- A value for the discrete output attribute $y \in \{1, \ldots, z\}$, where z is the number of classes for the discrete output attribute.

Note that a Boolean input or output attribute can be represented by a discrete attribute w_i or y, so that $v_i = 2$ or $z = 2$.

Let us consider a fuzzy classification model formed by M rules $R_1, \ldots, R_M$. At least one rule should exist for each of the z output classes, so that $M \geq z$. Each rule R_j ($j = 1, \ldots, M$) contains p fuzzy sets $\widetilde{A}_{ij}$ ($i = 1, \ldots, p$) associated to p real input attributes, q discrete values B_{ij} ($i = 1, \ldots, q$) associated to q discrete input attributes, and a discrete value C_j associated to the discrtete output attribute. A rule R_j has, therefore, the following structure:

$$\begin{array}{ll} R_j : If & x_1 \; is \; \widetilde{A}_{1j} \; and \; \ldots \; and \; x_p \; is \; \widetilde{A}_{pj} \; and \\ & w_1 \; is \; B_{1j} \; and \; \ldots \; and \; w_q \; is \; B_{qj} \\ \multicolumn{2}{l}{then \; y \; is \; C_j} \end{array} \tag{1}$$

Each fuzzy set $\widetilde{A}_{ij}$ $(i = 1, \ldots, p)$ $(j = 1, \ldots, M)$ can be described by a membership function $\mu_{\widetilde{A}_{ij}} : \mathcal{X}_i \to [0, 1]$, where $\mathcal{X}_i$ is the dominion of the real input attribute x_i. In our model, we use gaussian membership functions:

$$\mu_{\widetilde{A}_{ij}}(x_i) = \exp\left[-\frac{1}{2}\left(\frac{x_i - a_{ij}}{\sigma_{ij}}\right)^2\right]$$

where $a_{ij} \in [l_i, u_i]$ is the centre, and $\sigma_{ij} > 0$ is the variance. The output of the model $\psi(x, w)$ for a datum with a vector of real input attributes x and a vector of discrete input attributes w corresponds to the output $C \in \{1, \ldots, z\}$ whose model activation value $\lambda_C(x, w)$ is maximum, that is:

$$\psi(x, w) = \arg_C \max_{C=1}^{z} \lambda_C(x, w)$$

The activation value $\lambda_C(x, w)$ of the model for a datum (x, w) and an output class $C \in \{1, \ldots, z\}$ se is calculated by summing the firing degree $\varphi_j(x, w)$ of each rule R_j $(j = 1, \ldots, M)$ whose value for the discrete output attribute C_j is equal to C, that is:

$$\lambda_C(x, w) = \sum_{\substack{j = 1, \ldots, M \\ C_j = C}} \left((\phi_j(w) + 1) \prod_{i=1}^{p} \mu_{\widetilde{A}_{ij}}(x_i) \right)$$

where $\phi_j(w)$ is the number of discrete input attributes, so that $w_j = B_{ij}$.

2.2 Fuzzy Modeling Criteria

We consider three principal criteria: accuracy, transparency and compactness. Quantitative measures are defined for this criteria by the use of the suitable objective functions.

Accuracy. Classification Rate, $CR = \frac{\Phi}{N}$ where Φ is the number of data for which $\psi(x, w) = y$ and N is the total number of data.

Transparency. Similarity [14]:

$$S = \max_{\substack{i, k = 1, \ldots, M \\ j = 1, \ldots, p \\ \widetilde{A}_{ij} \neq \widetilde{A}_{kj}}} S(\widetilde{A}_{ij}, \widetilde{A}_{kj}) \tag{2}$$

where $S(\widetilde{A}, \widetilde{B}) = \max\left\{\frac{|\widetilde{A} \cap \widetilde{B}|}{|\widetilde{A}|}, \frac{|\widetilde{A} \cap \widetilde{B}|}{|\widetilde{B}|}\right\}$

Compactness. Number of rules (M), and the number of different fuzzy sets (L).

2.3 Optimisation Model

Based on the above observations, we propose the following optimisation model involving two objectives and one constraint:

$$\begin{array}{ll} Maximise & f_1 = CR \\ Minimise & f_2 = M \\ Subjet\ to & g_1 = S - g_s \leq 0 \end{array} \tag{3}$$

It is supposed that the number of different fuzzy sets (L) is less in those models with a lower number of rules (M), so that we only use M value in the optimisation model. Furthermore, we only wish to minimise the similarity (S) to a given threshold (g_s), since models with dissimilar fuzzy sets are not sufficiently accurate. In this way, the search space is reduced and the efficacy of the algorithms increases.

3 Multiobjetive Evolutionary Algorithms

We propose a hybrid learning system to search for multiple Pareto-optimal solutions simultaneously, taking into account criteria of precision, transparency and compactness. We study different multiobjective evolutionary algorithms to develop the structure and the parameters of the set of rules, and use a simplification operator of these rules to attain transparency and compactness.

Three multiobjective Pareto based evolutionary algorithms are used: pre-selection with niches, ENORA and NSGA-II. The first two are algorithms developed by the authors [7] and [13] respectively, while NSGA-II is the well known multiobjective evolutionary algorithm proposed by Deb in [3]. These three algorithms are adapted in this work to obtain fuzzy classification models.

3.1 Representation of Solutions

We use a representation involving the codification of real and discrete numbers, using the Pittsburgh approach. Each individual of a population contains a number of rules that can vary between min and max, where the values min and max are chosen by a decision-maker taking into account that there must be at least one rule per output class, that is, $z \leq min \leq M \leq max$.

For each rule R_j, $j = 1, \ldots, M$ we represent:

- The fuzzy sets associated to the real input attributes x_i, $i = 1, \ldots, p$, by real numbers $a_{ij} \in [l_i, u_i]$ and $\rho_{ij} = 2\sigma_{ij}^2 > 0$, which define the centres and variances, respectively.
- The discrete values associated to the discrete input attributes w_i, $i = 1, \ldots, q$, by integer numbers $b_{ij} \in \{1, \ldots, v_i\}$.
- The discrete value associated to the discrete output attribute by a integer number $c_j \in \{1, \ldots, z\}$.
- To carry out adaptive crossing and mutation, each individual has two discrete parameters $d \in (0, \delta)$ and $e \in (0, \epsilon)$ associated to crossing and mutation, where δ is the number of crossing operators and ε is the number of mutation operators. There are also two real parameters $e_c, e_v \in [0, 1]$ that define the amplitude of the mutation of the centres and variances of the fuzzy sets, respectively.

3.2 Initial Population

The individuals are initialised by randomly generating a uniform distribution within the search space limits. Another condition is included besides the limits related with the variance of the fuzzy Gaussian sets, so that:

$$\alpha_i = \frac{u_i - l_i}{\gamma} \leq \rho_{ij} \leq u_i - l_i = \beta_i,\ i = 1, \ldots, p$$

where γ is a value used to calculate the minimum variances.

One of the conditions that is maintained throughout the evolutive process is that there is at least one rule per output value of between 1 y z.

3.3 Handling the Constraints

The rule proposed in [10] is used to handle the constraints.

3.4 Variation Operators

To fully explore the search space the evolutive operators must work at the different levels of the individuals. Therefore different crossing or mutation operators are defined and applied adaptively with a probability of variation defined by the user p_v.

Fuzzy Set Crossover. Interchanges two random fuzzy sets.

Rule Crossover. Interchanges two random rules.

Rule Incremental Crossover. Add to each individual a random rule taken from the other individual.

Gaussian Set Centre Mutation. Mutate the centre of one or more random fuzzy sets.

Gaussian Set Variance Mutation. Mutate the variance of one or more random fuzzy sets.

Gaussian Set Mutation. Mutate a random fuzzy set, changing it for another random fuzzy set.

Rule Incremental Mutation. Add a new random fuzzy rule.

Discrete Value Mutation. Mutate a random discrete value.

3.5 Simplification of the Rule Set

The rule set simplification is performed according to the technique described in [7] and [15]. Those sets with $\eta_2 < S(\widetilde{A}, \widetilde{B})$ are combined and those with $\eta_1 < S(\widetilde{A}, \widetilde{B}) < \eta_2$.

3.6 Decision-Making Process

Let $S = \{s_1, \ldots, s_D\}$ be the set of non-dominated solutions so that $CR\,(s_i) \geq CR^{min}$, $(i = 1, \ldots, D)$ where CR^{min} is the minimum correct rate acceptable by the decision maker.

```
While S ≠ ∅ {
    Reach the solution s_i with the lowest number of rules and fuzzy sets.
    If s_i is sufficiently clear and accurate, return s_i.
}
There is no satisfactory solution.
```

4 Experiments and Results

This work is based on data gathered by the Clinical Information System of a hospital ICU in 1999 and 2002, selecting, following medical advice, a total of 99 patients with different infectious processes. From the medical records of each patient, the specialists

Table 1. Patient parameters considered: parameter name, description, type (R real, B Boolean), weight, associated input attribute and limits

Name	Description	Type	Weight	Attributes	Limits
Total	Total burnt surface %	R	0.164613	x_1	$0-85$
Prof	Deeply burnt surface %	R	0.115335	x_2	$0-85$
SAPS II	Severity score	R	0	–	
Weight	Patient's weight	R	0	–	
Age	Patient's age	R	0	–	
Pneumonia	Presence of pulmonary infection	B	0.207799	w_1	$1,2$
Sex	Patient's sex	B	0.069403	w_2	$1,2$
Inh	Use of inhibitors	B	0.067112	w_3	$1,2$
Wound-Infect	Infection by surgical wound	B	0.034936	w_4	$1,2$
aids-drugs	HVI drug consumption	B	0.025781	w_4	$1,2$
Co-Liver	Previous liver problems	B	0.024223	w_5	$1,2$
Bacteremia	Presence of bacteria in blood	B	0.023324	w_6	$1,2$
Co-Card	Previous cardiopathy	B	0.008141	w_7	$1,2$
Co-Respir	Previous respiratory problems	B	0.007035	w_8	$1,2$
HBP	High blood preassure	B	0.001650	w_9	$1,2$
Diabetes	Diabetic patient	B	0.000541	w_{10}	$1,2$
Co-Renal	Previous kidney problems	B	0	–	
Death	Prognosis of death	B	–	y	$1,2$

Table 2. Parameters used to run the algorithms and Non-dominated solutions (best results after 100 runs) obtained in this study for the problem of classifying mortality by infection in serious burnt patients. The solutions chosen for each algorithm after decision-making (section 3.6) are in bold.

Parameters	M	L	CR-Train	CR-Eval	S
$N = 100$	**Preselection with niches**				
$No\ Evaluations = 10^5$	**10**	**9**	**0.820225**	**0.8**	**0.065554**
$min = 10$	10	10	0.820225	0.8	0.045618
$max = 20$	14	11	0.853933	0.9	0.056930
$\gamma = 100$	11	11	0.842697	0.7	0.072503
$p_v = 0.1$	20	14	0.910112	0.8	0.099939
$g_s = 0.1$	**NSGA-II**				
$\eta_1 = 0.6$	10	8	0.910112	0.8	0.077349
$\eta_2 = 0.1$	**11**	**8**	**0.932584**	**0.8**	**0.077349**
$No\ Children = 10$ (Preselection with niches)	**ENORA**				
$No\ min\ Niches = 5$ (Preselection with niches)	**10**	**9**	**0.943820**	**0.8**	**0.083848**
$No\ max\ Niches = 35$ (Preselection with niches)	15	10	0.955056	0.8	0.093499

Table 3. Fuzzy model with 10 rules for classifying mortality by infection in serious burnt patients obtained by ENORA. For the discrete variables, the values 1 and 2 correspond to false and true Boolean values, respectively.

x_1	x_2	w_1	w_2	w_3	w_4	w_5	w_6	w_7	w_8	w_9	w_{10}	w_{11}	y
$(30.26; 23.72)$	$(19.25; 20.79)$	1	1	1	1	2	1	2	2	1	2	1	1
$(47.54; 9.02)$	$(19.25; 20.79)$	2	1	1	2	1	1	2	2	1	1	2	1
$(4.31; 10.11)$	$(19.25; 20.79)$	1	2	2	2	2	1	2	2	1	1	2	1
$(55.51; 8.50)$	$(42.32; 8.50)$	2	1	1	2	1	1	1	2	2	2	1	1
$(69.98; 10.95)$	$(19.25; 20.79)$	1	1	1	2	1	1	2	1	2	2	1	1
$(47.54; 9.02)$	$(51.39; 15.07)$	2	2	2	1	1	2	2	2	1	2	1	2
$(69.98; 10.95)$	$(19.25; 20.79)$	1	2	1	2	1	2	1	1	1	1	1	2
$(30.26; 23.72)$	$(19.25; 20.79)$	2	2	2	2	1	2	2	1	2	2	1	2
$(47.54; 9.02)$	$(19.25; 20.79)$	2	2	1	2	1	2	1	1	1	1	1	2
$(69.98; 10.95)$	$(32.21; 8.50)$	1	1	1	2	2	2	1	2	2	2	1	2

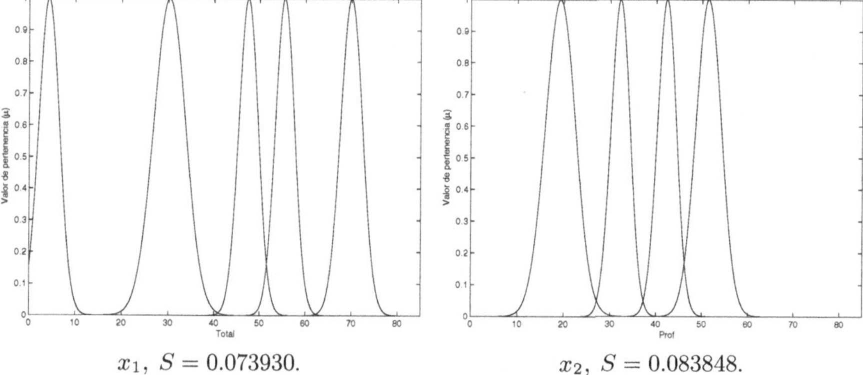

x_1, $S = 0.073930$. x_2, $S = 0.083848$.

consider that the parameters depicted in Table 1 contain the most clinically relevant evidences for establishing the survival estimation of individual patients.

For the study, 89 patients were randomly selected to form part of the training set, while the remaining 10 were used for the evaluation. Once the set had been established it was necessary to establish the importance of each parameter for the survival. This classification was carried out by reference to the medida de Mutual Information, based on Shannon's Entropy [4]. The weights depicted in Table 1 show the results obtained and, for this study, the parameters with weightings of 0: SAPS II, weight, age and renal comorbidity. We are left therefore with a problem of $p = 2$ real inputs and $q = 10$ discrete inputs. The two real inputs considered are within the range $[0, 85]$, while all the discrete inputs (and output) are boolean type.

One of the most common AI techniques used in the medical field for making decisions in the medical field are decision trees. With these techniques, the following classification rates have been obtained: 0.55 (ID3), 0.70 (J48) and 0.75 (ADT).

The preselection with niches, NSGA-II and ENORA was run 100 times with the parameters shown in Table 2 (left). Table 2 (right) shows the best non-dominated solutions obtained, while the solution chosen after the decision-making process (section 3.6) for

Table 4. Statistics and Box Plots for the hypervolume obtained with 100 runs of preselection with niches, ENORA and NSGA-II for classifying mortality through infection in serious burnt patients

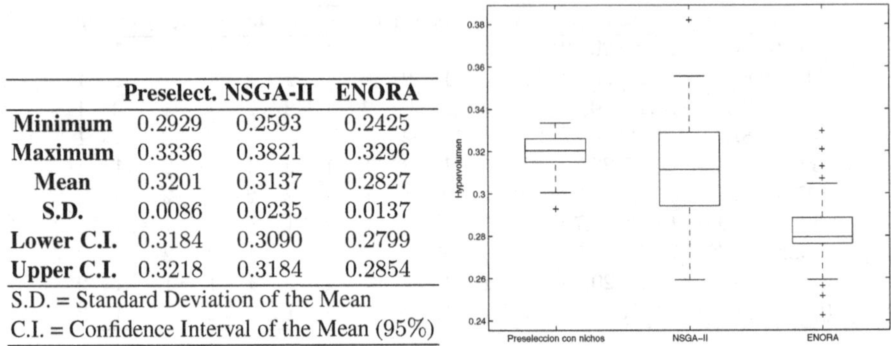

	Preselect.	NSGA-II	ENORA
Minimum	0.2929	0.2593	0.2425
Maximum	0.3336	0.3821	0.3296
Mean	0.3201	0.3137	0.2827
S.D.	0.0086	0.0235	0.0137
Lower C.I.	0.3184	0.3090	0.2799
Upper C.I.	0.3218	0.3184	0.2854

S.D. = Standard Deviation of the Mean
C.I. = Confidence Interval of the Mean (95%)

each algorithm is marked in bold. The definitive solution obtained by ENORA had 10 rules (Table 3).

To compare the algorithms, we use the hypervolume indicator (ν), which calculates the fraction of objective space not dominated by any of the solutions obtained by the algorithm [3,11,21].

The statistics and box plots depicted in Table 4 show that ENORA obtains less localisation and dispersal values than Preselection with niches and NSGA-II. Lastly, the confidence intervals of 95% for the mean obtained with the t-test show that ENORA provides lower values than Preselection with niches and NSGA-II. Therefore, the approximations obtained with ENORA are better than those obtained with Preselection with niches or NSGA-II, according to the hypervolume indicator ν. The t-test is robust with simples of more than 30 individuals, and so the results are significant, leading us to conclude that the differences between the hypervolume values obtained with the algorithms are statistically significant.

The statistical analysis shows that, for the type of multiobjective problem being considered, a Pareto search based on the partition of the search space into radial slots is more efficient than general search strategies based solely on diversity functions, such as NSGA-II, or diversity schemes involving the explicit formation of niches, such as Preselection with niches.

5 Conclusions

This article presents interesting results for the combination of multiobjective evolutionary algorithms and fuzzy modelling for classification purposes, using a theme of great relevance in clinical practice – the classification of mortality through infection in serious burnt patients. We propose a multiobjective optimisation model with constraints for fuzzy classification, taking into consideration the criteria accuracy, transparency and compactness. Three evolutionary multiobjective algorithms were implemented (Preselection with niches, ENORA an NSGA-II) combined with techniques of rule simplification.

The results obtained are better than those obtained with other techniques commonly used in medicine, with the added advantage that the proposed technique identifies alternative solutions. Statistical tests were carried out on the values hypervolume quality score to comare the algorithms; for the problem considered, ENORA provided better results than the other two algorithms, Preselection with niches and NSGA-II.

Acknowledgement

This work is partially founded by FEDER and Spanish MEC under the projects PET2006_0406 y TIN2006-15460-C04. Our gratitude to Dr. Francisco Palacios Ortega from the Hospital Universitario de Getafe for his collaboration and useful comments.

References

1. Casillas, J., Cordón, O., Herrera, F., Magdalena, L.: Interpretability improvements to find the balance interpretability-accuracy in fuzzy modeling: an overview. In: Casillas, J., Cordón, O., Herrera, F., Magdalena, L. (eds.) Interpretability Issues in Fuzzy Modeling. Studies in Fuzziness and Soft Computing, pp. 3–22. Springer, Heidelberg (2003)
2. Coello, C.A., Veldhuizen, D.V., Lamont, G.B.: Evolutionary Algorithms for Solving Multi-Objective Problems. Kluwer Academic/Plenum publishers, New York (2002)
3. Deb, K.: Multi-Objective Optimization using Evolutionary Algorithms. John Wiley and Sons, Ltd., Chichester (2001)
4. Klir, G.J., Folger, T.A.: Fuzzy Sets, Uncertainty, and Information. Prentice-Hall, Englewood Cliffs (1992)
5. Goldberg, D.E.: Genetic Algorithms in Search, Optimization, and Machine Learning. Addison-Wesley, Reading (1989)
6. Gómez-Skarmeta, A.F., Jiménez, F.: Fuzzy modeling with hibrid systems. Fuzzy Sets and Systems 104, 199–208 (1999)
7. Gómez-Skarmeta, A.F., Jiménez, F., Sánchez, G.: Improving Interpretability in Approximative Fuzzy Models via Multiobjective Evolutionary Algorithms. International Journal of Intelligent Systems 22, 943–969 (2007)
8. Ishibuchi, H., Murata, T., Turksen, I.: Single-objective and two-objective genetic algorithms for selecting linguistic rules for pattern classification problems. Fuzzy Sets and Systems 89, 135–150 (1997)
9. Ishibuchi, H., Nakashima, T., Murata, T.: Performance evaluation of fuzzy classifier systems for multidimensional pattern classification problems. IEEE Transactions on Systems, Man, and Cubernetics - Part B: Cybernetics 29(5), 601–618 (1999)
10. Jiménez, F., Gómez-Skarmeta, A.F., Sánchez, G., Deb, K.: An evolutionary algorithm for constrained multi-objective optimization. In: Proceedings IEEE World Congress on Evolutionary Computation (2002)
11. Laumanns, M., Zitzler, E., Thiele, L.: On the Effects of Archiving, Elitism, and Density Based Selection in Evolutionary Multi-objective Optimization. In: Zitzler, E., Deb, K., Thiele, L., Coello Coello, C.A., Corne, D.W. (eds.) EMO 2001. LNCS, vol. 1993, pp. 181–196. Springer, Heidelberg (2001)
12. Roubos, J.A., Setnes, M.: Compact fuzzy models through complexity reduction and evolutionary optimization. In: Proceedings of FUZZ-IEEE-2000, San Antonio, Texas, pp. 762–767 (2000)

13. Sánchez, G., Jiménez, J., Vasant, P.: Fuzzy Optimization with Multi-Objective Evolutionary Algorithms: a Case Study. In: IEEE Symposium of Computational Intelligence in Multicriteria Decision Making (MCDM), Honolulu, Hawaii (2007)
14. Setnes, M.: Fuzzy Rule Base Simplification Using Similarity Measures. M.Sc. thesis, Delft University of Technology, Delft, the Netherlands (1995)
15. Setnes, M., Babuska, R., Verbruggen, H.B.: Rule-based modeling: Precision and transparency. IEEE Transactions on Systems, Man and Cybernetics, Part C: Applications & Reviews 28, 165–169 (1998)
16. Spanish Society of Intensive-Critical Medicine and Coronary Units and Spanish Society of Emergency, Generalized infection mortality could be 20% off (in Spanish) (2007)
17. Takagi, T., Sugeno, M.: Fuzzy identification of systems and its application to modeling and control. IEEE Transactions on Systems, Man and Cybernetics 15, 116–132 (1985)
18. Valente de Oliveira, J.: Semantic constraints for membership function optimization. IEEE Transactions on Fuzzy Systems 19(1), 128–138 (1999)
19. Wang, L., Yen, J.: Extracting fuzzy rules for system modeling using a hybrid of genetic algorithms and Kalman filter. Fuzzy Sets and Systems 101, 353–362 (1999)
20. Yen, J., Wang, L.: Application of statistical information criteria for optimal fuzzy model construction. IEEE Transactions on Fuzzy Systems 6(3), 362–371 (1998)
21. Zitzler, E., Thiele, L., Laumanns, M., Fonseca, C.M., Grunert da Fonseca, V.: Performance Assessment of Multiobjective Optimizers: An Analysis and Review. IEEE Transactions on Evolutionary Computation 7(2), 117–132 (2003)

Knowledge Based Information Retrieval with an Adaptive Hypermedia System

Francisca Grimón[1], Josep Maria Monguet[2], and Jordi Ojeda[2]

[1] Departamento de Computación, FACYT, UC
[2] Universidad Politécnica de Cataluña
fgrimon@uc.edu.ve, jm.monguet@upc.edu, Jordi.Ojeda@upc.edu

Abstract. This paper describes research on information retrieval with an adaptive hypermedia system (AHS) used during three higher education courses taught in a blended learning (BL) environment. The system generates different work plans for each student, according to their profile. Work plans are adapted by means of an algorithm. AHS enable course contents to be adapted to the learning needs of each student and structured in a way that leads to many different learning paths. The case study method was used in this research. The results suggest that the AHS has a positive impact on the learning process. Further research is needed to confirm these results.

Keywords: adaptive hypermedia system, blended learning, adapted, algorithm.

1 Introduction

According to [10] "an AHS is a hypermedia system that adapts itself autonomously. This is, it monitors the user's behavior, records this behavior in a model of user and dynamically adapts the model current status."

The purpose of AHS is to create a learning space that can be adjusted to the particularities of each student, making it possible to set up educational environments where students reach learning objectives by means of contents and experiences in accordance with their aptitudes, interests, and preferences [1]. AHS, its methods and techniques have been frequently used in the area of education [2], hiding or showing information and links according to certain conditions to be met. Within this context, adaptation aims at rendering learning more efficient and effective. [13] As expressed in [3] and [4], adaptive hypermedia is an alternative for the traditional one-size-fits-all approach. These systems are useful to adapt the content to the needs of each user. According to [7], the term BL has gained ground in recent years as a specific way of describing education that uses technology. The definition of BL has created great controversy in scientific literature. Below are the given by the authors [14]:

1. An integrated combination of traditional learning with Web-based online approaches.

J. Mira et al. (Eds.): IWINAC 2009, Part I, LNCS 5601, pp. 457–463, 2009.

2. The combination of media and tools deployed in an e-learning environment.
3. The combination of a number of pedagogical techniques, irrespective of the learning technology used in each case. [9] Stated that the aim of a BL is to find a combination of resources and procedures that are adapted to the learning scenario, in order to attain better student performance.

2 Design of AHS

The elements of AHS with their different components: content, user, adaptation and evaluations are explained below.

- The contents component can be used to structure the knowledge that is to be transmitted. In addition, it allows information on contents to be stored and managed. The conceptual arrangement of contents is based on the model proposed in the ASK system [5][12] in which contents are presented via a network of concepts (nodes) that are predefined by the lecturer and should be learnt by the student. The arcs represent the transition from one concept to another, according to the knowledge that the student has gained. The transitions are presented in the form of evaluation questions.
- The user component contains the student's profile, which changes as the individual interacts with the system and acquires new knowledge. For each user, a model is saved that represents their knowledge of the course contents. This is called an overlay model (it defines the student's knowledge as a subset of an expert's knowledge). The asynchronous mode contains general and specific contents.
- The adaptation component involves a set of rules that enable contents to be adapted to a student's profile. The use of rules to make an adaptation is inspired by research into AHS, including "Modelling of an Adaptive Hypermedia System Based on Active Rule [11]". This research applied the following rules:

  When an event E occurs, if condition C is met, the system will execute action A.

  The adaptation is carried out via rules that specify which elements of the system will be shown and how they will behave, according to the user model [1]. The rules establish which contents the user should know and what knowledge they should gain before they can access specific contents [8]. In addition, they establish the way in which the user and content models are combined to carry out the adaptation [4]. In general, the rules enable the system to adaptively select content, taking into account the student's characteristics and what they need to know in order to meet the learning objectives.
- The evaluation component allows the evaluations to be adapted to the student's profile. There are two types of evaluations:

- ∘ Type A is an evaluation of knowledge on contents.
- ∘ Type B is an evaluation of the student's perception of the learning process.

To implement the AHS, the following aspects of each of the components are considered:

- Contents. This component is used to manage contents. Metadata for each content item in the AHS enables it to be described. The contents are stored in the content repository. Content management has the following functions: add a new content, modify, delete and search.
- Users. This component enables user profiles to be created, updated and managed. Metadata is available on each user, so that they can be identified in the system.
- Evaluations. This component is used to manage Type A (knowledge) and Type B (perception of learning) evaluations.
- Adaptation. This component defines the rules used to adapt the system to each user. The rules are as follows:
  - ∘ Evaluation rules. These indicate the criteria to be followed as a result of a student's performance in an evaluation on the contents that they have been assigned in their work plan. To carry out Type A evaluations, which enable the student to progress in the AHS, the student must first carry out Type B evaluations.

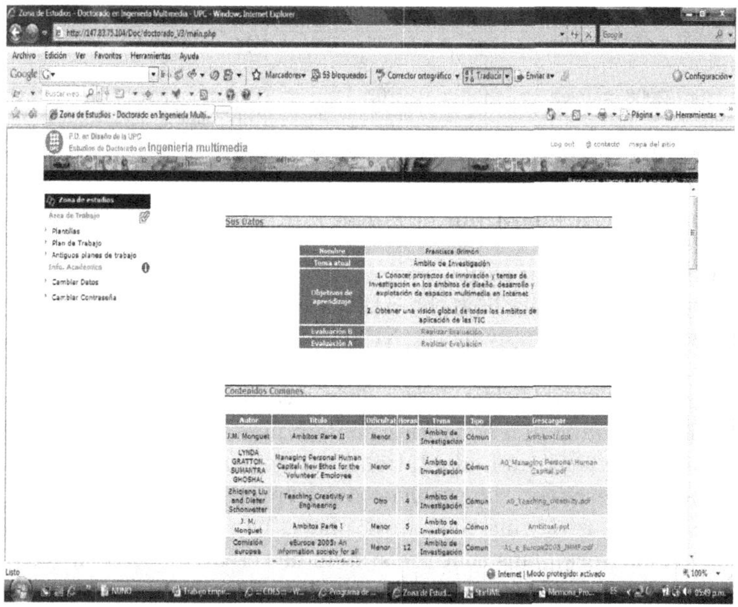

Fig. 1. Interface for the work plan

- Content rules. These indicate the criteria that should be followed in order to select contents, according to a student's profile. They include: discipline, preferences, objectives, the user's category of interest, the relation between the course subjects, and the key words associated with each subject.

Several interfaces are employed in the implementation of the synchronous and asynchronous modes of the model. Example is shown in Figure 1.

3 Methodology

The research strategy used was "Case Study" because it makes it possible to study a contemporaneous phenomenon in the real life's context [15]. The size of the samples and the instruments used for the empirical experiences conducted during this research work are presented in Table 1. Data analysis was performed once the information was obtained, processed and organized. To study the questionnaire open answers and the answers to the interview, the content analysis methodology was applied because it consists of classifying and/or coding the different elements of a message into categories [6].

Table 1. Empirical Experiences

Experiences	Number of participants	Instruments
Doctorate. (2005-2006 course)	16	Questionnaire 1, Assessment
Computer Science BA (2007 course)	26	Questionnaire 2, Interview 1, Assessment
Computer Science BA (2007 course)	26	Questionnaire 2, Interview 1, Assessment
Doctorate (2007-2008 course)	07	Questionnaire 2, Interview 1, Interview 2, Assessment

3.1 Experience Description

3.1.1 Software Description

For the purpose of knowing how the AHS-based learning process will progress, the system was designed applying the XP (eXtreme Programming) method, with a series of diagrams taken from the Rational Unified Process and the Unified Model Language as standard notation. The development was open source. The system has the following functionalities: content management, user management, assessment management, and adaptation management. It also has two kinds of users: teacher and student. The teacher can manage contents and monitor the students' evolution, reviewing their work plans. The students can see their current and previous work plans that the system has created based on their profile. They can also look for contents within the repository and see their academic information.

3.1.2 Exploratory Study

AHS creates a profile for each user, depending on the user's objectives, tastes and/or preferences concerning the course; this is carried out by means of an adaptation function. Profile refers to the particular conditions of each student concerning knowledge, academic experience/research/work activity, technological skills, interest/preferences in the course being studied, and progress. The system generates the profile based on the information supplied by the students when they access the tool and on previous knowledge. The profile is updated by means of the student-system interaction. Later, the system selects a repository of the contents that are shown to the student by means of a work plan that contains several features. The system includes two kinds of assessments. One self-assessment of the teaching-learning process (Type B Assessment) and a knowledge assessment (Type A Assessment). These assessments are related to the contents shown to the students in their respective work plans. The content progress will depend on whether the learning objectives have been covered, through the approval of the knowledge assessment.

3.1.3 Data Gathering Instruments

The instruments used for data gathering (questionnaires and interviews) were validated by a panel of experts from different knowledge areas. Interaction with experts was based on ICTs, because they were located in different countries:

Table 2. Results

Research questions	Test	Results
How will students rate the use of the AHS?	1	66.70 % Excellent 33.30 Good
	1	66.70 % Excellent 33.30 Good
	2	52.36 Excellent 39.12 Good 08.52 Regular
	3	66% Excellent 17% Good 17% Regular
How will students rate the contents provided by the system?	1	100 % Good
	2	64.00% Excellent 36.00 % Good
	3	33% Excellent 50%Good 17% Regular
How will students rate the content evaluations carried out by the system?	1	11.10% Excellent 66.70% Good 22.20% Regular
	2	35.00% Excellent 36.00% Good 29.00% Regular
	3	35.00% Excellent 36.00% Good 29.00% Regular
How will students rate their profile, which is generated by the system?	1	93.75% Excellent 06.25% Good
	2	50.00% Excellent 36.00% Good 14.00 % Regular
	3	33%Excellent 50% Good 17% Regular
Will the students consider that the AHS	1,2,3	100% Yes
has a positive effect on their learning process?		
How will the model influence the learning process of each individual?	1,2,3	100% Yes

Spain, Venezuela, and the US. Experts validated the appearance or content of the instruments and clarity review.

4 Results

The results indicate that the AHS had a positive effect on the learning process. The students gave a good rating to the AHS and stated that the contents were adapted to the profile of each user. The results of the different tests contribute to research into content personalisation in BL environments. In Table 2, Test 1 refers to the study undertaken on the subject "Research Methodology", in the 2006-2007 academic year. Test 2 shows the study undertaken on the subject "Information Systems", in the 2007 academic year. Finally, Test 3 refers to the study undertaken on the subject "Research Methodology", in the 2007-2008 academic year.

References

[1] Berlanga, A., García, J.: Sistemas Hipermedia Adaptativos en el Ámbito de la Educación. Technical Report. Departamento de Informática. Universidad de Salamanca (2004)
[2] Brusilovsky, P.: Adaptive Hypermedia: From Intelligent Tutoring Systems to Web-Based Education, pp. 1–7. Springer, Heidelberg (2000)
[3] Brusilovsky, P.: Aaptive Hypermedia. In: User Modeling and User-Adapted Interaction, vol. 11, pp. 87–110. Kluwer Academic Publisher, Netherlands (2001)
[4] De Bra, P.: Design Issues in Adaptive Web-Site Development. In: Proceedings of the 2nd Workshop on Adaptive Systems and User Modelling on the WWW at the 7th International Conference on User Modeling, pp. 29–39 (1999)
[5] Gaudioso, E.: Contribuciones al Modelado del Usuario en Entornos Adaptativos de Aprendizaje y Colaboración a través de Internet mediante técnicas de Aprendizaje Automático. Doctoral thesis. Universidad Nacional de Educación a Distancia. Madrid (2002)
[6] Gómez, M.: Análisis de contenido cualitativo y cuantitativo: Definición, clasificación y metodología. Revista de Ciencias Humanas 20, 103–113 (2000)
[7] Herrera, M.: Modelación del rendimiento estudiantil en ambientes de aprendizaje basados en blended learning y el método de casos de estudio. Proyecto de Thesis. Universidad Politécnica de Cataluña (2007)
[8] Medina, N., Molina, F., García, L.: An Author Tool Based on SEMHP for the Creation and Evolution of Adaptive Hypermedia Systems. In: ACM International Conference Proceeding Series, Workshop proceedings of the sixth international conference on Web engineering, vol. 155, Article No. 12 (2006)
[9] Monguet, J., Fábregas, J., Delgado, D., Grimón, F., Herrera, M.: Efecto del Blended Learning sobre el rendimiento y la motivación de los estudiantes. Journal Interciencia 31(3), 190–196 (2006)
[10] Parcus, N.: Software Engineering for Adaptive Hypermedia Systems Reference Model, Modeling Techniques and Development Process. Thesis Doctor. Universität München (2001)
[11] Raad, H., Causse, B.: Modelling of an Adaptive Hypermedia System Based on Active Rules, pp. 149–157. Springer, Heidelberg (2002)

[12] Schank, R., Cleary, C.: Engines for education. Lawrence Erlbaum Associates, Hillsdale (1995)

[13] Vélez, J.: Arquitectura para la Integración de las Dimensiones de Adaptación en un Sistema Hipermedia Adaptativo. Proyecto de Investigación. Universidad de Girona, Spain (2007)

[14] Whitelock, D., Jelfs, A.: Editorial: Journal of Educational Media Special Issue on Blended Learning. Journal of Educational Media 28(2-3), 99–100 (2003); Yin, R.: Case Study Research, Design and Methods, 3rd edn. Sage Publications, Newbury Park (2002)

Reveal the Collaboration in a Open Learning Environment

Antonio R. Anaya and Jesús G. Boticario

E.T.S.I.I. - UNED C/Juan del Rosal, 16, Ciudad Universitaria, 28040 Madrid, Spain
arodriguez@dia.uned.es, jgb@dia.uned.es

Abstract. The management and characterization of collaboration to improve students' learning is still an open issue, which needs standardized models and inferring methods for effective collaboration indicators, especially when online courses are based on open approaches where students are not following CSCL scripts. We have supplied our students with a scrutable (manageable and understandable) web application that shows an ontology, which includes collaborative features. The ontology structures collaboration context information, which has been obtained form explicit (based on questionnaires) and implicit methods (supported by several machine learning techniques). From two consecutive years of experiences with hundreds of students we researched students' interactions to find implicit methods to identify and characterize students' collaboration. Based on the outcomes of our experiments we claim that showing useful and structured information to students and tutors about students' collaborative features can have a twofold beneficial impact on students learning and on the management of their collaboration.

1 Introduction

It is really curious that although collaboration is a very common strategy in e-learning [1], modeling has not focused on it in depth. [2] noted the need to build a standard to analyze collaboration. They warned that research works on collaboration did not give enough information about the analysis process. Although describing the collaboration standard is not the scope of this research, we explain our analysis of the collaboration in detail.

We designed a long-term collaborative learning experience during 2006-07, 2007-08 and 2008-09 with fourth-year Artificial Intelligence (AI) and Engineering Based Knowledge students at UNED (UNED is the acronym for the Spanish University for Distance Education). This experience consisted of two main phases within a step-wise approach: the first phase was 3 weeks and the second phase was 10 weeks. In the second phase students were grouped into three-member teams. It was enough time for the students to complete the collaborative work and manage their collaborative process. In our educational environment we have found that UNED students, who are mainly adults with responsibilities other than learning, cannot be forced to collaborate in a typical CSCL environment because of the time restrictions of these environments [3]. We provided students

J. Mira et al. (Eds.): IWINAC 2009, Part I, LNCS 5601, pp. 464–475, 2009.

with an open collaborative learning experience, where students could manage their own collaborative learning process. Because of our research in first two years (2006-07 and 2007-08), we noted collaboration problems arose when the team members did not communicate much (lack of communication), worked at different rates (lack of coordination), or some members dropped out of the collaboration experience (abandonment). Obviously, the lack of information on the collaboration process in collaborative e-learning environments in comparison to face-to-face communication makes the learning process worse in these environments. The objective is to give useful and structured information on the collaboration process to students and tutors to be able to improve collaboration process management.

Accordingly, we developed a web application, whose objectives were to provide students and tutors with information on the collaboration context and process, and it works as self-regulation learning tool [4]. Collaboration context data were obtained from the questionnaire, which students had to answer at the beginning of the learning experience. Information on the collaboration process was achieved by analyzing the statistical indicators of student interaction during the collaboration learning process.

We researched the relation between student collaboration with the statistical indicators of student interactions during the academic years (2006-07 and 2007-08) with the objective of inferring student collaboration quantitatively and identifying the statistical indicators of interactions, which are related to collaboration. These objectives are summarized in a quantitative method that obtains the collaboration just after the collaboration process has finished. Finally, using a quantitative data mining method, we discovered that very collaborative students were the most active students and their activity caused more activity in others.

To cover the research objectives we have offered a web application to some students on the collaboration learning experience this academic year 2008-09. This web application is based on ontology, which includes collaborative features, and it can be navigated and managed by students. This web application also shows information on the collaboration context, the statistical indicators of student interaction in forums and inferred student collaboration. We have developed an open and scrutable web application [5], which means that the data displayed are structured, understandable and manageable by application users (students and tutors). These characteristics, nevertheless, restrict how the data are displayed and their format.

The next section collects some research works focusing on collaboration or scrutability. Then we describe the collaborative learning experience. After, we comment on the latest research to discover student collaboration according to their interaction and explain the data mining method used, the statistical indicators and inferred results. Then the web application is described and the ontology, on which the web application is based. Finally, we conclude with the discussion and future works.

2 Related Works

Although collaboration is one of the main strategies in e-learning [1], it has not been researched in depth compared with other user skills or characteristics [6,7]. However, research on collaboration can be divided, on the one hand, into building a collaboration model and, on the other hand, into identifying collaboration indicators.

Some research works have built a collaboration model. [8] used a collaboration model, which described student collaboration interaction on a structured course. After student interaction had finished, fuzzy logic was applied to compare the interaction with the collaboration model. Thus, student collaboration could be identified. An expert built the model describing student collaboration, which should have been done. [9] built a collaboration model consisting of a set of 25 meta-constraints representing ideal collaboration. Each meta-constraint consisted of a relevance condition, a satisfaction condition and a feedback message, which was presented when the constraint was violated. They studied the literature existing on characteristics of effective collaboration and also used their own experience in the collaborative work. [10] described a model for representing collaborative actions, aimed at supporting interaction analysis in CSCL. Its main contributions were the proposal of a context model, and of a generic taxonomy based on a bottom-up approach that focused on the actors intervening in each type of action. [11] proposed several models useful for describing Communities of Practice (CoP): community, actor, learner profile, competency, collaboration, process/activity, and lessons learnt. These models were built by adapting some existing models. The models proposed were structured in an ontology.

Other research works have used a set of collaboration indicators based on collaboration theories. [12,13] proposed 5 indicators based on the [14] theory. These indicators are: the use of strategies, intra-group cooperation, checking the success criteria, and monitoring. They noted that these five indicators did provide some insight into the collaborative work done by the groups. They could be used to detect group weaknesses in their collaborative learning process. [15] considered the six Collaborative Competency Levels defined in [16]. They decided to separate this type of competency because it defined important aspects in collaborative and cooperative student behavior. They were interested in modeling these user characteristics in order to establish their relation with learning process success. [17] proposed the Collaborative Learning (CL) Model, [18] described the potential indicators of effective collaborative learning teams based on a review of research in educational psychology and computer-supported collaborative learning, and empirical data. In addition, the CL Model provided a set of criteria for evaluating the system after development. [19] proposed items for assessing learners in web-based collaborative learning (WBCL). The learners assessed their colleagues (teammates) and themselves on the basis of interaction, collaborability, and accountability. Interaction can be measured by attendance, posting frequency to BBS, the number of on-line/offline chatting, and personal journal. Similarly, collaborability can be measured by personal journal and investigation, while self/peer assessment can be done for accountability. [20] described a rating scheme for assessing the quality of collaborative processes in computer-supported problem solving. The rating scheme assessed collaboration quality on nine process dimensions that integrate results from a qualitative analysis of transcribed collaboration dialogue with theoretical considerations based on the relevant literature. The nine dimensions of the rating scheme were: sustaining mutual understanding, dialogue management, information pooling, reaching consensus, task division, time management, technical coordination, reciprocal interaction, and individual task orientation.

There are still some open issues in collaborative e-learning environments to improve learning and collaboration. We are researching collaboration analysis with scrutability, because both concepts help students with special characteristics like those enrolled at UNED, who represent Lifelong Learning learners with partial dedication to their studies. We note that there are serious difficulties when creating a collaboration model that describes the best way of collaborating. We think that this is only possible in very structured and restrictive web environments, which we have not offered to our students. Showing students and tutors useful and understandable information on collaboration is the objective of this research. We go one step further than systems that only show the results of the collaboration analysis, because we have developed a web application showing the collaboration information using data mining techniques.

3 Educational Context

We offered AI students the possibility of taking part in a long-term collaborative learning experience during 2006-07, 2007-08 and 2008-09. The learning experience consisted of practical collaborative work, which covered 3 months (see Figure 1). The activity structure was divided into two main phases within a step-wise approach. The collaborative learning experience was offered to all students enrolled in the subject. The number of students that were initially interested in participating in the experience was 260 the first year (2006-07), 239 the second year (2007-08) and 207 the third year (2008-09). The 2008-09 learning experience has not finished yet.

The learning experience has two parts or phases: first, where the learners do individual work; second, where three-member teams do collaborative work. We

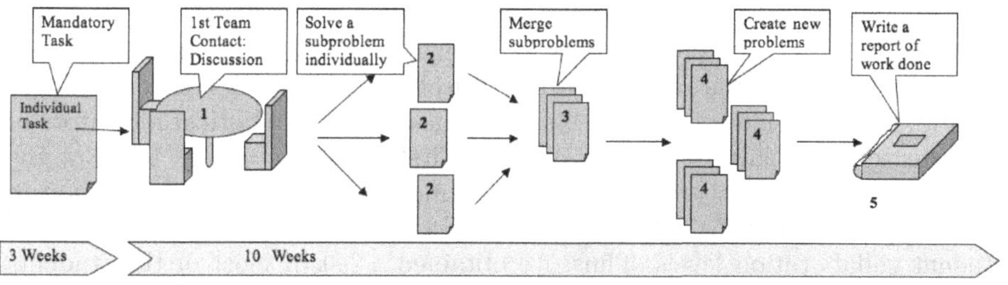

Fig. 1. Description of the open collaborative learning experience

provided a learning platform dotLRN (`http://dotlrn.org/`), which supports all learning experience activities and stores all the interactions, which take place on the platform, in a relation database. The first phase goes on for 3 consecutive weeks. Here the student answers an initial questionnaire and solves an individual task. The questionnaire requests personal data and information on the possibilities of the learners collaborating with others. From this questionnaire the collaboration context is obtained. The answers are used to group the learners into three-member teams. The individual task is mandatory and the learners have to solve a very simple problem. Only the learners who pass the individual task can start the collaborative or second phase. During this phase a general virtual environment is open for all students of the subject with common services (FAQs, news, surveys, calendar and forums). The number of students who finished this phase was 125 in 2006-07, 140 in 2007-08 and 115 in 2008-09.

The second phase covered 10 weeks and the students were grouped in teams with 3 members. We chose 3 members because learning is more efficient in small groups (mainly groups with 3 or 4 members) [21]. The team members had to follow 6 consecutive tasks throughout the collaborative experience. (1) In the forums the team discusses which problem they are going to address from the ones that are given to them. (2) This phase is mainly individual work and consists of each team member solving one of the three different subproblems. (3) The team members have to integrate their previously generated individual solutions. (4) The experimental task takes place in this phase and here the team has to create other related planning problems that are based on the original one. (5) Finally, the team has to provide a report that covers all the activities and their corresponding results. At the end (6), the learners are asked to fill in a final questionnaire, which includes several topics with valuable information associated with the collaboration results. 104 students finished this phase in 2006-07, 122 in 2007-08, and there are no available data for 2008-09 yet. During the second phase virtual spaces for each three-member team were opened, where the teams could perform the tasks. The specific virtual spaces include documents, surveys, news, a task manager and forums.

4 Inferring Collaboration Process Information

We have analyzed the collaboration interaction in the previous two collaborative learning experiences (academic years 2006-07 and 2007-08). We have researched student collaboration and interactions with the objective of obtaining a quantitative relation between both aspects and measuring the collaboration before or just after the collaboration process has finished. If there is a relation between collaboration and interaction, we can only measure the statistical indicators of the interaction to know the collaboration. First of all, we needed to know the students' collaboration and an expert identified the students' collaboration in the first two experiences. The expert read all the forum messages and measured student collaboration levels. Thus, we obtained a list of most of the students labeled according to their collaboration level. The expert used a scale of 8 values (1, high collaboration level; 8, low collaboration level).

Apart from this, we built datasets containing statistical indicators of student interaction in forums. The statistical indicators considered in the dataset were: number of threads or conversations that the learner started (num_thrd), and their average, square variance and the number of threads divided by their variance; the number of messages sent (num_msg), and their average, square variance and the number of messages divided by their variance; the number of replies in the thread started by the user (num_reply_thrd), and divided by the number of user threads; the number of replies to messages sent by the user (num_reply_msg), and divided by the number of user messages. According to [22], the number of threads (num_thrd) represents user initiative, the number of messages (num_msg) represents user activity, and the number of replies represents the activity caused by the user.

We ran the cluster algorithm EM with the dataset to obtain clusters of students. These clusters were compared with the list of students according to their collaboration level to identify relations between clusters and the collaboration level [23]. Therefore, the process just described to discover the level of collaboration was: (1) we ran the EM cluster algorithm with the dataset. We used the data mining software WEKA [24] and the clustering algorithm EM [25]. (2) We looked for the main statistical indicators used by EM to build the cluster. (3) We compared the cluster with the expert's list to identify the relation between the cluster and the collaboration level. (4) We identified learner collaboration levels according to their cluster. We explain the research results below.

5 Results

We ran an EM cluster algorithm with the 2006-07 and 2007-08 datasets. The EM cluster algorithm was configured to obtain three clusters. Figure 2 shows the relation between the clusters and the statistical indicators *num_msg* (number of messages sent by a student) and *num_reply_msg* (number of replies to messages sent by a student). We have selected these indicators, because the relation with the clusters is the clearest.

The X-axis shows the values of the statistical indicator (num_msg second column and num_reply_msg third column) and the Y-axis shows the three clusters (cluster-0 at down, cluster-1 in the middle and cluster-2 on the top). The clusters are ordered according to their average value. Although the cluster borders are not clear, because the clusters overlap, the cluster center value is always different in every cluster. We conclude that the EM cluster algorithm has identified three clear clusters in the datasets.

We conclude that the EM cluster algorithm has identified three clear clusters in the datasets, and the statistical indicator "num_msg" and "num_reply_thrd" identify the relation between the clusters better.

Following the research plan, we compared the cluster with lists of collaboration levels, which the expert supplied. Table 1 describes the results. The collaboration level for each cluster is the average of the estimated collaboration level given by the expert to every instance/student in the cluster. We note that the highest

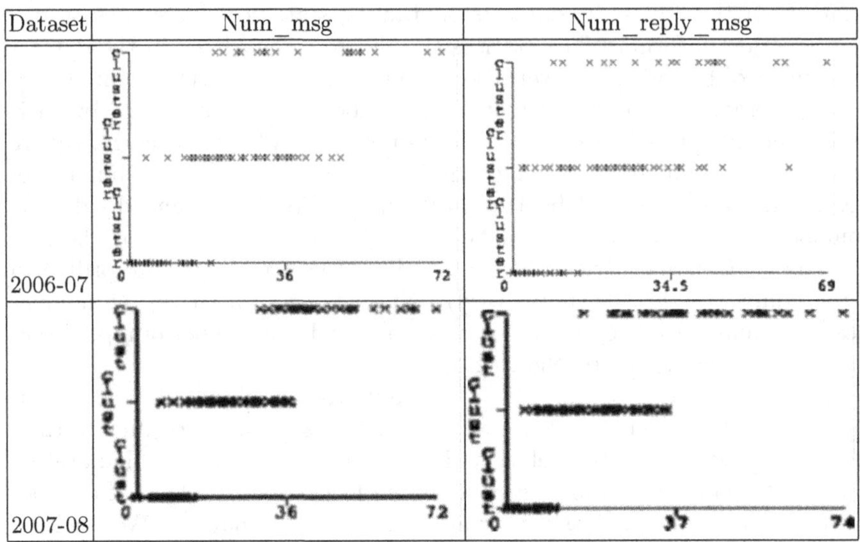

Fig. 2. Graphics of statistical indicators versus clusters

Table 1. Cluster collaboration level average

Dataset	Cluster-0	Cluster-1	Cluster-2
D-I-2006-07	5.62	3.85	3.26
D-I-2007-08	5.21	4.39	3.89

values in the table indicate the lowest collaboration, and the lowest values indicate the highest collaboration.

Now we can connect the clusters to the collaboration level. The cluster-0, which groups students who sent few messages and these messages received few replies, collects not very collaborative students. The cluster-2, which groups students who sent a lot of messages and these messages received a lot of replies, collects very collaborative students. The cluster-1 collects students that are not very collaborative or very collaborative.

From this research we conclude that measuring student collaboration in a quantitative way during the collaboration process is possible and it depends on student activity and the activity caused by students (indicators *num_ msg* and *num_ reply_ msg*). It has been possible to measure this information during the current collaboration learning experience and it is being shown to students.

6 Web Application

One of the problems of e-learning is the lack of information available to students and tutors. Students use e-mail, forum or chat to talk with others. However, from all these communication media less information is obtained than face-to-face communication. The lack of information makes it difficult to manage the collaboration process.

We have developed a web application as a self-regulation learning tool to give students and tutors structured information on the student collaboration context and process so that they can improve collaboration process management. Since the web application shows information to students and tutors, this information has to be understandable. We have followed the ideas of open and scrutable ontology and model [5,26,27]to develop the web application, which shows the collaboration context and collaboration process information and lets information owners manage some data. Figure 3 shows the web application. We note that to cover the objectives we have based the web application on ontology, which includes collaborative features.

The web application shows data, which students gave when they answered the initial questionnaire, the statistical indicator of student interaction, and the student collaboration level according to the EM cluster algorithm results (see section Inferring Collaboration Process Information). The web application is based on ontology on collaboration, which stores and structures data in a simple and understandable way. All information is easily navigable by common navigators. The information has been structured according to the hierarchical order of the ontology. The information collected in the class Static_Data, which is explained in the next section, can be managed by the data owner, who is the student described by the data.We have grouped all students in the same web application instance so that they can navigate their own data, teammate data, and others' data.

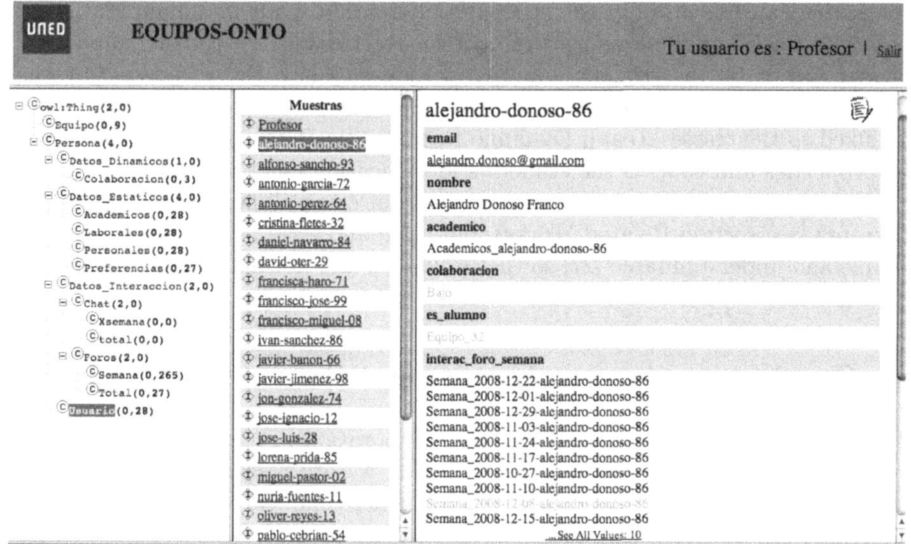

Fig. 3. User information page

This has given them enough information about the collaboration context, their own collaboration, and others' collaboration. The students and tutors can use all the information to improve collaboration process management. Information on interactions and collaboration is updated once a week.

7 An Ontology on Collaboration

The web application is based on ontology. We propose using ontology in this case, because ontology is a standard, structured, understandable and reusable way of describing concepts and relating concepts with one another [28]. For this reason ontology is suitable for describing the learner model [27], which can be used in a scrutable way by students [26]. The ontology objectives have been to compile data of the collaboration context and process.

To cover the first objective, students were asked collaboration context information (when they could connect to the web platform and do collaborative tasks, address, job, etc.) and student context information (personal data, academic data, etc.). The second objective was covered inferring student collaboration levels according to their statistical indicators of interactions. We explained this point earlier in the section Inferring Collaboration Process Information.

Collaboration ontology is a set of classes ordered hierarchically. Two classes hang from the main class (Thing): Team and Person. The class Team collects the collaborative three-member team. The class Person has four subclasses: Dynamic Data, Static Data, Interaction Data and User. Dynamic Data has data that can change over time and collects the inferring value of the student's collaboration level. Static Data has data that cannot change over time like personal data, academic data, occupation data and preferences. Interaction Data collects the statistical indicators of the student's interaction. Only the main statistical indicators are shown: number of the thread started by the student, number of messages sent by the student, number of messages in a thread started by the student, and number of replies to messages sent by the student. The class User is linked to the classes Team, Dynamic Data, Static Data and Interaction Data by properties and collects the student's identification data.

We know that the ontology proposed does not cover all possible collaboration concepts (see section Related Works), but open and scrutable ontology must be simple and understandable [27] so that students, and not machines, can use and manage it.

8 Conclusion, Current Results and Future Works

This research focuses on improving the collaboration process in e-learning environments. The e-learning experience, which we are researching, is aimed at UNED's students, who have responsibilities other than learning. These kinds of students cannot be forced to collaborate in a typical CSCL environment because of the time restrictions of these environments [3]. However it is advisable to use collaboration in this context, because it can solve the isolation problem of

UNED's students. For this reason we offered fourth-year Artificial Intelligence (AI) and Engineering Based Knowledge students the possibility of doing tasks in a collaborative e-learning environment.

We propose a web application as self-regulation learning tool, whose objectives were to compile collaboration context information to make collaboration easier, to show information on the collaboration process to improve management, and let students manage their own data in the ontology.

The collaboration context and process information was obtained from three different sources: (1) student answers to the initial questionnaire; (2) the statistical indicators of student interaction; (3) the student collaboration level according to the EM cluster algorithm results. From comparing the clusters with a list of the collaboration levels supplied by an expert we have found that collaboration is related to student interaction. We note that the number of messages sent and the number of replies to those messages are rough indicators of student collaboration. We conclude that the student who is more active and causes more activity is the most collaborative.

We cannot say whether we have achieved our objectives yet. We are waiting for the end of the collaboration experience to analyze the students' answers to the final questionnaire, the web application log analysis and the student subject exam ranking. After this, we shall have all the data to evaluate the web application.

There are some open issues in this research. We do not have a clear explanation on why the clustering algorithm has provided the best results for this research. To clarify this issue we are carrying out parallel research where the inferring method relies on decision tree algorithms [29]. We are currently collecting results from the datasets so that we can subsequently compare the new results from the application of decision tree algorithms with the results reported in this paper. Another open issue is ontology. We have said that creating a collaboration standard is not our objective, but we should work in this direction updating our ontology and model according to user model standards such as IMS Learner Information standards.

References

1. Soller, A., Martinez, A., Jermann, P., Muehlenbrock, M.: From mirroring to guiding: A review of state of the art technology for supporting collaborative learning. International Journal of Artificial Intelligence in Education 15, 261–290 (2005)
2. Strijbos, J.W., Fischer, F.: Methodological challenges for collaborative learning research. Learning and Instruction 17 (2007)
3. Gaudioso, E., Santos, O.S., Rodríguez, A., Boticario, J.G.: A proposal for modelling a collaborative task in a web-based learning environment. In: Brusilovsky, P., Corbett, A.T., de Rosis, F. (eds.) UM 2003. LNCS, vol. 2702, Springer, Heidelberg (2003)
4. Steffens, K.: Self-regulation and computer based learning. Anuario de Psicología 32(2), 77–94 (2001)

5. Bull, S., Kay, J.: Student models that invite the learner in: The smili open learner modelling framework. IJAIED, International Journal of Artificial Intelligence in Education 17(2), 89–120 (2007)
6. Kobsa, A.: Generic user modeling systems. In: Brusilovsky, P., Kobsa, A., Nejdl, W. (eds.) The Adaptive Web: Methods and Strategies of Web Personalization (2001)
7. Brusilovsky, P., Millan, E.: User models for adaptive hypermedia and adaptive educational systems. In: Brusilovsky, P., Kobsa, A., Nejdl, W. (eds.) Adaptive Web 2007. LNCS, vol. 4321, pp. 3–53. Springer, Heidelberg (2007)
8. Redondo, M.A., Bravo, C., Bravo, J., Ortega, M.: Applying fuzzy logic to analyze collaborative learning experiences in an e-learning environment. USDLA Journal 17(2), 19–28 (2003)
9. Baghaei, N., Mitrovic, A.: From modelling domain knowledge to metacognitive skills: Extending a constraint-based tutoring system to support collaboration. In: Conati, C., McCoy, K., Paliouras, G. (eds.) UM 2007. LNCS (LNAI), vol. 4511, pp. 217–227. Springer, Heidelberg (2007)
10. CSCL 2003: An Xml-Based Representation Of Collaborative Interaction, CSCL 2003 (2003)
11. Vidou, G., Dieng-Kuntz, R., Ghadi, A.E., Evangelou, C., Giboin, A., Tifous, A., Jacquemart, S.: Towards an ontology for knowledge management in communities of practice. In: Reimer, U., Karagiannis, D. (eds.) PAKM 2006. LNCS(LNAI), vol. 4333, pp. 303–314. Springer, Heidelberg (2006)
12. Collazos, C.A., Guerrero, L.A., Pino, J.A., Ochoa, S.F.: Evaluating collaborative learning processes. In: Haake, J.M., Pino, J.A. (eds.) CRIWG 2002. LNCS, vol. 2440, pp. 203–221. Springer, Heidelberg (2002)
13. Collazos, C.A., Guerrero, L.A., Pino, J.A., Renzi, S., Klobas, J., Ortega, M., Redondo, M.A., Bravo, C.: Evaluating collaborative learning processes using system-based measurement. Educational Technology and Society 10(3), 257–274 (2007)
14. Johnson, D., Johnson, R.: Cooperative, competitive, and individualistic learning. Journal of Research and Development in Education 12, 8–15 (1978)
15. Baldiris, S., Santos, O.C., Barrera, C., Boticario, J.G., Velez, J., Fabregat, R.: Linking educational specifications and standards for dynamic modelling in adaptaplan. In: International Workshop on Representation models and Techniques for Improving e-Learning: Bringing Context into Web-based Education (ReTIeL 2007), Denmark (August 2007)
16. VI International Symposium on Educative Informatics (SIIE 2004): Supporting a collaborative task in a web-based learning environment with Artificial Intelligence and User Modelling techniques, VI International Symposium on Educative Informatics, SIIE 2004 (2004)
17. Soller, A.: Supporting social interaction in an intelligent collaborative learning system. International Journal of Artificial Intelligence in Education 12(1), 40–62 (2001)
18. Fourth International Conference on Intelligent Tutoring Systems (ITS 1998): Promoting effective peer interaction in an intelligent collaborative learning environment, San Antonio, TX, Fourth International Conference on Intelligent Tutoring Systems, ITS 1998 (1998)
19. Park, C.J., Hyun, J.S.: Comparison of two learning models for collaborative e-learning. In: Pan, Z., Aylett, R.S., Diener, H., Jin, X., Göbel, S., Li, L. (eds.) Edutainment 2006. LNCS, vol. 3942, pp. 50–59. Springer, Heidelberg (2006)
20. Meier, A., Spada, H., Rummel, N.: A rating scheme for assessing the quality of computer-supported collaboration processes. Computer-Supported Collaborative Learning (2), 63–86 (2006)

21. Johnson, D., Johnson, F.: Learning Together: Group Theory and Group Skills (1975)
22. Santos, O.C., Rodríguez, A., Gaudioso, E., Boticario, J.G.: Helping the tutor to manage a collaborative task in a web-based learning environment. In: AIED 2003: Supplementary Proceedings, pp. 153–162 (2003)
23. Talavera, L., Gaudioso, E.: Mining student data to characterize similar behavior groups in unstructured collaboration spaces. In: Proceedings of the Workshop on Artificial Intelligence in CSCL, Valencia, Spain, 16th European Conference on Artificial Intelligence (ECAI 2004), pp. 17–23 (2004)
24. Witten, I.H., Frank, E.: Data Mining. Morgan Kaufmann, San Francisco (2005)
25. Gama, J., Gaber, M.M.: Learning from Data Streams: Processing Techniques in Sensor Networks. Springer, Heidelberg (2007)
26. Kay, J.: Ontologies for reusable and scrutable student models. In: Mizoguchi, R. (ed.) AIED Workshop W2: Workshop on Ontologies for Intelligent Educational Systems, pp. 72–77 (1999)
27. Kay, J., Lum, A.: Ontology-based user modelling for semantic web. In: Ardissono, L., Brna, P., Mitrović, A. (eds.) UM 2005. LNCS, vol. 3538, pp. 11–19. Springer, Heidelberg (2005)
28. Chen, W., Mizoguchi, R.: Leaner model ontology and leaner model agent. In: Kommers, P. (ed.) Cognitive Support for Learning - Imagining the Unknown, pp. 189–200 (2004)
29. Berikov, V., Litvinenko, A.: Methods for statistical data analysis with decision trees, Novosibirsk, Sobolev Institute of Mathematics (2003)

Reasoning on the Evaluation of Wildfires Risk Using the Receiver Operating Characteristic Curve and MODIS Images

L. Usero[1] and M. Xose Rodriguez-Alvarez

[1] Departamento de Ciencias de la Computación, Universidad de Alcalá
[2] Universidad de Santiago de Compostela
luis.usero@uah.es, mariajose.rodriguez.alvarez@usc.es

Abstract. This paper presents a method to evaluate the wildfires risk using the Receiver Operating Characteristic (ROC) curve and Terra moderate resolution imaging spectroradiometer (MODIS) images. To evaluate the wildfires risk fuel moisture content (FMC) was used, the relationship between satellite images and field collected FMC data was based on two methodologies; empirical relations and statistical models based on simulated reflectances derived from radiative transfer models (RTM). Both models were applied to the same validation data set to compare their performance. FMC of grassland and shrublands were estimated using a 5-year time series (2001-2005) of Terra moderate resolution imaging spectroradiometer (MODIS) images. The simulated reflectances were based on the leaf level PROSPECT coupled with the canopy level SAILH RTM. The simulated spectra were generated for grasslands and shrublands according to their biophysical parameters traits and FMC range. Both RTM-based models, empirical and statistical, offered similar accuracy with better determination coefficients for grasslands. In this work, we have evaluated the accuracy of (MODIS) images to discriminate between situations of high and low fire risk based on the FMC, by using the Receiver Operating Characteristic (ROC) curve. Our results show that none of the MODIS bands have a good discriminatory capacity (0.9984) when used separately, but the joint information provided by them offer very small misclassification errors.

1 Introduction

Fuel moisture content (FMC) is one of the variables that drive fire danger. FMC is a quotient of these two variables, EWT and DM, that can be estimated independently. This variable conditions fire, since the drier the vegetation the easier fires ignite and propagate [2]. Water stress causes changes in the spectral reflectance and transmittance of leaves [16]. Direct estimation by field sampling provides the most accurate method to obtain FMC [15], commonly using gravimetric methods, namely the weight difference between fresh and dry samples [12]. However, this approach is very costly and the generalization to regional or global scales results unfeasible. The use of meteorological indices is widespread,

J. Mira et al. (Eds.): IWINAC 2009, Part I, LNCS 5601, pp. 476–485, 2009.

since they provide an easy spatial and diachronic estimation of FMC [3], but they also present operational difficulties since the weather stations are often located far from forested areas and maybe scarce in fire prone regions. Furthermore, these estimations are reasonably well suited for dead fuels, because their water content is highly related to atmospheric conditions. However, in live fuels, species physiological characteristics and adaptation to drought imply a great diversity of moisture conditions with the same meteorological inputs [18]. Radiative transfer models are theoretical models with a strong physical basis, generally based on the theory of radiative transfer is its name. To calculate the solar irradiance reaching Earth's surface, simulating the processes that suffers through the atmosphere. Radiative transfer models show that these spectral measurements are related to equivalent water thickness (EWT), water content per area unit, and dry matter content (DM), matter content per area unit [5]; [10]. The PROSPECT model is a radiative transfer model designed and developed by S. Jacquemoud and F. Baret in 1990 and is one of the models used in many scientific experiments. This model is based on the model proposed by Allen, which represents the optical properties of leaves from 400nm to 2500nm wavelength. The dispersion is described by a spectral index of refraction (n) and a parameter that characterizes the internal structure of the leaf (N). The absorption is modeled using the concentration of pigments (chlorophyll a + b), water content (Cw) and their corresponding spectral absorption coefficients.

SAILH model is a simulation model of the reflectivity at the canopy based on the Kubelka-Munk theory used to describe the interaction of radiation with plant canopies in a fast and under controlled conditions. This model is a variant of the SAIL model developed by Wout Verhoef in 1984 belongs to the type of canopy-level models as models of turbidity. As SAIL, SAILH model is a model of a turbidity level of canopy, so it treats the vegetation as an infinitely extended horizontal layer, flat and uniform, consisting of elements of vegetation randomized in parallel layers make small reflective particles that absorb, depending on their optical properties (reflectivity and transmissivity). SAILH can simulate the canopy reflectance at given the reflectivity and transmissivity of the leaf, the reflectivity of the soil, leaf area index (LAI) The Receiver Operating Characteristic (ROC) analysis comes from statistical decision theory [Green and Swets, 1966] and was originally used during World War II for the analysis of radar images. The first applications of this theory within the medical area occurred during the late 1960s. Today the ROC analysis is a widespread method in the medical field and many textbooks and articles, for example [Kraemer, 1992, Bamber, 1975,Metz, 1978, DeLong et al., 1988, Hanley, 1989, McClish, 1989, Armitage and Berry, 1994],have descriptions of it. From the computer science point of view, ROC analysis has been increasingly used as a tool to evaluate discriminate effects among different methods. The ROC curve relies heavily on notations as sensitivity and specificity and these values depend on the specific data set. Even though the values for sensitivity and specificity in theory lie in the interval [0; 1], in practice the borders are decided by the data set. If the QROC curve is considered instead a better way to compare different tests are given since all values have the same interval. This is accomplished by using two quality measures that transforms sensitivity and specificity values from different data sets to a comparable interval. MODIS (or Moderate Resolution Imaging Spectroradiometer) is a key instrument aboard the Terra (EOS AM) and Aqua (EOS PM) satellites. Terra's orbit around the Earth is timed so that it passes from north to south across the equator in the morning, while Aqua passes south to north over the equator in the afternoon. Terra MODIS and Aqua MODIS are viewing the entire Earth's surface every 1 to 2 days, acquiring data in 36 spectral bands, or groups of wavelengths (see MODIS Technical Specifications). These data will improve our understanding of global dynamics and processes occurring on the land, in the oceans, and in the lower atmosphere. MODIS is playing a vital role in the development of validated, global, interactive Earth system models able to predict global change accurately enough to assist policy makers in making sound decisions concerning the protection of our environment.

2 Methods

The empirical approach was derived from multivariate linear regression (MLR) analysis between field collected FMC data and reflectance values derived from the moderate resolution imaging spectroradiometer (MODIS). The field samples were divided in two sets: 60% for calibrating the model and the remaining 40% for the validation. Two different models were built for grasslands and shrublands. The simulation approach was derived from RTM that were parameterized using field data, auxiliary information derived from MODIS products and the knowledge of the type of canopy architecture that define which RTM is appropriate [7]. Once the simulated reflectance values for grasslands and shrublands were obtained for the whole solar spectrum, they were convolved to the MODIS spectral wavelengths and band widths. Finally, separate MLR models between the simulated reflectances and the grassland and shrubland FMC values were built, in a similar way to the empirical method. Those equations were applied to the MODIS data for the same validation dataset as the empirical model to compare the performances of both approaches.

2.1 Field Sampling

A field campaign has been carried out by our research group since 1996 to the present in the Cabaneros National Park (Central Spain; Fig. 1) to collect samples of different Mediterranean species for field FMC estimation. Three plots of grassland and two of shrubland (Cistus ladanifer L., Rosmarinus officinalis L., Erica arborea L. and Phyllirea angustifolia L.) sized 30m 30 m, were collected in gentle slopes (¡5%) and homogeneous patches. For this paper, FMC values of C. ladanifer L. were selected as representative for shrubland plots since it is very common in Mediterranean siliceus areas.

It appears in a 29.79% of the study area covering a radius of 100 km from the National Park being the dominant species in more than 6% versus less than

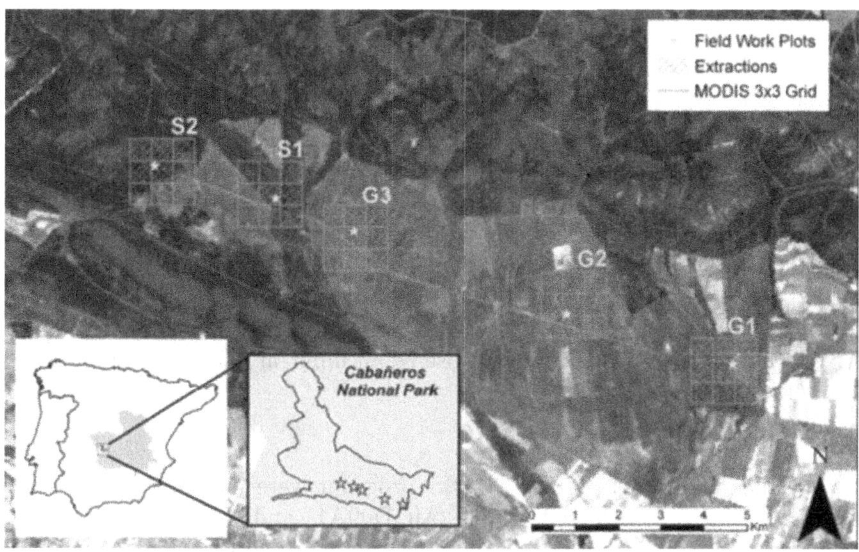

Fig. 1. Map of Spain showing the location of Cabaeros National Park, as well as a false color composite Landsat image showing the midpoint of the shrubland (S1 and S2) and grassland (G1, G2 and G3) plots used in this analysis. The grey boxes indicate the 3 T 3 MODIS grid (1.5 km T 1.5 km) centered at the plots. Shaded boxes indicate the window adapted to the shrub shape plot.

16% of appearance and 1% of dominance of the other three species together in the same area. In addition to this, it is a typical pioneer species that regenerates easily by seeds after diverse types of handlings and disturbances (Nuez Olivera, 1988), so it is the primary colonizer in areas with recurrent wildfires, which are of special interest in this study. The sampling protocol followed standard methods described in [6] and was repeated every 8 days during the spring and summer seasons from 1996 to 2002 and every 16 days from 2003 on. For this paper, FMC measurements taken from 2001 to 2005 have been used to correspond with the temporal series of the MODIS images. Cm2 FMC was computed from the difference of fresh and dry weight as follows:

$$FMC(\%) = \frac{W_f - W_d}{W_d} \times 100 \quad (1)$$

where W_f is fresh weight of leaves and small terminal branches (in the case of shrub species) or the whole plant (in the case of grassland), and Wd is dry weight, after oven drying the samples for 48 h at 60 8C. After 2004, FMC field sampling incorporated the collection of variables that are critical for running the RTM at leaf level, such as dry matter content (DM), equivalent water content (EWT) and chlorophyll content (Ca + b). DM and EWT were computed as follows:

$$DM = \frac{W_d}{A} \quad (2)$$

and

$$EWT = \frac{W_f - W_d}{A} \tag{3}$$

where A is the leaf area.

C. ladanifer L. leaf area was measured with an image analysis Delta system (Delta Devices LTD, Cambridge. England). Ca + b was measured by means of destructive sampling and measurement of leaf concentration in laboratory with the dimethyl sulfoxide (DMSO) method and spectrophotometric readings, according to [19]. For grasslands, DM and Ca + b measurements were provided by a field ecologist working in similar environments (Valladares, personal communication). Spectral soil reflectance was also measured with a GER 2600 (GER Corp., Millbrook, NY) radiometer to use as an input at canopy level model.

2.2 MODIS Data

Two standard products of the MODIS program were chosen for this study: the MODIS/Terra surface reflectance (MOD09A1) and the MODIS-Terra leaf area index (LAI) and fraction of photosynthetically active radiation (FPAR) (MOD15A2). The first is an 8-day composite product of atmospherically corrected reflectance for the first seven spectral bands of the MODIS sensor at a spatial resolution of 500m (Fig. 2). This product includes ancillary information, such as sun and sensor angles [17]. The standard MOD15A2 product was selected to take into account the strong effect of LAI variations on reflectance as well as to parametrize the RTM. This product is generated daily at 1 km spatial resolution and composited over an 8-day period based on the maximum value of the FPAR for that period (Knyazikhin, 1999). The original products were downloaded from the Land Processes Distributed Active Archive Center (LP DAAC) of the United States Geological Survey (USGS) (http://edcimswww.cr.usgs.gov/pub/imswelcome/) and reprojected from sinusoidal to UTM 30 T Datum European 1950 (ED50) using nearest neighbour interpolation resampling. MOD15A2 data were resampled to 500m to match the resolution of the MOD09A1 product using the same interpolation algorith. The values of a given plot for comparing with the field data were extracted from each composited image using the median value of a 3 X 3 pixel kernel located at the center of the field plot. A 3 X 3 window was used in order to reduce the potential noise due to residual atmospheric effects and georeferencing errors. In the case of shrublands, extraction windows were adapted to the shape of shrub patches to avoid including mixed pixels. To verify this approach the coefficient of variation (CV) was computed for reflectances for a Landsat image (30m X 30m pixel size) within the extraction windows. The CV decreased from 0.052 and 0.255 of the 3×3 windows in the near infrared band (NIR) and the short wave infrared (SWIR) bands, respectively, to 0.050 and 0.195 with the adapted window. The extractions of reflectance data of each pixel were derived from the 8-day composite that had a closest selected day to the field collections. A wide range of vegetation indexes were calculated to be included as independent variables in the empirical MLR model. Only one form of the NDII using band 6

(1628-1652 nm) was calculated based on previous studies which show stronger correlations between this band and field measured FMC values than other MODIS bands in the SWIR region [14];[24]. The first five indices in Table 1 measure greenness variations, which are only indirectly related to leaf water content. The other indices included in Table 1 are more directly related to water content, by combining water absorption in the SWIR wavelengths with other bands that are insensitive to water content [9]. Although greenness indices do not include water absorption bands, they can be used as an indirect estimation of water content, since moisture variations affect chlorophyll activity, leaf internal structure and LAI of many Mediterranean plants (Bowyer and Danson, 2004). In this sense, as the plant dries, changes in leaf internal structure cause a decrease in the reflectance in the NIR and an increase in the visible region, as a result of reducing photosynthetic activity and LAI values. However, this relation cannot be generalized for all ecosystems because, for example, variations on chlorophyll content can also be caused by plant nutrient deficiency, disease, toxicity and phonological stage [4].

3 Data Analysis and Conclusions

Appropriate statistical techniques for evaluating the accuracy of a given (continuous) classifier in distinguishing between two states (S1 and S2) are based on ROC curve analysis [1, 2]. Let Y a continuous variable. Based on the values of Y, the clasification of an observation as belonging to the state S1 or S2, can be made by choosing a threshold value c: if the observation is classified as S1 and if as S2. In this situation, the ROC curve is defined as the plot of the true-positive rate (TPR, the probability of correct classification for S1) versus the false-positive rate (FPR, the probability of misclassification for S2), across all possible threshold values. Related to the ROC curve, several indexes, as the area under the curve (AUC) or the index of Youden, are considered as summaries of the discrimination capability of the classifier. The AUC is the most commonly used one, taking values between 0.5 (no discrimination power) and 1 (perfect discrimination power).

In many situations, however, the classification rule based on the values of Y that minimize the overall misclassification error, is not necessarily the criterio used in ROC analysis. Moreover, in this context it is known that the best classifier based on Y with a threshold as classification decision is that based on the conditional probability of one of the states (e.g. S1) given the values of Y [3, 4]. Therefore, the best classifier can be expressed as:

$$\tilde{Y} = f(Y)P[S1|Y] \subset (0,1) \tag{4}$$

In practice, however, the function $f()$ of 4 is not known, and it is estimation would be required. In this work, we have evaluated the accuracy of several MODIS bands to classify situations of high and low risk of fire. From our data set, we classified as high risk of fire those observations with values of FMC under

60% and as fire low risk, those with values above this threshold. For this analysis, it was considered as classifier, not the values of the MODIS bands, but their values transformed from equation 4. To estimate the function $f()$ generalized additive models (GAM) [20] have been applied. Generalized additive models are flexible non parametric regression models that enable much more accurate fitting of the real data than the parametric linear models usually used. For each of the MODIS bands considered (from 1 to 7), the following logistic regression model were fitted

$$f(MB) = P[RF = 1|MB] = g^{-1}(h(MB)) \subset (0,1) \tag{5}$$

where MB denotes de MODIS band variable, RF is a binary variable taking the value 1 for high risk of fire and 0 for low risk, g is the logit function (known), and h is a smooth unknown function. The dataset was divided in two sets: 60% for fitting the logistic regression model (2) (calibration set), and the remaining 40% to estimate the ROC curve (validation set). Once the model (2) was fitted, we have used the estimated probabilities on the validation set to obtain the ROC curve and the AUC. Figures 2 and 3 show, for MODIS bands 1 and 2, the estimated transformation $\hat{f}()$, along with the ROC curve and AUC. As can be seen from these figures, none of the MODIS bands have a good accuracy in distinguishing between situations of high and low fire risk, showing AUC values close to 0.5 in all but the MODIS 2 band, with a value of 0.7. Therefore, we have conducted another analysis to evaluate if combining different MODIS bands to yield a composite classifier, we can obtain better classification capability. The methodology in this case has been the same as for the analysis above. First, we have estimated the conditional probability of high fire risk given the different MODIS bands values, by using flexible multivariate logistic regression [20]:

$$f(MB1, MB2, ..., MB7) = P[RF = 1|MB1, ..., MB7] = g^{-1}(\sum_{i=1}^{7} h_i(MB_i)) \subset (0,1) \tag{6}$$

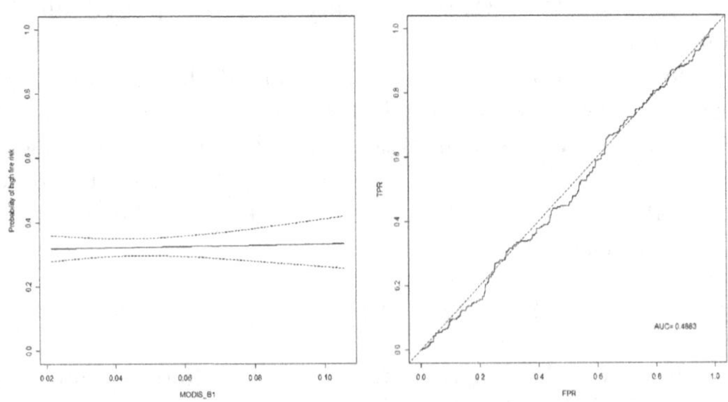

Fig. 2. Left: Estimated probability of high risk of fire as a function of MODIS band 1 values, along with 95% confidence interval. Right: Estimated ROC curve and AUC.

where MB_i denotes the MODIS band i variable, RF is a binary variable taking the value 1 for high risk of fire and 0 for low risk, g is the logit function (known), and h_i are smooth unknown functions. Not all the MODIS bands were finally included in model (3) . The MODIS bands 3 and 7 were excluded, since they did not provide an improvement in the deviance explained. Once the model (3) was fitted, we have used the probabilities estimated on the validation set to obtain the ROC curve. The result is shown in Figure 4. As can be seen, the accuracy of the combination of several MODIS bands in distinguishing between high and

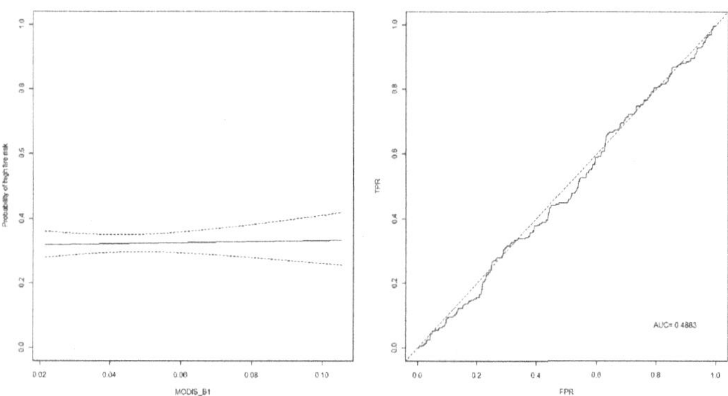

Fig. 3. Left: Estimated probability of high risk of fire as a function of MODIS band 2 values, along with 95% confidence interval. Right: Estimated ROC curve and AUC.

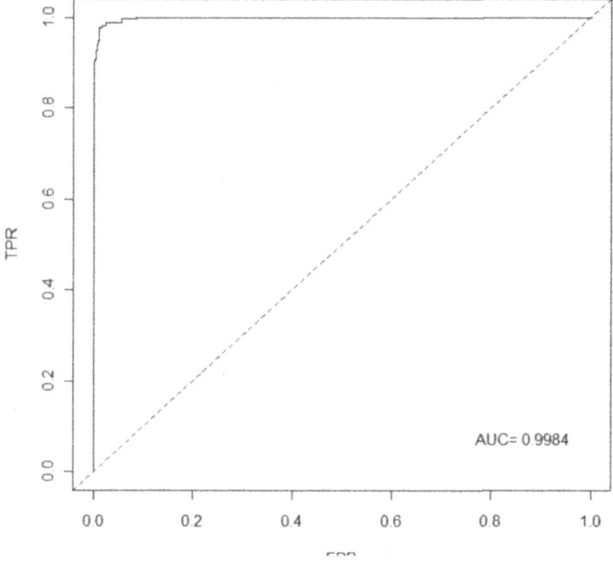

Fig. 4. Estimated ROC curve and for the combination of several MODIS bands

low risk of fire is almost perfect, with a AUC value near 1. All statistical analysis were carried out in R [8]. Thin plate splines smoothers [22] have been used to estimate the function(s) h[i] in (2) and (3), with optimal smoothing parameters chosen automatically by use of the Un-Biased Risk Estimator criterion (UBRE) [21]. All models have been fitted by use of the gam function of the mgcv package [8]. As regard to the estimation of the ROC curve and AUC, the R package ROCR has been used.

References

1. Bowyer, P., Danson, F.M.: Sensitivity of spectral reflectance to variation in live fuel moisture content at leaf and canopy level. Remote Sens. Environ. 92, 297–308 (2004)
2. Burgan, R.E., Rothermel, R.C.: BEHAVE: Fire Behavior Prediction and Fuel Modeling System. Fuel Subsystem. GTR INT-167, USDA Forest Service, Ogden, Utah (1984)
3. Camia, A., Bovio, G., Aguado, I., Stach, N.: Meteorological fire danger indices and remote sensing. In: Chuvieco, E. (ed.) Remote Sensing of Large Wildfires in the European Mediterranean Basin, pp. 39–59. Springer, Berlin (1999)
4. Ceccato, P., Flasse, S., Tarantola, S., Jacquemoud, S., Grégoire, J.M.: Detecting vegetation leaf water content using reflectance in the optical domain. Remote Sens. Environ. 77, 22–33 (2001)
5. Ceccato, P., Gobron, N., Flasse, S., Pinty, B., Tarantola, S.: Designing a spectral index to estimate vegetation water content from remote sensing data. Part 1, Theoretical approach. Remote Sens. Environ. 82, 188–197 (2002)
6. Chuvieco, E., Aguado, I., Cocero, D., Riaño, D.: Design of an empirical index to estimate fuel moisture content from NOAA-AVHRR analysis in forest fire danger studies. Int. J. Remote Sens. 24(8), 1621–1637 (2003a)
7. Combal, B., Baret, F., Weiss, M., Trubuil, A., Mace, D., Pragne're, A., Myneni, R., Knyazikhin, Y., Wang, L.: Retrieval of canopy biophysical variables from bidirectional reflectance Using prior information to solve the ill-posed inverse problem. Remote Sens. Environ. 84, 1–15 (2002)
8. R Development Core Team: A language and environment for statistical computing. R Foundation for Statistical Computing, Vienna, Austria (2008) ISBN 3-900051-07-0, `http://www.R-project.org`
9. Fourty, T., Baret, F.: Vegetation water and dry matter contents estimated from top-of-the atmosphere reflectance data: a simulation study. Remote Sens. Environ. 61, 34–45 (1997)
10. Jacquemoud, S., Baret, F.: PROSPECT: a model of leaf optical properties spectra. Remote Sens. Environ. 34, 75–91 (1990)
11. Knyazikhin, Y., et al.: MODIS leaf area index (LAI) and fraction of photosynthetically active radiation absorbed by vegetation (FPAR) product (MOD15). Algorithm Theoretical Basis Document (1999), `http://eospso.gsfc.nasa.gov/atbd/modistables.html`
12. Lawson, B.D., Hawkes, B.C.: Field evaluation of moisture content model for medium-sized logging slash. In: Proceedings of the 10th Conference on Fire and Forest Meteotology, Ottawa, Canada, pp. 247–257 (1989)
13. Nuñez Olivera, E.: Ecología del jaral de Cistus ladanifer, Universidad de Extremadura (1988)

14. Roberts, D.A., Peterson, S., Dennison, P.E., Sweeney, S., Rechel, J.: Evaluation of airborne visible/infrared imaging spectrometer (AVIRIS) and moderate resolution imaging spectrometer (MODIS) measures of live fuel moisture and fuel condition in a shrubland ecosystem in southern California. J. Geophy. Res. 111, G04S02 (2006), doi:10.1029/2005JG000113
15. Usero, R., Zarco, U.: Estimación del contenido de agua del combustible mediante el uso de redes neuronales artificiales CEDI Granada (2005)
16. Ustin, S.L., Jacquemoud, S., Zarco-Tejada, P.J., Asner, G.: Remote Sensing of Environmental Processes: State of the Science and New Directions. In: Ustin, S.L. (ed.) Manual of Remote Sensing. Remote Sensing for Natural Resource Management and Environmental Monitoring, New York, vol. 4, pp. 679–730 (2004)
17. Vermote, E.F., Vermeulen, A.: Atmospheric correction algorithm: spectral reflectances (MOD09). NASA (1999)
18. Viegas, D.X., Piñol, J., Viegas, M.T., Ogaya, R.: Estimating ive fine fuels moisture content using meteorologically-based indices. Int. J. Wildland Fire 10, 223–240 (2001)
19. Wellburn, A.R.: The spectral determination of chlorophylls a and b, as well as total carotenoids, using various solvents with spectrophotometers of different resolution. J. Plant Physiol. 144, 307–313 (1994)
20. Wood, S.: Generalized Additive Models. Chapman and Hall/CRC, Boca Raton (2006)
21. Wood, S.: Stable and efficient multiple smoothing parameter estimation for generalized additive models. Journal of the American Statististical Association 99, 549–673 (2004)
22. Wood, S.: Thin plate regression splines. Journal of the Royal Statistical Society: Series B 65, 95–114 (2003)
23. Yebra, M., Chuvieco, R.: Estimation of live fuel moisture content from MODIS images for fire risk assessment agricultural and forest meteorology, vol. 148, pp. 523–536. Elsevier, Amsterdam (2008)
24. Yebra, M., de Santis, A., Chuvieco, E.: Estimación del peligro de incendios a partir de teledetección y variables meteorológicas: variación temporal del contenido de humedad del combustible. Recursos Rurais 1, 9–19 (2005)

Optimised Particle Filter Approaches to Object Tracking in Video Sequences

Artur Loza[1], Fanglin Wang[2], Miguel A. Patricio, Jesús García, and José M. Molina[3]

[1] University of Bristol, United Kingdom
artur.loza@bristol.ac.uk
[2] Shanghai Jiao Tong University, China
hardegg@sjtu.edu.cn
[3] Universidad Carlos III de Madrid, Spain
{mpatrici,jgherrer}@inf.uc3m.es, molina@ia.uc3m.es

Abstract. In this paper, the ways of optimising a Particle Filter video tracking algorithm are investigated. The optimisation scheme discussed in this work is based on hybridising a Particle Filter tracker with a deterministic mode search technique applied to the particle distribution. Within this scheme, an extension of the recently introduced structural similarity tracker is proposed and compared with the approach based on separate and combined colour and mean-shift tracker. The new approach is especially applicable to real-world video surveillance scenarios, in which the presence of multiple targets and complex background pose a non-trivial challenge to automated trackers. The preliminary results indicate that a considerable improvement in tracking is achieved by applying the optimisation scheme, at the price of a moderate computational complexity increase of the algorithm.

1 Introduction

Recently there has been an increased interest in object tracking in video sequences supplied by either a single camera or a network of cameras [1, 2, 3]. Reliable tracking methods are of crucial importance in many surveillance systems as they enable human operators to remotely monitor activity across areas of interest, enhance situation awareness and help the surveillance analyst with the decision-making process. One of the important tracking applications are the surveillance systems, utilised in a wide range of environments such as: transport systems, public spaces (shopping malls, car parks, etc.), industrial environments, government or military establishments. The objects tracked are usually moving in an environment characterised by a high variability that requires sophisticated algorithms for video acquisition, camera calibration, noise filtering, motion detection; capable of learning and adapting to the changing conditions.

A realistic surveillance scenario is considered in this work, in which complex multi-object scene is monitored by means of video tracking. The video sequence representing such scenario is highly challenging for any tracking technique, due

J. Mira et al. (Eds.): IWINAC 2009, Part I, LNCS 5601, pp. 486–495, 2009.

to the the presence of spurious or similarly-coloured background and interacting or occluded targets. It is generally acknowledged that it is not possible to perform successful tracking in such complex scenes, for extended period of time, using only one type of a specialised tracker [4]. Improved tracking can be achieved by using a small number low complexity complementary algorithms, or a single hybrid solution obtained by combining several techniques.

The existing hybrid PF tracking approaches concentrate on optimising the sparse sampling of the state space. The predicted particles are usually modified by using an external optimisation procedure that uses the available measurement as an input. Recently proposed hybrid-PF schemes combine the MS and PF trackers [5,6,7]. MS is usually applied to each particle in order to move it to more optimal position. Such a hybrid solution requires less particles than when using PF alone and results in improved performance of the tracker. The most consistent implementation of this scheme has been proposed in [8], where the data association problem is formulated and MS algorithm is 'embedded seamlessly' into the PF algorithm. Therein, the deterministic MS-induced particle bias with a superimposed Gaussian distribution is considered a new proposal distribution.

In this correspondence, a PF tracker extension within this scheme is investigated. Unlike in the colour-based PF-MS combination, the optimisation scheme proposed is applied to the recently introduced SSIM-PF tracker [9]. Moreover, a different deterministic procedure is utilised in the new method, as the SSIM-PF is optimised by a gradient ascent procedure, based on the SSIM-cue. The preliminary results obtained with this method show that not only a more stable performance of the SSIM-PF tracker is achieved, but also the number of particles required is reduced. This secondary effect off-sets, to some extent, the computational cost of the gradient search procedure.

The remaining part of the paper is organised as follows. Probabilistic solutions to the video tracking problem are reviewed in Section 2, with special emphasis on the state estimation techniques. The proposed optimisation scheme is described in Section 3. In Section 4, the evaluation of video tracking simulation is presented, together with discussion of the results. Finally, Section 5 presents the conclusions of the study and suggests areas for future work.

2 Probabilistic Solutions to Video Tracking Problem

2.1 Bayesian Tracking Framework

Bayesian inference methods have gained a strong reputation for tracking and data fusion applications, because they avoid simplifying assumptions that may degrade performance in complex situations and have the potential to provide a optimal or sub-optimal solutions [10]. In case of the sub-optimal solution, the proximity to the theoretical optimum depends on the computational capability to execute numeric approximations and the feasibility of probabilistic models for target appearance, dynamics, and measurements likelihoods.

Table 1. The particle filter algorithm

---

0. Initialisation: Generate samples $\{\boldsymbol{x}_0^{(\ell)}\}_{\ell=1}^N \sim p(\boldsymbol{x}_0)$. Set weights $W_0^{(\ell)} = 1/N$.
For $k = 1, 2, \ldots, K$ do:
1. Prediction: Sample $\boldsymbol{x}_k^{(\ell)} \sim p(\boldsymbol{x}_k|\boldsymbol{x}_{k-1}^{(\ell)})$ from the motion model.
2. Update: Compute the weights $W_k^{(\ell)} \propto W_{k-1}^{(\ell)} \mathcal{L}(\boldsymbol{z}_k|\boldsymbol{x}_k^{(\ell)})$ and normalise them $\widehat{W}_k^{(\ell)} = W_k^{(\ell)} / \sum_{\ell=1}^{N} W_k^{(\ell)}$.
3. Resample: Duplicate / eliminate samples $\boldsymbol{x}_k^{(\ell)}$ with high / low weights and set $W_k^{(\ell)} = \widehat{W}_k^{(\ell)} = 1/N$.

In the Bayesian tracking framework the best posterior estimate of $\boldsymbol{x}_k$ is inferred from available measurements, $\boldsymbol{Z}_k$, based on derivation the a posteriori pdf $p(\boldsymbol{x}_k|\boldsymbol{Z}_k)$. Assuming that the posterior pdf at time $k-1$ is available, the prior pdf of the state at time k is obtained via the Chapman-Kolmogorov equation:

$$p(\boldsymbol{x}_k|\boldsymbol{Z}_{k-1}) = \int_{\mathbb{R}^{n_x}} p(\boldsymbol{x}_k|\boldsymbol{x}_{k-1}) p(\boldsymbol{x}_{k-1}|\boldsymbol{Z}_{k-1}) d\boldsymbol{x}_{k-1} \tag{1}$$

where $p(\boldsymbol{x}_k|\boldsymbol{x}_{k-1})$ is the state transition probability. Once a measurement $\boldsymbol{Z}_k$ is available, $p(\boldsymbol{x}_k|\boldsymbol{Z}_k)$ is recursively obtained according to Bayes update rule

$$p(\boldsymbol{x}_k|\boldsymbol{Z}_k) = \frac{p(\boldsymbol{Z}_k|\boldsymbol{x}_k) p(\boldsymbol{x}_k|\boldsymbol{Z}_{k-1})}{p(\boldsymbol{Z}_k|\boldsymbol{Z}_{k-1})} \propto p(\boldsymbol{Z}_k|\boldsymbol{x}_k) p(\boldsymbol{x}_k|\boldsymbol{Z}_{k-1}) \tag{2}$$

where $p(\boldsymbol{Z}_k|\boldsymbol{Z}_{k-1})$ is a normalising constant and $p(\boldsymbol{Z}_k|\boldsymbol{x}_k)$ is the measurement likelihood. Commonly used estimators of X_k, include the maximum a posteriori (MAP), $\hat{X}_k = \arg\max_{X_k} p(\boldsymbol{x}_k|\boldsymbol{Z}_k)$, and the minimum mean squared error (MMSE), $\hat{X}_k = \int \boldsymbol{x}_k p(\boldsymbol{x}_k|\boldsymbol{Z}_k) d\boldsymbol{x}_k$, which is equivalent to the expected value of the state.

2.2 Particle Filtering for State Vector Estimation

Particle filtering [1,10,11,12] is a method relying on sample-based reconstruction of probability density functions. The aim of sequential particle filtering is to evaluate the *posterior* pdf $p(\boldsymbol{x}_k|\boldsymbol{Z}_k)$ of the state vector $\boldsymbol{x}_k \in \mathbb{R}^{n_x}$, given a set $\boldsymbol{Z}_k = \{\boldsymbol{z}_i\}_{i=1}^k$ of sensor measurements up to time k. The quality (importance) of the ℓth particle (sample) of the state, $\boldsymbol{x}_k^{(\ell)}$, is measured by the weight associated with it, $W_k^{(\ell)}$.The pseudo-code description of a generic PF tracking algorithm is shown in Table 1.

Two major stages can be distinguished in the particle filtering method: *prediction* and *update*. During prediction, each particle is modified according to the state model of the region of interest in the video frame, including the perturbation of the particle's state according to the motion model $p(\boldsymbol{x}_k|\boldsymbol{x}_{k-1}^{(\ell)})$.

For a particle representation, the filtered posterior density is approximated as:

$$p(\boldsymbol{x}_k|\boldsymbol{Z}_k) \approx \sum_{\ell=1}^{N} \widehat{W}_k^{(\ell)} \delta(\boldsymbol{x}_k - \boldsymbol{x}_k^{(\ell)}) \tag{3}$$

based on the likelihood $\mathcal{L}(\boldsymbol{z}_k|\boldsymbol{x}_k^{(\ell)})$ (9) of the measurement and particle weights. Here, $\delta(.)$ is the Dirac delta function. Consequently, the posterior mean state is computed using the collection of particles as $\hat{\boldsymbol{x}}_k = \sum_{\ell=1}^{N} \widehat{W}_k^{(\ell)} \hat{\boldsymbol{x}}_k^{(\ell)}$.

An inherent problem with particle filters is degeneracy (the case when only one particle has a significant weight). A *resampling* procedure helps to avoid this by eliminating particles with small weights and replicating the particles with larger weights. In this work, the systematic resampling method [13] was used.

For completeness of the presentation the remaining constituents of the tracking algorithm, motion and likelihood model, are briefly described below, in the way they are implemented in this work.

The motion of the moving object is modelled by the random walk model, $\boldsymbol{x}_k = \boldsymbol{F}\boldsymbol{x}_{k-1} + \boldsymbol{v}_{k-1}$, with a state vector $\boldsymbol{x}_k = (x_k, y_k, \dot{x_k}, \dot{y_k}, s_k)^T$ comprising the pixel coordinates of the centre of the region surrounding the object and the region scale s_k. $\boldsymbol{F}$ is the transition matrix and $\boldsymbol{v}_k$ is the process noise assumed to be white, Gaussian, with a covariance matrix $\boldsymbol{Q} = \text{diag}(\sigma_x^2, \sigma_y^2, \sigma_{\dot{x}}^2, \sigma_{\dot{y}}^2, \sigma_s^2)$.

The normalised distance between the two regions $\boldsymbol{t}_{\text{ref}}$ (reference region) and $\boldsymbol{t}_k$ (current region), for particle ℓ is substituted into the likelihood function, modelled as an exponential:

$$\mathcal{L}(\boldsymbol{z}_k|\boldsymbol{x}_k^{(\ell)}) \propto \exp\left(-D^2(\boldsymbol{t}_{\text{ref}}, \boldsymbol{t}_k)/D_{\min}^2\right), \tag{4}$$

where $D_{\min} = \min_{\boldsymbol{x}}\{D(\boldsymbol{t}_{\text{ref}}, \boldsymbol{t}_k)\}$. This smooth likelihood function, although chosen empirically by the authors of [1], has been in widespread use for a variety of cues ever since. The Bhattacharyya distance D is commonly used to calculate similarity between target and reference objects, described by their colour histograms h:

$$D(\boldsymbol{t}_{\text{ref}}, \boldsymbol{t}_k) = \left(1 - \sum_{i=1}^{B} h_{\text{ref},i} h_{k,i}\right)^{0.5}. \tag{5}$$

2.3 Importance Sampling and Proposal Distributions

The particle weights in (3) are updated based on the principle of importance sampling [10]

$$w_k^i \propto w_{k-1}^i \frac{p(\mathbf{z}_k|\mathbf{x}_k^i)p(\mathbf{x}_k^i|\mathbf{x}_{k-1}^i)}{q(\mathbf{x}_k^i|\mathbf{x}_{k-1}^i, \mathbf{z}_{1:k})}, \tag{6}$$

where $q(\mathbf{x}_k^i|\mathbf{x}_{k-1}^i, \mathbf{z}_{1:k})$ is a proposal, called an *importance density* and $p(\mathbf{z}_k|\mathbf{x}_k^i)$ is modelled by a likelihood function. The most popular choice of the importance density is the prior, $p(\mathbf{x}_k|\mathbf{x}_{k-1})$. This choice results in a simple implementation of the weight update stage (cf. (6))

$$w_k^i \propto w_{k-1}^i p(\mathbf{z}_k|\mathbf{x}_k^i) . \tag{7}$$

However, using the transition information alone may not be sufficient to capture the complex dynamics of some targets. It has been shown that an optimal importance density is defined as function of the state and a new measurement/additional information $q(\mathbf{x}_k^i|\mathbf{x}_{k-1}^i, \mathbf{z}_{1:k})$. Therefore, in this work, the use of a mixture distribution containing additional information as the importance density is proposed

$$q(\mathbf{x}_k^i|\mathbf{x}_{k-1}^i, \mathbf{z}_{1:k}) = \sum_{m=1}^{M} \alpha_m f_m(\mathbf{x}_{1:k}^i, \mathbf{z}_{1:k})$$

where $\alpha_m, \sum_{m=1}^{M} \alpha_m = 1$ are normalised weights of M components of the mixture. Among candidates for f_m, proposed in this work, are the prior, blob detection and data association distributions. For $M = 1$ and $f_1(\mathbf{x}_{1:k}^i, \mathbf{z}_{1:k}) = p(\mathbf{x}_k^i|\mathbf{x}_{k-1}^i)$ the generic PF is obtained. Example of such a mixture importance density is two densities proposed in [8, 14, 15], resulting from inclusion of the Adaboost detection information.

3 Optimised Structural Similarity-Based Tracker

3.1 Structural Similarity Measure

A region-based tracking algorithm typically compares the current frame region, I, with the object template, J, by means of a distance or similarity measure. A recently proposed image metrics, Structural SIMilarity Image Quality Index (SSIM), used in our method, is defined as follows [16]

$$S = \left(\frac{2\mu_I\mu_J + C_1}{\mu_I^2 + \mu_J^2 + C_1}\right)\left(\frac{2\sigma_{IJ} + C_2}{\sigma_I^2 + \sigma_J^2 + C_2}\right), \tag{8}$$

where μ, σ and σ_{IJ} denote the sample mean, standard deviation and covariance, respectively, and $C_{1,2}$ are small positive constants used for the numerical stability purposes. S is symmetric and maps the similarity between two images to the interval $(-1, 1] : S = 1$ iff $I = J$. This similarity measure is converted into a dissimilarity (distance) measure by taking $D = 1 - S$.

This similarity measure has been selected based on its good ability to capture perceptual similarity of images. The SSIM measure simulates the perceptual process of the human visual system by measuring the luminance, contrast and structural similarity of the two images. Another important feature of the SSIM index is that the normalised measurements in SSIM are sensitive to the relative rather than absolute image distortions [16], thus making this measure suitable to video tracking in varying conditions. The SSIM measure was first successfully applied to particle-filter video object tracking in [9]. Therein it was also demonstrated that the structure comparison is more reliable in scenarios when spurious (e.g. camouflaged) objects appear in the scene or when there is not enough discriminative colour information available.

3.2 Likelihood Model

The normalised distance between the two regions J (reference region) and I (current region), for particle ℓ is substituted into the likelihood function, modelled as an exponential:

$$\mathcal{L}(\boldsymbol{z}_k|\boldsymbol{x}_k^{(\ell)}) \propto \exp\left(-D^2(I,J)/D^2_{\min}\right), \tag{9}$$

where $D_{\min} = \min_{\boldsymbol{x}}\{D(I,J)\}$. The structural properties of the region are extracted through SSIM (8) and are used directly to calculate the distance D in (9). The likelihood function is then used to evaluate the importance weights of the particle filter, to update the particles and to obtain the overall estimate of the centre of the current region $\boldsymbol{t}_k$, as described in Table 1.

3.3 Gradient SSIM-PF Algorithm

An extension to SSIM-PF, by deterministically modifying each particle according to the local structural similarity surface, referred to as DSSIM, is proposed in this correspondence. In general terms, the estimated target location, $\mathbf{x}_k^0$ is moved along the direction the structural similarity gradient by one pixel in each iteration until no further improvement is achieved, or the limit of iterations is reached (20 in our simulations). This step is performed for each particle, following its prediction (see 1. in Table 1). The prior and proposal distributions are then approximated as Gaussian distributions centred on each respective prediction and the resulting particle weight is re-calculated as shown in (6). It should be noted, that the gradient ascent procedure can be applied directly to tracking, outside of the probabilistic framework, as proposed in [17]. The resulting tracking algorithm, although deterministic, is much faster than its probabilistic counterpart, at the price of loosing some flexibility offered by the PF-based solutions.

4 Experiments

In this section, we present a performance analysis and a comparison of the colour- and the structural similarity-based PF algorithms, applied to tracking of targets (humans) in a real-world video sequence. Additionally, we have tested a colour-based PF tracker optimised by MS procedure as proposed in [15]. Moreover, the tracking results by applying the optimising procedures (MS and SSIM gradient search) directly to the video have been included for the completeness.

The performance of the six algorithms (SSIM-PF, COL-PF, their optimised versions, SSIM-PF-DSSIM and COL-PF-MS, and the optimisation procedures themselves, DSSIM and MS) was evaluated with the use of the sequence *cross* (5 sec duration), taken from our multimodal database [18]. The sequence contains

Fig. 1. Pedestrian tracking test results

three people walking rapidly in front of a stationary camera. The main challenges posed by this sequence are: the colour similarity between the tracked object and the background or other passing people, and a temporal near-complete occlusion of the tracked person by a passer-by.

In order to assess the preliminary tracking results and compare some of the features of the trackers tested, a discussion of based on the visual observation of their outputs is provided below. In order to clearly illustrate our point, we have concentrated on tracking one of the pedestrians in a fragment of the sequence. Figure 1 presents the extracted frames of the output of the six trackers. The graphs therein are a superposition of the video frame and the corresponding object region contour as estimated by the trackers.

Based on the observation of the estimated target regions in Figure 1, it can be concluded that the gradient structural similarity procedure locates the object precisely in majority of the frames. It fails, however, to recover from an occlusion towards the end of the sequence. The performance of the SSIM-PF tracker is very unstable, due to a very low number of particles used (see the discussion below) and the tracker looses the object half-way through the sequence. On the other hand, the combined algorithm, SSIM-PF-DSSIM, tracks the object successfully throughout the sequence. The MS tracker has completely failed to track the object. Since the MS algorithm is a memory-less colour-based tracker, its poor performance in these two sequences is due to the object's fast motion and its similarity to the surrounding background. The colour-based tracker, COL-PF, performs similarly to SSIM-PF, however, it locates the object somewhat more precisely. Finally, the combined COL-PF-MS tracker, appears to be more stable than its non-optimised version. Nevertheless, the objects is eventually lost as a result of the occlusion.

It should be noted that in order to illustrate the benefit of using the optimisation procedures, a very low number of the particles for PF-based methods has been chosen (20 for SSIM-based and 30 for colour-based PF). Consequently, it allowed us to observe that the resulting tracking instability and failures are partially mitigated by the use of the optimisation procedures. Moreover, since the optimisation procedures are faster than PFs, such a combination does not increase the computational load considerably. On the contrary, the appropriate combination of the two methods, results in a lower number of the particles required and thus reducing the processing time. Conversely, it can be shown that, in some cases, non-optimised tracker can achieve a similar performance to the optimised tracker utilising a larger number of particles and thus being more computationally demanding.

5 Conclusion

In this paper, hybrid probabilistic and deterministic approaches to video target tracking have been investigated, and their advantages and mutual complementarities have been identified. The particular issue addressed herein is concerned with tracking pedestrians in the presence of spurious or similarly-coloured targets, which may interact or become temporarily occluded. The tracking methods considered in this work include the colour- and structural similarity-based Particle Filters algorithms. Their performance is contrasted with the hybrid PF method, combined with the mean-shift and gradient deterministic optimisation method.

The tracking results show that the structural similarity-based PF trackers are more accurate compared to the colour-based trackers, in the video sequence tested. The main finding of the work presented in this correspondence is that the best tracking performance among the techniques tested is achieved by a combination of the deterministic gradient optimisation method and the SSIM-PF method, without increasing significant the computational complexity of the system.

Among the research issues that will be the subject of further investigation is the speed and reliability improvement of the proposed optimised hybrid technique. It is envisaged that this could be achieved by replacing the simple gradient search with a more efficient optimisation procedure and by more accurate modelling of the resulting proposal density.

Acknowledgment

This work was partially supported by Projects CICYT TIN2008-06742-C02-02/TSI, CICYT TEC2008-06732-C02-02/TEC, SINPROB, CAM MADRINET S-0505/TIC/0255 and DPS2008-07029-C02-02.The authors would also like to thank Dr Henry Knowles, University of Bristol, for making his implementation of the Mean-Shift algorithm available to us.

References

1. Pérez, P., Vermaak, J., Blake, A.: Data fusion for tracking with particles. Proceedings of the IEEE 92(3), 495–513 (2004)
2. Hampapur, A., Brown, L., Connell, J., Ekin, A., Haas, N., Lu, M., Merkl, H., Pankanti, S.: Smart video surveillance: Exploring the concept of multiscale spatiotemporal tracking. IEEE Signal Processing Magazine 22(2), 38–51 (2005)
3. Yilmaz, A., Javed, O., Shah, M.: Object tracking: A survey. ACM Comput. Surv. 38(4) (2006)
4. Shearer, K., Wong, K.D., Venkatesh, S.: Combining multiple tracking algorithms for improved general performance. Pattern Recognition 34(6), 1257–1269 (2001)
5. Maggio, E., Cavallaro, A.: Hybrid particle filter and mean shift tracker with adaptive transition model. In: IEEE International Conference on Acoustics, Speech, and Signal Processing, 2005. Proceedings (ICASSP 2005), vol. 2, pp. 221–224 (2005)
6. Shan, C., Tan, T., Wei, Y.: Real-time hand tracking using a mean shift embedded particle filter. Pattern Recogn. 40(7), 1958–1970 (2007)
7. Bai, K., Liu, W.: Improved object tracking with particle filter and mean shift. In: Proceedings of the IEEE International Conference on Automation and Logistics, Jinan, China, vol. 2, pp. 221–224 (2007)
8. Cai, Y., de Freitas, N., Little, J.J.: Robust visual tracking for multiple targets. In: Leonardis, A., Bischof, H., Pinz, A. (eds.) ECCV 2006. LNCS, vol. 3954, pp. 107–118. Springer, Heidelberg (2006)

9. Loza, A., Mihaylova, L., Bull, D.R., Canagarajah, C.N.: Structural similarity-based object tracking in multimodality surveillance videos. Machine Vision and Applications 20(2), 71–83 (2009)
10. Arulampalam, M., Maskell, S., Gordon, N., Clapp, T.: A tutorial on particle filters for online nonlinear/non-Gaussian Bayesian tracking. IEEE Trans. on Signal Proc. 50(2), 174–188 (2002)
11. Isard, M., Blake, A.: Condensation – conditional density propagation for visual tracking. Intl. J. of Computer Vision 28(1), 5–28 (1998)
12. Ristic, B., Arulampalam, S., Gordon, N.: Beyond the Kalman Filter: Particle Filters for Tracking Applications. Artech House, Norwood (2004)
13. Kitagawa, G.: Monte Carlo filter and smoother for non-Gaussian nonlinear state space models. J. Comput. Graph. Statist. 5(1), 125 (1996)
14. Okuma, K., Taleghani, A., de Freitas, N., Little, J., Lowe, D.: A boosted particle filter: Multitarget detection and tracking. In: Pajdla, T., Matas, J(G.) (eds.) ECCV 2004. LNCS, vol. 3021, pp. 28–39. Springer, Heidelberg (2004)
15. Lu, W.L., Okuma, K., Little, J.J.: Tracking and recognizing actions of multiple hockey players using the boosted particle filter. Image Vision Comput. 27(1-2), 189–205 (2009)
16. Wang, Z., Bovik, A., Sheikh, H., Simoncelli, E.: Image quality assessment: from error visibility to structural similarity. IEEE Transactions on Image Processing 13(4), 600–612 (2004)
17. Wang, F., Loza, A., Yang, J., Xue, Y., Yu, S.: Surveillance video object tracking with differential SSIM. In: Proc. IEEE Intl. Conf. on Multimedia and Expo., Cancun, Mexico (June 2009)
18. Lewis, J.J., Nikolov, S.G., Loza, A., Canga, E.F., Cvejic, N., Li, J., Cardinali, A., Canagarajah, C.N., Bull, D.R., Riley, T., Hickman, D., Smith, M.I.: The Eden Project Multi-Sensor Data Set. Technical Report TR-UoB-WS-Eden-Project-Data-Set (2006), `http://www.imagefusion.org`

Towards Interoperability in Tracking Systems: An Ontology-Based Approach

Juan Gómez-Romero, Miguel A. Patricio, Jesús García, and José M. Molina

Applied Artificial Intelligence Group, University Carlos III of Madrid
{jgromero,mpatrici,jgherrer}@inf.uc3m.es, molina@ia.uc3m.es

Abstract. Current video surveillance systems are expected to allow for management of video sequences recorded by distributed cameras. In order to obtain a complete view of the scene, it is necessary to apply data fusion techniques to the acquired data. This increases the need of communicating various and probably heterogeneous components in a vision system. A solution is to use a common and well-defined vocabulary to achieve understanding between them. In this work, we present an OWL ontology aimed at the symbolic representation of tracking data. The ontological representation of tracking data improves system interoperability and scalability, and facilitates the development of new functionalities. The ontology can be used to express the tracking information contained in the messages exchanged by the agents in CS-MAS, a multi-agent framework for the development of multi-camera vision systems.

1 Introduction

In recent years, the interest in computer vision systems has grown considerably. Some notable examples are automated security and surveillance systems (to detect intrusions or abnormal activities in facilities), road traffic control (for the management of traffic lights and driver advertising), plane identification in airports (for logging of plane takeoffs and landings), and monitoring of players in sports (to study of the movements and activities of players during training or matches). In the simplest case, vision systems have a single sensor, which provides a sequence of frames to the image processing algorithms. These algorithms aim to automatically detect, identify, and predict the actions that are being performed in the observation area. If watched entities of the system are controlled, other sensor technologies can be used to get additional and/or more accurate data, such as RFID or ultra-wideband.

An expected feature of video-surveillance systems in real applications [1] is the capability to track and maintain identity (ID) of all detected objects in real-time. Tracking algorithms is concerned with the estimation of the number of objects in a scene, together with their instantaneous locations, kinematic states and any other characteristics required. Tracking is a complex problem, since several difficulties can arise as a result of the imperfect functioning of video sensors and the wide range of possible scenes. Numerous soft computing techniques have been proposed to improve the results of tracking systems [2].

J. Mira et al. (Eds.): IWINAC 2009, Part I, LNCS 5601, pp. 496–505, 2009.

The decreasing price of video camera hardware and the development of wireless network technologies have given rise to the proliferation of multi-camera systems. Although the existence of multiple cameras inevitably increases the complexity of the systems, it also improves the results, since multiple fields of view are considered. However, it is necessary to implement suitable procedures to integrate data generated in each camera, i.e. to apply data fusion techniques. Data fusion is the process of combining data to refine state estimates and predictions [3]. Data fusion in tracking involves three steps: (i) the creation of a common spatiotemporal coordinate space; (ii) the transmission of the information acquired in each node; (iii) the alignment of observations to assign labels to objects. Data fusion systems are usually organized by following the guidelines of the JDL model [4], which classifies data fusion processes according to the abstraction and the refinement of the entities considered.

A promising approach to the development of third generation surveillance applications is to rely on the well-known multi-agent paradigm. Accordingly, Patricio et al. proposed CS-MAS (Cooperative Surveillance Multi-Agent System), a BDI agent-based framework for the development of multi-camera systems specially focused on solving tracking issues [5]. This framework provides a reference architecture to organize, communicate, and coordinate all the procedures carried out by a distributed vision system, including data fusion. Nodes in CS-MAS are implemented as agents that proactively interchange data (particularly, tracking data). Communication between agents in the CS-MAS is performed by interchanging messages. The content of these messages is the direct serialization of the data structures used by the agents. Two main components of CS-MAS are camera agents, which execute a tracking algorithm to process data acquired from visual sensors, and fusion agents, which carry out data fusion processes.

In this paper, we propose a semantic model to represent the visual information shared in distributed artificial vision systems. Specifically, we present an ontology to describe the tracking information contained in the messages interchanged among the camera and the fusion agents of the CS-MAS. The ontology behaves as an agreed vocabulary that allow tracking data to be represented in an abstract, common and understandable way. Ontologies are a well-known formalism to represent structured knowledge and have proved to be valid in several scenarios that require interoperation between heterogeneous entities, e.g. the Semantic Web [6]. In this document, we overview the structure of this ontology and show an example of annotation of the tracking data handled by CS-MAS.

The remainder of this paper is organized as follows. In section 2, we study the advantages of using ontologies to represent tracking data in artificial vision systems. Section 3 reviews some related work on the use of ontologies in (distributed) visual systems. Next, in section 4, we describe the formulation of the ontology for the representation of tracking data. In section 5, we illustrate the use of the ontology with a brief example on the annotation of the outputs of the tracking system. Finally, the paper concludes with a discussion of our proposal and plans for future research work.

2 Ontologies for Visual Data Representation

Ontologies are a widely extended knowledge representation formalism that offers interesting features, such as standardization, sharing, and reutilization of knowledge [7,8]. Ontologies are based on Description Logics, a family of decidable logics for representing structured knowledge [9]. The most extended ontology language is the standard Web Ontology Language OWL [10], which has a successor in the in-progress specification OWL 2 [11]. The description of tracking data with ontologies, which have well-defined semantics, provides support for better interoperability within the artificial vision system. This results in the following advantages:

- *Obtaining implicit knowledge.* In principle, the ontology does not provide more information than the data acquired by the image-processing procedure. However, since this information is represented in a formal language, further reasoning tasks can be easily performed, and additional facts can be deduced.
- *Support for high-level information management and fusion.* On top of the ontological representation, further functionalities, dealing with high-level abstractions, can be implemented. Our objective is to extend this approach to context-aware reasoning, with a view on activity recognition. More abstract objects (e.g. people, moving items) with special features or behaviors can be defined by relying on the ontology, as well as scene interpretation rules. The ontology acts as an intermediate between the high and low-level descriptions.
- *Decoupling of internal and external representations.* The ontology allows the camera agents to define their own data structures, as long as they use the ontology to communicate with other agents.
- *Understanding among agents.* Different tracking techniques, with different algorithms and data, can be implemented in each camera agent. When communication is performed, the visual data managed by each agent is represented by using the common ontology. As a result, understanding is accomplished more easily.
- *Extensibility of the architecture.* New modules with enhanced capabilities can be easily introduced in the system. The processing in these modules can be completely encapsulated, while communication is carried out through well-defined interfaces by interchanging messages with an ACL envelope and content expressed with the ontology.
- *Improved data manipulation and querying.* Ontologies are logic-based knowledge representation formalisms with associated reasoning procedures. Visual data expressed with the ontology can be easily queried and transformed.
- *Implementation of mash-up applications.* New applications can be developed by relying on the formal representation of visual data. For instance, an off-line tool for the visualization of the frames and tracks of a video sequence could be implemented. The input of this tool would be the annotated visual data generated by the tracking procedure, whereas the output would be a visual presentation of them. This application does not depend on the data representation of the tracking algorithms, which can change without having to modify it.

3 Related Work

Formal knowledge representation languages are receiving notable attention from applied researchers in very different areas in the last years. Since its very beginning, the multi-agent paradigm has considered that knowledge representation models are crucial to accomplish interoperability in distributed systems. Vision systems, in turn, are being enhanced with cognitive capabilities by implementing knowledge-based reasoning procedures that allow them to achieve context-awareness and interpret scene activities.

CS-MAS has been implemented with JADEX [12], a framework for the development of BDI agent-based systems compliant with the FIPA[1] standard. FIPA provides mechanisms in the transport language (FIPA-ACL) to define ontologies for specifying the contents of the messages exchanged by the agents. The content language in FIPA is SL (Semantic Language), a first order logic-based language for specifying domain knowledge. FIPA-SL has been used for agent communication for years, but it has very important disadvantages. On the one hand, it is undecidable in its general form (it is not guaranteed that all inferences are computable in a finite time). On the other hand, there are very few tools supporting FIPA-SL. Hence, it has been proposed to use Semantic Web ontology languages to represent domain knowledge in agent communication [13,14]. OWL overcomes SL drawbacks: (i) it is decidable (it is based on well-studied Description Logics); (ii) it has a wide support (APIs, reasoners, etc.). We have used OWL to create the tracking data ontology, in line with other current developments.

Ontologies are being applied to solve vision problems at different abstraction levels (according to the JDL classification). In the lowest level, one of the most important contributions is COMM (Core Ontology for MultiMedia) [15], an ontology to represent MPEG-7 data with OWL. COMM does not aim at representing high-level entities, such as people, events, or activities occurring in the scene. Instead, it identifies the components of a MPEG-7 video sequence in order to link them with (Semantic) Web resources. Similarly, the Media Annotations Working Group of the W3C is working in an OWL-based language for adding metadata to Web images and videos [16].

Conversely, other approaches do aim at modeling video content knowledge. François et al. present a framework for video event representation and annotation [17]. In this framework, VERL (Video Event representation Language) defines the concepts to describe processes (entities, events, time, composition operations, etc.), and VEML (Video Event Markup Language) is a XML-based vocabulary to markup video sequences (scenes, samples, streams, etc.). VEML 2.0 has been expressed in OWL-DL, but only partially, because it imports VERL elements that need a more expressive language. The limitation in the number of entities represented in VEML 2.0 reduces its usefulness, as it is discussed in [18], which presents a framework that supports representation of uncertain knowledge.

---

[1] `http://www.fipa.org`

In a more abstract level, scene interpretation issues are being dealt with ontologies as well. In [19], it is presented a proposal for scene interpretation based on Description Logics and supported by the reasoning features of RACER[2] inference engine. The problem of representing high-level semantics of situations with a computable formalism is also faced in [20]. The authors present an OWL ontology (STO, Situation Theory Ontology) that encodes Barwise's situation semantics. Both research works tackle the problem of transforming numeric data to symbolic objects, because scene interpretation must eventually take raw video data as an input. Our representation has been purposely designed to solve this problem and could be used in combination with these other approaches.

4 The Tracking Entities Description Ontology

The tracking entities description ontology (Trend, in short) has been developed in OWL with the Protégé editor[3], version 4, which supports some of the extensions of the OWL 2 specification. The ontology is publicly available at the authors' web page[4].

OWL ontologies are developed from the following primitive elements: (i) concept or classes, which represent the basic ideas of the domain that must be understood; (ii) instances or individuals, which are concrete occurrences of concepts; (iii) relations, which represent binary connections between individuals or individuals and typed values. Domain knowledge is represented by asserting axioms, which establish restrictions over concepts, instances, and relations, describing their attributes by delimiting their possible interpretation.

Concepts in TREND represent video-processing objects, e.g. *track*, *frame*, *sequence*, and *camera*. Relations are established between these concepts, e.g. *track is active in frame*, and *sequence has been recorded by camera*. The axioms in TREND define restrictions on these properties, e.g. *a sequence is composed by several frames*, and *a track has only one combination of colors in a given frame*. The individuals of the TREND ontology are the concrete values calculated by the tracking algorithm, e.g. *track 1 in position (120, 240)*. Axioms defining concepts and relations are general and valid in every case (i.e. extensive knowledge), whereas the values of the instances are introduced in each execution of the algorithm (i.e. intensive knowledge). Figure 1 shows schematically an excerpt of the classes and relations in the TREND ontology.

We have imported two base ontologies in TREND to represent temporal and geometrical entities, respectively the TIME and the GEOM ontologies. TIME is an adaptation of the OWL-Time ontology [21], which is a well-known proposal to represent time stamps. GEOM has been built by relying on the proposal described in [22], and contains geometric concepts, such as Point, PointSet, Curve (a subclass of PointSet), or Polygon (a kind of Curve). We do not consider spatial relations in our geometry ontology, although RCC-8-like semantics could be

[2] `http://www.racer-systems.com`

[3] `http://protege.stanford.edu`

[4] `http://www.giaa.inf.uc3m.es/miembros/jgomez/ontologies/trend.owl`

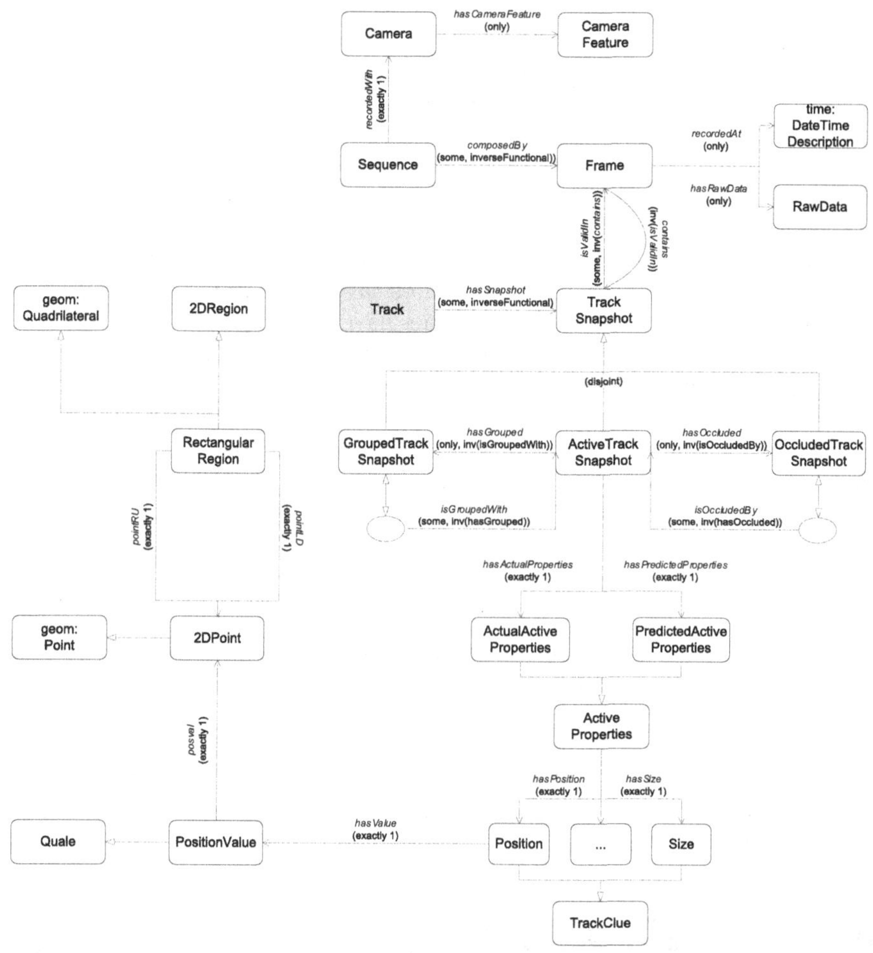

Fig. 1. Excerpt of the TREND ontology

introduced by using rules or other mechanism [23]. Additional concepts and restrictions have been defined in the TREND ontology to match graphical concepts, e.g. 2DPoint (a subclass of geom:Point, related to pair of real values), 2DRegion (a kind of geom:Polygon), RectangularRegion (a kind of 2DRegion, defined by the upper-left and the down-right points). Concepts and axioms to represent Camera, CameraProperty, and video Sequence have been also created. At this moment we have not developed the elements to describe track trajectories, but it can be realized that these ontologies can be easily included to augment the vocabulary.

The core concepts in TREND are Frame and Track. A frame is identified by a frame ID and may be marked with a time stamp to know when it was recorded. The time stamp is a date description created with the OWL-Time vocabulary.

The definition of tracks is more complex. We want to keep all the information related to a track (activity, groupings, occlusions, position, size, velocity, etc.), which changes between frames, and not only its lastly updated values. Therefore, we must connect tracks, frames, and track properties at each frame, which is a ternary relation. To solve this issue, we have followed a design pattern proposed by the W3C Semantic Web Best Practices and Deployment Working Group to define ternary relations in OWL ontologies [24]. Thus, we have associated a set of TrackSnapshots to each Track. Each TrackSnapshot is asserted to be valid in various Frames. We have created three disjoint subclasses of track snapshot: Active, Grouped, and Occluded. Only active snapshots have visual properties (TrackClue), which can be actual or predicted. Properties have been represented by following the *qualia* approach (used in the upper ontology DOLCE[5]), which distinguishes between properties themselves and the value space in which they take values. Thus, we have an ActiveProperties structure associated to each ActiveTrackSnapshot, with clues such as as velocity, color, position, etc. TrackClues take values from Quale spaces. All these properties are applicable during the frames in which the snapshot has been asserted to be valid.

5 Use Case

In this section, we provide an example of annotation of tracks processed by the tracking algorithm executed by a camera agent of the CS-MAS. The tracking process analyzes different aspects of the input image stream and provides as a result the tracking data. The video sequence used in this example is the APIDIS basketball dataset[6], generated by the APIDIS European project. This dataset is especially suitable for data fusion problems, although in this example we use frames recorded by a single camera.

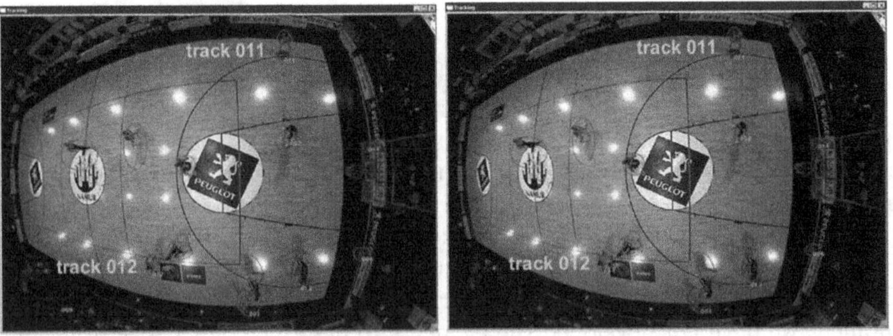

Fig. 2. Two frames of a video sequence of the APIDIS dataset

[5] http://www.loa-cnr.it/DOLCE.html
[6] http://www.apidis.org/Dataset/

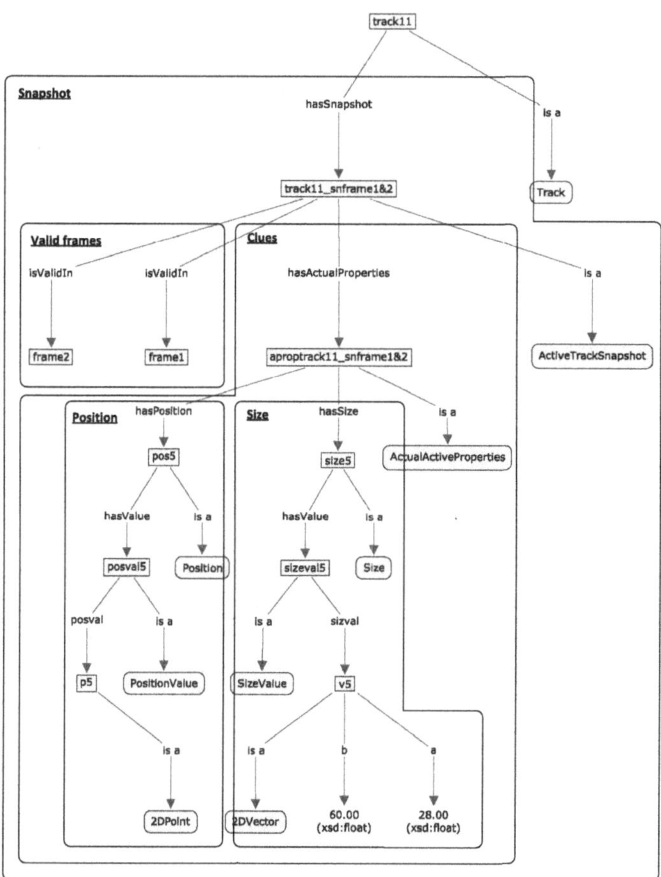

Fig. 3. Representation of the snapshot of track 11 valid in frames 1 and 2 (not moving)

We have described two tracks (track 11 and track 12) appearing in two frames (frame 1 and frame 2) of the video sequence (Fig. 2). The difference between frame 1 and frame 2 is quite subtle, since only minor position change can be appreciated visually. The tracking algorithm does detect changes in track 12 (associated to the player that has the ball), but not in track 11 (associated to a player in a peripheral position). For space reasons, we only show an excerpt of the ontological representation of track 11 in frames 1 and 2 (Fig. 3). This example is fully deployed in the authors' web page[7]. The use of the annotated data in a FIPA-compliant multi-agent system such as CS-MAS is quite straightforward, as shown in [14].

[7] `http://www.giaa.inf.uc3m.es/miembros/jgomez/ontologies/trend-instances.owl`

6 Conclusions and Future Work

In this paper, we have proposed an ontology to represent tracking data in artificial vision systems. The ontology is used in the CS-MAS system to communicate agents by means of FIPA messages with contents described semantically. We have presented the structure of the ontology, which is publicly available, and an example of annotation of tracking data.

Endorsing data with semantics has several advantages. Communication is easily achieved among agents, since the contents of messages are expressed in the same well-defined language. Systems are more flexible, extensible, and independent of the implementation technologies. The ontology also facilitates the development of extended functionalities for the vision system built on top of it, as well as the publication of tracking data.

We plan to continue this research work various directions. First of them, we will fully integrate the ontological representation with the tracking software and CS-MAS. This may imply further refinements or simplifications of the current ontology, since it has been developed with a very broad scope. Other interesting contribution would be the implementation of software tool for visualizing and exporting the data obtained by tracking algorithms and annotated with the ontology. Such a program would be useful to review and test the accuracy of tracking processes, both for expert and non-expert users. Moreover, we plan to use the ontology as a basis for further high-level processing of visual data. We are developing an extension of the tracking system that uses context-knowledge to infer additional scene information. This context layer is built on top of the tracking data ontology, which is the first step in the evolution from numeric to symbolic information. Last but not least, it is interesting to highlight that the representation can be used in data fusion processes at different levels, from object refinement to situation assessment. We believe that formal knowledge representations can be valuable to accomplish successful data fusion, and consequently, studying the applicability of this and other eventual OWL ontologies is a very promising research direction.

Acknowledgement. This work was supported in part by Projects CICYT TIN2008-06742-C02-02/TSI, CICYT TEC2008-06732-C02-02/TEC, SINPROB, CAM MADRINET S-0505/TIC/0255 and DPS2008-07029-C02-02.

References

1. Yilmaz, A., Javed, O., Shah, M.: Object tracking: A survey. ACM Computing Surveys 38(4), 13 (2006)
2. Patricio, M.A., Castanedo, F., Berlange, A., Pérez, O., García, J., Molina, J.M.: Computational Intelligence in Visual Sensor Networks: Improving Video Processing Systems. In: Computational Intelligence in Multimedia Processing: Recent Advances, pp. 351–377. Springer, Heidelberg (2008)
3. Steinberg, A.N., Bowman, C.L., White, F.E.: Revisions to the JDL data fusion model. In: Proceedings of SPIE, vol. 3719, pp. 430–441. SPIE (1999)

4. Steinberg, A.N., Bowman, C.L.: Revisions to the JDL data fusion model. In: Handbook of Multisensor Data Fusion, pp. 45–67. CRC Press, Boca Raton (2009)
5. Patricio, M.A., Carbó, J., Pérez, O., García, J., Molina, J.M.: Multi-agent framework in visual sensor networks. Journal of Applied Signal Processing (1), 226–247 (2007)
6. Horrocks, I.: Ontologies and the Semantic Web. Communications of the ACM 51(12), 58–67 (2008)
7. Gruber, T.R.: A translation approach to portable ontology specifications. Knowledge Acquisition 5(2), 199–220 (1993)
8. Chandrasekaran, B., Josephson, J., Benjamins, V.: What are ontologies, and why do we need them? IEEE Int. Systems and Their Applications 14(1), 20–26 (1999)
9. Baader, F., Horrocks, I., Sattler, U.: Description Logics. In: Handbook of Knowledge Representation, pp. 135–180. Elsevier, Amsterdam (2008)
10. McGuinness, D.L., van Harmelen, F.: OWL Web Ontology Language Overview (2004), `http://www.w3.org/TR/owl-features/`
11. Bao, J., Kendall, E.F., McGuinness, D.L., Wallace, E.K.: OWL 2 Web Ontology Language: Quick reference guide (2008), `http://www.w3.org/TR/owl2-quick-reference/`
12. Braubach, L., Pokahr, A., Lamersdorf, W.: Jadex: A BDI-Agent System Combining Middleware and Reasoning. In: Software Agent-Based Applications, Platforms and Development Kits, pp. 143–168. Birkhäuser, Basel (2005)
13. Hendler, J.: Agents and the Semantic Web. IEEE Intelligent Systems 16(2), 30–37 (2001)
14. Schiemann, B., Schreiber, U.: OWL-DL as a FIPA-ACL content language. In: Proceedings of the Workshop on Formal Ontology for Communicating Agents, Malaga, Spain (2006)
15. Arndt, R., Troncy, R., Staab, S., Hardman, L., Vacura, M.: COMM: designing a Well-Founded multimedia ontology for the web. In: Aberer, K., Choi, K.-S., Noy, N., Allemang, D., Lee, K.-I., Nixon, L., Golbeck, J., Mika, P., Maynard, D., Mizoguchi, R., Schreiber, G., Cudré-Mauroux, P. (eds.) ASWC 2007 and ISWC 2007. LNCS, vol. 4825, pp. 30–43. Springer, Heidelberg (2007)
16. Lee, W., Bürger, T., Sasaki, F.: Use cases and requirements for ontology and API for media object 1.0 (2009), `http://www.w3.org/TR/media-annot-reqs/`
17. François, A.R., Nevatia, R., Hobbs, J., Bolles, R.C., Smith, J.R.: VERL: an ontology framework for representing and annotating video events. IEEE Multimedia 12(4), 76–86 (2005)
18. Westermann, U., Jain, R.: Toward a common event model for multimedia applications. IEEE Multimedia 14(1), 19–29 (2007)
19. Neumann, B., Möller, R.: On Scene Interpretation with Description Logics. Image and Vision Computing 26, 82–101 (2008)
20. Kokar, M.M., Matheus, C.J., Baclawski, K.: Ontology-based situation awareness. Information Fusion 10(1), 83–98 (2009)
21. Hobbs, J., Pan, F.: Time ontology in OWL (2006), `http://www.w3.org/TR/owl-time/`
22. Maillot, N., Thonnat, M., Boucher, A.: Towards ontology-based cognitive vision. Machine Vision and Applications 16(1), 33–40 (2004)
23. Grütter, R., Scharrenbach, T., Bauer-Messmer, B.: Improving an RCC-Derived geospatial approximation by OWL axioms. In: Proceedings of the 7th International Semantic Web Conference (ISWC 2008), pp. 293–306 (2008)
24. Noy, N., Rector, A.: Defining n-ary relations on the semantic web (2006), `http://www.w3.org/TR/swbp-n-aryRelations/`

Multimodal Agents in Second Life and the New Agents of Virtual 3D Environments

A. Arroyo, F. Serradilla, and O. Calvo

Dpto. Sistemas Inteligentes Aplicados, Universidad Politecnica de Madrid, Spain
aarroyo@eui.upm.es, fserra@eui.upm.es, ocalvo@gmail.com

Abstract. The confluence of 3D virtual worlds with social networks imposes to software agents, in addition to his conversational functions, the same behaviors as those common to human-driven avatars. In this paper we explore the possibilities of the use of metabots (metaverse robots) in virtual 3D worlds and we introduce the concept of AvatarRank as a measure of the avatar's popularity and the concept of the extended Turing test to assess the anthropomorphness of the metaverse's elements.

1 Introduction

We are living the development of so-called Web 2.0 and a good number of applications are being introduced into many users' lives and are profoundly changing the roots of society by creating new ways of communication and cooperation. Similarly, in recent years the first virtual worlds have emerged in which humans, through their avatars, "cohabit" with other users. This is a new interaction model based on a technology born from multiplayer online games, a 3D virtual world imagined by Neil Stephenson in his novel "Snow Crash" [1] and previously by Vernor Vinge on the short story "True Names" [2], a world in which humans feel at home, a metaverse. This new model is more humane because it simulates the real environment's characteristics in which the human being is and has become. If our senses create the reality in which we are developing ourselves, to experience virtual reality worlds we will have to live in accordance with the capabilities of our senses, to live in digital extensions of our physical world.

These virtual worlds or metaverses are in fact true social networks and they are useful for interaction between people in different locations. Likewise, in the three-dimensionality context it is very appropriate to develop virtual robots with the same appearance as that of the human-driven avatars. These new virtual robots are called metabots –term coined from the contraction of the terms metaverse and robot–. A metabot will therefore be a fully capable software completely able to interact in one or more metaverses through one or more avatars. The representation of a metabot should be much more intuitive to humans than the traditional one as it simulates reality and then it does not require a specific learning. Additionally, the metabot can use any common channel based on information brokers, databases, indexing engines, social networks, etc... to serve as a bidirectional link between the "in-metaverse" and "out-metaverse".

J. Mira et al. (Eds.): IWINAC 2009, Part I, LNCS 5601, pp. 506–516, 2009.

As already mentioned, these virtual environments are useful for human interaction. However, we don't need to leave aside the great possibilities provided for research in human-machine communication and machine-machine communication. To humanize the software systems, conversational agents or systems which allow people to carry out at least part of their work through some spoken dialogue were developed [3].

For the moment, there is no machine able to pass the Turing test, but today there are conversational robots that they have almost surpassed this test. Each year there is a call for Loebner Prize in AI in the British University of Reading, in which participant robots answer a series of questions asked by human interrogators. In the eighteenth call for this award in October 2008, all conversational agents were able to deceive to at least one of the interrogators. The winner of this edition has cheated 25% of the interrogators. Communication with this bot is one of the experiments discussed in this article. The metabots could improve those communication process in 3D virtual spaces in many ways: gestures, glances, facial expression, movement...

On the other hand, direct manipulation interfaces [4] must have more flexible interaction methods and powerful man-machine integration [5] if we want the productive capacities of individuals exceeds the current limit imposed by current interfaces. Therefore, the field of intelligent agents, particularly agents that focus on the World Wide Web, was have a strong growth, mainly in the MIT [6]. Communication between proactive software systems is part of the research on multiagent systems and, more specifically, in the languages of communication among agents (ACL).

Ubiquitous computing aims to build a hardware and software infrastructure that enables a new type of relation between man and computers through flexible communication interfaces and, of course, incorporating more intelligence to software systems. The technology of intelligent agents will be the key to this ease of use and power that finally will put at our disposal in this new computing paradigm. Again the highest regards for this line of research is MIT, and especially the Media Lab of this institution [7].

In this article we show the results of a series of experiments in this new field of experimentation. First, we present the software base developed to support these new experiences and the concepts used (the AvatarRank and the extended Turing test). Then we will discuss one to one our experiences with metabots. We show several examples of metabots as information agents: The usher, a receptionist metabot implemented as a classical agent; the chatterbot, a multiagent metabot and; the pretender metabot, a classic information agent and an indexation engine.

2 Related Works

One of the main challenges of AI since his early days has been to achieve the man-machine communication through natural language. We are still far from having a computer program that is capable of recognizing the complexity of natural language, reasoning about the content of communication and respond appropriately to a human user. However, the use of programs designed to simulate an intelligent conversation without an understanding of the contents of it is growing in many areas. These programs are known as conversational agents or more typically as Chatterbots, Chatbots or Talkbots.

Although in many cases may appear to intelligently interpret the input provided by the user, the truth is that most of the currently available chatterbots simply look for keywords in the message and select the response from all those stored in its database data by applying a set of rules.

During the decades of 60 and 70 of the past century the first chatterbots appeared, including: ELIZA, that simulates a psychologist, using pattern recognition over data inputs provided by user to establish the answer, and SHRDLU, more focused in understanding the natural language through AI techniques and less focused on the appearance of human conversation. ELIZA allowed conversation on any topic with messages that mimic those provided by a human but without any real content. SHRDLU was able to "live" in an environment (known in the field of AI as "cube world") and establish a conversation with the user in a manner consistent with the world that it lives in. Currently, it is not surprising to find that actual chatterbots have features of both of them.

In the first category, we highlight the chatterbot ALICE (Artificial Linguistic Internet Computer Entity) initially developed by Richard Wallace in 1995, actually having a strong community of partners being an open source project. The main contribution of this scheme is the usage of a modelling language called AIML (Artificial Intelligence Markup Language) [8], based on XML, to specify the rules that apply heuristic conversational chatterbot. A promising line for this type of chatterbots is to use the technologies of the Semantic Web and multi-platforms to extend its functionality [9].

In the second category we find a set of proposals that reflect the variety of AI techniques employed in the natural language recognition. As an example we can cite the case of chatterbot called Jabberwacky that stores the contents of all conversations that maintains and uses it to find the most appropriate response using contextual pattern recognition. Chatterbots also have expertise in specific tasks in natural language processing, such as those provided by Google Talk to chat with us about automatic translators in 30 different languages as if they were one more of our list of contacts.

The most advanced chatterbots, as we have said, are still far from overcoming the famous Turing Test and the Imitation Game [10]. We can still observe their rules of operation if we focus on the responses obtained, but it is not difficult to find examples of humans who where "tricked" by a chatterbot to the point of arriving at an appointment in the real world [11].

Moreover, the new virtual worlds constitute a new area in which chatterbots can incorporate new features. In those three-dimensional virtual spaces in which users establish contact through avatars that represent them, conversational agents must show active behavior. These behaviors (gestures, movements...) have to emulate a human-driven avatar and reflect the personality of the profile or type of user they represent.

The confluence of 3D virtual worlds with social networks imposes to virtual agents, in addition to his conversational functions, the need of an "inhabitant" of these environments. There are currently chatterbots in virtual worlds like Second Life based in AIML with disparate forms and behaviors (vendors, tour guides...). The movements of the robots that inhabit 3D environments are based on the same principles that are used in mobile robotics for decades so the models used in this field [12] are easy to implement in the simulations conducted in virtual worlds. In the present paper we will discuss on different experiences that explore all these aspects.

3 Concepts on Metabots

3.1 Definition

A metabot is a software virtual robot with the same appearance as that of human-driven avatars. It is a fully capable software that interact in one or more metaverses through one or more avatars. It is a new kind of user interface which is much more intuitive to humans as it simulates human behavior and thus not require a specific learning. For a human, a metabot behaves as a conversational agent and, in the digital space, a metabot behaves like a human-driven avatar with the ability to communicate with other software agents and with any other common channel based on information retrieval system, databases, indexing engines, social networks, recommendations system, etc.

A metabot is a mobile, proactive and perceptive agent. It moves in a 3D environment like human-driven avatars; it can also learn from them. It communicates with them as a conversational agent since it is one of them. It communicates with other agents by using ACL and/or multiagent platforms. To live in 3D-persistent worlds, the metabot must have concepts such as space and time. Ontologies used by these agents, therefore, must rely on these concepts and metabots must have a isomorphic mental map with the 3D virtual world loaded and interpretable by a human.

3.2 AvatarRank

One of the first problems that arise when implementing metabots is to define the parameters within which we humans communicate in metaverse and how to make metabots sensitive to these parameters. For example, how a metabot can know when an avatar is interested in it?.

Humans have millions of years of evolution to help us in these tasks, and yet sometimes it is difficult for us to know how to behave at a party. In any case, humans instinctively know what a conversation circle is or who is the most popular in a classroom. These behaviors related to human consciousness of the space environment are moved to metaverses with almost no difference. Even more, the main goal of a metaverse is to minimize this difference. In the virtual space also exist the concepts of closeness, touching, attention, etc. So in order to build metabots that are able to integrate these absolutely anthropomorphic behaviors we must isolate and define certain parameters and calculations in order to measure a concept which we called Avatar Rank: How popular is an avatar?

When we use the word "popular" we mean the amount of attention that is receiving an avatar in a single moment. Therefore the higher the rank of an avatar, the more attention from other avatars (whether human or robot), and as a consequence, any activity has more potential to be received. The Avatar Rank concept is also built over a more comprehensible measure: the Score. The concept is the same as the Avatar Rank avatars but only works unidirectionally and between two avatars. Therefore the score of A over B [score (A, B)] is defined as the amount of attention that B is receiving from A.

The calculation of the Avatar Rank is derived from a matrix of scores for all the Avatars to all the Avatars and accumulating the value of the columns from that matrix. In turn, the score is calculated by taking two immediate trigonometric measures: the distance between A and B and the face angle of A on B. In short, the more B is looking to A, and the more A and B are close, the higher is the Score. Following equations defines the value for an Avatar Rank and Score functions.

$$score(AoverB) = score(A \leftarrow B) = S_{a,b} = score(a,b)$$

$$score(a,b) = \frac{Dmax - (distance(a,b) - Dmin)}{Dmax} \cdot \frac{\Pi - |atan(b-a) - rotation(b)|}{\Pi}$$

$$scoreMatrix_{NxN} = \begin{bmatrix} S_{1,1} & S_{1,2} & .. & S_{1,N} \\ S_{2,1} & S_{2,2} & .. & S_{2,N} \\ .. & .. & .. & .. \\ S_{N,1} & S_{N,2} & .. & S_{N,N} \end{bmatrix} \quad \forall i \in N, S_{i,i} = 0$$

$$avatarRank(X) = \frac{1}{N-1} \cdot \sum_{i=1}^{N} S_{X,i}$$

In the following pics there's a graphical representation of the Score function.

The Score function is a mathematically formulated version of the human personal space concept as we will discuss later. In the following lines we will comment some examples of how avatars interact themselves with the Score and Avatar Rank functions.

Example 1. A and B are near, but C is out of conversation.
In this case, A and B have an average AvatarRank since they are in the "sensitive zone" of each other. In the other hand, C has a near zero AvatarRank, since is in no body's zone.

Example 2. The follower. A is following B.
A has a near zero Avatar Rank and B has a very high Avatar Rank. When an avatar follows another, is placing the target avatar in the middle of the sensitive zone.

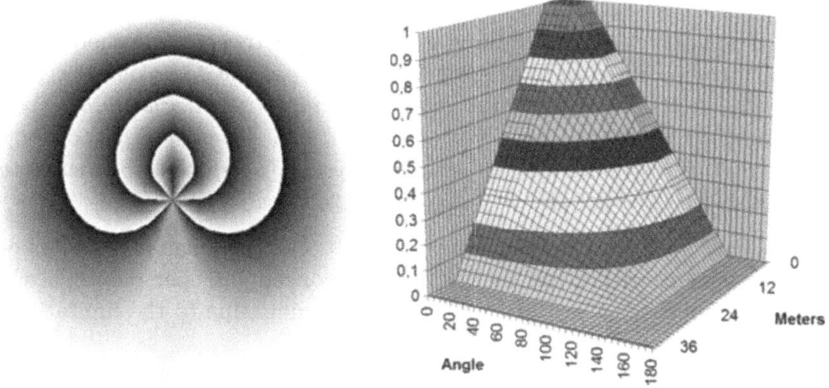

Fig. 1 and 2. Score function. Left: Spatial representation. Right: Linear scale representation.

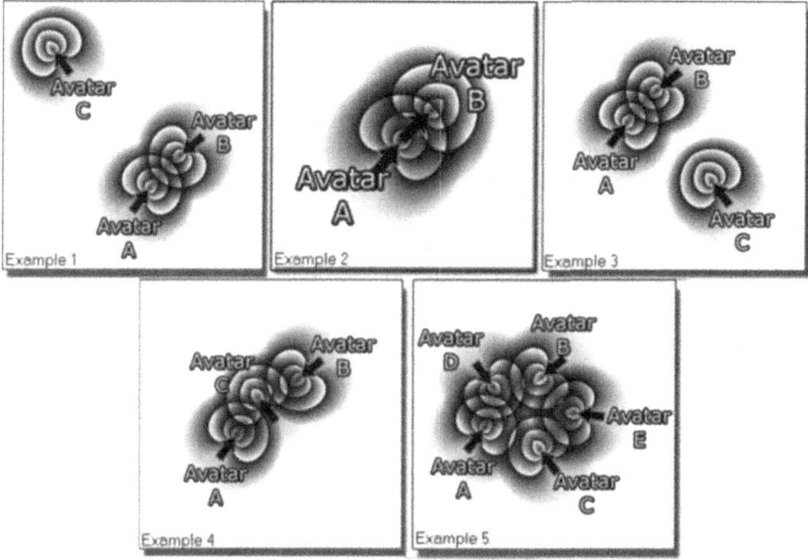

Fig. 3. Avatar Rank examples

Example 3. A and B are near, but C is somewhat interested in A and B
In this case, A and B have also an average Avatar Rank, but it's a bit higher than in the case above, since C is "looking" at them.

Example 4. Alpha avatar. A and B are about C, but C is not paying attention to A and B *Now A and B have a medium-low Avatar Rank, since they are a bit far than in case above. But in this case C has a very high Avatar Rank, as A and B have C in their hottest part of the sensitive zone.*

Example 5. The party. Common talking group. Everybody sees everybody. *This is a very common anthropomorphic behavior. Some avatars come together to have a little chat time. All avatars have a medium-high Avatar Rank, even when no avatar is completely in a hot sensitive zone. When multiple avatars are together and their sensitive zones are overlapped a "hotspot" is created. Any avatar placed in this hotspot will be a top avatar rank automatically.*

So, with the Score functions we are modeling an absolutely human and social behavior to implement within the metaverse: the concept of "personal space"[1] However, these approaches are still rather crude, mainly due to the limitations of current metaverses.

Conclusions. The word anthropomorphism surrounds this paper and is the common nexus of the tree main ideas we are handling: metabots, metaverses and Turing Test. In the way to achieve a real artificial intelligence we will need to implement with the metaverse tools the anthropomorphic rules of our humanity.

3.3 Metabots and the Extended Turing Test

One constant idea in this paper is that metaverse are the first information system built by humans for humans. Previously, with more limited resources to keep and manage the information, humans have to learn to use concepts not close to our reality, for example, a mouse that moves across a screen is something that we learn to do as a way to reach information. However in metaverses the system is built to adapt to humans, that is, designed and planned from anthropomorphic principles. Moreover, the more anthropomorphic objects and services are, the better metaverse experience will be. That last paragraph is well known by all metaverse developers. Even in this early stage of metaverse evolution, we all know instinctively that "it should be that way". When we formalised this idea we find a new paradigm for the building of everything related to metaverses, from objects to avatars: the maximum credible metaphor paradigm.

This new paradigm says that anything that is to be done into a metaverse should be performed with the maximum possible details and quality to be credible for humans. The implications and details of that paradigm are so vast that should be discussed in another paper, but we need to stand on this concept to reach another one, the extended Turing test.

If we take some pictures from real world and some from a metaverse and offer them to a couple of judges, how many of them will know which shot is real and which not?.

Until now, the Turing Test was only intended to be applied to "speaking" artificial intelligences. Since the Turing Test is also anthropomorphic based, why not to extend the test schema to anthropomorphic spaces, services, avatars, objects, etc.?

Even more, if not only a bunch of judges, but thousands of people wandering around metaverses, using services, speaking with metabots, buying objects or

[1] http://en.wikipedia.org/wiki/Personal_space

building islands, isn't that a continuous evaluation of the surrounding anthropomorphness?. In fact, in the very moment one human gets into a metaverse, is evaluating everything. Probably, the future success of the metaverses will be conditioned to who credible are their metaphors.

Conclusions. Turing Test, Metaverse and Metabot share tied-together by anthropomorphism. One of the critics to the Turing Test is about its impractically [13] is "the anthropomorphism of the test prevents it from being truly useful for the task of engineering intelligent machines". This means that the only environment where the Turing Test can be ran with guarantees must be anthropomorphic, then, the metaverse.

4 Experimentation in Metabots: Some Use Cases

4.1 Metabot Running a Classic Information Agent: The Usher Metabot

One of the most common Information Agent use is to retrieve information from a standard web page and, after formatting and filtering the desired data, return it to the query origin. As discussed above in this paper, the UPM owns an island in Second Life called TESIS. On that island the EUI built its Virtual facilities in which numerous educational activities are performed.

Standing in the front desk of that blue crystal building there's a metabot that works as a virtual usher. Her work is very simple: if an avatar walks in and speak she'll listen the words and search in a web page[2] where all the information of teachers and administrative staff is. Once gathered all contact information, the usher metabot will say over the public audible channel.

Conclusions. This metabot is almost the simplest bot that can be built, but despite of this fact, it's easy to see its potential as communication channel between massive load if information that we can find on the web and, by now huge but almost empty metaverse. Another important goal is to give an anthropomorphic user interface to web access, following the maximum credible metaphor paradigm.

4.2 Metabot Running a Multiagent: The Chatterbox Metabot

At the 18th Loebner Prize for artificial intelligence, held on the weekend of 11-12 October, 2008, Elbot convinced three of the 12 human interrogators he was indistinguishable from human, beating the other contestants and taking the Bronze Prize. But the most important characteristic for us is that anybody can chat with Elbot through its webpage www.elbot.com. That feature gave us the idea of making up a fancy body for Elbot within Second Life: the Chatterbox Metabot. But Elbot only speaks English and, unfortunately, most of people that works and learn in TESIS island are mostly Spanish speakers. So we get another step: using the online Google translator services we translate the English

[2] http://www.eui.upm.es/Escuela/InfoGeneral

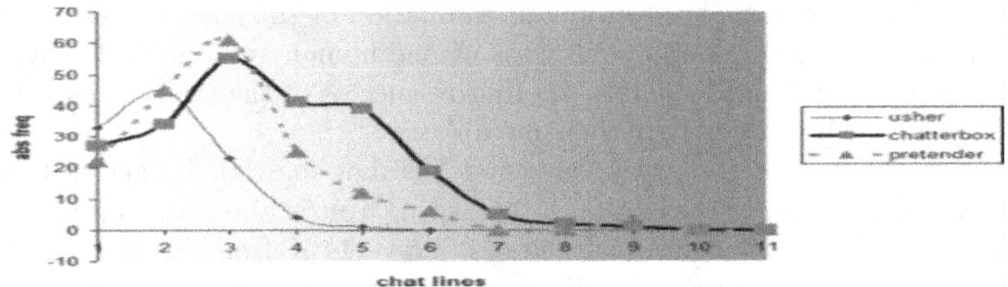

Fig. 4. Chat lines by metabot model: usher, chatterbox or pretender

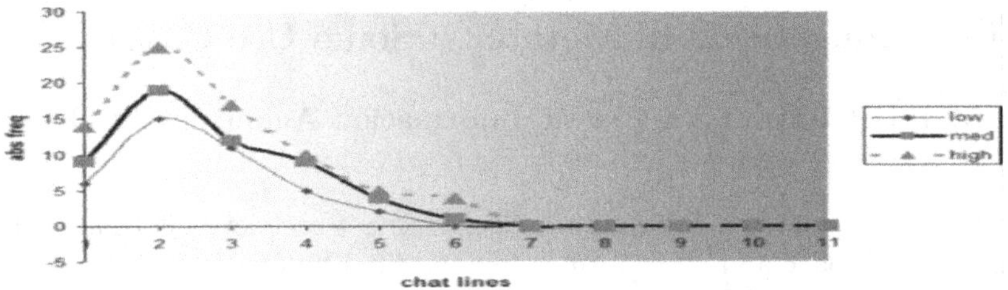

Fig. 5. Chat lines by human-likeness avatar

paragraphs of Elbot to Spanish, the Chatterbox Metabot speaks in Spanish with any avatar and when that avatar replies to the Metabot, we translate it back to English before send response to Elbot. We have to say that sometimes the final quality of the conversation is diminished with all this translation.

Conclusions. Using the pure web 2.0 concept "mashup" we can condense all the Internet range intelligence in one only metabot. Bringing the known concepts of wild computing or distributed digital consciousness [14] to the metaverse and its implicit anthropomorphism could be the kick-off to a new path for investigation in artificial intelligence, like the Extended Turing Test.

4.3 Metabot Running Information Agent + Index Server: The Pretender Metabot

For this metabot we have used a mixed solution of classic information agent and an indexation engine. The information agent is able to fetch a famous quote from the www.tusfrases-celebres.net website and then index it with an Apache Lucene engine. Every time someone comes near to our "pretender" metabot and speaks, it performs a search on the quotations index, and pronounces the phrase with the best score on the audible channel. If no one sentence gives satisfactorily results, the metabot will fetch one randomly.

This experiment aims to investigate an approach based on pure imitation, working with large amounts of data and tools to access this information quickly. The difference between the Chatterbox metabot and the Pretender metabot is this one also get information from within the Metaverse, acquiring information about its nearly environment. Conclusions. We called "pretender" to this metabot, because this one just pretends to be intelligent, but it's only using inverse index algorithms and a lot of curiosity. But here we have the final question, if it only looks like intelligence but works, is it finally some kind of intelligence?

5 Conclusions

The metabots mentioned above have been running for several months on TESIS island in Second Life metaverse, and now we are able to expose the different kinds of reaction that SL users have to metabots, depending on its type, look, behaviour. Measures: we have logged the number of paragraphs (chat lines) interchanged in every chat time between a metabot and a human. We called this "chat length". In the first test, we placed the three metabots and logged all chats for two months.

In this graphic we can see that the Usher metabot is the less spoken, the pretender has better chat length and the chatterbox bot is the winner. The conclusion is that the better chat skills, the longer chat we will have. Based on the results of the previous study we made another one. We pick usher metabot and repeated the test with three different avatars. A human like one, a barely human, and a blatantly robotic look (high, med and low respectively).

Conclusion: The simple image of the metabot can affect severely to the final results. This validates the maximum credible paradigm. When performing test number 1 and checking the logs with the questions and the answers, we realized that some people never realized that wasn't speaking with a machine, this is, passed our "continuous Turing test". The results were about 15% for Chatterbox and 4% for Pretender metabot.

The first clear conclusion obtained with the presented experiences is that metabots are a very promising field of experimentation for the implementation of the models used in Artificial Intelligence and the development of man-machine communication in all its aspects and without the need for a specific learning by humans. An immediate goal of our work is the development of an infrastructure that allows easy aggregation of new Information Agents, so that a single centralised metabot can, with growing capabilities of information gathering and information injection, put it into the metaverses taking it from various sources.

Acknowledgements

This work has been partially supported by Optenet, project P086130230: "BABY-BOT: Creating a 'virtual child' as a conversational agent, artificial intelligence techniques for the detection of criminal behaviour in virtual environments and social networking" and by 2008 UPM competition call.

References

1. Stephenson, N.: Snow Crash. Bantam Books (1992)
2. Vinge, V.: True Names. Dell Binary Star #5 (1981)
3. Bernsen, N.O., Dybkjaer, L., Dybkjaer, H.: A dedicated task-oriented dialogue theory in support of spoken language dialogue systems design. In: ICSLP 1994, pp. 875–878 (1994)
4. Lieberman, H.: Autonomous interface agents. In: Proceedings of CHI 1997, Atlanta, Georgia, March 1997, pp. 67–74. ACM Press, New York (1997)
5. Lieberman, H., Van-Dyke, N., Vivacqua, A.: Let's Browse: a collaborative browsing agent. Knowledge-Based Systems 12(8), 427–431 (1999)
6. Media Lab Projects List, MIT Media Laboratory (October 2008)
7. Dertouzos, M.L.: The future of computing. Scientific American (July 1999)
8. Wallace, R.: The Elements of AIML Style. ALICE A.I. Foundation (2003)
9. Freese, E.: Enhancing AIML Bots using Semantic Web Technologies. In: Proceedings of Extreme Markup Languages, Montreal-Quebec (Agosto 7-10, 2007)
10. Turing, A.M.: Computing machinery and intelligence. Mind, New Series 59(236), 433–460 (1950)
11. Epstein, R.: From Russia, with Love. Scientific American Mind, (October/November 2007)
12. Breazeal, C.: Designing Sociable Robots. MIT Press, Cambridge (2003)
13. Russell, S.J., Norvig, P.: Artificial Intelligence: A Modern Approach. Prentice Hall, Englewood Cliffs (2003)
14. Goertzel, B.: Creating Internet Intelligence: Wild Computing, Distributed Digital Consciousness, and the Emerging Global Brain. Springer, Heidelberg (2002)

Application of Artificial Neural Networks to Complex Dielectric Constant Estimation from Free-Space Measurements

Antonio Jurado, David Escot, David Poyatos, and Ignacio Montiel

Laboratorio de Detectabilidad
Instituto Nacional de Técnica Aeroespacial (INTA)
Ctra. Ajalvir Km. 4, 28850, Torrejón de Ardoz, Spain
{juradola,escotbd,poyatosmd,montielsi}@inta.es,
http://www.inta.es

Abstract. Adequate characterization of materials allows the engineer to select the best option for each application. Apart from mechanical or environmental characterization, last decades' rise in the exploitation of the electromagnetic spectrum has made increasingly important to understand and explain the behavior of materials also in that ambit. The electromagnetic properties of non-magnetic materials are governed by their intrinsic permittivity or dielectric constant and free-space measurements is one of the various methods employed to estimate this quantity at microwave frequencies. This paper proposes the application of Artificial Neural Networks (ANNs) to extract the dielectric constant of materials from the reflection coefficient obtained by free-space measurements. In this context, two kind of ANNs are examined: Multilayer Perceptron (MLP) and Radial Basis Function (RBF) networks. Simulated materials are utilized to train the networks with and without noise and performance is tested using an actual material sample measured by the authors in an anechoic chamber.

Keywords: Materials characterization, complex permittivity, free-space measurements, multilayer perceptron, radial basis function.

1 Introduction

For many years, evaluation of electromagnetic properties of materials has been a fundamental aspect and a challenging problem with an important variety of applications [1]. The increased exploitation of the electromagnetic spectrum implies a need for material characterization in different scenarios and for new applications. In this sense, the authors' background is related with radar systems and technologies, and particularly with Radar Cross Section (RCS) analysis. RCS represents the reflectivity of a given target and is the ratio between the power scattered by the target and the power of the plane wave signal impinging on it. It depends on the shape and size of the target but also on its intrinsic electromagnetic properties (relative complex permittivity and permeability (ϵ_r^*, μ_r^*)) and the frequency and polarization of the incident wave [2].

J. Mira et al. (Eds.): IWINAC 2009, Part I, LNCS 5601, pp. 517–526, 2009.

Nowadays, numerical techniques permit the realization of RCS analysis of complex targets with software programs. Thanks to this, tough electromagnetic problems can be faced. For instance, these prediction codes contribute to the design process of *stealth* aircrafts (platforms with reduced or low observable radar signature [3]) by allowing to extensively characterize a Computer Aided Design (CAD) model prior to making a physical prototype. Also, computers can be used to estimate the RCS of aircrafts and populate with it the data base needed in Non Cooperative Target Identification (NCTI) techniques [4]. In either case, to prepare a trustworthy simulation, it is necessary to have good knowledge of the dielectric properties of all media involved.

There are different ways to obtain the complex electromagnetic constants of samples (probe method [5], resonant cavity method [6], free-space measurements [7],...) but accounting for the applications and frequencies of interest for the authors, their background, the available facilities at INTA and the preparation of the samples, the technique chosen is the one based on free-space measurements. In this method either one or two antennas are used and then, the reflection and/or transmission coefficients are measured. From these data the constants can be extracted.

Traditionally, an iterative process needs to be implemented to find the roots of the error function and determine the complex relative permittivity and permeability (ϵ_r^*, μ_r^*) from the measured quantity(s). This paper is the continuation of a previous work in which soft-computing techniques were applied to obtain the dielectric constant of non-magnetic materials from the reflection coefficient [8]. Genetic Algorithms (GAs), Particle Swarm Optimization (PSO) and Artificial Neural Networks (ANNs) were used with satisfactory results in that work, but when the input data was contaminated with noise (which is a more realistic situation), better results were obtained with ANNs than with the optimization techniques (GAs, PSO). This preliminary result encouraged to study the behavior of neural networks trained with noisy inputs, so the next sections will examine the performance of two kind of networks (Multilayer Perceptron (MLP) and Radial Basis Function (RBF) networks) in this situation.

2 Problem Formulation

The reflection coefficient of a sample can be measured in free-space using a single antenna following the set-up sketched in Fig. 1a. The PC controls the positioner and triggers the Vector Network Analyzer (VNA). The sample needs to be metal-backed (Fig. 1b) so that enough response is received in the antenna. The anechoic chamber diminishes the effect of multiple reflections and well-known measurement techniques like *software gating* or *background subtraction* can be employed to improve the signal-to-noise ratio [9].

From transmission line theory, reflection coefficient is related to complex permittivity and permeability via the following general equations (no assumptions or approximations for low losses materials have been made):

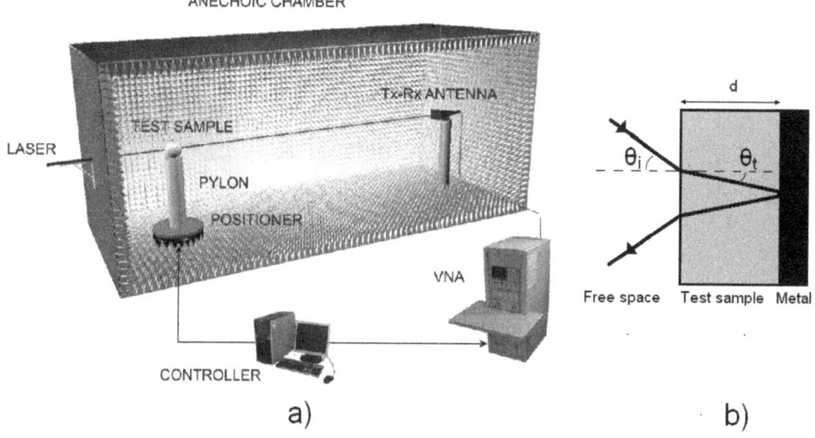

Fig. 1. Measurement of samples: a) Set-up in an anechoic chamber. b) Behavior of a wave impinging on a metal-backed sample.

$$\Gamma_{\perp} = \frac{\sqrt{\frac{\mu_r^*}{\epsilon_r^*}} cos(\theta_i) tanh(jk_0 d\sqrt{\mu_r^* \epsilon_r^*} cos(\theta_t)) - cos(\theta_t)}{\sqrt{\frac{\mu_r^*}{\epsilon_r^*}} cos(\theta_i) tanh(jk_0 d\sqrt{\mu_r^* \epsilon_r^*} cos(\theta_t)) + cos(\theta_t)} . \quad (1)$$

$$\Gamma_{\parallel} = \frac{\sqrt{\frac{\mu_r^*}{\epsilon_r^*}} cos(\theta_t) tanh(jk_0 d\sqrt{\mu_r^* \epsilon_r^*} cos(\theta_t)) - cos(\theta_i)}{\sqrt{\frac{\mu_r^*}{\epsilon_r^*}} cos(\theta_t) tanh(jk_0 d\sqrt{\mu_r^* \epsilon_r^*} cos(\theta_t)) + cos(\theta_i)} . \quad (2)$$

where $\Gamma_{\perp}$ and $\Gamma_{\parallel}$ are the perpendicular and parallel reflection coefficients, d is the sample thickness, $k_0 = \frac{2\pi}{\lambda}$ is the free-space wavenumber, θ_i is the incidence angle, θ_t the transmitted angle (Fig. 1b) and ϵ_r^* and μ_r^* are the relative complex permittivity and permeability:

$$\epsilon_r^* = \epsilon_r^{'} - j\epsilon_r^{''} . \quad (3)$$

$$\mu_r^* = \mu_r^{'} - j\mu_r^{''} . \quad (4)$$

From equations (1) and (2) complex constants ϵ_r^* and μ_r^* can be extracted.

3 Application of Artificial Neural Networks

The networks are trained with a set of simulated materials with random constants and thickness. The performance is tested with a 20x20 cm actual sample of Arlon© CuClad 250GX-0620 55 11 whose thickness is $d = 1.7mm$ and whose complex dielectric constant is provided by the manufacturer. This sample is a non-magnetic material (μ_r^*=1) so the problem is reduced to the estimation of the complex permittivity and there is no need to measure off-normal, meaning that $\theta_i = 0$ and (1) and (2) are the same and only one of them is enough to extract

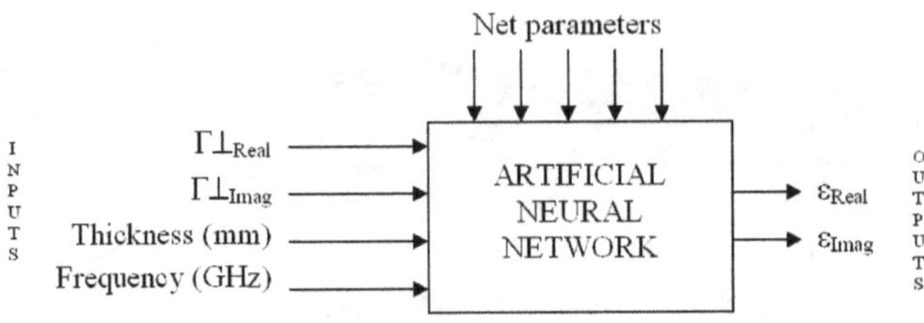

Fig. 2. Proposed ANN structure

Table 1. Network parameters used in this study

MLPs			RBFs	
Name	Hidden layer 1	Hidden layer 2	Name	Spread parameter
MLP1	20 nodes	-	RBF1	1
MLP2	20 nodes	10 nodes	RBF2	0.75
MLP3	10 nodes	5 nodes	RBF3	0.5

the real ($\epsilon_r^{'}$) and imaginary parts ($\epsilon_r^{''}$) of ϵ_r^*. This simplification diminishes the complexity of the problem but does not limit its utility as the conclusions can be easily extrapolated to oblique incidence and μ_r^* determination.

Therefore, ninety different materials are defined with random values of $\epsilon_r^{'}$, $\epsilon_r^{''}$ and d ranging from 1 to 20, 0 to 20 and 0.5 to 2 mm respectively. For each simulated material, the reflection coefficient is computed according to (1) at 121 frequencies equally spaced from 9 to 11 GHz. These are the data utilized to train different networks based on the MLP or the RBF concepts. The synthetic materials are randomly shuffled and, for the MLP networks, 1/2 of data are used for training, 1/4 for validation and 1/4 for test, whereas for the RBF, 3/4 are used for training and 1/4 for test. All the networks follow the structure of Fig. 2 with four inputs (real and imaginary parts of the reflection coefficient, frequency and thickness) and two outputs (real and imaginary parts of the complex relative permittivity). The input parameters are normalized between -1 and +1 in both kind of nets.

The networks are designed using the *Neural Network Toolbox* provided by MATLAB. For the multilayer perceptrons three different topologies are investigated, one with one hidden layer (MLP1) and two with two (MLP2 and MLP3), according to Table 1. Similarly, for the radial basis function networks, other three configurations are studied, all of them using gaussian functions but with different spread parameter, as shown in Table 1.

4 Results

The following figures show the performance of the aforementioned networks compared with the values provided by Arlon for this band of frequencies: $\epsilon_r^{'} = 2.55$

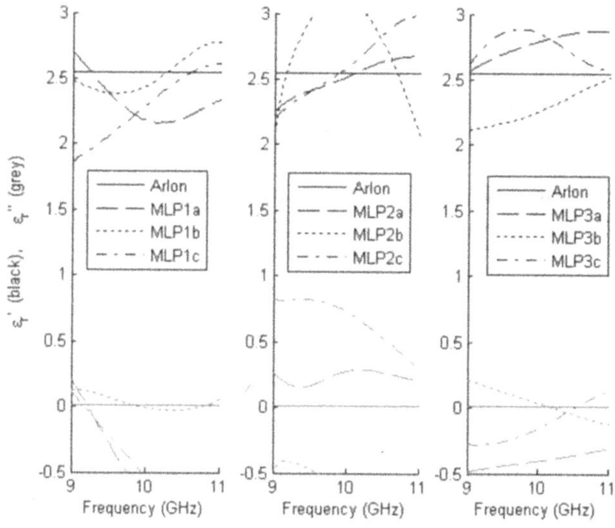

Fig. 3. MLP networks

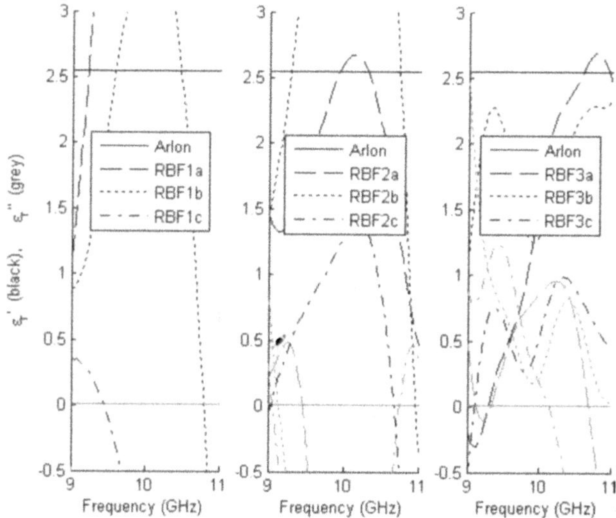

Fig. 4. RBF networks

and $\epsilon_r^{''} = 0.0056$. Both multilayer perceptrons and radial basis function networks are slightly different each time the network is trained so, to account for this variability, three instances (a, b and c) have been trained for each network of Table 1. Up to 200 epochs have been allowed each time and under these circumstances the training processes converge.

Table 2. Dielectric constant estimation error

		$\sigma_{\epsilon_r^{'}}$	$\sigma_{\epsilon_r^{''}}$			$\sigma_{\epsilon_r^{'}}$	$\sigma_{\epsilon_r^{''}}$			$\sigma_{\epsilon_r^{'}}$	$\sigma_{\epsilon_r^{''}}$
MLP1	a	0.2802	0.8300	MLP2	a	0.1299	0.2284	MLP3	a	0.2450	0.4059
	b	0.1441	0.0576		b	0.4019	0.6760		b	0.2878	0.1054
	c	0.3522	0.6129		c	0.2618	0.6642		c	0.2272	0.1874
	η	**0.2588**	**0.5001**		η	**0.2645**	**0.5229**		η	**0.2533**	**0.2329**
RBF1	a	7.4019	7.7911	RBF2	a	1.0121	5.0017	RBF3	a	1.5354	0.6137
	b	1.8167	10.183		b	2.2968	4.5114		b	0.7653	0.7961
	c	4.7425	14.384		c	2.4059	2.9184		c	1.9987	1.1617
	η	**4.6537**	**10.7860**		η	**1.9049**	**4.1483**		η	**1.4331**	**0.8572**

Figures 3 and 4 compare the results obtained with the three MLPs and the three RBFs studied respectively. Black lines correspond to the real part of ϵ_r^* and gray lines to the imaginary part. While RBFs results look unacceptable, the three topologies of multilayer perceptrons behave better (it seems that MLP3 is the best fitted topology), but, in order to provide another mean of evaluating the goodness of an instance, the following error σ is calculated, both for the real and imaginary parts of the dielectric constant,

$$\sigma = \sqrt{\frac{\sum_{i=1}^{N} (\epsilon_i - \epsilon_{manufacturer})^2}{N}} \tag{5}$$

where N is the number of frequencies. The computations of this error are collected in Table 2 where η is the arithmetic mean of the three instances a, b and c for each case. These data confirm that, under these circumstances, the RBF networks tried are not able to reach a good result and the MLPs topologies have similar good performance, proving that artificial neural networks can be used for dielectric constant estimation of materials.

However actual measurements are not free of noise (caused by diverse sources: misalignments, thermal noise, instrumentation errors,...) and consequently, better performance could be expected if the synthetic materials employed to train the networks were contaminated with noise. This idea was preliminary announced in [8], and will be developed in the following subsections.

4.1 Noisy Training Data

To emulate the measurement error and evaluate its influence in the determination of $\epsilon_r^{'}$ and $\epsilon_r^{''}$, the reflection coefficients obtained for the ninety defined materials are contaminated with gaussian noise of zero mean and a variance of 0.5 dB in modulus and 0.1° in phase. With these new data the same topologies are trained again. Figure 5 depicts the results obtained with the MLPs

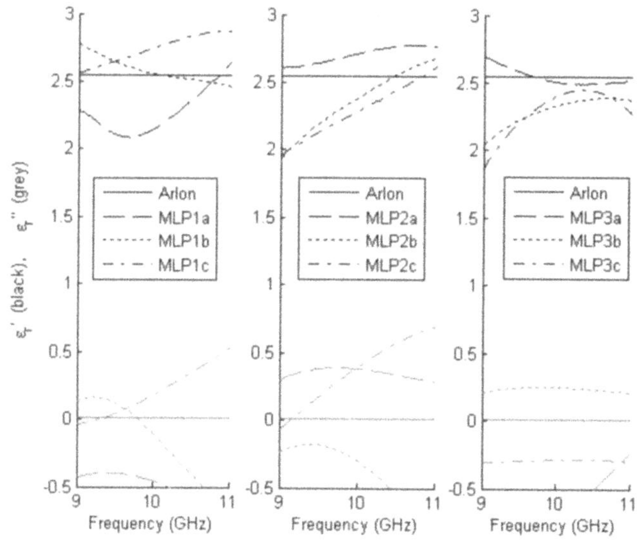

Fig. 5. MLP networks with noisy inputs

Table 3. Dielectric constant estimation error with noisy inputs

		$\sigma_{\epsilon_r'}$	$\sigma_{\epsilon_r''}$			$\sigma_{\epsilon_r'}$	$\sigma_{\epsilon_r''}$			$\sigma_{\epsilon_r'}$	$\sigma_{\epsilon_r''}$
MLP1	a	0.3220	0.5139	MLP2	a	0.1692	0.3461	MLP3	a	0.0571	0.8621
	b	0.0917	0.3452		b	0.2768	0.4639		b	0.2692	0.2284
	c	0.2207	0.2790		c	0.3130	0.4416		c	0.2980	0.2989
	η	**0.2115**	**0.3794**		η	**0.2530**	**0.4172**		η	**0.2081**	**0.4631**
RBF1	a	1.4148	6.7279	RBF2	a	4.8909	7.1193	RBF3	a	2.8467	3.3500
	b	2.3297	5.5645		b	2.4537	9.4395		b	0.8964	1.0378
	c	1.6430	4.5300		c	1.9278	6.6730		c	2.2232	3.4506
	η	**1.7958**	**5.6075**		η	**3.0908**	**7.7439**		η	**1.9888**	**2.6128**

alternatives. The statistics shown in Table 3 support the hypothesis that improved estimations can be obtained if the training data include the unavoidable measurement error.

Unfortunately, once again, networks based on the RBF concept produce estimations much worse than MLPs when the training data is contaminated with noise. Most probably, due to the range width and the number of synthetic materials employed, the input data is too scattered for the networks tested.

4.2 Narrow Range

It is expected that if the range for the dielectric constant (both real and imaginary parts) is reduced, the results could be better because the distribution of the training data would be less sparse. Taking this into account the ranges for ϵ_r'

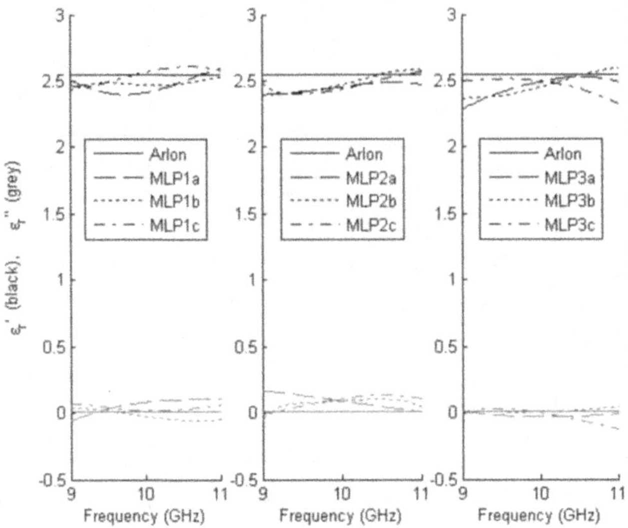

Fig. 6. MLP networks: narrow range and noisy inputs

Table 4. Dielectric constant estimation error: narrow range and noisy inputs

		$\sigma_{\epsilon_r'}$	$\sigma_{\epsilon_r''}$			$\sigma_{\epsilon_r'}$	$\sigma_{\epsilon_r''}$			$\sigma_{\epsilon_r'}$	$\sigma_{\epsilon_r''}$
MLP1	a	0.0685	0.0416	MLP2	a	0.1034	0.0935	MLP3	a	0.1125	0.0283
	b	0.0555	0.0378		b	0.0870	0.0742		b	0.1131	0.0169
	c	0.0418	0.0632		c	0.1054	0.0929		c	0.1066	0.0607
	η	**0.0553**	**0.0485**		η	**0.0986**	**0.0869**		η	**0.1107**	**0.0353**
RBF1	a	0.1097	0.5023	RBF2	a	0.1139	0.1071	RBF3	a	0.0615	0.0965
	b	0.0853	0.4720		b	0.1052	0.1240		b	0.1071	0.1459
	c	0.0392	0.2603		c	0.0595	0.0642		c	0.0864	0.0621
	η	**0.0781**	**0.4115**		η	**0.0929**	**0.0984**		η	**0.0850**	**0.1015**

and ϵ_r'' are adjusted from [1 20] and [0 20] to [1 5] and [0 4] respectively. Within this margins, ninety new synthetic materials are defined and their reflection coefficients are calculated with (1). Then, gaussian noise is added like in the above section, and with this data the three MLPs and the three RBFs networks are trained.

Figures 6 and 7 show the results obtained when the measured reflection coefficient is fed into the two kind of networks respectively. It is worth noting that the estimations have notably improved. It is also appreciated in the statistics presented in Table 4 that both of them behave better than in the case of a wider range, confirming the expectations. It is remarkable the improvement achieved with the RBFs networks that were unable to produce any useful results in the previous sections.

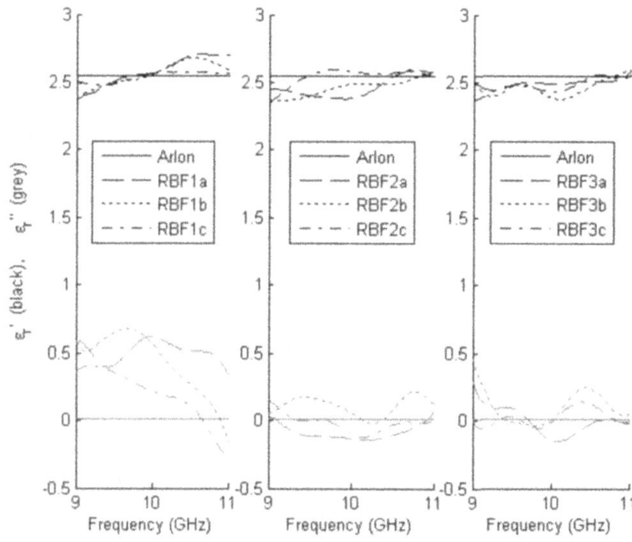

Fig. 7. RBF networks: narrow range and noisy inputs

5 Conclusions and Future Work

The results presented confirm that artificial neural networks can be useful to extract the complex relative permittivity of a material from a free-space measured sample. Firstly, the networks need to be trained, and for this purpose a set of artificial materials with thickness and real and imaginary parts of the dielectric constant randomly chosen within a predefined range must be defined. It has been shown that the width of these ranges have influence on the performance of the networks, so a multi-stage calculation procedure is established. In a first stage, and if no other *a priori* information is known, the ranges for the dielectric constant should be kept wide.

Then, the reflection coefficient for those materials is computed and with these data a first realization of any of the proposed MLPs examined in this paper can be trained. With them, a coarse estimation of ϵ_r^* is easily reached. As expected, the addition of noise emulates the measurement error and better estimations are obtained.

In a second stage, the ranges should be adjusted according the results obtained in the first one. A new set of materials must be defined within those limits and the reflection coefficient must be calculated and contaminated again. Now, either MLP o RBF networks must be trained again and good estimations can be expected when the measured reflection coefficient is fed into them. If fine results are needed this second stage can be repeated over and over again.

Additional improvements could be reached following the next proposed lines of work. First, for each kind of concept (MLP or RBF) all the topologies tried in Table 1 produce similar outputs but, if needed, a Monte-Carlo simulation with hundred of instances could be run in order to obtain a detailed statistical behavior and select the best topology. Second, the structure of the networks could be duplicated so to have a network specialized in obtaining the real part of ϵ_r^* and another specialized in the imaginary part.

Finally, the best option would be training the networks with actual materials with known electromagnetic properties instead of using simulated synthetic materials, but those kind of materials are not easily available.

References

1. Von Hippel, A.: Dielectric materials and applications. Technology Press of MIT, Cambridge (1954)
2. Knott, E.F., Tuley, E.F., Shaeffer, J.F.: Radar cross section, 2nd edn. Scitech Publishing (2004)
3. Lynch Jr., D.: Introduction to RF stealth. Scitech Publishing (2004)
4. Montiel, I., Poyatos, D., González, I., Escot, D., García, C., Diego, E.: FASCRO code and the synthetic database generation problem. In: NATO-RTO Symposium on Target Identification and Recognition using RF systems, pp. 15.1–15.16 (2004)
5. Misra, D.K.: On the measurement of the complex permittivity of materials by an open-ended coaxial probe. IEEE Microwave Guided Wave Letter 5, 161–163 (1995)
6. Meng, B., Booske, J., Cooper, R.: Extended cavity perturbation technique to determine the complex permittivity of dielectric materials. IEEE Trans Microwave Theory Tech MTT-43, 2633–2636 (1995)
7. Musil, J., Zacek, F.: Microwave Measurements of Complex Permittivity by Free-Space Methods and Their Applications. Elsevier, New York (1986)
8. Poyatos, D., Escot, D., Montiel, I., Olmeda, I.: Complex permittivity estimation by bio-inspired algorithms for target identification improvement. In: Mira, J., Álvarez, J.R. (eds.) IWINAC 2007. LNCS, vol. 4528, pp. 232–240. Springer, Heidelberg (2007)
9. Escot-Bocanegra, D., Poyatos-Martínez, D., Fernández-Recio, R., Jurado-Lucena, A., Montiel-Sánchez, I.: New benchmark radar targets for scattering analysis and electromagnetic software validation. Progress In Electromagnetics Research, PIER 88, 39–52 (2008)

Author Index